MICROCONTROLLERS

HIGH-PERFORMANCE SYSTEMS AND PROGRAMMING

MICROCONTROLLERS

HIGH-PERFORMANCE SYSTEMS AND PROGRAMMING

Julio Sanchez
Eastern Florida State College

Maria P. Canton
Brevard Public Schools

CRC Press
Taylor & Francis Group
Boca Raton London New York

CRC Press is an imprint of the
Taylor & Francis Group, an **informa** business

CRC Press
Taylor & Francis Group
6000 Broken Sound Parkway NW, Suite 300
Boca Raton, FL 33487-2742

© 2014 by Taylor & Francis Group, LLC
CRC Press is an imprint of Taylor & Francis Group, an Informa business

No claim to original U.S. Government works

Printed on acid-free paper
Version Date: 20130923

International Standard Book Number-13: 978-1-4665-6665-1 (Hardback)

Library of Congress Cataloging-in-Publication Data

Sanchez, Julio, 1938-
   Microcontrollers : high-performance systems and programming / Julio Sanchez, Maria P. Canton.
      pages cm
   Includes bibliographical references and index.
   ISBN 978-1-4665-6665-1 (hardback)
   1. Microcontrollers. 2. Microcontrollers--Programming. 3. Programmable controllers I. Canton, Maria P. II. Title.

TJ223.P76S362 2013
629.8'95--dc23   2013036871

Visit the Taylor & Francis Web site at
http://www.taylorandfrancis.com

and the CRC Press Web site at
http://www.crcpress.com

Table of Contents

Preface

Microcontrollers: High-Performance Systems and Programming can be considered a continuation of and a complement to our previous two titles on the subject of microcontroller programming. In the present book we focus on the line of high-perforance microcontrollers offered by Microchip. In addition to their enhanced features, extended peripherals, and improved performance, there are several practical factors that make the high-performance PIC series a better choice than their mid-range predecessors for most systems:

- The possibility of programming high-performance microcontrollers in a high-level language (C language).

- Source code compatibility with PIC16 microcontrollers, which facilitates code migration from mid-range to PIC18 devices.

- Pin compatibility of some PIC18 devices with their PIC16 predecessors. This makes possible the reuse of PIC16 controllers in circuits originally designed for mid-range hardware. For example, the PIC18F442 and PIC18F452 in 40-pin DIP configuration are pin-compatible with the popular PIC16F877. Similarly, the PIC18F242 and PIC18F252, in 28-pin DIP format, are pin compatible with the PIC16F684.

- Microchip pricing policy makes available the high-performance chips at a lower cost than their mid-range equivalents. Recently we have priced the 18F452 at $6.32 while the 16F877 sells from the same source at $6.72.

Expanded functionality, high-level programmability, architectural improvements that simplify hardware implementation, code and pin-layout compatibility, and lower cost make it easy to select a high-performance PIC over its mid-range counterpart. One consideration that is sometimes mentioned in favor of the mid-range devices is the abundance of published application circuits and code samples. Our book attempts to correct this. Although it should also be mentioned that some PIC16 processors with small footprints have no PIC18 equivalent, which explains why some mid-range devices continue to hold a share of the microcontroller marketplace.

Like our preceding titles in this field, the book is intended as a reference and resource for engineers, scientists, and electronics enthusiasts. The book focuses on the needs of the working professional in the fields of electrical, electronic, computer, and software engineering. In developing the material for this book, we have adopted the following rules:

1. The use of standard or off-the-shelf components such as input/output devices, integrated circuits, motors, and programmable microcontrollers, which readers can easily duplicate in their own circuits.

2. The use of inexpensive or freely available development tools for the design and prototyping of embedded systems, such as electronic design programs, programming languages and environments, and software utilities for creating printed circuit boards.

3. Our sample circuits and programs are not copyrighted or patented so that readers can freely use them in their own applications.

Our book is designed to be functional and hands-on. The resources furnished to the reader include sample circuits with their corresponding programs. The circuits are depicted and labeled clearly, in a way that is easy to follow and reuse. Each circuit includes a parts list of the resources and components required for its fabrication. For the most important circuits, we also provide tested PCB files. The sample programs are matched to the individual circuits but general programming techniques are also discussed in the text. There are appendices with useful information and the book's online software contains a listing of all the sample programs developed in the text.

Julio Sanchez

Maria P. Canton

Chapter 1

Microcontrollers for Embedded Systems

1.1 Embedded Systems

An embedded system is a computer with specific control functions. It can be part of a larger computer system or a stand-alone device. Most embedded systems must operate within real-time constraints. Embedded systems contain programmable processors that are either microcontrollers or digital signal processors (DSPs). The embedded system is sometimes a general-purpose device, but more often it is used in specialized applications such as washing machines, telephones, microwave ovens, automobiles, and many different types of weapons and military hardware.

A microcontroller or DSP usually includes a central processor, input and output ports, memory for program and data storage, an internal clock, and one or more peripheral devices such as timers, counters, analog-to-digital converters, serial communication facilities, and watchdog circuits. More than two dozen companies in the United States and abroad manufacture and market microcontrollers. Mostly they range from 8- to 32-bit devices. Those at the low end are intended for very simple circuits and provide limited functions and program space, while the ones at the high end have many of the features associated with microprocessors. The most popular microcontrollers include several from Intel (such as the 8051), from Zilog (derivatives of their famous Z-80 microprocessor) from Motorola (such as the 68HC05), from Atmel (the AVR), the Parallax (the BASIC Stamp), and many from Microchip. Some of the high-end Microchip microcontrollers and DSPs are the topic of this book.

1.2 Microchip PIC

The names PIC and PICmicro are trademarks of Microchip Technology. Microchip prefers the latter designation because PIC is a registered trademark in some European countries. It is usually assumed that PIC stands for Peripheral Interface Controller, although the original acronym was Programmable Interface Controller. More recently, Microchip has stated that PIC stands for Programmable Intelligent Computer, a much nicer, albeit not historically true version of the acronym.

The original PIC was built to complement a General Instruments 16-bit CPU designated the CP-1600. The first 8-bit PIC was developed in 1975 to improve the performance of the CP-1600 by offloading I/O tasks from the CPU. In 1985, General Instrument spun off its microelectronics division. At that time, the PIC was re-designed with internal EPROM to produce a programmable controller. Today, hundreds of versions and variations of PIC microcontrollers are available from Microchip. Typical on-board peripherals include input and output ports, serial communication modules, UARTs, and motor control devices. Program memory ranges from 256 words to 64k words and more. The word size varies from 12 to 14 or 16 bits, depending on the specific PIC family.

1.2.1 PIC Architecture

PIC microcontrollers contain an instructions set that varies in length from 35 instructions for the low-end devices to more than 70 for the high end. The accumulator, which is known as the work register in PIC documentation, is part of many instructions because the low- and mid-range PICs contain no other internal registers accessible to the programmer. The PICs are programmable in their native Assembly Language. C language and BASIC compilers have also been developed. Open-source Pascal, JAL, and Forth compilers are also available, although not very popular.

It is often mentioned that one of the reasons for the success of the PIC is the support provided by Microchip. This support includes development software, such as a professional-quality development environment called MPLAB, which can be downloaded free from the company's website (www.microchip.com). The MPLAB package includes an assembler, a linker, a debugger, and a simulator. Microchip also sells an in-circuit debugger called MPLAB ICD 2. Other development products intended for the professional market are also available from Microchip.

In addition to the development software, the Microchip website contains a multitude of free support documents, including data sheets, application notes, and sample code. Furthermore, the PIC microcontrollers have gained the support of many hobbyists, enthusiasts, and entrepreneurs who develop code and support products and publish their results on the Internet. This community of PIC users is a treasure trove of information and know-how easily accessible to the beginner and useful even to the professional. One such Internet resource is an open-source collection of PIC tools named GPUTILS, which is distributed under the GNU General Public License. GPUTILS includes an assembler and a linker. The software works on Linux, Mac OS, OS/2, and Windows. Another product, called GPSIM™, is an open source simulator featuring PIC hardware modules.

1.2.2 Programming the PIC

Stand-alone programming a PIC microcontroller requires the following tools and components:

- An Assembler or high-level language compiler. The software package usually includes a debugger, simulator, and other support programs.

- A computer (usually a PC) on which to run the development software.

- A hardware device called a programmer that connects to the computer through the serial, parallel, or USB line. The PIC is inserted in the programmer and "blown" by downloading the executable code generated by the development system. The hardware programmer usually includes the support software.

- A cable or connector for connecting the programmer to the computer.

- A PIC microcontroller.

Alternatively, some PIC microcontrollers can be programmed while installed in their applications boards. Although this option can be very useful as a production and distribution tool, for reasons of space it is not discussed in this book.

PIC Programmers

The development system (assembler or compiler) and the programmer driver are the software components. The computer, programmer, and connectors are the hardware elements. Figure 6.1 shows a commercial programmer that connects to the USB port of a PC. The one in the illustration is made by MicroPro.

Figure 1.1 *USB PIC programmer made by MicroPro.*

Many other programmers are available on the market. Microchip offers several high-end models with in-circuit serial programming (ICSP) and low-voltage programming (LVP) capabilities. These devices allow the PIC to be programmed in the target circuit. Some PICs can write to their own program memory. This makes possible the use of so-called bootloaders, which are small resident programs that allow loading user software over the RS-232 or USB lines. Programmer/debugger combinations are also offered by Microchip and other vendors.

Development Boards

A development board is a demonstration circuit that usually contains an array of connected and connectable components. Their main purpose is as a learning and experiment tool. Like programmers, PIC development boards come in a wide range of prices and levels of complexity. Most boards target a specific PIC microcontroller or a PIC family of related devices. Lacking a development board, the other option is to build the circuits oneself, a time-consuming but valuable experience. Figure 1.2 shows the LAB-X1 development board for the 16F87x PIC family.

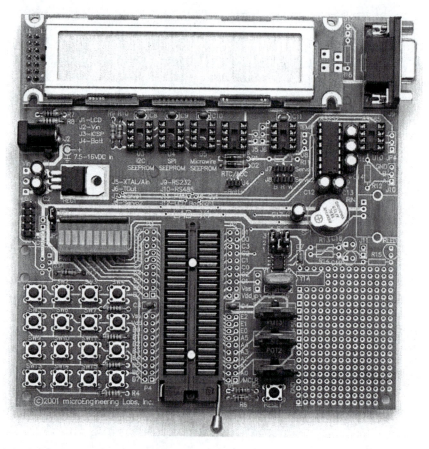

Figure 1.2 *LAB-X1 development board.*

The LAX-X1 board, as well as several other models, are products of microEngineering Labs, Inc. Development boards from Microchip and other vendors are also available.

1.3 PIC Architecture

PIC microcontrollers are roughly classified by Microchip into three groups: baseline, mid-range, and high-performance. Figure 1.3 shows the components of each PIC family at the time of this writing.

Microchip PIC and dsPIC Families

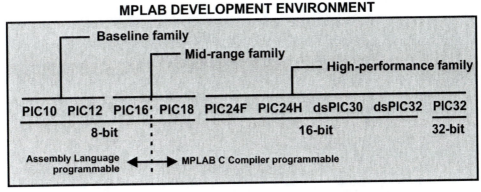

Figure 1.3 Microchip PIC and dsPIC families.

Within each of the groups the PIC are classified based on the first two digits of the PIC's family type. However, the sub-classification is not very strict, as there is some overlap. In fact, we find PICs with 16X designations that belong to the baseline family and others that belong to the mid-range group. In the following sub-sections we describe the basic characteristics of the various sub-groups of the three major PIC families with 8-bit architectures.Table 1.1 shows the principal hardware characteristics of each of the four 8-bit PIC families

Table 1.1

8-bit PIC Architectures Comparison Chart

	BASELINE	MID-RANGE	ENHANCED	PIC18
Pin Count	6-40	8-64	8-64	18-100
Interrupts	No	Single interrupt	Single interrupt Context saved	Multiple Interrupts Context saved
Performance	5 MIPS	5 MIPS	8 MIPS	Up to 16 MIPS
Instructions	33, 12-bit	35, 14-bit	49, 14-bit	83, 16-bit
Program Memory	Up to 3 KB	Up to 14 KB	Up to 28 KB	Up to 128 KB
Data Memory	138 Bytes	368 Bytes	1,5 KB	4 KB
Hardware Stack	2 level	8 level	16 level	32 level
Total Number of Devices	16	58	29	193
Families	PIC10 PIC12 PIC14 PIC16	PIC12 PIC16	PIC12FXXX PIC16F1XX	PIC18

1.3.1 Baseline PIC Family

This group includes members of the PIC10, PIC12, PIC14, and PIC16 families. The devices in the baseline group have 12-bit program words and are supplied in 6- to 28-pin packages. The microcontrollers in the baseline group are described as being suited for battery-operated applications because they have low power requirements. The typical member of the baseline group has a low pin count, flash program memory, and low power requirements. The following types are in the Baseline group:

- PIC10 devices

- PIC12 devices

- PIC14 devices

- Some PIC16 devices

We present a short summary of the functionality and hardware types of the baseline PICs in the sections that follow, although these devices are not covered in this book.

PIC10 devices

The PIC10 devices are low-cost, 8-bit, flash-based CMOS microcontrollers. They use 33 single-word, single-cycle instructions (except for program branches, which take two cycles. The instructions are 12-bits wide. The PIC10 devices feature power-on reset, an internal oscillator mode which saves having to use ports for an external oscillator. They have a power-saving SLEEP mode, A Watchdog Timer, and optional code protection.

The recommended applications of the PIC10 family range from personal care appliances and security systems to low-power remote transmitters and receivers. The PICs of this family have a small footprint and are manufactured in formats suitable for both through hole or surface mount technologies. Table 1.2 lists the characteristics of the PIC10F devices.

Table 1.2

PIC10F Devices

	10F200	10F202	10F204	10F206
Clock:				
Maximum Frequency of Operation (MHz)	4	4	4	4
Memory:				
Flash Program Memory	256	512	256	512
Data Memory (bytes)	16	24	16	24
Peripherals:				
Timer Module(s)	TMR0	TMR0	TMR0	TMR0
Wake-up from Sleep	Yes	Yes	Yes	Yes
Comparators	0	0	1	1
Features:				
I/O Pins	3	3	3	3
Input Only Pins	1	1	1	1
Internal Pull-ups	es	Yes	Yes	Yes
In-Circuit Serial Programming	Yes	Yes	Yes	Yes
Instructions	33	33	33	33
Packages:	-------------------------------- 6-pin SOT-23 ----------------------------			
	-------------------------------- 8-pin PDIP ----------------------------			

Two other PICs of this series are the 10F220 and the 10F222. These versions include four I/O pins and two analog-to-digital converter channels. Program memory is 256 words on the 10F220 and 512 in the 10F222. Data memory is 16 bytes on the F220 and 23 in the F222.

PIC12 Devices

The PIC12C5XX family are 8-bit, fully static, EEPROM/EPROM/ROM-based CMOS microcontrollers. The devices use RISC architecture and have 33 single-word, single-cycle instructions (except for program branches that take two cycles). Like the PIC10 family, the PIC12C5XX chips have power-on reset , device reset, and an internal timer. Four oscillator options can be selected, including a port-saving internal oscillator and a low-power oscillator. These devices can also operate in SLEEP mode and have watchdog timer and code protection features.

The PIC12C5XX devices are recommended for applications ranging from personal care appliances, security systems, and low-power remote transmitters and receivers. The internal EEPROM memory makes possible the storage of user-defined codes and passwords as well as appliance setting and receiver frequencies. The various packages allow through-hole or surface mounting technologies. Table 1.3 lists the characteristics of some selected members of this PIC family.

Table 1.3

PIC 12CXXX and 12CEXXX Devices

	12C508(A) 12C509A 12CR509A	12C518	12CE519	12C671 12C672 12C673	12CE674
Clock: Maximum Frequency of Operation (MHz)	4	4	4	10	10
Memory: EPROM Program Memory (bytes)	25/41/41	25	41	128	128
Peripherals: EEPROM Data Memory (bytes)	—	16	16	0/0/16	16
Timer Module(s)	TMR0	TMR0	TMR0	TMR0	TMR0
A/D Converter (8-bit) Channels	—	—	—	4	4
Features: Interrupt Sources	—	—	—	4	4
I/O Pins	5	5	5	5	5
Input Pins	1	1	1	1	1

(continues)

Table 1.3

PIC 12CXXX and 12CEXXX Devices **(continued)**

	12C508(A) 12C509A 12CR509A	12C518	12CE519	12C671 12C672 12C673	12CE674
Internal Pull-ups	Yes/Yes/No	Yes	Yes	Yes	Yes
In-Circuit Serial Programming	Yes/No	Yes	Yes	Yes	Yes
Number of Instructions	33	33	33	35	35
Packages	8-pin DIP SOIC	8-pin DIP JW,SOIC	8-pin DIP JW. SOIC	8-pin DIP SOIC	8-pin DIP JW

Two other members of the PIC12 family are the 12F510 and the 16F506. In most respects these devices are similar to the ones previously described, except that the 12F510 and 16F506 both have flash program memory. Table 1.4 lists the most important features of these two PICs.

Table 1.4

PIC12F510 and 12F675

	12F629	12F675
Clock:		
Maximum Frequency of Operation (MHz)	20	20
Memory:		
Flash Program Memory	1024	1024
Data Memory (SRAM bytes)	64	64
Peripherals:		
Timers 8/16 bits	1/1	1/1
Wake-up from Sleep on Pin Change	Yes	Yes
Features:		
I/O Pins	6	6
Analog comparator module	Yes	Yes
Analog-to-digital converter	No	10-bit
In-Circuit Serial Programming	Yes	Yes
Enhanced Timer1 module	Yes	Yes
Interrupt capability	Yes	Yes
Number of Instructions	35	35
Relative addressing	Yes	Yes
Packages	8-pin PDIP, SOIC, DFN-S	8-pin PDIP SOIC, DFN-S

Two other members of the PIC12F are the 12F629 and 12F675. The only difference between these two devices is that the 12F675 has a 10-bit analog-to-digital converter while the 629 has not A/D converter. Table 1.5 lists some important features of both PICs.

Table 1.5

PIC12F629 and 12F675

	12F629	12F675
Clock:		
Maximum Frequency of Operation (MHz)	20	20
Memory:		
Flash Program Memory	1024	1024
Data Memory (SRAM bytes)	64	64
Peripherals:		
Timers 8/16 bits	1/1	1/1
Wake-up from Sleep on Pin Change	Yes	Yes
Features:		
I/O pins	6	6
Analog comparator module	Yes	Yes
Analog-to-digital converter	No	10-bit
In-circuit serial programming	Yes	Yes
Enhanced Timer1 module	Yes	Yes
Interrupt capability	Yes	Yes
Number of instructions	35	35
Relative addressing	Yes	Yes
Packages	8-pin PDIP	8-pin PDIP
	SOIC	SOIC
	DFN-S	DFN-S

Several members of the PIC12 family, 12F635, 12F636, 12F639, and 12F683, are equipped with special power-management features (called nanowatt technology by Microchip). These devices were especially designed for systems that require extended battery life.

PIC14 Devices

The single member of this family is the PIC14000, which is built with CMOS technology. This makes the PIUC14000 fully static and gives it industrial temperature range. The 14000 is recommended for battery chargers, power supply controllers, power management system controllers, HVAC controllers, and for sensing and data acquisition applications.1.3.2

1.3.3 Mid-range PIC Family

The mid-range PICs includes members of the PIC12 and PIC16 groups as well as the PIC 18 group. According to Microchip the mid-range PICs all have 14-bit program words with either flash or OTP program memory. Those with flash program memory also have EEPROM data memory and support interrupts. Some members of the mid-range group have USB, I2C, LCD, USART, and A/D converters. Implementations range form 8 to 64 pins.

PIC16 Devices

This is by far the largest mid–range PIC group. Currently over 80 versions of the PIC16 are listed in production by Microchip. Although we do not cover the mid-range devices in this book, we have selected a few of its most prominent members of the PIC16 family to list their most important features. These are found in Table 1.6.

Table 1.6

PIC16 Devices

	16C432	16C58	16C770	16F54	16F84A	16F946
Clock:						
Maximum Frequency MHz	20	40	20	20	20	20
Memory:						
Program memory type	OTP	OTP	OTP	Flash	Flash	Flash
K-bytes	3.5	3	3.5	0.75	1.75	14
K-words	2	2	2	0.5	1	8
Data EEPROM	0	0	0	0	64	256
Peripherals:						
I/O channels	12	12	16	12	13	53
ADC channels	0	0	6	0	0	8
Comparators	0	0	0	0	0	2
Timers	1/8-bit	1/8-bit 1/16-bit	2/8-bit	1/8-bit	1/8-bit	2/8-bit 1/16-bit
Watchdog timer	Yes	Yes	Yes	Yes	Yes	Yes
Features:						
ICSP	Yes	No	Yes	No	Yes	Yes
ICD	No	No	No	No	0	1
Pin count	20	18	20	18	18	64
Communications	-	-	MPC/SPI	-	-	AUSART
Packages	20/CERDIP, 20/SSOP 208mil	18/CERDIP 18/PDIP 18/SOIC 300mil	20/CERDIP 20/PDIP 20/SOIC 300mil	18/PDIP 18/SOIC 300mil	18/PDIp 18/SOIC 300mil	64/TQFP

Microchip documentation refers to an enhanced mid-range family composed of PIC12FXXX and PIC16F1XX devices. These devices maintain compatibility with the previous members of the mid-range family while providing additional performance. Their most important new features include multiple interrupts, fourteen additional instructions, up to 28 KB program memory, and additional peripheral modules.

1.3.3 High-Performance PICs and DSPs

The high-performance PICs belong to the PIC18 and PIC32 groups. The motivation for expanding the PIC arquitecture and modifying the core of the mid-range PICs relate to the following limitations:

- Small-size stack

- Single interrupt vector

- Limited instruction set

- Small memory size

- Limited number of peripherals

- No high-level language programmability

The devices in the PIC16 group have 16-bit program words, flash program memory, a linear memory space of up to 2 Mbytes, as well as protocol-based communications facilities. They all support internal and external interrupts and a much larger instruction set than members of the baseline and mid-range families. The PIC18 family is also a large one, with over seventy different variations currently in production. These devices are furnished in 18 to 80 pin packages. Microchip describes the PICs in this family as high-performance with integrated A/D converters.

Digital Signal Processor

The notion of digital signal processing starts with the conversion of analog signal information such as voice, image, temperature, or pressure primitive data to digital values that can be stored and manipulated by a computing device. Converting the data from its primitive analog form to a digital format makes it much easier to analyze, display, store, process, or convert the data to another format. Digital signal processing is based on the fact that computing and data processing operations are easier to perform on digital data than on raw analog signals.

The concept of digital signal processing can be illustrated by means of a satellite-based Earth imagining system (such as the Landsat EROS) shown in Figure 1.4.

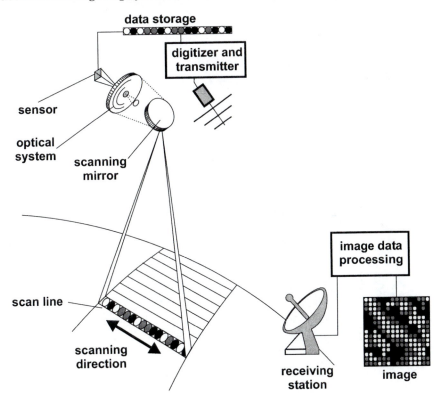

Figure 1.4 *Schematic of a space-borne imaging system.*

The optical-mechanical instrument onboard a spacecraft, shown in Figure 1.4, consists of several subsystems. The scanning mirror collects the radiation, which is imaged by an optical system onto a sensor device. The sensor performs an analog-to-digital conversion and places the digital values in a temporary storage structure. During its orbit, the satellite reaches a location in space from which it can communicate with an Earth receiving station. At this time, the transmitter and support circuitry send the digital data to the receiving station. The receiving station processes this data and formats it into an image. In this scheme, digital signal processing can take place as the image data is sensed by the instrument and temporarily stored on board the satellite, or when the raw data received by the Earth station is converted into an image that can be manipulated, viewed, stored, or re-processed.

Analog-to-Digital

Conversion from analog-to-digital form and vice versa are not formally operations of a DSP. However, these conversions are so often required during signal processing that most DSP devices include the analog-to-digital and digital-to-analog conversion hardware.

Analog-to-digital conversion is usually performed by sampling the signal at uniform time intervals and using the sampled value as representative of the region between the intervals. Figure 1.5 shows an example of analog-to-digital conversion by sampling.

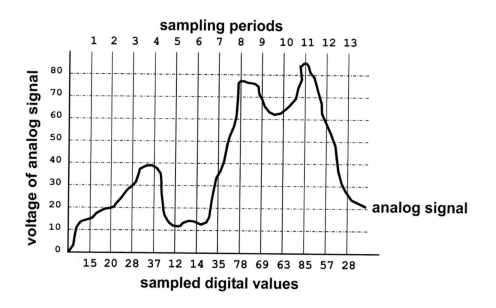

Figure 1.5 *Analog-to-digital conversion by sampling.*

In Figure 1.5 we see that the sampled values are actually an approximation of the analog curve, as the variations between each interval are lost in the conversion process. Therefore, the more sampling periods, the more accurate the approximation. On the other hand, too small a sampling rate tends to reduce the significance of the data by producing repeated values in the digital record.

Chapter 2

PIC18 Architecture

2.1 PIC18 Family Overview

The PIC18 family was designed to provide ease of use (programmable in C), high performance, and effortless integration with previous 8-bit families. In addition to the standard modules found in the PIC16 and previous families, the PIC18 includes several advanced peripherals, such as CAN, USB, Ethernet, LCD and CTMU. Its principal features are

- Nanowatt technology ensures low power consumption
- 83 instructions (16-bit wide)
- C language optimized
- Up to 2 MB addressable program memory
- 4KB maximum RAM
- 32-level hardware stack
- 8-bit file select register
- Integrated 8x8 hardware multiplier

The performance of the PIC18 series is the highest in the Microchip 8-bit architecture. Figure 2.1 is a block diagram of the PIC18 architecture.

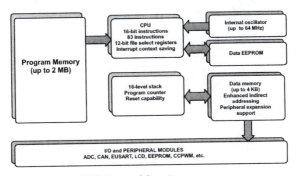

Figure 2.1 *Block diagram of PIC18 architecture.*

Although the PIC16 series has been very successful in the microcontroller marketplace, it also suffers from limitations and constraints. Perhaps the most significant limitation is that the devices of the PIC16 family can only be programmed in Assembly language. Other limitations result from the device's RISC design. For example, the absence of certain types of opcodes, such as the Branch instruction, make it necessary to combine a skip opcode followed by a goto operation in order to provide a conditional, targeted jump. Other limitations relate to the hardware itself: small stack and a single interrupt vector. As the complexity, memory size, and the number of peripheral modules increased, the limitations of the PIC16 series became more evident.

In the PIC18 series, Microchip reconsidered its PIC16 design rules and produced a completely new style microcontroller, with a much more complex core, while limiting the changes to the peripheral modules. The degree of change can be deduced from the expansion of the instruction set from 35 14-bit to 83 16-bit operation codes. Memory has gone from 14 to 128 KB; the stack from 8 levels to 32 levels. These changes made it possible to optimize the PIC18 series for C language programming.

2.1.1 PIC18FXX2 Group

At the present time, Microchip lists 193 different devices in the PIC18 family. These devices are available with pin counts from 28 to 100 and in the SOIC, DIP, PLCC, SPDIP, QFN, SSOP, TQFP, QFN, and LQFP packages. For consistency with the tutorial nature of this book, we have selected the PIC18F4X2 group with identical DIP and SOIC pinouts. Figure 2.2 shows the pin diagram for the PIC18F4X2 devices.

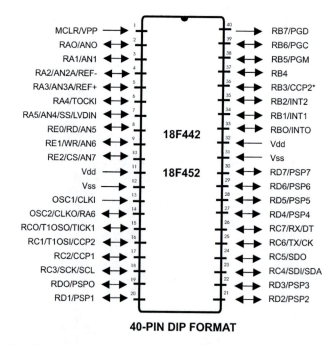

40-PIN DIP FORMAT

Figure 2.2 *Pin diagram for PIC18F4X2 devices.*

For learning and experimentation the devices in DIP packages are more convenient because they can be easily inserted in the ZIF (zero insertion force) sockets found in most programming devices, development boards, and breadboards. The devices in Figure 1.1 and Figure 1.2 are so equipped. A PLCC (plastic leaded chip carrier) package with 44 pins is also available for 18F442 and 18F452 devices. We do not cover this option.

2.1.2 PIC18FXX2 Device Group Overview

These devices come in 28-pin and 40-pin packages, as well as in a 44-pin PLCC package previously mentioned. The 28-pin devices do not have a Parallel Slave Port (PSP). Also, the number of analog-to-digital (A/D) converter input channels is reduced to 5. An overview of features is shown in Table 2.1

Table 2.1

Principal Features of Devices in the PIC18FXX2 Family

FEATURES	PIC18F242	PIC18F252	PIC18F442	PIC18F452
Operating Frequency	DC - 40 MHz	DC - 40 MHz	DC - 40 MHz	DC - 40 MHz
Program Memory (Bytes)	16K	32K	16K	32K
Program Memory (Instructions)	8192	16384	8192	16384
Data Memory (Bytes)	768	1536	768	1536
Data EEPROM Memory (Bytes)	256	256	256	256
Interrupt Sources	17	17	18	18
I/O Ports	A, B, C	A, B, C	A, B, C, D, E	A, B, C, D, E
Timers	4	4	4	4
Capture/Cornpare /PWM Modules	2	2	2	2
Serial Communications	----------------------- MSSP --- Addressable USART			
Parallel Communications	-	-	PSP	PSP
10-bit Analog-to-Digital Module	5 channels	5 channels	8 channels	8 channels
RESETS (and Delays)	-------------------------- POR, BOR, Reset ---------------------------- Instruction, Stack Full, Stack Underflow, (PWRT, OST)			
Programmable Low Voltage Detect	Yes	Yes	Yes	Yes
Programmable Brown-out Reset	Yes	Yes	Yes	Yes
Instruction Set	75 Instructions	75 Instructions	75 Instructions	75 Instructions
Packages	28-pin DIP 28-pin SOIC SOIC	28-pin DIP 28-pin SOIC SOIC	40-pin DIP PLCC 44-pin SOIC	40-pin DIP QFP PLCC 44-pin SOIC

From Table 2.1 the following general features of the PIC18FXX2 devices can be deduced:

1. Operating frequency is 40 MHz for all devices. They all have a 75 opcode instruction set.

2. Program memory ranges from 16K (8,192 instructions) in the PIC18F2X2 devices to 32K (16,384 instructions) in the PIC18F4X2 devices.

3. Data memory ranges for 768 to 1,536 bytes.

4. Data EEPROM is 256 bytes in all devices.

5. The PIC18F2X2 devices have three I/O poerts (A, B, and C) and the PIC18F4X2 devices have five ports (A, B, C, D, and E).

6. All devices have four timers, two Capture/Compare/PWM modules, MSSP and adressable USART for serial communications and 10-bit analog-to-digital modules.

7. Only PIC18F4X2 devices have a parallel port.

2.1.3 PIC18F4X2 Block Diagram

The block diagram of the 18F4X2 microcontrollers, which correspond to the 40-pin devices of Figure 2.2, is shown in Figure 2.3.

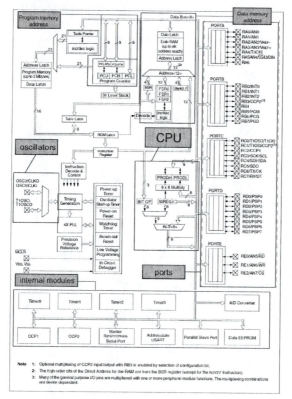

Figure 2.3 *PIC18F4X2 block diagram.*

2.1.4 Central Processing Unit

In Figure 2.3 the dashed rectangle labeled CPU (central processing unit) contains the 8-bit Arithmetic Logic Unit, the Working register labeled WREG, and the 8-bit-by-8-bit hardware multiplier, described later in this chapter. The CPU receives the instruction from program memory according to the value in the Instruction register and the action in the Instruction Decode and Control block. An interrupt mechanism with several sources (not shown in Figure 2.3) is also part of the PIC18FXX2 hardware.

The Status Register

The Status register, not shown in Figure 2.3, is part of the CPU and holds the individual status bits that reflect the operating condition of the individual elements of the device. Figure 2.4 shows the bit structure of the Status register.

bits:	7	6	5	4	3	2	1	0
	−	−	−	N	OV	Z	DC	C

```
bit 4 N:    Negative bit
            1 = Arithmetic result is negative
            0 = Arithmetic result is positive
bit 3 OV:   Overflow bit
            1 = Overflow in signed arithmetic
            0 = No overflow occurred
bit2 Z:     Zero bit
            1 = The result of an operation is zero
            0 = The result of an operation is not zero
bit 1 DC:   Digit carry/borrow bit for ADDWF, ADDLW, SUBLW,
            and SUBWF instructions. For borrow the polarity
            is reversed.
            1 = A carry-out from the 4th bit of the result
            0 = No carry-out from the 4th bit of the result
            For rotate instructions (RRF and RLF) this bit
            is loaded with either bit 4 or bit 3 of the
            source register.
bit 0 C:    Carry/borrow bit for ADDWF, ADDLW, SUBLW, and
            SUBWF instructions. For borrow the polarity
            is reversed.
            1 = A carry-out from the most significant bit
            0 = No carry-out from the most significant bit
            For rotate instructions (RRF and RLF) this bit
            is loaded with either bit 4 or bit 3 of the
            source register.
```

Figure 2.4 *Status register bitmap.*

Program Counter Register

The 21-bit wide Program Counter register specifies the address of the next instruction to be executed. The register mapping of the Program Counter register is shown in Figure 2.5.

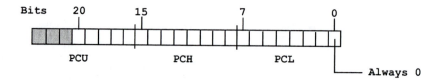

Figure 2.5 *Register map of the Program Counter.*

As shown in Figure 2.5, the low byte of the address is stored in the PCL register, which is readable and writeable. The high byte is stored in the PCH register. The upper byte is in the PCU register, which contains bits <20:16>. The PCH and PCU registers are not directly readable or writeable. Updates to the PCH register are performed through the PCLATH register. Updates to the PCU register are performed through the PCLATU register.

The Program Counter addresses byte units in program memory. In order to prevent the Program Counter from becoming misaligned with word instructions, the LSB of PCL is fixed to a value of '0' (see Figure 2.5). The Program Counter increments by 2 to the address of the next sequential instructions in the program memory.

The CALL, RCALL, GOTO, and program branch instructions write to the Program Counter directly. In these instructions, the contents of PCLATH and PCLATU are not transferred to the program counter. The contents of PCLATH and PCLATU are transferred to the Program Counter by an operation that writes PCL. Similarly, the upper 2 bytes of the Program Counter will be transferred to PCLATH and PCLATU by an operation that reads PCL.

Hardware Multiplier

All PIC18FXX2 devices contain an 8 x 8 hardware multiplier in the CPU. Because multiplication is a hardware operation it completes in a single instruction cycle. Hardware multiplication is unsigned and produces a 16-bit result that is stored in a 16-bit product register pair labeled PRODH (high byte) and PRODL (low byte).

Hardware multiplication has the following advantages:

- Higher computational performance
- Smaller code size of multiplication algorithms

The performance increase allows the device to be used in applications previously reserved for Digital Signal Processors.

Interrupts

PIC18FXX2 devices support multiple interrupt sources and an interrupt priority mechanism that allows each interrupt source to be assigned a high or low priority level. The high-priority interrupt vector is at OOOOO8H and the low-priority interrupt vector is at 000018H. High-priority interrupts override any low-priority interrupts that may be in progress. Ten registers are related to interrupt operation:

- RCON
- INTCON
- INTCON2
- INTCON3
- PIR1, PIR2
- PIE1, PIE2
- IPR1, IPR2

Each interrupt source (except INTO) has three control bits:

- A Flag bit indicates that an interrupt event has occurred.
- An Enable bit allows program execution to branch to the interrupt vector address when the flag bit is set.
- A Priority bit to select high-priority or low priority for an interrupt source.

Interrupt priority is enabled by setting the IPEN bit {mapped to the RCON<7> bit}. When interrupt priority is enabled, there are 2 bits that enable interrupts globally. Setting the GIEH bit (1NTCON<7>) enables all interrupts that have the priority bit set. Setting the GIEL bit (INTCON<6>) enables all interrupts that have the priority bit cleared. When the interrupt flag, the enable bit, and the appropriate global interrupt enable bit are set, the interrupt will vector to address OOOOO8h or 000018H, depending on the priority level. Individual interrupts can be disabled through their corresponding enable bits.

When the IPEN bit is cleared (default state), the interrupt priority feature is disabled and the interrupt mechanism is compatible with PIC mid-range devices. In this compatibility mode, the interrupt priority bits for each source have no effect and all interrupts branch to address OOOOO8H.

When an interrupt is handled, the Global Interrupt Enable bit is cleared to disable further interrupts. The return address is pushed onto the stack and the Program Counter is loaded with the interrupt vector address, which can be OOOOO8H or 000018H. In the Interrupt Service Routine, the source or sources of the interrupt can be determined by testing the interrupt flag bits. To avoid recursive interrupts, these bits must be cleared in software before re-enabling interrupts. The "return from interrupt" instruction, RETFIE, exits the interrupt routine and sets the GIE bit {GIEH or GIEL if priority levels are used), which re-enables interrupts.

Several external interrupts are also supported, such as the INT pins or the PORTB input change interrupt. In these cases, the interrupt latency will be three to four instruction cycles. Interrupts and interrupt programming are the subject of Chapter 8.

2.1.5 Special CPU Features

Several CPU features are intended for the following purposes:

- Mmaximize system reliability
- Minimize cost through the elimination of external components
- Provide power-saving operating modes
- Offer code protection

These special features are related to the following functions and components:

- SLEEP mode
- Code protection
- ID locations
- In-circuit serial programming
- SLEEP mode

SLEEP mode is designed to offer a very low current mode during which the device is in a power-down state. The application can wakeup from SLEEP through the following mechanisms:

1. External RESET
2. Watchdog Timer Wake-up
3. An interrupt

The Watchdog Timer is a free running on-chip RC oscillator, that does not require any external components. This RC oscillator is separate from the RC oscillator of the OSC1/CLKI pin. That means that the WDT will run, even if the clock on the OSC1/CLKI and OSC2/CLKO/ RA6 pins of the device has been stopped, for example, by execution of a SLEEP instruction.

Watchdog Timer

A Watchdog Timer time-out (WDT) generates a device RESET. If the device is in SLEEP mode, a WDT causes the device to wakeup and continue in normal operation (Watchdog Timer Wake-up). If the WDT is enabled, software execution may not disable this function. When the WDTEN configuration bit is cleared, the SWDTEN bit enables/disables the operation of the WDT. Values for the WDT postscaler may be assigned using the configuration bits.

The CLRWDT and SLEEP instructions clear the WDT and the postscaler (if assigned to the WDT) and prevent it from timing out and generating a device RESET condition. When a CLRWDT instruction is executed and the postscaler is assigned to the WDT, the postscaler count will be cleared, but the postscaler assignment is not changed.

The WDT has a postscaler field that can extend the WDT Reset period. The postscaler is selected by the value written to 3 bits in the CONFIG2H register during device programming.

Wake-Up by Interrupt

When global interrupts are disabled (the GIE bit cleared) and any interrupt source has both its interrupt enable bit and interrupt flag bit set, then one of the following will occur:

When an interrupt occurs before the execution of a SLEEP instruction, then the SLEEP instruction becomes a NOP. In this case, the WDT and WDT postscaler will not be cleared, the TO bit will not be set, and PD bits will not be cleared.

If the interrupt condition occurs during or after the execution of a SLEEP instruction, then the device will immediately wakeup from SLEEP. In this case, the SLEEP instruction will be completely executed before the wake-up. Therefore, the WDT and WDT postscaler will be cleared, the TO bit will be set, and the PD bit will be cleared.

Even if the flag bits were checked before executing a SLEEP instruction, it may be possible for these bits to set before the SLEEP instruction completes. Code can test the PD bit in order to determine whether a SLEEP instruction executed. If the PD bit is set, the SLEEP instruction was executed as a NOP. To ensure that the WDT is cleared, a CLRWDT instruction should be executed before a SLEEP instruction.

Low Voltage Detection

For many applications it is desirable to be able to detect a drop in device voltage below a certain limit. In this case, the application can define a low voltage window in which it can perform housekeeping tasks before the voltage drops below its defined operating range. The Low Voltage Detect feature of the PIC18FXX2 devices can be used for this purpose. For example, a voltage trip point for the device can be specified so that when this point is reached, an interrupt flag is set. The program will then branch to the interrupt's vector address and the interrupt handler software can take the corresponding action. Because the Low Voltage Detect circuitry is completely under software control, it can be "turned off" at any time, thus saving power.

Implementing Low Voltage Detect requires setting up a comparator that reads the reference voltage and compares it against the preset trip-point. This trip-point voltage is software programmable to any one of sixteen values by means of the 4 bits labeled LVDL3:LVDLO. When the device voltage becomes lower than the preselected trip-point, the LVDIF bit is set and an interrupt is generated.

Device Configuration

Several device configurations can be selected by programming the configuration bits. These bits are mapped, starting at program memory address 300000H. Note that this address is located in the configuration memory space (300000H to 3F0000H), which is only accessed using table read and table write operations. When the configuration bits are programmed, they will read as '0'; when left unprogrammed they will read as '1'.

MPLAB development tools provide an __CONFIG directive, together with a set of device-specific operands, that simplify selecting and setting the desired configuration bits. This topic is explored in the book's chapters related to programming.

2.2 Memory Organization

Devices of the PIC18FXX2 family contain three independent memory blocks:

- Program Memory
- Data Memory
- Data EEPROM

Because the device uses a separate buss, the CPU can concurrently access the data and program memory blocks.

2.2.1 Program Memory

The Program Counter register is 21 bit wide and therefore capable of addressing a maximum of 2-Mbyte program memory space. Accessing a location between the physically implemented memory and the 2-Mbyte maximum address will read all zeroes. The PIC18F242 and PIC18F442 devices can store up to 8K of single-word instructions. The PIC18F252 and PIC18F452 devices can store up to 16K of single-word instructions. The RESET vector address is at OOOOH and the interrupt vector addresses are at 0008H and 0018H. Figure 2.6 shows the memory map for the PIC18FXX2 family.

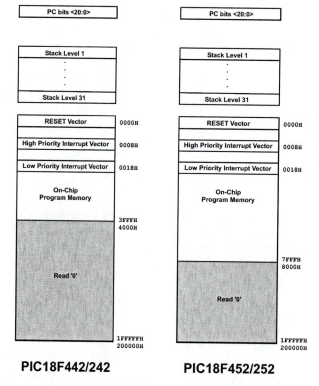

PIC18F442/242   **PIC18F452/252**

Figure 2.6 *Program memory map for the PIC18FXX2 family.*

2.2.2 18FXX2 Stack

The PIC18FXX2 stack is 31 address deep and allows as many combinations of back-to-back calls and interrupts to occur. When a CALL or RCALL instruction is executed, the Program Counter is pushed onto the stack. When a CALL or RCALL instruction is executed, or an interrupt is acknowledged, the Program Counter is pulled off the stack. This also takes place on a RETURN, RETLW, or RETFIE instruction. PCLATU and PCLATH registers are not affected by any of the RETURN or CALL instructions.

The stack consists of a 31-word deep and 21-bit wide RAM structure. The current stack position is stored in a 5-bit Stack Pointer register labeled STKPTR. This register is initialized to OOOOOB after all RESETS. There is no RAM memory cell associated with Stack Pointer value of OOOOOB. When a CALL type instruction executes (PUSH operation), the stack pointer is first incremented and the RAM location pointed to by STKPTR is written with the contents of the PC. During a RETURN type instruction (POP operation), the contents of the RAM location pointed to by STKPTR are transferred to the PC and then the stack pointer is decremented.

The stack space is a unique memory structure and is not part of either the program or the data space in the PIC18FXX2 devices. The STKPTR register is readable and writeable, and the address on the top of the stack is also readable and writeable through SFR registers. Data can also be pushed to or popped from the stack using the top-of-stack SFRs. Status bits indicate if the stack pointer is at, or beyond the 31 levels provided.

Stack Operations

Figure 2.7 shows the bit structure of the STKPTR register. The STKPTR register contains the stack pointer value, as well as a stack full and stack underflow) status bits. The STKPTR register can be read and written by the user. This feature allows operating system software to perform stack maintenance operations. The 5-bit value in the stack pointer register ranges from 0 through 31, which correspond to the available stack locations. The stack pointer is incremented by push operations and decremented when values are popped off the stack. At RESET, the stack pointer value is set to 0.

bits:	7	6	5	4	3	2	1	0
	STKOVF	STKUNF		SP4	SP3	SP2	SP1	SP0

bit 7   **STKOVF:**
    1 = Stack became full or overflowed
    0 = Stack has not overflowed
bit 6   **STKUNF:**
    1 = Stack underflow occurred
    0 = No stack underflow occurred
bit 5   **Unimplemented:** Read as 0
bit 4-0   **SP4:SP0:** Stack Pointer location

Figure 2.7 *STKPTR register bit map.*

The STKOVF bit is set after the program counter is pushed onto the stack 31 times without popping any value off the stack. Notice that some Microchip documentation refers to a STKFUL bit, which appears to be a synonym for the STKOVF bit. To avoid confusion, we only use the STKOVF designation in this book.

The STKOVF bit can only be cleared in software or by a Power-On Reset (POR) operation. The action that takes place when the stack becomes full depends on the state of the STVREN (Stack Overflow Reset Enable) configuration bit. The STVREN bit is bit 0 of the CONFIG4L register. If the STVREN bit is set, a stack full or stack overflow condition will cause a device RESET. Otherwise, the RESET action will not take place. When the stack pointer has a value of 0 and the stack is popped, a value of zero is entered to the Program Counter and the STKUNF bit is set. In this case, the stack pointer remains at 0. The STKUNF bit will remain set until cleared in software or a POR occurs. Returning a value of zero to the Program Counter on an underflow condition has the effect of vectoring the program to the RESET vector. User code can provide logic at the RESET vector to verify the stack condition and take the appropriate actions.

Three registers, labeled TOSU, TOSH and TOSL, hold the contents of the stack location pointed to by the STKPTR register. The address mapping of these registers is shown in Figure 2.8.

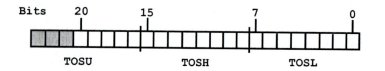
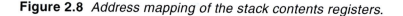

Figure 2.8 *Address mapping of the stack contents registers.*

Users can implement a software stack by manipulating the contents of the TOSU, TOSH, and TOSL registers. After a CALL, RCALL, or interrupt, user software can read the value in the stack by reading the TOSU, TOSH, and TOSL. These values can then be placed on a user-defined software stack. At return time, user software can replace the TOSU, TOSH, and TOSL with the stored values. At this time, global interrupts should have been disabled in order to prevent inadvertent stack changes.

Fast Register Stack

A fast return from interrupts is available in the PIC18FXX2 devices. This action is based on a Fast Register Stack that saves the STATUS, WREG, and BSR registers. The fast version of the stack is not readable or writable and is loaded with the current value of the three registers when an interrupt takes place. The FAST RETURN instruction is then used to restore the working registers and terminate the interrupt.

The fast register stack option can also be used to store and restore the STATUS, WREG, and BSR registers during a subroutine call. In this case, a fast call and fast return instruction are executed. This is only possible if no interrupts are used.

Instructions in Memory

Program memory is structured in byte-size units but instructions are stored as two bytes or four bytes. The Least Significant Byte of an instruction word is always stored in a program memory location with an even address, as shown in Figure 2.5. Figure 2.9 shows three low-level instructions as they are encoded and stored in program memory

LOCATIONS IN PROGRAM MEMORY:

INSTRUCTIONS:		LSB = 1	LSB = 0	Word Address
				00000H
				00002H
				00004H
				00006H
MOVLW	055H	0FH	55H	00008H
GOTO	000006H	EFH	03H	0000AH
		00H	00H	0000CH
MOVFF	12H, 456H	C1H	23H	0000EH
		04H	56H	00010H
				00012H
				00014H

Figure 2.9 *Instruction encoding.*

The CALL and GOTO instructions have an absolute program memory address embedded in the instruction. Because instructions are always stored on word boundaries, the data contained in the instruction is a word address. This word address is written to Program Counter bits <20:1>, which accesses the desired byte address. Notice in Figure 2.9 that the instruction

```
GOTO      000006H
```

is encoded by storing the number of single-word instructions that must be added to the Program Counter (03H). All program branch instructions are encoded in this manner.

2.2.3 Data Memory

Data memory is implemented as static RAM. Each register in the data memory has a 12-bit address, allowing up to 4096 bytes of data memory in the PIC18FXX2 devices. Data memory consists of Special Function Registers (SFRs) and General Purpose Registers (GPRs). The SFRs are used for control and status operations and for implementing the peripheral functions. The GPRs are for user data storage. Figure 2.10 is a map of data memory in the PIC18FXX2 devices.

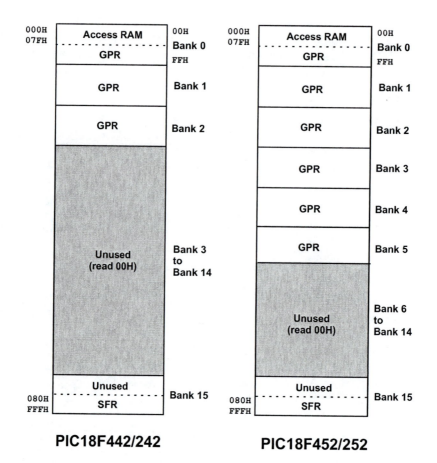

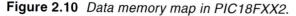

Figure 2.10 *Data memory map in PIC18FXX2.*

In Figure 2.10, GPRs start at the first location of Bank 0 and grow to higher memory addresses. Memory is divided into 255-byte units called banks. Seven banks are implemented in the PIC18F452/252 devices and four banks in the PIC18F442/242 devices. A read operation to a location in an unimplemented memory bank always returns zeros.

The entire data memory may be accessed directly or indirectly. Direct addressing requires the use of the BSR register. Indirect addressing requires the use of a File Select Register (FSRn) and a corresponding Indirect File Operand (INDFn). Addressing operations are discussed in Chapter 11 in the context of LCD programming.

Each FSR holds a 12-bit address value that can be used to access any location in the Data Memory map without banking. The SFRs start at address F80H in Bank 15 and extend to address 0FFFH in either device. This means that 128 bytes are assigned to the SFR area although not all locations are implemented. The individual SFRs are discussed in the context of their specific functionality. Figure 2.11 shows the names and addresses of the Special Function Registers.

Address	Name	Address	Name	Address	Name	Address	Name	Address	Name
FFFh	TOSU	FDFh	INDF2[3]	FBFh	CCPR1H	F9Fh	IPR1		
FFEh	TOSH	FDEh	POSTINC2[3]	FBEh	CCPR1L	F9Eh	PIR1		
FFDh	TOSL	FDDh	POSTDEC2[3]	FBDh	CCP1CON	F9Dh	PIE1		
FFCh	STKPTR	FDCh	PREINC2[3]	FBCh	CCPR2H	F9Ch	—		
FFBh	PCLATU	FDBh	PLUSW2[3]	FBBh	CCPR2L	F9Bh	—		
FFAh	PCLATH	FDAh	FSR2H	FBAh	CCP2CON	F9Ah	—		
FF9h	PCL	FD9h	FSR2L	FB9h	—	F99h	—		
FF8h	TBLPTRU	FD8h	STATUS	FB8h	—	F98h	—		
FF7h	TBLPTRH	FD7h	TMR0H	FB7h	—	F97h	—		
FF6h	TBLPTRL	FD6h	TMR0L	FB6h	—	F96h	TRISE[2]		
FF5h	TABLAT	FD5h	T0CON	FB5h	—	F95h	TRISD[2]		
FF4h	PRODH	FD4h	—	FB4h	—	F94h	TRISC		
FF3h	PRODL	FD3h	OSCCON	FB3h	TMR3H	F93h	TRISB		
FF2h	INTCON	FD2h	LVDCON	FB2h	TMR3L	F92h	TRISA		
FF1h	INTCON2	FD1h	WDTCON	FB1h	T3CON	F91h	—		
FF0h	INTCON3	FD0h	RCON	FB0h	—	F90h	—		
FEFh	INDF0[3]	FCFh	TMR1H	FAFh	SPBRG	F8Fh	—		
FEEh	POSTINC0[3]	FCEh	TMR1L	FAEh	RCREG	F8Eh	—		
FEDh	POSTDEC0[3]	FCDh	T1CON	FADh	TXREG	F8Dh	LATE[2]		
FECh	PREINC0[3]	FCCh	TMR2	FACh	TXSTA	F8Ch	LATD[2]		
FEBh	PLUSW0[3]	FCBh	PR2	FABh	RCSTA	F8Bh	LATC		
FEAh	FSR0H	FCAh	T2CON	FAAh	—	F8Ah	LATB		
FE9h	FSR0L	FC9h	SSPBUF	FA9h	EEADR	F89h	LATA		
FE8h	WREG	FC8h	SSPADD	FA8h	EEDATA	F88h	—		
FE7h	INDF1[3]	FC7h	SSPSTAT	FA7h	EECON2	F87h	—		
FE6h	POSTINC1[3]	FC6h	SSPCON1	FA6h	EECON1	F86h	—		
FE5h	POSTDEC1[3]	FC5h	SSPCON2	FA5h	—	F85h	—		
FE4h	PREINC1[3]	FC4h	ADRESH	FA4h	—	F84h	PORTE[2]		
FE3h	PLUSW1[3]	FC3h	ADRESL	FA3h	—	F83h	PORTD[2]		
FE2h	FSR1H	FC2h	ADCON0	FA2h	IPR2	F82h	PORTC		
FE1h	FSR1L	FC1h	ADCON1	FA1h	PIR2	F81h	PORTB		
FE0h	BSR	FC0h	—	FA0h	PIE2	F80h	PORTA		

Note 1: Unimplemented registers are read as '0'.
2: This register is not available on PIC18F2X2 devices.
3: This is not a physical register.

Figure 2.11 *PIC18FXX2 Special Function Registers map.*

2.2.4 Data EEPROM Memory

EEPROM stands for Electrically Erasable Programmable Read-Only Memory. This type of memory is used in computers and embedded systems as a nonvolatile storage. You find EEPROM in flash drives, BIOS chips, and in memory facilities such flash memory and EEPROM data storage memory found in PICs and other microcontrollers.

EEPROM memory can be erased and programmed electrically without removing the chip. The predecessor technology, called EPROM, required that the chip be removed from the circuit and placed under ultraviolet light in order to erase it. In embedded systems, the typical use of serial EEPROM on-board memory, and EEPROM ICs, is in the storage of passwords, codes, configuration settings, and other information to be remembered after the system is turned off.

Data EEPROM is readable and writable during normal operation. EEPROM data memory is not directly mapped in the register file space. Instead, it is indirectly addressed through the SFRs. Four SFRs used to read and write the program and data EEPROM memory. These registers are

- EECON1
- EECON2
- EEDATA
- EEADR

In operation, EEDATA holds the 8-bit data for read/write and EEADR holds the address of the EEPROM location being accessed.

All devices of the PIC18FXX2 family 256 bytes of data EEPROM with an address range from Oh to FFh. EEPROM access and programming are discussed in Chapter 10.

2.2.5 Indirect Addressing

The instruction set of most processors, including the PICs, provide a mechanism for accessing memory operands indirectly. Indirect addressing is based on the following capabilities:

1. The address of a memory operand is loaded into a register. This register is called the pointer.

2. The pointer register is then used to indirectly access the memory location at the address it "points to."

3. The value in the pointer register can be modified (usually incremented or decremented) so as to allow access to other memory operands.

Indirect addressing is useful in accessing data tables in manipulating software stacks.

In the PIC18FXX2 architecture, indirect addressing is implemented using one of three Indirect File Registers (labeled INDFx) and the corresponding File Select Register (labeled FSRx). Any instruction using an INDFx register actually accesses the register pointed to by the FSRx. Reading an INDF register indirectly (FSR = 0) will read OOH. Writing to the INDF register indirectly, results in a no operation.

The INDFx registers are not physical registers in the sense that they cannot be directly accessed by code. The FSR register is the pointer register that is initialized to the address of a memory operand. Once a memory address is placed in FSRx, any action on the corresponding INDFx register takes place at the memory location pointed at by FSR. For example, if the FSR0 register is initialized to memory address 0x20, then clearing an INDF0 register has the effect of clearing the memory location at address 0x20. In this case, the action on the INDF0 register actually takes place at the address contained in the FSR0 register. Now if FSR (the pointer register) is incremented and INDF is again cleared, the memory location at address 0x21 will be cleared. Indirect addressing is covered in detail in the programming chapters later in the book.

2.3 PIC18FXX2 Oscillator

In the operation of any microprocessor or microcontroller, it is necessary to provide a "clock" signal in the form of a continuously running, fixed-frequency, square wave. The operation and speed of the device are entirely dependent on this clock frequency. In addition to the fetch/execute cycle of the CPU, other essential timing functions are also derived from this clock signal ranging from timing and counting operations pulses required in communications. In many PIC microcontrollers, the internal or external component that generates this clock signal is called the oscillator.

Every microcontroller or microprocessor must operate with a clock signal of a specified frequency. The principal clock signal is divided internally by a fixed value, thus creating a lower-frequency signal. Each cycle of this slower signal is called an instruction cycle by Microchip. The instruction cycle can be considered the primary unit of time in the action of the CPU because it determines how long an instruction takes to execute. The original clock signal is also used to create phases or time stages within the instruction cycle or in other microcontroller operations. In PIC18FXX2 devices, the main oscillator signal is divided by four. For example, a clock signal frequency of 40 MHz produces an instruction cycle frequency of 10 MHz. Many microcontrollers, including PICs, provide an internal oscillator signal; however, this is not the case with the PIC18FXX2 devices, which require an external device to provide the clock signal. The pins labeled OSC1 and OSC2 (see Figure 2.2) are used with the oscillator function.

2.3.1 Oscillator Options

The PIC18FXX2 can be operated in eight different oscillator modes. The configuration bits labeled FOSC2, FOSC1, and FOSCO allow selecting one of these eight modes during start-up. Table 2.2 shows the designations and description of the eight oscillator modes.

Table 2.2

Oscillator Types

CODE	TYPE
LP	Low-Power Crystal
XT	Crystal/Resonator
HS	High-Speed Crystal/Resonator
HS + PLL	High-Speed Crystal/Resonator with PLL enabled
RC	External Resistor/Capacitor
RCIO	External Resistor/Capacitor withI/O pin enabled
EC	External Clock
ECIO	External Clock with I/O pin enabled

Crystal Oscillator and Ceramic Resonator

The designations XT, LP, HS or HS+PLL in Table 2.2 refer to modes in which a crystal or ceramic resonator os connected to the OSC1 and OSC2 pins to establish a clock signal for the device. The PIC18FXX22 requires that crystals be parallel cut because serial cut crystals can give frequencies outside the manufacturer's specifications. Figure 2.12 shows the wiring and components required for oscillator modes LP, XT, and HS.

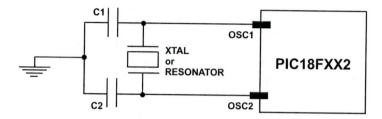

CAPACITOR SELECTION (C1 and C2)
FOR CERAMIC RESONATORS

Mode	Freq	C1 and C2
XT	455 kHz	68 - 100 pF
	2.0 MHz	15 - 68 pF
	4.0 MHz	15 - 68 pF
HS	8.0 MHz	10 - 68 pF
	16.0 MHz	10 - 22 pF

CAPACITOR SELECTION (C1 and C2)
FOR CRYSTAL OSCILLATOR

Mode	Freq	C1 and C2
LP	32.0 kHz	33 pF
	200 kHz	15 pF
XT	200 kHz	22 - 68 pF
	1.0 MHz	15 pF
	4.0 MHz	15 pF
HS	4.0 MHz	15 pF
	8.0 Mhz	15 - 33 pF
	20.0 MHz	15 - 33 pF
	25.0 Mhz	15 - 33 pF

Figure 2.12 *Oscillator schematics for LP, XT, and HS modes.*

An external clock may also be used in the HS, XT, and LP oscillator modes. In this case, the clock is connected to the device's OSC1 pin while the OSC2 pin is left open.

RC Oscillator

The simplest and least expensive way of providing a clocked impulse to the PIC is with an external circuit consisting of a single resistor and capacitor. This circuit is usually called an RC oscillator. The major drawback of the RC option is that the frequency of the pulse depends on the supply voltage, the normal variations in the actual values of the resistor and capacitor, and the operating temperature. This makes the RC oscillator option only suitable for applications that are timing insensitive. Figure 2.13 shows the circuit required for the RC and RCIO oscillator modes.

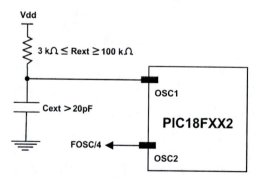

Figure 2.13 *RC and RCIO oscillator modes.*

The two variations of the RC option are designated RC and RCIO. In the RC option, the OSC2 pin is left open. In the RCIO variation, the OSC2 pin provides a signal with the oscillator frequency divided by 4 (FOSC/4 in Figure 2.13). This signal can be used for testing or to synchronize other circuit components.

External Clock Input

The EC and ECIO Oscillator modes are used with an external clock source connected to the OSC1 pin. Figure 2.14 shows the circuit for the EC oscillator mode.

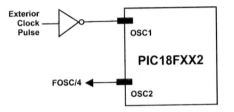

Figure 2.14 *External clock oscillator mode.*

In the EC mode (Figure 2.14), the oscillator frequency divided by 4 is available on the OSC2 pin. This signal may be used for test purposes or to synchronize other circuit components.

The ECIO oscillator is similar to the EC mode except that the OSC2 pin becomes an additional general-purpose I/O source; specifically, the OSC2 pin becomes bit 6 of PORTA (RA6).

Phase Locked Loop Oscillator Mode

With the Phase Locked Loop (PLL) a circuit is provided as a programmable option. This is convenient for users who want to multiply the frequency of the incoming crystal oscillator signal by 4, as in Figure 2.12. For example, if the input clock frequency is 10 MHz and the PLL oscillator mode is selected, the internal clock frequency will be 40 MHz. The PLL mode is selected by means of the FOSC<2:0> bits. This requires that the oscillator configuration bits are programmed for the HS mode. Otherwise, the system clock will come directly from OSC1.

2.4 System Reset

The PIC18FXX2 documentation refers to the following eight possible types of RESET.

1. Power-On Reset (POR)
2. Master Clear Reset during normal operation (MCLR)
3. Reset during SLEEP (MCLR)
4. Watchdog Timer Reset
5. Programmable Brown-Out Reset
6. Action of the RESET Instruction
7. Stack Full Reset
8. Stack Underflow Reset

The status of most registers is unknown after Power-on Reset (POR) and unchanged by all other RESETS. The remaining registers are forced to a "RESET state" on Power-on Reset, MCLR, WDT Reset, Brownout Reset, MCLR Reset during SLEEP, and by the RESET instruction.

2.4.1 Reset Action

Most registers are not affected by a WDT wake-up, as this is viewed as the resumption of normal operation. Status bits from the RCON register, RI, TO, PD, POR, and BOR, are set or cleared differently in the various RESET actions. Software can read these bits to determine the type of RESET. Table 2.3 shows the RESET condition for some Special Function Registers.

Table 2.3

RESET State for some SFRs

Condition	Program Counter	RCON Register	$\overline{RI}$	$\overline{TO}$	$\overline{PD}$	$\overline{POR}$	$\overline{BOR}$	STKFUL	STKUNF
Power-On Reset	0000h	0--1 1100	1	1	1	0	0	u	u
MCLR Reset during normal operation	0000h	0--u uuuu	u	u	u	u	u	u	u
Software Reset during normal operation	0000h	0--0 uuuu	0	u	u	u	u	u	u
Stack Full Reset during normal operation	0000h	0--u uu11	u	u	u	u	u	u	1
Stack Underflow Reset during normal operation	0000h	0--u uu11	u	u	u	u	u	1	u
MCLR Reset during SLEEP	0000h	0--u 10uu	u	1	0	u	u	u	u
WDT Reset	0000h	0--u 01uu	1	0	1	u	u	u	u
WDT Wake-up	PC + 2	u--u 00uu	u	0	0	u	u	u	u
Brown-out Reset	0000h	0--1 11u0	1	1	1	1	0	u	u
Interrupt wake-up from SLEEP	PC + 2[1]	u--u 00uu	u	1	0	u	u	u	u

Legend: u = unchanged, x = unknown, - = unimplemented bit, read as '0'

Note 1: When the wake-up is due to an interrupt and the GIEH or GIEL bits are set, the PC is loaded with the interrupt vector (0x000008h or 0x000018h).

Some circuits include a hardware reset mechanism that allows the user to force a RESET action, usually by activating a switch that brings low the MCLR line. The same circuit holds high the MCLR line during device operation by tying it to the Vdd source. Figure 2.15 shows a possible schematic for a pushbutton reset switch on the MCLR line.

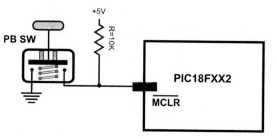

Figure 2.15 *RESET switch on the MCLR line.*

Power-On Reset (POR)

A Power-on Reset pulse is generated on-chip when a Vdd rise is detected. Users can take advantage of the POR circuitry by tying the MCLR pin to Vdd either directly or through a resistor, as shown in Figure 2.15. This circuit eliminates external RC components usually needed to create a Power-on Reset delay.

Power-Up Timer (PWRT)

The Power-up Timer (PWRT) provides a fixed nominal time-out from POR. The PWRT operates on an internal RC oscillator. The chip is kept in RESET as long as the PWRT is active. This action allows the Vdd signal to rise to an acceptable level. A configuration bit is provided to enable/disable the PWRT.

Oscillator Start-Up Timer (OST)

The Oscillator Start-up Timer (OST) provides a delay of 1024 oscillator cycles from the time of OSC1 input until after the PWRT delay is over. This ensures that the processor fetch/execute cycle does not start until the crystal oscillator or resonator has started and stabilized. The OST time-out is invoked only for XT, LP, and HS modes and only on Power-on Reset or wake-up from SLEEP.

PLL Lock Time-Out

When the Phase Locked Loop Oscillator Mode is selected, the time-out sequence following a Power-on Reset is different from the other oscillator modes. In this case, a portion of the Power-up Timer is used to provide a fixed time-out that is sufficient for the PLL to lock to the main oscillator frequency. This PLL lock time-out (TPLL) is typically 2 ms and follows the oscillator start-up time-out (OST),

Brown-Out Reset (BOR)

A temporary reduction in electrical power (brown-out condition) can activate the chip's brown-out reset mechanism. A configuration bit (BOREN) can be cleared to disable or set to enable the Brown-out Reset circuitry. If Vdd falls below a predefined value for a predetermined period, the brown-out situation will reset the chip. The chip will remain in Brown-out Reset until Vdd rises above the predefined value.

Time-Out Sequence

On power-up, the time-out sequence follows this order:

1. After the Power-on Reset (POR) time delay has expired, the Power-up Time (PWRT) time-out is invoked

2. The Oscillator Start-up Time (OST) is activated

The total time-out will vary based on the particular oscillator configuration and the status of the PWRT. In RC mode with the PWRT disabled, there will be no time-out at all. Because the time-outs occur from the POR pulse, if MCLR is kept low long enough, the time-outs will expire. Bringing MCLR high will begin execution immediately. This is useful for testing purposes or to synchronize more than one PIC18FXXX device operating in parallel.

2.5 I/O Ports

PIC18FXX2 devices come equipped with either five or three ports. PIC18F4X2 devices have five ports and PIC18F2X2 devices have three ports. Ports provide access to the outside world and are mapped to physical pins on the device. Some port pins are multiplexed with alternate functions of peripheral modules. When a peripheral module is enabled, that pin ceases to be a general-purpose I/O.

Ports are labeled with letters of the alphabet and are designated as port A (PORTA) to port E (PORTE). Port pins are bi-directional, that is, each pin can be configured to serve either as input or output. Each port has three registers for its operation. These are

- A TRIS register that determines data direction

- A PORT register used to read the value stored in each port pin or to write values to the port's data latch

- A LAT register that serves as a data latch and is useful in read-modify-write operations on the pin values

The status of each line in the port's TRIS register determines if the port's line is designated as input or output. Storing a value of 1 in the port's line TRIS register makes the port line an input line, while storing a value of 0 makes it an output line. Input port lines are used in communicating with input devices, such as switches, keypads, and input data lines from hardware devices. Output port lines are used in communicating with output devices, such as LEDs, seven-segment displays, liquid-crystal displays (LCDs), and data output line to hardware devices.

Port pins are bit mapped, however, they are read and written as a unit. For example, the PORTA register holds the status of the eight pins possibly mapped to port A, while writing to PORTA will write to the port latches. Write operations to ports are actually read-modify-write operations. Therefore, the port pins are first read, then the value is modified, and then written to the port's data latch.

As previously mentioned, some port pins are multiplexed; for example, pin RA4 is multiplexed with the Timer0 module clock input, labeled T0CKI. In Figure 2.2 the port pin is shown as RA4/T0CKI. Other port pins are multiplexed with analog inputs and with other peripheral functions. The device data sheets contain information regarding the functions assigned to each device pin.

2.5.1 Port Registers

In PIC18FXX2 devices, ports are labeled PORTA, PORTB, PORTC, PORTD, and PORTE. PORTD and PORTE are only available in PIC18F4X2 devices. The characteristics of each port are detailed in the device's data sheet. For example, PORTA is a 7-bit wide, bi-directional port. The corresponding Data Direction register is TRISA. If software sets a TRISA bit to 1, the corresponding PORTA pin will serve as an input pin. Clearing a TRISA bit (= 0) will make the corresponding PORTA pin an output. It is easy to remember the function of the TRIS registers because the number 1 is reminiscent of the letter I and the number 0 of the letter O.

Reading the PORTA register reads the status of the pins, whereas writing to it will write to the port latch. The Data Latch register (LATA) is also memory mapped. Read-modify-write operations on the LATA register read and write the latched output value for PORTA. The RA4 pin is multiplexed with the Timer0 module clock input to become the RA4/TOCKI pin. This pin is a Schmitt Trigger input and an open drain output. All other RA port pins have TTL input levels and full CMOS output drivers.

All other PORTA pins are multiplexed with analog inputs and the analog VREF+ and VREF- inputs. The operation of each pin is selected by clearing/setting the control bits in the ADCON1 register (A/D Control Register). The TRISA register controls the direction of the PORTA pins. This is so even when the port pins are being used as analog inputs.

2.5.2 Parallel Slave Port

The Parallel Slave Port is implemented on the 40-pin devices only, that is, those with the PIC18F4X2 designation. In these devices, PORTD serves as an 8-bit wide Parallel Slave Port when the control bit labled PSPMODE (at TRISE<4>) is set. It is asynchronously readable and writable through the RD control input pin (REO/RD) and WR control input pin (RE1/WR).

The Parallel Slave Port can directly interface to an 8-bit microprocessor data bus. In this case, the microprocessor can read or write the PORTD latch as an 8-bit latch. Programming and operation of the Parallel Slave Port is discussed later in this book.

2.6 Internal Modules

In electronics a module can be loosely defined as an assembly of electronic circuits or components that performs as a unit. All PIC microcontrollers contain internal modules to perform specific functions or operations. In this sense we can refer to the Timer module, the Capture/Compare/PWM module, or the Analog-to-Digital Converter module. By definition, a module is an internal component.

A peripheral or peripheral device, on the other hand, is an external component, such as a printer, a modem, or a Liquid Crystal Display (LCD). Microcontrollers often communicate with peripheral devices through their I/O ports or through their internal modules. We make this clarification because sometimes in the literature we can find references to the "peripheral components" or the "peripherals" of a microcontroller when actually referring to modules.

2.6.1 PIC18FXX2 Modules

Most PIC microcontrollers contain at least one internal module, and many devices contain ten or more different modules. The following are the standard modules of the PIC18FXX2 family of devices:

- Timer0 module: 8-bit/16-bit timer/counter with 8-bit programmable prescaler
- Timer1 module: 16-bit timer/counter
- Timer2 module: 8-bit timer/counter with 8-bit period register

- Timer3 module: 16-bit timer/counter
- Two Capture/Compare/PWM (CCP) modules
- Master Synchronous Serial Port (MSSP) module with two modes of operation
- Universal Receiver and Transmitter (USART)
- 10-bit Analog-to-Digital Converter (A/D)
- Controller Area Network (CAN)
- Comparator Module
- Parallel Slave Port (PSP) module

The structure and details of the internal modules are discussed in the programming chapters later in this book.

Chapter 3

Programming Tools and Software

3.1 Environment

In order to learn and practice programming microcontrollers in embedded systems, you will require a development and testing environment. This environment will usually include the following elements:

1. A software development environment in which to create the program's source file and generate an executable program that can later be loaded into the hardware device. This environment often includes debuggers, library managers, and other auxiliary tools.

2. A hardware device called a "programmer" that transfers the executable program to the microcontroller itself. In the present context, this process is usually called "burning" or "blowing" the PIC.

3. A circuit or demonstration board in which the program (already loaded onto a PIC microcontroller) can be tested in order to locate defects or confirm its functionality.

In the present chapter, we discuss some of the possible variations in these elements for PIC18F programming and system development.

3.1.1 Embedded Systems

An embedded system is designed for a specific purpose, in contrast to a computer system, which is a general-purpose machine. The embedded system is intended for executing specific and predefined tasks, for example, to control a microwave oven, a TV receiver, or to operate a model railroad. In a general-purpose computer, on the other hand, the user may select among many software applications. For example, the user may run a word processor, a Web browser, or a database management system on the desktop. Because the software in an embedded system is usually fixed and cannot be easily changed, it is called "firmware."

At the heart of an embedded system is a microcontroller (such as a PIC), some-times several of them. These devices are programmed to perform one, or, at most, a few tasks. In the most typical case an embedded system also includes one or more "peripheral" circuits that are operated by dedicated ICs or by functionality con-tained in the microcontroller itself. The term "embedded system" refers to the fact that the programmable device is often found inside another one; for instance, the control circuit is embedded in a microwave oven. Furthermore, embedded systems do not have (in most cases) general-purpose devices such as hard disk drives, video controllers, printers, and network cards.

The control for a microwave oven is a typical embedded system. The controller includes a timer (so that various operations can be clocked), a temperature sensor (to provide information regarding the oven's condition), perhaps a motor (to option-ally rotate the oven's tray), a sensor (to detect when the oven door is open), and a set of pushbutton switches to select the various options. A program running on the embedded microcontroller reads the commands and data input through the key-board, sets the timer and the rotating table, detects the state of the door, and turns the heating element on and off as required by the user's selection. Many other daily devices, including automobiles, digital cameras, cell phones, and home appliances, use embedded systems and many of them are PIC-based.

3.1.2 High- and Low-Level Languages

All microcontrollers can be programmed in their native machine language. The term "machine language" refers to the primitive codes, internal to the CPU, that execute the fundamental operations that can be performed by a particular processor. The proces-sor's fetch/execute cycle retrieves the machine code from program memory and per-forms the necessary manipulations and calculations. For example, the instruction represented by the binary opcode

```
00000000 00000100
```

clears the watchdog timer register in the PIC18FXX devices.

Programming, loosely defined, refers to selecting, configuring, and storing in pro-gram memory a sequence of primitive opcodes so as to perform a specific function or task. The machine language programmer has to manually determine the opera-tion code for each instruction in the program and places these codes, in a specific order, in the designated area of program memory.

Assembly language is based on a software program that recognizes a symbolic language where each machine code is represented by a mnemonic instruction and a possible operand. A program, called an "assembler," reads these instructions and operands from a text file and generates the corresponding machine codes. For ex-ample, in order to encode the instruction that clears the watchdog (00000000 00000100 binary in the previous example), the assembly language programmer in-serts in the text file the keyword

```
CLRWDT
```

The assembler program reads the programmer's text file, parses these mnemonic keywords and their possible operands, and stores the binary opcodes in a file for later execution. Because assembly language references the processor opcodes, it is a machine-specific language. An assembler program operates only on devices that have a common machine language, although some minor processor-specific variations can, in some cases, be selectively enabled. Because of their association with the hardware, machine language and assembly language are usually referred to as "low-level languages."

High-level programming languages, such as C, Pascal, and Fortran, provide a stronger level of abstraction from the hardaware device. It is generally accepted that compared to low-level languages, high-level programming is more natural, easier to learn, and simplifies the process of developing program languages. The result is a simpler and more understandable development environment that comes at some penalties regarding performance, hardware control, and program size.

Rather than dealing with registers, memory addresses, and call stacks, a high-level language deals with variables, arrays, arithmetic or Boolean expressions, subroutines and functions, loops, threads, locks, and other abstract concepts. In a high-level language, the design focuses on usability rather than optimal program efficiency. Unlike low-level assembly languages, high-level languages have few, if any, elements that translate directly into the machine's native opcodes.

The term "abstraction penalty" is sometimes used in the context of high-level languages in reference to limitations that are evident when computational resources are limited, maximum performance is required, or hardware control is mandated. In some cases, the best of both worlds can be achieved by coding the noncritical portions of a program mostly in a high-level language while the critical portions are developed in assembly language. This results in mixed-language programs, which are discussed later in this book.

It should be noted that many argue that modern developments in high-level languages, based on well-designed compilers, produce code comparable in efficiency and control to that of low-level languages. Another advantage of high-level languages is that their design is independent of machine structures, and the hardware features of a specific device result in code that can be easily ported to different systems. Finally we should observe that the terms "low-level" and "high-level" languages are not cast in stone: to some, assembly language with the use of macros and other tools becomes a high-level language, while C is sometimes categorized as low-level due to its compact size, direct memory addressing, and low-level operands.

As previously mentioned, the major argument in favor of high-level languages is their ease of use and their faster learning curve. The advantages of assembly language, on the other hand, are better control and greater efficiency. It is true that arguments that favor high-level languages find some justification in the computer world, but these reasons are not always valid in the world of microcontroller programming. In the first place, the microcontroller programmer is not always able to avoid complications and technical details by resorting to a high-level language because the programs relate closely to hardware devices and to electronic circuits. These devices and circuits must be understood at their most essential level if they are to be controlled and operated by software. For example, consider a microcontroller program that must provide some sort of control baseded on the action of a thermostat. In this case, the programmer must become familiar with temperature sensors, analog-to-digital conversions, motor controls, and so on. This is true whether the program will be written in a low- or a high-level language. For these reasons we have considered both high-level and low-level programming of the microcontrollers is discussed in this book.

3.1.3 Language-Specific Software

Developing programs in a particular programming language requires a set of matching software tools. These software development tools are either generic, that is, suitable for any programming language, or multi-language. Fortunately, for PIC programming, all the necessary software tools are furnished in a single development environment that includes editors, assemblers, compilers, debuggers, library managers, and other utilities. This development package, called MPLAB, discussed in the following sections.

3.2 Microchip's MPLAB

MPLAB is the name of the PIC assembly language development system provided by Microchip. The package is furnished as an integrated development environment (IDE) and can be downloaded from the company's website at www.microhip.com. The MPLAB package is furnished for Windows, Linux, and Mac OS systems. At the time of this writing, the current MPLAB version is 8.86.

3.2.1 MPLAB X

A new implementation of MPLAB is named MPLAB X. This new package, available free on the Microchip website, is described by Microchip as "an integrated environment to develop code for embedded microcontrollers." This definition matches the one for the conventional MPLAB; however, the MPLAB X package brings many changes to the conventional MPLAB environment. In the first place, MPLAB X is based on the open source NetBeans IDE from Oracle. This has allowed Microchip to add many features and to be able to quickly update the software in the context of a more extensible architecture. Microchip also states that MPLAB X provides many new features that will be especially beneficial to users of 16- and 32-bit microprocessor families, where programs can quickly become extremely complex.

Because MPLAB X is still considered "work in progress," we have not used it in developing the programs that are part of this book. Furthermore, the expanded features of this new environment have added complications in learning and using this package. For the processors considered in this book, and the scope of the developed software, we have considered these new features an unnecessary complication.

3.2.2 Development Cycle

The development cycle of an embedded system consists of the following steps

1. Define system specifications. This step includes listing the functions that the system is to perform and determining the tests that will be used to validate their operations.

2. Select system components according to the specifications. This step includes locating the microcontroller that best suits the system.

3. Design the system hardware. This step includes drawing the circuit diagrams.

4. Implement a prototype of the system hardware by means of breadboards, wire boards, or any other flexible implementation technology.

5. Develop, load, and test the software.

6. Implement the final system and test hardware and software.

The commercial development of an embedded system is hardly ever the work of a single technician. More typically, it requires the participation of computer, electrical, electronic, and software engineers. Note that, in the present context, we consider computer programmers as software engineers. In addition, professional project managers are usually in charge of the development team.

3.3 An Integrated Development Environment

The MPLAB development system, or integrated development environment, consists of a system of programs that run on a PC. This software package is designed to help develop, edit, test, and debug code for the Microchip microcontrollers. Installing the MPLAB package is usually straightforward and simple. The package includes the following components:

1. MPLAB editor. This tool allows creating and editing the assembly language source code. It behaves very much like any Windows™ editor and contains the standard editor functions, including cut and paste, search and replace, and undo and redo functions.

2. MPLAB assembler. The assembler reads the source file produced in the editor and generates either absolute or relocatable code. Absolute code executes directly in the PIC. Relocatable code can be linked with other separately assembled modules or with libraries.

3. MPLAB linker. This component combines modules generated by the assembler with libraries or other object files, into a single executable file in .hex format.

4. MPLAB debuggers. Several debuggers are compatible with the MPLAB development system. Debuggers are used to single-step through the code, breakpoint at critical places in the program, and watch variables and registers as the program executes. In addition to being a powerful tool for detecting and fixing program errors, debuggers provide an internal view of the processor, which is a valuable learning tool.

5. MPLAB in-circuit emulators. These are development tools that allow performing basic debugging functions while the processor is installed in the circuit.

Figure 3.1 is a screen image of the MPLAB program. The application on the editor window is one of the programs developed later in this book.

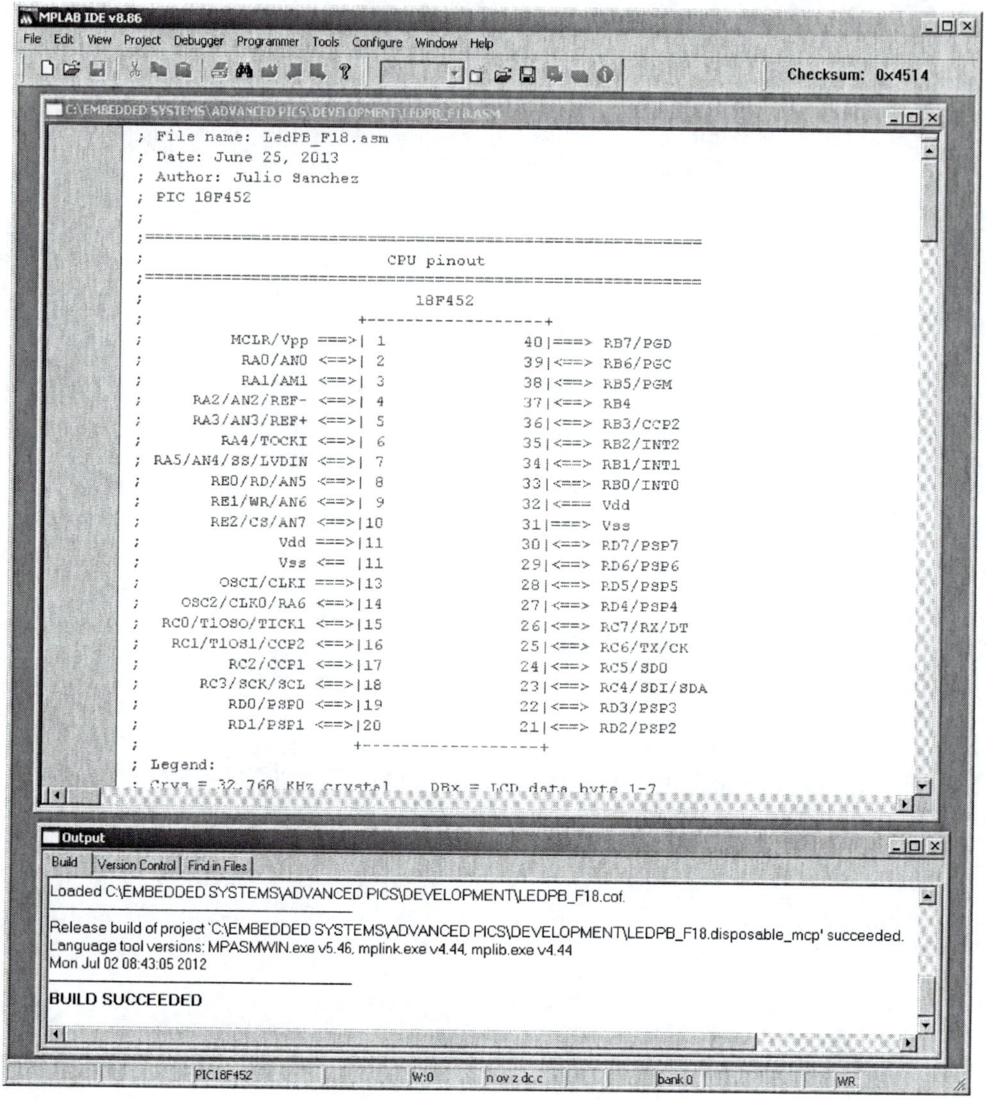

```
MPLAB IDE v8.86                                                          _ |□| x
File  Edit  View  Project  Debugger  Programmer  Tools  Configure  Window  Help

 □ ☞ 🗎 | ✂ ▯ ▮ | 🖨 ▦ ▭ ▮ ▮ ? |              ▾ ▯ ☞ 🗎 ▯ ▮ ▯ ①        Checksum: 0x4514

 C:\EMBEDDED SYSTEMS\ADVANCED PICS\DEVELOPMENT\LEDPB_F18.ASM                      _ |□| x
    ;  File name: LedPB_F18.asm
    ;  Date: June 25, 2013
    ;  Author: Julio Sanchez
    ;  PIC 18F452
    ;
    ;=====================================================================
    ;                          CPU pinout
    ;=====================================================================
    ;                           18F452
    ;               +-------------------+
    ;      MCLR/Vpp ===>| 1          40 |===> RB7/PGD
    ;      RA0/AN0  <==>| 2          39 |<==> RB6/PGC
    ;      RA1/AM1  <==>| 3          38 |<==> RB5/PGM
    ;   RA2/AN2/REF- <==>| 4         37 |<==> RB4
    ;   RA3/AN3/REF+ <==>| 5         36 |<==> RB3/CCP2
    ;      RA4/TOCKI <==>| 6         35 |<==> RB2/INT2
    ; RA5/AN4/SS/LVDIN <==>| 7       34 |<==> RB1/INT1
    ;      RE0/RD/AN5 <==>| 8        33 |<==> RB0/INT0
    ;      RE1/WR/AN6 <==>| 9        32 |<=== Vdd
    ;      RE2/CS/AN7 <==>|10        31 |===> Vss
    ;           Vdd ===>|11          30 |<==> RD7/PSP7
    ;           Vss <== |11          29 |<==> RD6/PSP6
    ;      OSCI/CLKI ===>|13         28 |<==> RD5/PSP5
    ;   OSC2/CLKO/RA6 <==>|14        27 |<==> RD4/PSP4
    ;   RC0/T1OSO/TICKI <==>|15      26 |<==> RC7/RX/DT
    ;   RC1/T1OSI/CCP2 <==>|16       25 |<==> RC6/TX/CK
    ;      RC2/CCP1 <==>|17          24 |<==> RC5/SDO
    ;      RC3/SCK/SCL <==>|18       23 |<==> RC4/SDI/SDA
    ;      RD0/PSP0 <==>|19          22 |<==> RD3/PSP3
    ;      RD1/PSP1 <==>|20          21 |<==> RD2/PSP2
    ;               +-------------------+
    ;  Legend:
    ;    Crys = 32.768 KHz crystal     DBx = LCD data byte 1-7

 Output                                                                          _ |□| x
 Build | Version Control | Find in Files |
 Loaded C:\EMBEDDED SYSTEMS\ADVANCED PICS\DEVELOPMENT\LEDPB_F18.cof.
 ─────────────────────────────────────────────────
 Release build of project `C:\EMBEDDED SYSTEMS\ADVANCED PICS\DEVELOPMENT\LEDPB_F18.disposable_mcp' succeeded.
 Language tool versions: MPASMWIN.exe v5.46, mplink.exe v4.44, mplib.exe v4.44
 Mon Jul 02 08:43:05 2012
 ─────────────────────────────────────────────────
 BUILD SUCCEEDED

 PIC18F452              W:0         n ov z dc c          bank 0              WR
```

Figure 3.1 *Screen snapshot of MPLAB IDE version 8.64.*

3.3.1 Installing MPLAB

In the normal installation, the MPLAB executable will be placed in the following path:

```
C:\Program Files\Microchip\MPASM Suite
```

Although the installation routine recommends that any previous version of MPLAB be removed from the system, we have found that it is unnecessary, considering that several versions of MPLAB can peacefully coexist.

Once the development environment is installed, the software is executed by clicking the MPLAB IDE icon. It is usually a good idea to drag and drop the icon onto the desktop so that the program can be easily activated.

With the MPLAB software installed, it may be a good idea to check that the applications were placed in the correct paths and folders. Failure to do so produces assembly-time failure errors with cryptic messages. To check the correct path for the software, open the Project menu and select the Set Language Tool Locations command. Figure 3.2 shows the command screen.

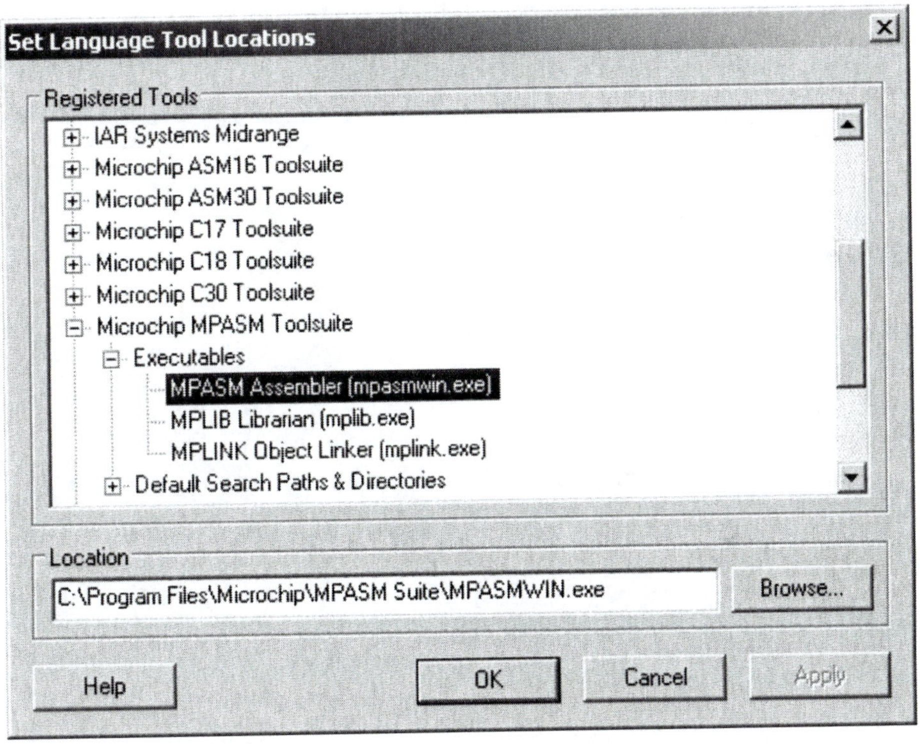

Figure 3.2 *MPLAB 8.64 set language tool locations screen.*

In the Set Language Tool Locations window, make sure that the file location coincides with the actual installation path for the software. If in doubt, use the <Browse> button to navigate through the installation directories until the executable program is located. In this case, mpasmwin.exe. Follow the same process for all the executables in the tool packages that will be used in development. For assembly language programs this is the Microchip MPASM Toolsuite shown in Figure 3.2

3.3.2 Creating the Project

In MPLAB, a project is a group of files generated or recognized by the IDE. Figure 3.3 shows the structure of an assembly language project.

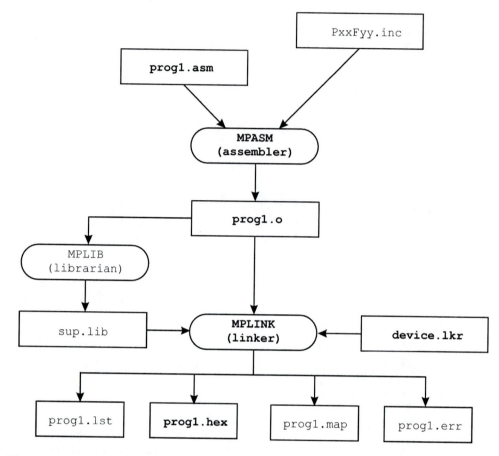

Figure 3-3 *MPLAB project files.*

Figure 3.3 shows an assembly language source file (prog1.asm) and an optional processor-specific include file that are used by the assembler program (MPASM) to produce an object file (prog1.o). Optionally, other sources and other include files may form part of the project. The resulting object file, as well as one or more optional libraries, and a device-specific script file (device.lkr), are then fed to the linker program (MPLINK), which generates a machine code file (prog1.hex) and several support files with listings, error reports, and map files. It is the .hex file that is used to blow the PIC.

Other files, in addition to those in Figure 3.3, may also be produced by the development environment. These vary according to the selected tools and options. For example, the assembler or the linker can be made to generate a file with the extension .cod, which contains symbols and references used in debugging.

Projects can be created using the <New> command in the Project menu. The programmer then proceeds to configure the project manually and add to it the required files. An alternative option, much to be preferred when learning the environment is using the <Project Wizard> command in the project menu. The wizard will prompt you for all the decisions and options that are required, as follows

1. Device selection. Here the programmer selects the PIC hardware for the project, for example, 18F452.

2. Select a language toolsuite. The purpose of this screen is to make sure that the proper development tools are selected and their location is correct.

3. Next, the wizard prompts the user for a name and directory. The Browse button allows for navigating the system. It is also possible to create a new directory at this time.

4. In the next step, the user is given the option of adding existing files to the project and renaming these files if necessary. Because most projects reuse a template, an include file, or other prexisting resources, this can be a useful option.

5. Finally the wizard displays a summary of the project parameters. When the user clicks on the <Finish> button the project is created and programming can begin.

Figure 3.4 is a snapshot of the final wizard screen.

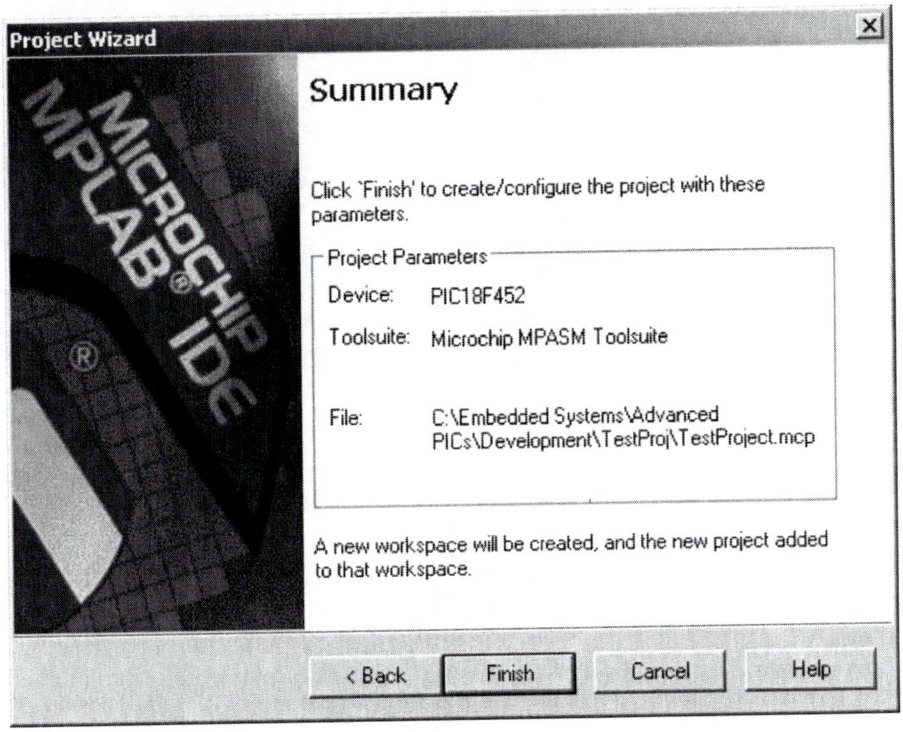

Figure 3-4 *Final screen of the Project Creation Wizard.*

3.3.3 Setting the Project Build Options

The <Build Options: Project> command in the Project menu allows the user to customize the development environment. For projects to be coded in assembly language, the MPASM Assembler Tab on the Build Options for Project screen is one of the the most used. The screen is shown in Figure 3.5.

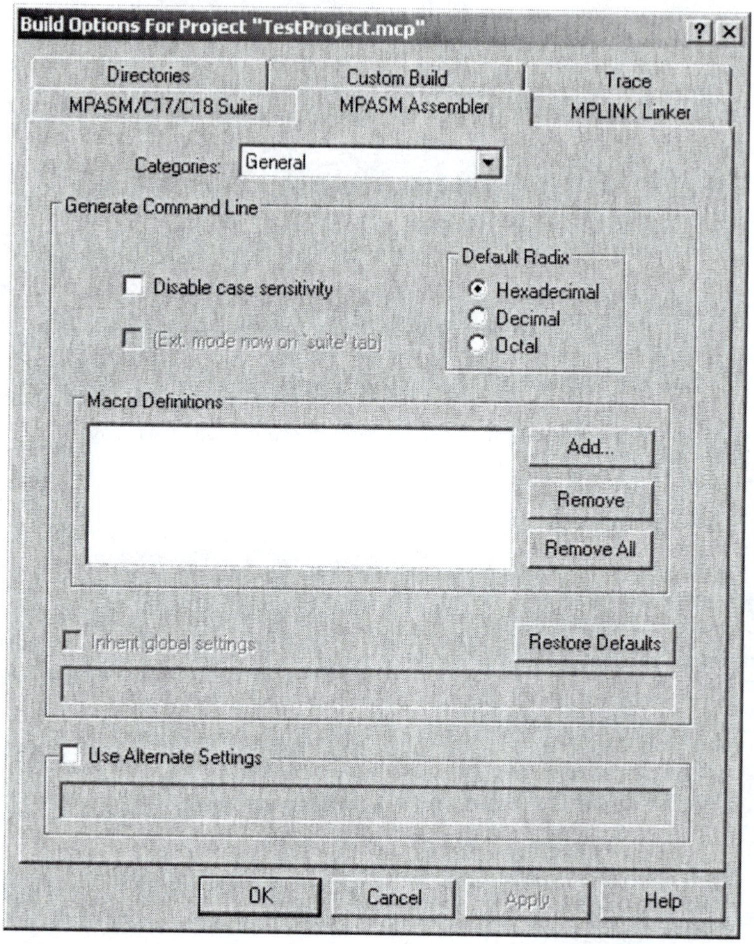

Figure 3-5 *MPASM assembler tab in the build options screen.*

The MPASM Assembler tab allows performing the following customizations:

1. Disable/enable case sensitivity. Normally, the assembler is case-sensitive. Disabling case sensitivity turns all variables and labels to uppercase.

2. Select the default radix. Numbers without formatting codes are assumed to be hex, decimal, or octal according to the selected option. Assembly language programmers usually prefer the hexadecimal radix.

3. The macro definition windows allows adding macro directives. The use of macros is not discussed in this book.

4. The Use Alternate Settings textbox is provided for command line commands in non-GUI environments.

5. The Restore Defaults box turns off all custom configurations.

6. Selecting Output in the Categories window provides command line control options in the output file.

3.3.4 Adding a Source File

The code and structure of an assembly language program is contained in a text file referred to as the source file. The name of the source file is usually descriptive of its purpose or function. At this point in the project, the programmer will usually import an existing file or a template that serves as a skeleton for the project's source file, the name of the project or a variation on this name. Alternatively, the source file can be coded from scratch, although the use of a template saves considerable effort and avoids errors.

Click on the Add New File to Project command in the Project menu to create a new source file from scratch. Make sure that the new file is given the .asm extension and the development environment will automatically save it in the Source Files group of the Project Directory. At this time, a blank editor window will be opened and you can start typing the source code for your program.

Select the Add Files to Project command in the Project menu to import an existing file or a template into the project. In either case, you may have to rename the imported file and remove the old one from the project. If this precaution is not taken, an existing file may be overwritten and its contents lost. For example, to use the template file names PIC18F_Template.asm in creating a source file named PIC18F_Test1.asm, proceed as follows:

1. Make sure that the Project window is displayed by selecting the Project command in the View menu.

2. Right-click the Source Files option in the Project window and select the Add Files...

3. Find the file PIC18F_Template.asm dialog and click on the file name and then the Open button. At this point, the selected file appears in the Source Files box of the Project window.

4. Double click the file name (PIC18F_Template.asm) and MPLAB will open the Editor Window with the file loaded.

5. Rename the template file (PIC18F_Template.asm) by selecting the Editor window, then the Save As command in the MPLAB File menu. Enter the name under which the file is to be saved (PIC18F_Test1.asm in this walkthrough). Click on the Save button.

6. At this point, the file is renamed but not inserted in the Project. Right-click the Source Files option in the Project window and select the Add Files... Click on the PIC18F_Test1.asm file name and on the Open button. The file PIC18F_Test1.asm now appears in the Source Files option of the Project window.

7. Right-click on the PIC18F_Template.asm filename in the Project window and select the Remove command. This removes the template file from the project.

In many cases the MPLAB environment contains duplicate commands that provide alternative ways for achieving the same results, as is the case in the previous walkthrough and in others listed in this book. When there are several ways to obtain the same result, we have tried to select the more intuitive, simpler, and faster one.

3.3.5 Building the Project

Once all the options have been selected, the installation checked, and the assembly language source file written or imported, the development environment can be made to build the project. Building consists of calling the assembler, the linker, and any other support program in order to generate the files shown in Figure 3.3 and any others that may result from a particular project or IDE configuration. The build process is initiated by selecting the <Build All> command in the Project menu. Once the building concludes, a screen labeled Output is displayed showing the results of the build operation. If the build succeeded, the last line of the Output screen will show this result. Figure 3.6 shows the MPLAB program screen after a successful build in the preceding walkthrough.

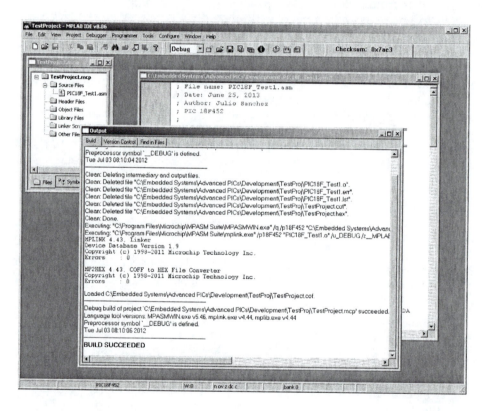

Figure 3-6 *MPLAB program screen showing the Build All command result.*

3.3.6 .hex File

The build process results in several files, depending on the options selected during project definition. One of these files is the executable, which contains the machine codes, addresses, and other parameters that define the program. This is the file that is "blown" in the PIC. The location of the .hex file depends on the option selected during project creation (see Figure 3.5). The Directories tab of the Build Options for Project dialog box contains a group labeled Build Directory Policy, as shown in Figure 3.7.

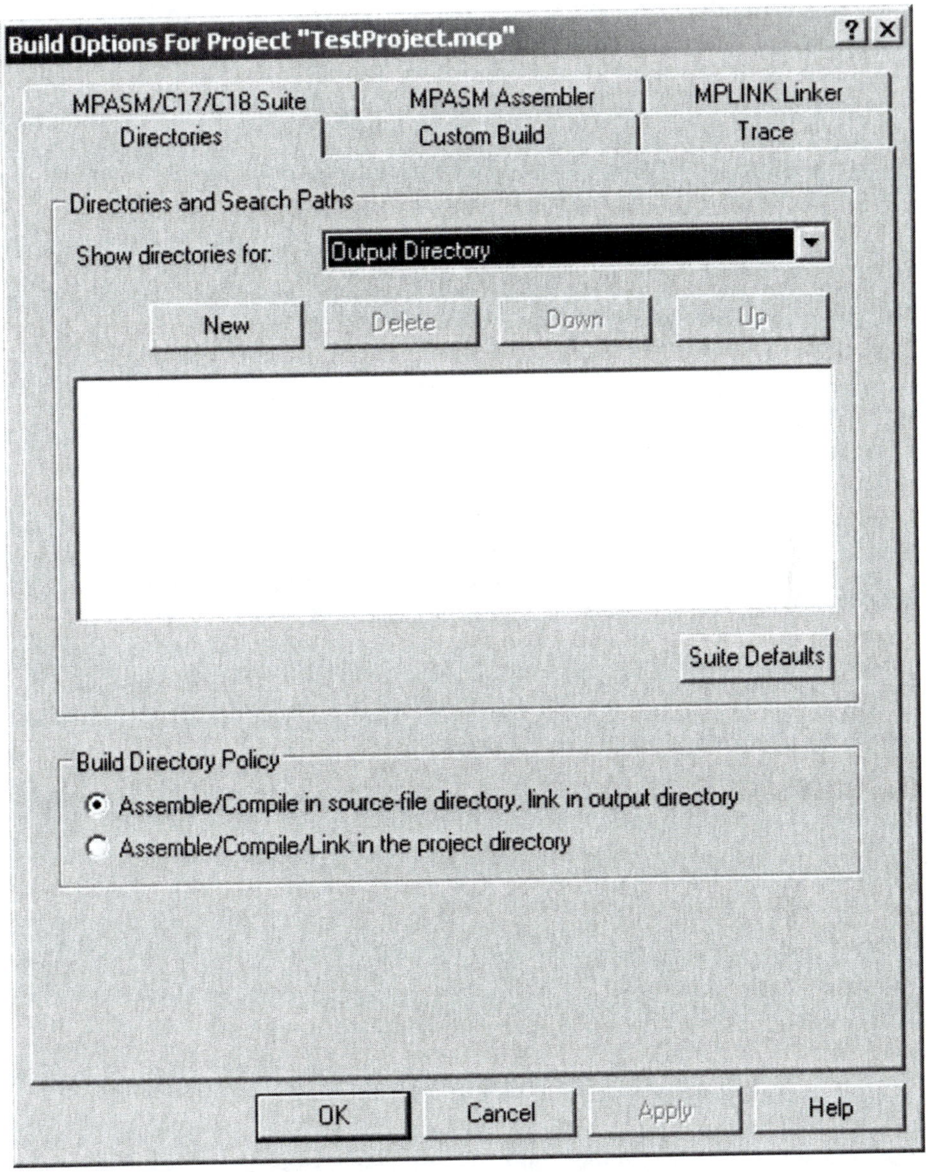

Figure 3-7 *The Directories Tab in Build Options for Project.*

The radio buttons in this group allow selecting two options:

1. Assemble and compile in the source-file directory and link in the output directory.

2. Assemble, compile, and link in the project directory.

The output directory can be selected by clicking the New button in the Directories and Search Path box, and then clicking the filename continuation button, labeled with ellipses. The executable (.hex) file and all other files generated by the build process will be placed accordingly.

3.3.7 Quickbuild Option

While learning programming or during prototype development, we sometimes need to create a single executable file without dealing with the complexity of a full-fledged project. For these cases, MPLAB provides a Quickbuild option in the Project menu. The resulting .hex file can be used to blow a PIC in the conventional manner and contains all the necessary debugging information. We have used the Quickbuild option for most of the PIC18F assembly language programs developed in this book. The typical command sequence for using the Quickbuild option is as follows:

1. Make sure the correct PIC device is selected by using the Configure>Select Device command.

2. Make sure the Quickbuild option is active by selecting Project>Set Active Project>None. This places the environment in the Quickbuild mode.

3. Open an existing assembly language file in the editor window using the File>Open command sequence. If creating a new program from scratch, use the File>New command to open the editor window and type in the assembly code. Then use the File>Save option to save the file with the .asm extension. In either case the file must have the .asm extension.

4. Select Project>Quickbuild file.asm to assemble your application, where file.asm is the name of your active assembly file.

5. If the assembly proceeds without errors, the .hex file will be placed in the same directory as the .asm source file.

3.4 MPLAB Simulators and Debuggers

In the context of MPLAB documentation, the term "debugger" is reserved for hardware debuggers while the software versions are called "simulators." Although this distinction is not always clear, we will abide by this terminology in order to avoid confusion. The reader should note that there are MPLAB functions in which the IDE considers a simulator as a debugger.

The MPLAB standard simulator is called MPLAB SIM. SIM is part of the Integrated Development Environment and can be selected at any time. The hardware debuggers currently offered by Microchip are ICD 2, ICE 2000, ICE 4000, and MPLAB REAL ICE. In addition, Microchip offers several hardware devices that serve simultaneously as a programmer and a debugger. The ones compatible with the PIC18F family are called PICKit2 and PICKit3.

A simulator, as the term implies, allows duplicating the execution of a program one instruction at a time, and also viewing file registers and symbols defined in the code. Debuggers, on the other hand, allow executing a program one step at a time or to a predefined breakpoint while the PIC is installed in the target system. This makes possible not only viewing the processor's internals, but also the state of circuit components in real-time.

The best choice for a debugger usually depends on the complexities and purpose of the program under development. As debuggers get more sophisticated and powerful, their price and complexity increase. In this sense, the powerful debugger that may be suitable for a commercial project may be overkill for developing simple applications or for learning programming. The software debugger furnished with MPLAB (MPLAB Sim) is often suitable for all except the most complex and elaborate circuits. In the section that follows we present an overview of MPLAB Sim. Later in this chapter we present a brief overview of some pupolar hardware debuggers.

3.4.1 MPLAB SIM

Microchip documentation describes the SIM program as a discrete-event simulator. SIM is part of the MPLAB IDE and is selected by clicking on the Select Tool command in the Debugger menu. The command offers several options, one of them being MPLAB SIM. Once the SIM program is selected, a special debug toolbar is displayed. The toolbar and its functions is shown in Figure 3.8.

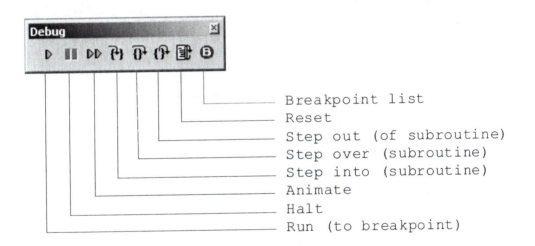

Figure 3.8 *SIM toolbar.*

In order for the simulator to work the program must first be successfully built. The most commonly used simulator methods are single-stepping though the code and breakpoints. A breakpoint is a labeled location at a program line at which the simulator will stop and wait for user actions.

Using Breakpoints

Breakpoints provide a way of inspecting program results at a particular place in the code. Single-stepping is executing the program one instruction at a time. The three step buttons are used in single-stepping. The first one allows breaking out of a subroutine or procedure. The second one is for bypassing a procedure or subroutine while in step mode. The third one single-steps into whatever line follows.

Breakpoints are set by double-clicking at the desired line while using the editor. The same action removes an existing breakpoint. Lines in which breakpoints have been placed are marked, on the left document margin, by a letter "B" enclosed in a red circle. Right-clicking while the cursor is on the program editor screen provides a context menu with several simulator-related commands. These include commands to set and clear breakpoints, to run to the cursor, and to set the program counter to the code location at the cursor.

Watch Window

The View menu contains several commands that provide useful features during program simulation and debugging. These include commands to program memory, file registers, EEPROM, and special function registers. One command in particular, called <Watch>, provides a way of inspecting the contents of FSRs and GPRs on the same screen. The <Watch> command displays a program window that contains reference to all file registers used by the program. The user then selects which registers to view and these are shown in the Watch window. The Watch window is shown in Figure 3.9.

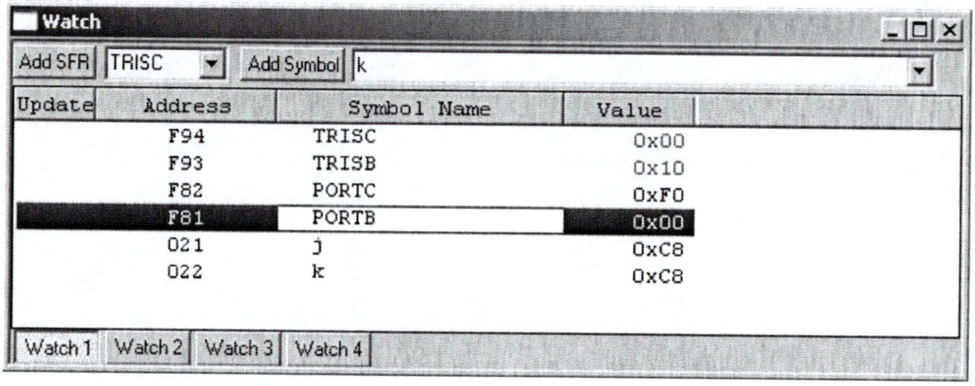

Update	Address	Symbol Name	Value
	F94	TRISC	0x00
	F93	TRISB	0x10
	F82	PORTC	0xF0
	F81	PORTB	0x00
	021	j	0xC8
	022	k	0xC8

Watch 1 | Watch 2 | Watch 3 | Watch 4

Figure 3-9 *Use of Watch window in MPLAB SIM.*

The contents of the various registers can be observed in the Watch window when the program is in the single-step or breakpoint mode. Those that have changed since the last step or breakpoint are displayed in red (not seen in Figure 3.9). The user can click on the corresponding arrows in the Watch window to display all the symbols or registers. The <Add Symbol> or <Add FSR> button is then used to display the selected register. Four different Watch windows can be enabled, labeled Watch 1 to Watch 4 at the bottom of the screen in Figure 3.9.

Simulator Trace

Another valuable tool available from the View menu is the one labeled <Simulator Trace>. The Simulator Trace provides a way for recording the step-by-step execution of the program so it can be examined in detail. To make sure that the system is set up for simulator tracing,you can select the Degugger>Settings>OSC/Trace command in the Simulator Trace option of the View menu. The default settings for simulator tracing is shown in Figure 3.10.

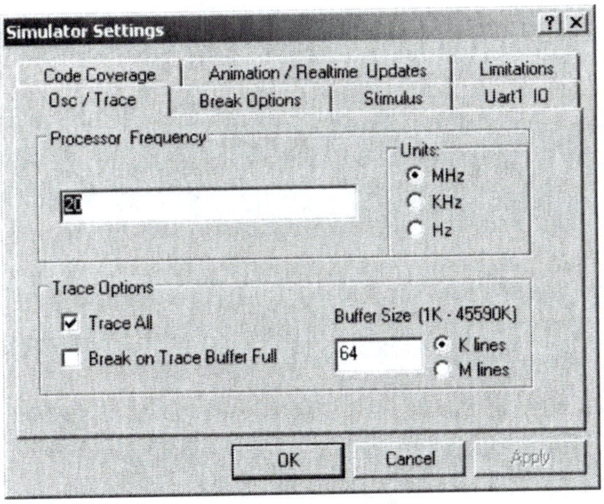

Figure 3.10 *Default oscillator settings for Trace command.*

To set up for tracing, right-click on the white space of the editor window and select Add Filter-in Trace or the Add Filter-out Trace by right-clicking the desired option. In the Filter-In Trace mode you select the code that is traced, and therefore displayed in the Trace window, by right-clicking on the editor window and selecting the Add Filter-in Trace option. You may also select a portion of the program code by first highlighting the desired code area. In the filter-out trace mode, you exclude text from trac3e by highlighting the text to be excluded before selecting the Add Filter-out Trace option. A section of the code designated for tracing can be removed by highlighting it and the selecting the Remove Filter Trace option. To remove the entire code designated for tracing, you right-click and then select the Remove All Filter Traces option. Figure 3.11 is a screen snapshot of the Simulator Trace window.

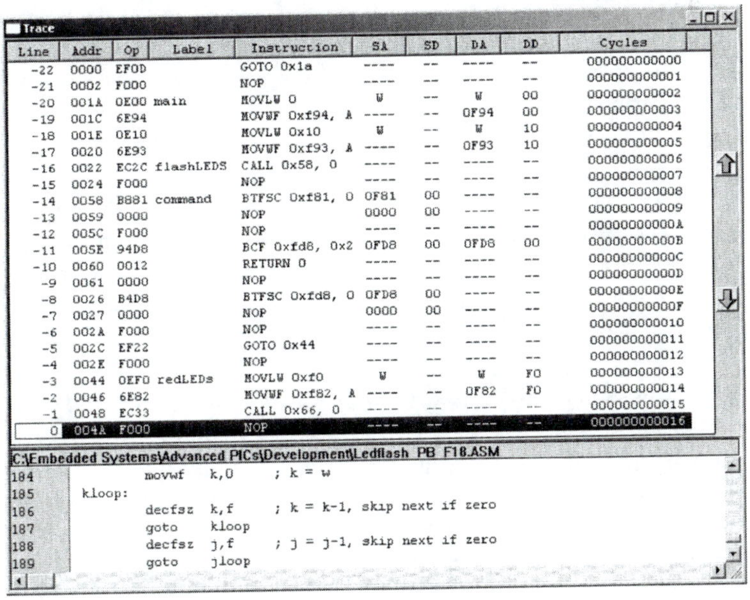

Figure 3.11 *Simulator Trace window snapshot.*

3.4.2 MPLAB Stimulus

Most microcontroller applications interact with hardware devices for input or output operations. For example, a simple program or program section uses an LED to report the state (pressed or un-pressed) of a pushbutton switch. If this program is being de-bugged with a hardware debugger, it is possible to mechanically activate the pushbutton switch in order to test its operation. When using a software debugger, there is no such thing as a pushbutton switch to press at execution time. The MPLAB Stimulus feature provides a way of simulating hardware inputs or changes in values in registers or memory while debugging with Sim.

The action or change detected by the software can be one of the following:

1. A change in level or a pulse to an I/O pin of a port.

2. A change in the values in an SFR (Special Function Register) or other data mem-ory.

3. A specific instruction is reached or a certain time has passed in the simulation.

4. A predefined condition is satisfied.

Microchip defines two types of stimulus:

1. Asynchronous: For example, a one-time change to the I/O pin or register is trig-gered by a firing button on the stimulus GUI within the MPLAB IDE.

2. Synchronous: For example, a predefined series of signal/data changes to an I/O pin, SFR or GPR, such as a clock cycle.

Stimulus Dialog

Use the Stimulus Dialog window to define the terms and conditions in which an exter-nal stimuli takes place. This dialog is used for creating both asynchronous and syn-chronous stimulus on a stimulus workbook. The workbook is the internal MPLAB file where a set of stimulus settings are saved. Advanced users can export stimulus work-book settings to an SCL file for later access.

To create a new workbook, select the Stimulus command in the Debugger menu, then select New Workbook. To open an existing workbook to modify it or add new entries select the Open Workbook option. For this command to work, a stimulus workbook must have been previously created and saved. The new stimulus window is shown in Figure 3.12.

The Asynch tab is used to enter asynchronous, user-generated stimulus to the workbook. Stimulus are entered row-by-row. Usually, the Pin/SFR column is first ac-tivated and a port, pin, or SFR selected from the drop-down list. For example, a pro-gram that uses a circuit in which a pushbutton switch is wired to port B, line 0, will select RB0 from the list. Then the action for that specific stimulus must be selected in the Action column. The choices are Set High, Set Low, Toggle, Pulse High, and Pulse Low. Once the Action has been selected, the Fire button displays a > symbol to indicate that the action can be triggered.

Stimulus - [Untitled]

| Asynch | Pin / Register Actions | Advanced Pin / Register | Clock Stimulus | Register Injection | Register Trace |

Fire	Pin / SFR	Action	Width	Units	Comments / Message

| Advanced... | Apply | Remove | Delete Row | Save | Exit | Help |

Figure 3.12 *A blank Stimulus workbook.*

The Width and Units columns are used to define the number of units and the unit type for "Pulse" actions. The Units column allows selecting instruction cycles, nanoseconds, microseconds, milliseconds, and seconds. In regular stimulus, the Comments column is used to specify a comment for the specific entry in the workbook.

An sample debugging session with MPLAB Sim is offered in Appendix B.

3.4.3 MPLAB Hardware Debuggers

A more powerful and versatile debugging tool is a hardware or in-circuit debugger. Hardware debuggers allow tracing, breakpointing, and single-stepping through code while the PIC is installed in the target circuit. The typical in-circuit debugger requires several hardware components, as shown in Figure 3.13.

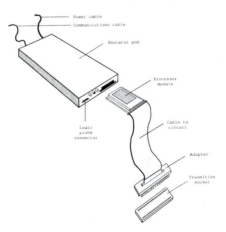

Figure 3.13 *Components of a typical hardware debugger.*

The emulator pod with power supply and communications cable provides the basic communications and functionality of the debugger. The communications line between the PC and the debugger can be an RS-232, a USB, or a parallel port line. The processor module fits into a slot at the front of the pod module. The processor is device specific and provides these functions to the debugger. A flex cable connects the processor module to an interchangeable device adapter that allows connecting to the several PICs supported by the system. The transition socket allows connecting the device adapter to the target hardware. A separate socket allows connecting logic probes to the debugger.

Microchip provides two models of in-circuit hardware debuggers, which they call In Circuit Emulators, or ICEs. The ICE 2000 is designed to work with most PICs of the mid-range and lower series, while the ICE 4000 is for the PIC18x high-end family of PICs. Microchip also furnishes in-circuit debuggers designated as ICD 2 and ICD 3. The purpose of these devices is to provide similar features as their full-fledged in-circuit emulators (ICE) at much reduced prices. However, a disadvantage of the ICD 2 and ICD 3 systems is that they require the exclusive use of some hardware and software resources in the target and that the system be fully functional. The ICEs, on the other hand, provide memory and clock so that the processor can run code even if it is not connected to the application board.

Some Microchip in-circuit debuggers (such as ICD 2 and ICD 3 and their less-expensive versions PICkit 2 and PICkit 3) also serve as in-circuit programmers. With these devices, the firmware in the target system can be programmed and reprogrammed without any other hardware components or connections. For this reason we discuss the Microchip in-circuit debuggers, previously mentioned, in the context of development programmers later in this chapter.

3.4.4 An Improvised Debugger

The functionality of an actual hardware debugger can sometimes be replaced with a little ingenuity and a few lines of code. Most PICs are equipped with EEPROM memory. Programmers (covered in the following section) have the ability to read all the data stored in the PIC, including EEPROM. These two facts can be combined in order to obtain runtime information without resorting to the cost and complications of a hardware debugger. For example, if a defective application is suspected of not finding the expected value in a PIC port, the developer can write a few lines of code to store the port value on an EEPROM memory cell. An endless loop following this operation makes sure that the stored value is not changed. Now the PIC is inserted in the circuit and the application executed. When the endless loop is reached, the PIC is removed from the circuit and placed back in the programmer. The value stored in EEPROM can now be inspected so as to determine the runtime state of the machine. In many cases, this simple trick allows us to locate an otherwise elusive bug.

3.5 Development Programmers

In the context of microcontroller technology, a programmer is a device that allows transferring the program onto the chip. This is usually called "burning" a PIC or, more commonly, "blowing" a PIC. Development programmers are used for configuring and testing an application and for any other non-production uses. Production programmers are used in manufacturing. In the present context, we refer to development programmers.

Most programmers have three components:

1. A software package that runs on the PC.
2. A cable connecting the PC to the programmer.
3. A hardware programming device.

Dozens of PIC development programmers are available on the Internet. The programmer "cottage industry" started when Microchip released to the public the programming specifications of the PIC without requiring contracts or a nondisclosure agreement. The commercial development programmers available online range from a "no parts" PIC programmer that has been around since 1998, to sophisticated devices costing hundreds of dollars and providing many additional features and refinements. Some PIC programmers are designed to interact with the MPLAB development environment while others are stand-alone gadgets. For the average PIC user, a nice USB programmer with a ZIF socket and the required software can be purchased for about $50.00. Build-it-yourself versions are also available for about half this amount. Figure 3.14 shows the hardware components of a USB PIC programmer from Ett Corporation.

Figure 3.14 *ET-PGM PIC programmer.*

An interesting feature of the Ett programmer in Figure 3.14 is its compatibility with the PICkit 2 and PICkit 3 packages from Microchip.

3.5.1 Microchip PICkit 2 and PICkit 3

The PICkit 2 and PICkit 3 packages from Microchip are low-cost debuggers that also serve as development programmers. They are compatible with the following device families:

1. Baseline devices such as PIC10FXX, PIC12FXX, and PIC16FXX

2. Midrange devices such as PIC12FXXX, PIC16FXXX, and PIC16HVXXX

3. PIC18F, PIC18F_J_, and PIC18F_K_ devices

4. PIC24 devices

5. dsPIC33 devices

The PICkit 2/3 devices allow debugging and programming of PIC and dsPIC microcontrollers using the MPLAB graphical user interface. The PICkit 2/3 programmers are connected to a PC via a USB interface, and to the target circuit via a Microchip debug (RJ-11) connector. Figure 3.15 shows the PICkit 2 programmer.

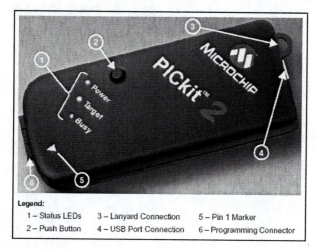

Figure 3.15 *PICkit 2 programmer from Microchip.*

The connector uses two device I/O pins and the reset line to implement in-circuit debugging and In-Circuit Serial Programming.

The following features are claimed for the PICkit 3 device:

- Full-speed USB at 12 Mbits/s
- Real-time execution
- MPLAB IDE compatible
- Built-in overvoltage/short circuit monitor
- Firmware upgradeable from PC/Web download
- Fully enclosed
- Supports low voltage to 2.0 volts (2.0V to 6.0V range)

- Diagnostic LEDs (power, busy, error)
- Read/write program and data memory of microcontroller
- Erase program memory space with verification
- Freeze peripherals at breakpoint

The PICkit 2/3 interface executes when the debugger/programmer is selected. The interface software allows the following operations:

1. Selecting the processor family
2. Reading the hex file in the device
3. Checking that the device is blank
4. Erasing the device
5. Verifying the device
6. Writing hex code to the hardware
7. Importing and exporting hex files
8. Reading EEPROM Data in the device
9. Enabling code and data protection

Figure 3.16 is a screen snapshot of the PICkit 2 programmer application.

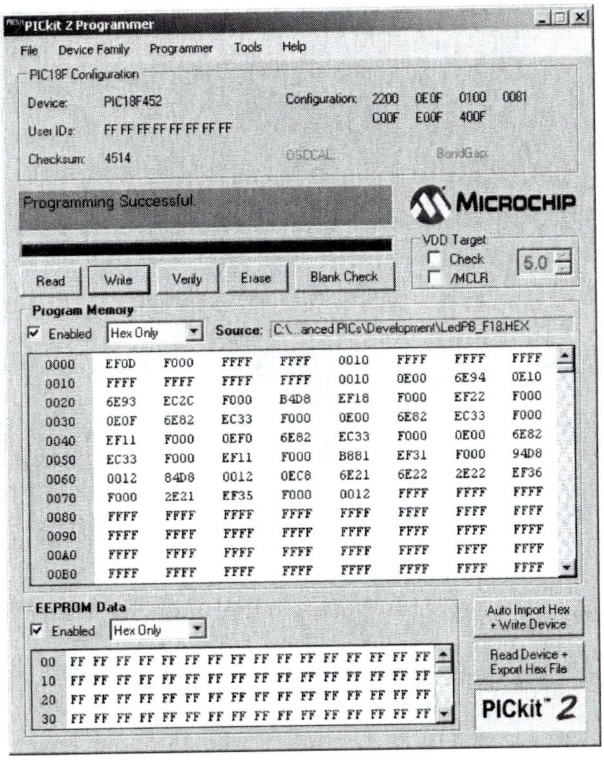

Figure 3.16 *PICkit 2 application.*

By design, the PICkit 2/3 programmers require a direct connection to the target system via an RJ-11 telephone connector and cable. In the circuit, two lines from the RJ-11 must be wired to the appropriate PIC pins. While in the development process, this requirement often becomes an inconvenience because we must provide the required hardware in every circuit. Some PIC programmers (such as the ET PGM PIC programmer in Figure 3.14) allow blowing a PIC directly by inserting it into a socket provided in the programmer hardware. By also using a ZIF socket in the target device, it is easy to remove the PIC, blow it in the programmer, and then return it to the target circuit for testing. Nevertheless, it is a developer's choice whether to move the PIC from the programmer to the target circuit or to use in circuit programming by wiring the circuit board directly to the programmer.

3.5.2 Micropro USB PIC Programmer

Another popular PIC programmer is Model 3128 from Micropro. This device connects to the USB port and comes equipped with a 40-pin ZIF socket, as shown in Figure 3.17.

Figure 3.17 *USB PIC programmer by Micropro.*

3.5.3 MPLAB ICD 2 and ICD 3 In-Circuit Debuggers/Programmers

At a higher end than the PICkit 2 and PICkit 3 devices, Microchip provides the ICD 2 and its upgrade ICD 3 in-circuit debuggers and programmers. These products are furnished in kits that include all the necessary components, such as cables, PIC hardware, power supply, and demonstration board. Figure 3.18 shows the main module of the ICD 3 package.

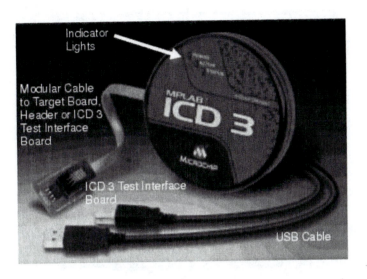

Figure 3.18 *Microchip ICD 3 main module.*

Like all in-circuit debuggers/programmers, the ICD 2 and ICD 3 devices require that the target circuit be configured for in-circuit operations. This means that the PIC in the target circuit must support in-serial operations, and that the circuit itself must contain an RJ-11 female connector wired to the processor's PGC, PGD, and MCLR lines. This means that these pins cannot be used as port pins for lines RB6 and RB7. Several other hardware resource on the PIC are reserved for in-circuit operations with ICD 2 and ICD 3 devices, and some circuit design restrictions must be followed.

3.6 Test Circuits and Development Boards

Microcontroller programming does not take place in a vacuum. In developing a PIC application that turns on an LED when a pushbutton switch is pressed, we would probably want to test the code in a circuit that contains all the hardware. This includes the PIC itself, the required support components and wiring to make the PIC operate, as well an LED and a pushbutton switch connected to the corresponding PIC port lines. This circuit can be constructed on a breadboard, on a printed circuit board, or found in a standard demonstration board. Demonstration boards can be purchased commercially or developed in-house. Lacking a development board, the other option is to build the circuits oneself, usually on a breadboard or perf board, a time-consuming but valuable experience.

3.6.1 Commercial Development Boards

A development or demonstration board is a circuit, typically on a printed circuit board. The board usually contains an array of connected and connectable components whose main purpose is to serve as an experimental tool. Like programmers, PIC development boards come in a wide range of prices and levels of complexity. Most boards target a specific set of PIC microcontroller families. Figure 3.19 shows the LAB-X1 development board.

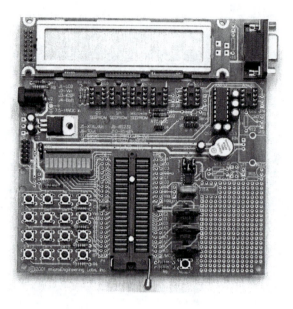

Figure 3.19 *LAB-X1 development board.*

The LAX-X1 board as well as several other models are products of microEngineering Labs, Inc. Some of the sample programs developed for this book were tested on a LAB-X1 board. Several development boards are available from Microchip and other vendors.

It is also possible to develop experiment/demo boards in house to suit the particular hardware options under development or for learning purposes. The advantage of these custom boards is that they serve to test, not only the software, but also the circuit design. Figure 3.20 shows a demo board developed to test the PIC 18F452 microcontroller as well as many of the hardware elements covered in the text.

Figure 3.20 *Custom demo board for the PIC 18F452.*

Appendix C contains circuit diagrams and PCB files for readers who may wish to reproduce this board. Notice that we have glued a paper strip to the microcontroller so as to make visible some minimal information regarding the pin functions. We have found this method particularly useful in developing breadboard circuits, discussed later in this chapter.

3.6.2 Circuit Prototype

A microcontroller is a circuit element that, by itself, serves no useful purpose. In our previous titles (*Microcontroller Programming the Microchip PIC* and *Embedded System Circuits and Programming*), we devote considerable attention to circuit components and their integration into electronic devices. Because of the greater complexity of the microcontrollers discussed, the present book concentrates on programming. However, because microcontrollers live in electronic circuits, it is impossible to completely ignore the hardware element.

The development of an electronic circuit typically follows several well-defined stages during which the product progresses through increasingly refined phases. This progression ensures that the final product meets all the design requirements and performs as expected. The designer or the engineer cannot be too careful in avoiding manufacturing errors that later force the scrapping of multiple components or, at best, forces costly or unsightly circuit repairs. A common norm followed by electronic firms is not to proceed to fabrication until a finished and unmodified prototype has been exhaustively tested and evaluated. Even the text and labels on the circuit board should be checked for spelling errors and to make sure that the final placement of hardware components will not hide some important information printed on its surface. The possible defects and errors in an electronic circuit includes the following types:

- Incorrect selection of components
- Defective product design or component placement
- Defective microcontroller programming
- Defects in wiring and connectivity

The methodology that uses a cyclic process of prototyping, testing, analyzing, and refining a product is known as iterative design. The practicality of iterative design results from the assumption that changes to a product are easier and less expensive to implement during the early stages of the development cycle. The iterative design model, also called the spiral model, can be described by the following steps:

1. The circuit is designed on paper, usually with the support of software, as previously described.

2. The paper circuit is checked and evaluated by experts other than the designer or designers.

3. A primitive hardware prototype is developed and tested. Breadboarding and wire-wrapping are the most common technologies used for this first-level prototype.

4. The breadboarded or wire-wrapped prototype is tested and evaluated. If changes to design are required, the development process restarts at Step 1.

5. A second-level prototype is developed, usually by means of printed circuit boards. This PCB prototype is evaluated and tested. If major modifications are required, the development process is restarted at Step 1.

6. A final, third-level, prototype is developed using the same technology and number of signal layers as will be used in the final product. If modifications and changes are detected during final testing, the development process is restarted at Steps 1, 3, or 5 according to the nature of the defect that must be remedied.

7. If the final prototype passes all tests and evaluations, a short production run is ordered. This short run allows finding problems in the manufacturing stage that can sometimes be remedied by making modifications to the original design or by changing the selected component or components.

The preceding steps assume a very conventional circuit and a simple and limited development process. The mass production of electronic components must consider many other factors.

3.6.3 Breadboard

One of the most useful tools for the experimenter and developer of embedded systems is a breadboard. The name originated in the early days of radio amateurs who would use a wooden board (sometimes an actual bread board) with inserted nails or thumbtacks to test the wiring and components of an experimental circuit. The modern breadboard is usually called a solderless breadboard because components and wires can be connected to each other without soldering them. The term "plugboard" is also occasionally used. Figure 3.21 shows a populated breadboard for a motor driver circuit.

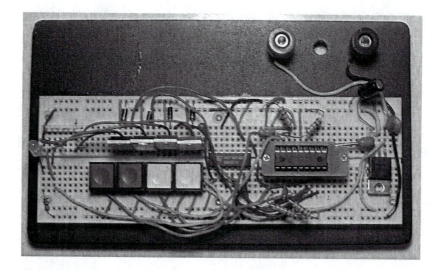

Figure 3.21 *Breadboard with a motor driver circuit.*

The main component of a modern solderless breadboard is a plastic block with perforations that contact internal, tin-plated, spring clips. These clips provide the contact points. The spacing between holes is usually 0.1 inches, which is the standard spacing for pins in many nonminiaturized electronic components and in ICs in dual inline (DIP) packages. Capacitors, switches, resistors, and inductors can be inserted in the board by cutting or bending their contact wires. Most boards are rated for 1 amp at 5 volts. Solderless breadboards come in different sizes and designs. The one in Figure 3.22 has the plastic interconnection component mounted on a metal base and includes female banana plug terminals suitable for providing power to the board. Simpler boards consist of a perforated plastic block with the corresponding spring clips under the perforations. These can sometimes serve as modules that can be attached to each other in order to accommodate a more complex circuit.

On the other hand, more sophisticated devices, sometimes called digital laboratories, include with the breadboard several circuits for providing power in different intensities as well as signals with the corresponding adjusters and selectors, as well as switches of several types. Figure 3-22 shows the IDL_800 Digital Lab.

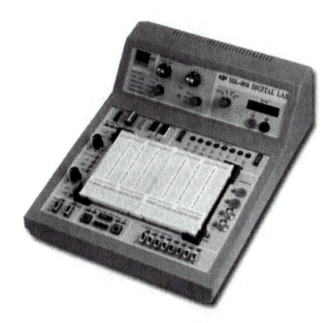

Figure 3.22 *IDL-800 Digital Lab breadboard.*

Limitations of Breadboards

Although breadboards are valuable tools in prototyping the circuit, the design community is divided regarding their value. One of the common problems with breadboards is faulty contacts. The spring-loaded clips, designed to provide connection between components, often fail. This can lead the operator to believe that there is something wrong with the circuit, when it is just a faulty contact. Another problem is that wire-based connectors are often longer than necessary, which introduces electrical problems that may not be related to the circuit itself. In some sensitive circuits, the routing required by the board hardware may produce interference or spurious signals. The circuit developer must take all these limitations into account when testing breadboard circuits.

Another mayor limitation is that solderless breadboard cannot easily accommodate surface-mount components. Furthermore, many other standard components are not manufactured to meet the 0.1 inch spacing of a standard breadboard and are also difficult to connect. Sometimes the circuit developer can build a breakout adapter as small as a PCB that includes one or more rows of 0.1 inch pins. The board-incompatible component can then be soldered to the adapter and the adapter plugged into the board. But the need to solder components to the adapter negates some of the advantages of the breadboard. However, with components that are likely to be reused, the adapter may be a viable option.

For example, a 6P6C male telephone plug connector, often used in in-circuit debugging and programming of PIC microcontroller, does not fit a standard breadboard. In this case we can build a breakout adapter such as the one shown in Figure 3.23.

Figure 3.23 *Breadboard adapter for 6P6C connector*

Breadboarding Tools and Techniques

Several off-the-shelf tools are available to facilitate breadboarding and others can be made in-house. One of the most useful ones is a set of jumper wires of different lengths and color. These jumper kits are usually furnished in a plastic organizer. Longer, flexible connectors are also available and come in handy when wiring large or complex circuits.

Wiring mistakes with microcontroller pins are easy to make in breadboarding integrated circuit components. In this case the operator must frequently look up the circuit diagram or the component schematics to determine the action on each IC pin. One solution is to use a drawing program to produce a labeled drawing of the component's pinout. The drawing is scaled to the same size as the component and then printed on paper or cardboard. A cutout is then glued, preferably with a nonpermanent adhesive, to the top part of the IC so that each pin clearly shows a logo that is reminiscent of its function. Figure 3.24 shows a portion of a breadboard with a labeled IC used in developing a circuit described later in this book.

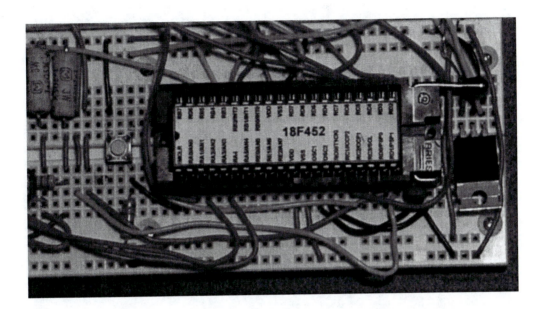

Figure 3.24 *Labeled IC in a breadboard.*

Notice in Figure 3.24 that the PIC 18F452 microcontroller is inserted in a device called a ZIF (zero insertion force) socket. When the ZIF socket handle is lifted, the IC can be easily removed. ZIF sockets are often used in prototypes and demo boards so that the component can be easily replaced or reprogrammed.

3.6.4 Wire Wrapping

Another popular technique used in the creation of prototypes and individual boards is wire wrapping. In wire-wrapped circuits, a square, gold-plated post is inserted in a perforated board. A silver-plated wire is then wrapped seven turns around the post, which results in twenty-eight contact points. The silver- and the gold-plated surfaces cold-weld, producing connections that are more reliable than the ones on a printed circuit board, especially if the board is subject to vibrations and physical stress. The use of wire-wrapped boards is common in the development of telecommunications components but solderless breadboards have replaced wire wrapping in conventional prototype development.

3.6.5 Perfboards

Thin sheets of isolating material with holes at 0.1-inch spacing are also used in prototype development and testing and in creating one-of-a-kind circuits. The holes in the perfboard contain round or square copper pads on either one or both sides of the board. In the perfboard, each pad is electrically insulated. Boards with interconnected pads are sometimes called stripboards. Components including ICs, sockets, resistors, capacitors, connectors, and the like, are inserted into the perfboard holes and soldered or wire wrapped on the board's back side. Figure 3.25 shows a circuit on a perfboard.

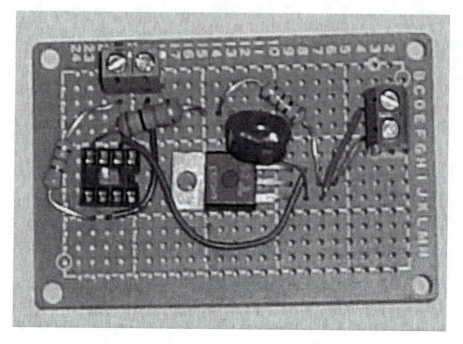

Figure 3.25 *Circuit on a perfboard.*

In this section we have omitted mentioning circuit prototyping methods and techniques that have become obsolete, such as point-to-point wiring and through-hole construction.

3.6.6 Printed Circuit Boards

The methods and techniques described so far, including breadboards, wirewrapping, and perfboards, are used in developing and testing the electronics of the circuit itself. Once the circuit prototype has been tested and evaluated, the next step is usually the production of a circuit board that can house the components in a permanent manner and thus becomes a prototype of the final product. This typically requires a printed circuit board, or PCB, where the components can be mechanically housed and electrically connected. The PCB is also called a printed wiring board (PWB) or a printed circuit assembly (PCA). Very few commercially made electronic devices do not contain at least one PCB.

The base of a conventional PCB is a nonconductive laminate made from an epoxy resin, with etched conductive copper traces that provide pathways for signals and power. PCBs can be produced economically in large or small volumes, and even individually. Production operations on the PCB include etching, drilling, routing, creating solder-resistant layers, screen printing, and installation of components. All these operations can be automated in a production setting or done by hand by the hobbyist or when creating a prototype. PCB technology has flourished because the final product is inexpensive and reliable. Standards regarding PCB design, assembly, and quality control are published by IPC. Figure 3.26 shows two images of a PCB.

Figure 3-26 *PCB drawing and populated board.* (Image from Wikimedia Commons.)

The image on the left-hand side of Figure 3.26 shows the PCB as it appears in the design software. In this example, the board is double sided, although the image for the reverse side is not shown in the illustration. The board shown on the right-hand side of Figure 3.26 shows the finished product, populated with surface-mount and through-the-hole components. Appendix C describes, step-by--step, the creation of a PCB.

Chapter 4

Assembly Language Program

4.1 Assembly Language Code

A program in PIC assembly language consists of an ASCII text file that includes the following elements:

- Machine opcodes (with possible operands) representing the hardware instructions that are part of the processor's instruction set, including the keywords, symbols, and the mathematical and logical operators recognized by the assembler.

- Assembler directives (including macro directives) in the form of keywords recognized by the assembler program or by other components of the development software.

- Labels indicating a location in the assembly language code.

- References to other files that can be loaded or included at assembly time.

- Comments in the form of text lines, or portions of text lines, that are ignored by the assembler but that serve to explain, document, organize, or embellish the code.

Programming in any language can be made easier by using pre-developed and pre-tested code fragments or subroutines to perform common or frequent operations. The collection of these fragments, sample programs, and subroutines constitute a programmer's equity. The better organized, coded, and tested this personal library is the easier it will be to develop a new application. In programming, reusability fosters productivity, as is the case in so many other fields. Programmers who fail to discover this simple truth do not last very long in the profession.

4.1.1 A Coding Template

Of the many code samples in the programmer's toolbox, the first one usually needed is a general coding template to start the program. This coding template will vary considerably according to the program requirements and the programmer's style. Furthermore, companies often provide standard coding templates that the employed programmers are required to use. In any case, using a code template to start a programming project is always a good idea. The following is a very simple coding template for applications that use the PIC18F452:

```
; File name: PIC18F_Template.asm
; Date: June 25, 2012
; Copyright 2012 by Julio Sanchez and Maria P. Canton
; Processor: PIC 18F452
;
;=========================================================
;                   definition and include files
;=========================================================
      processor       18F452         ; Define processor
      #include  <p18F452.inc>
; =========================================================
;                   configuration bits
;=========================================================
      config WDT = OFF
      config OSC = HS
      config LVP = OFF
      config DEBUG = OFF
      config PWRT = ON
;
; Turn off banking error messages
      errorlevel      -302
;
;=========================================================
;                   variables in PIC RAM
;=========================================================
; Access RAM locations from 0x00 to 0x7F
;var1           equ        0x000          ; Sample variable declarations
;var2           equ        0x002
;=========================================================
;                        program
;=========================================================
      org        0      ; start at address
      goto     main
; Space for interrupt handlers
;=============================
;     interrupt vectors
;=============================
      org        0x08       ; High-priority vector
      retfie     0x01       ; Fast return
      org        0x18       ; Low-priority vector
      retfie
;=========================================================
;                  main program entry point
;=========================================================
main:
; Template assumes that program uses data in the access bank
; only. Applications that use other banks will reset the
; BSR bits and should erase these lines
      movlb         0          ; Access bank to BSR register
; Program code
;

            end        ; END OF PROGRAM
```

Program Header

Most programming templates begin with a header that provides information about the project, the author or authors, the development environment, and a copyright notice. Professional headers often extend over several pages and may include a mini-manual on the program, the project's development history, a listing of all found and fixed bugs and other defects, and even circuit diagrams in ASCII graphics.

Program Environment Directives

Frequently found, following the header, is a listing of definition directives for the processor and the #include directive for the processor's include file. The processor directive allows defining the CPU in code. In this case, if the processor selected with the MPLAB <Configure><Device> command does not match, it is superseded by the code line.

The included file or files can be located in the current working directory, in the source file directory, or in the MPASM executable directory. The syntax for the include statement can be as follows:

```
#include include_file
#include "include_file"
#include <include_file>
```

The file defined by the #include statement is read in as source code and becomes part of the program. If the filename includes spaces, then the filename must be enclosed in quotes or brackets.

Configuration Bits

It is convenient to set the configuration bits using the config directive rather than writing to the high memory area where the configuration bits are stored. Table 4.1 shows the operands recognized by the config directive on the PIC18FXX devices.

Table 4.1

18F452 Configuration Options

DIRECTIVE	ACTION:
OSC =	LP, XT, HS, RC
OSC =	EC (OSC2 as Clock Out)
OSC =	HSPLL (HS-PLL Enabled)
OSC =	RCIO (RC-OSC2 as RA6)
OSCS =	Oscillator switch ON/OFF
PWRT =	Power-on timer ON/OFF
BOR =	Brown-out reset ON/OFF
BORV =	45, 42, 27, 25 (4.5 to 2.5V)
WDT =	Watchdog timer ON/OFF
WDTPS =	Watchdog postscaler 1, 2, 4, 8, 16, 32, 64, 128
CCP2MUX =	CCP2 MUX ON/OFF
STVR =	Stack overflow reset ON/OFF
LVP =	Low-voltage ICSP ON/OFFOFF
DEBUG =	Background debugger ON/OFF
CP0 =	Code protection ON/OFF

continues

Table 4.1

*18F452 Configuration Options (**continued**)*

DIRECTIVE	ACTION:
CPx =	Code protection block x = 1/2/3 ON/OFF
CPB =	Boot block code protection ON/OFF
CPD =	Data EEPROM code protection ON/OFF
WRTx =	Write block x protection x = 1/2/3 ON/OFF
WRTx =	Boot block x protection x = B/C/D ON/OFF
EBTRx =	Table Read Protection x = 0/1/2/3 ON/OFF
EBTRB =	Boot block table read protection ON/OFF

The template file sets the oscillator, watchdog timer, low voltage protection, and background debugger bits.

Error Message Level Control

Errors cannot be disabled but the errorlevel directive can be used to control the types of messages that are displayed or printed. The most common use of the errorlevel directive is to suppress the display of banking errors. In the template code, this is accomplished with the following statement:

```
errorlevel -302
```

Variables and Constants

The template does not declare variables or constants but does contain a message area that defines two sample local variables named var1 and var2 respectively. The 127 RAM locations from 0x00 to 0x7f (in access RAM) are sufficient for many programs. Because of the simpler coding and faster execution speed the access RAM area should be used preferably for program variables. The template uses the equ directive to assign variable names to addresses in RAM. Several other ways of reserving memory and using RAM for variables are explored later in this book.

Code Area and Interrupts

The 18FXX2 PICs have multiple interrupt sources: a high-priority interrupt with a vector at 008H and a low-priority interrupt with a vector at 018H, while the start-up and reset vector is at address 00H. In order to accommodate this structure, the template file provides a jump over the interrupt vectors as well as a code location for interrupt handlers. The label "main" marks the entry point for non-interrupt program code. The template concludes with the end directory, which is required to indicate the end of a build operation.

4.1.2 Programming Style

The program developer's main challenge is writing code that performs the task at hand and that is both simple and elegant. In the present context, this means writing a PIC assembly language program that assembles without errors (usually after some tinkering), that is understandable, and can be maintained by its author and by other programmers. How we attempt to achieve these goals defines our programming style. The task is by no means trivial, especially when programming in a low-level language.

Source File Comments

One of the first realizations of a beginning programmer is how quickly one forgets the reasoning that went into our code. A few weeks, even a few hours, after we developed a routine that, at the time, appeared obvious, is now cryptic and undecipherable and the logic behind it defies our understanding. The only solution to this volatility of program logic is to write simple code and good program comments that explain, not only the elementary, but also the more elaborate trains of thoughts and ideas behind our code.

In PIC assembly language, the comment symbol is the semicolon (;). The presence of a semicolon indicates to the assembler that everything that follows, to the end of the line, must be ignored. Using comments judiciously and with good taste is the mark of the expert software engineer. Programs with few or confusing comments fall into the category of "spaghetti code." In the programmer's lingo, "spaghetti code" refers to a coding style that cannot be deciphered or understood, reminiscent of the entanglement in a bowl of cooked spaghetti. The worse offense that can be said about a person's programming style is that he or she produces spaghetti code.

How we use comments to explain our code or even to decorate it is a matter of personal preference. However, there are certain common-sense rules that should always be considered:

- Do not use program comments to explain the programming language or reflect on the obvious.

- Abstain from humor in comments. Comedy has a place in the world but it is not in software. By the same token, stay away from vulgarity, racial or sexist remarks, and anything that could be offensive or embarrassing. You can never anticipate who will read your code.

- Write short, clear, readable comments that explain how the program works.

- Decorate or embellish your code using comments according to your taste.

Clearly commented bitmaps, banners, and many other code embellishments do not add to the quality and functionality of the code. It is quite possible to write very sober and functional programs without using these gimmicks. The decision of how to comment and how much to decorate our programs is one of style.

4.2 Defining Data Elements

Most PIC programs require the use of general-purpose file registers. These registers are allocated to memory addresses reserved for this purpose in the PIC architecture. Data memory structure was discussed in Section 2.2.2 and can be seen in Figure 2.10. Because the areas at these memory locations are already reserved for use as GPRs, the program can access the location either by coding the address directly or by assigning to that address a symbolic name.

4.2.1 equ Directive

The equ (equate) directory assigns a variable name to a location in RAM memory. For example:

```
Var1 equ 0x0c      ; Name var1 is assigned to location 0x0c
```

Actually the name (in this case var1) becomes an alias for the memory address to which it is linked. From that point on the program accesses the same variable if it references var1 or address 0x0c, as follows:

```
movf      var1,w,A  ; Contents of var1 to w in access RAM
```
or
```
movf      0x0c,w,A  ; Same variable to w
```

In addition to the equ directive, PIC assembly language recognizes the C-like #define directive, so the name assignation could also have been done as follows:

```
#define  var1      0x0c
```

In addition, the set directive is an alternative to equ, with the distinction that addresses defined with set can be changed later in the code.

4.2.2 cblock Directive

Another way of defining memory data is by using one of the data directives available in PIC assembly language. Although there are several of these, perhaps the most useful is the cblock directive. The cblock directive specifies an address for the first item, and other items listed are allocated consecutively by the assembler. The group ends with the endc directive. The following code fragment shows the use of the cblock/endc directives:

```
; Reserve 20 bytes for string buffer
    cblock    0x20
    strData
    endc
; Reserve three bytes for ASCII digits
    cblock    0x34
    asc100
    asc10
    asc1
    endc
```

Actually, the cblock directive defines a group of constants that are assigned consecutive addresses in RAM. In the previous code fragment, the allocation of 20 bytes for the buffer named strData is illusory because no memory is reserved. The illusion works because the second cblock starts at address 0x34, which is 20 bytes after strData. The coding also assumes that the programmer will abstain from allocating other variables in the buffer space.

Although most of the time named variables are to be preferred to hard-coded addresses, there are times when we need to access an internal element of some multi-byte structure. In these cases, the hard-coded form could be more convenient.

4.2.3 Access to Banked Memory

Having to deal with memory banks is one of the aggravations of PIC mid-range programming. The designers of the PIC 18FXX2 devices have simplified banking by creating an access bank that is more convenient to use and provides faster execution.

In the 18FXX2 PICs banks are designated starting with bank 0. All PICs of this family contain either three or six memory banks, as shown in Figure 2.10. In all devices bank 0 is designated as access RAM and comprises 128 bytes of General Purpose Registers (range 0x00 to 0x7f) as well as 128 bytes of Special Function Registers (range 0x80 to 0xff). This leaves memory for two additional GPR banks in the PIC 18F442/242 devices and four additional GPR banks in the PIC18F452/252 devices.

We have mentioned that many applications require no more memory than the 128 bytes available in the GPR access bank. In these cases, code will make sure that the "a" bit in the instruction word is set to 0, as in the instruction

```
movf     var1,w,0  ; Contents of var1 to w in access RAM.
```

Because the include file for the 18FXX2 processors defines that the constant A (mnemonic for access) is equal to 0, we could have also coded:

```
movf     var1,w,A
```

By the same token, bits 0 to 3 in the BSR register hold the upper 4 bits of the 12-bit RAM address. The movlb instruction is provided so that a literal value can be moved directly into the bank select register (BSR) bits; for example,

```
movlb     2          ; Select bank 2
```

At this point, any instructions that access RAM with the A bit set to 1 will reference data in bank 2; for example,

```
movwf     0x10,1   ; Move w to register at bank 2:0x10
```

Note that instructions that use the a bit to define bank access default to a = 1, that is, to using the BSR bits in computing the data address. For this reason, programs that only use the access bank can prevent errors and shorten the coding by setting the BSR register to this bank, as follows:

```
movlb     0          ; Select access bank
```

We have included this line in the template with the note that applications that use other banks will eventually supersede this statement.

4.3 Naming Conventions

One of the style issues that a programmer must decide relates to the naming conventions followed for program labels and variable names. The MPLAB assembler is case sensitive by default; therefore MY_PORT and my_port refer different registers.

4.3.1 Register and Bit Names

We have seen that a programmer can define all the standard and program-specific registers (SFRs and GPRs) using equ or #define directives. A safer approach is to import an include file (.inc extension) furnished in the MPALB package for each different PIC. The include files have the names of all SFRs and bits used by a particular device. These include files can be found in the MPLAB/MPASM Suite directory of the installed MPLAB software. The following is a segment of the P18F452.INC file.

```
;========================================================================
;         Verify Processor
;========================================================================
          IFNDEF __18F452
            MESSG "Processor-header file mismatch.  Verify selected
                  processor."
          ENDIF


;========================================================================
;         18Fxxx Family         EQUates
;========================================================================
FSR0                EQU  0
FSR1                EQU  1
FSR2                EQU  2
FAST                EQU  1
W                   EQU  0
A                   EQU  0
ACCESS              EQU  0
BANKED          EQU  1
;========================================================================
;         16Cxxx/17Cxxx Substitutions
;========================================================================
  #define DDRA    TRISA           ; PIC17Cxxx SFR substitution
  #define DDRB    TRISB           ; PIC17Cxxx SFR substitution
  #define DDRC    TRISC           ; PIC17Cxxx SFR substitution
  #define DDRD    TRISD           ; PIC17Cxxx SFR substitution
  #define DDRE    TRISE           ; PIC17Cxxx SFR substitution
;========================================================================
;
;         Register Definitions
;
;========================================================================
;----- Register Files ----------------------------------------------------
TOSU                EQU  H'0FFF'
TOSH                EQU  H'0FFE'
TOSL                EQU  H'0FFD'
STKPTR              EQU  H'0FFC'
PCLATU              EQU  H'0FFB'
PCLATH              EQU  H'0FFA'
PCL                 EQU  H'0FF9'
TBLPTRU             EQU  H'0FF8'
TBLPTRH             EQU  H'0FF7'
TBLPTRL             EQU  H'0FF6'
TABLAT              EQU  H'0FF5'
PRODH               EQU  H'0FF4'
PRODL               EQU  H'0FF3'

INTCON              EQU  H'0FF2'
INTCON1             EQU  H'0FF2'
INTCON2             EQU  H'0FF1'
INTCON3             EQU  H'0FF0'

INDF0               EQU  H'0FEF'
POSTINC0            EQU  H'0FEE'
```

```
POSTDEC0        EQU    H'0FED'
PREINC0         EQU    H'0FEC'
PLUSW0          EQU    H'0FEB'
FSR0H           EQU    H'0FEA'
FSR0L           EQU    H'0FE9'
WREG            EQU    H'0FE8'


INDF1           EQU    H'0FE7'
POSTINC1        EQU    H'0FE6'
POSTDEC1        EQU    H'0FE5'
PREINC1         EQU    H'0FE4'
PLUSW1          EQU    H'0FE3'
FSR1H           EQU    H'0FE2'
FSR1L           EQU    H'0FE1'
BSR             EQU    H'0FE0'


INDF2           EQU    H'0FDF'
POSTINC2        EQU    H'0FDE'
POSTDEC2        EQU    H'0FDD'
PREINC2         EQU    H'0FDC'
PLUSW2          EQU    H'0FDB'
FSR2H           EQU    H'0FDA'
FSR2L           EQU    H'0FD9'
STATUS          EQU    H'0FD8'


TMR0H           EQU    H'0FD7'
TMR0L           EQU    H'0FD6'
T0CON           EQU    H'0FD5'
.
.
.
```

Notice that all names in the include file are defined in capital letters. It is probably a good idea to adhere to this style instead of creating alternate names in lowercase. The C-like #include directive is used to refer, the include files at assembly time; for example, the template file contains the following line:

```
#include <p18f452.inc>
```

4.4 PIC 18Fxx2 Instruction Set

The PIC18FXX2 instruction set added several features to preexisting mid-range instructions. Thirty six new instructions were added, status bit operation were changed, some register bit locations and names were changed, and one instruction (clrw) is no longer supported. The total of seventy-six instructions is divided as follows:

- 31 byte-oriented instructions
- 5 bit-oriented instructions
- 22 control instructions
- 10 literal operations
- 8 data and program memory operations

Three instructions require two program memory locations. The remaining seventy-three instructions are encoded in a single memory word (16 bits). Each single word instruction consists of a 16-bit word divided into an OPCODE, which specifies the instruction type and one or more OPERANDS, which further specify the instruction.

4.4.1 Byte-Oriented Instructions

Most byte-oriented instructions have three operands:

- The file register (specified by the 'f' descriptor)
- The destination of the result (specified by the 'd' descriptor)
- The accessed memory (specified by the 'a' descriptor)

The file register descriptor 'f determines the file register to be used by the instruction. The destination descriptor 'd' specifies where the result of the operation is to be placed. If 'd' is zero, the result is placed in the WREG register. If 'd' is one, the result is placed in the file register specified in the instruction. The descriptor 'a' determines whether the operand is in banked or access memory. Zero represents access memory, and one is used for banked memory.

4.4.2 Bit-Oriented Instructions

Bit-oriented instructions have three operands:

- The file register (specified by the 'f' descriptor)
- The bit in the file register (specified by the 'b' descriptor)
- The accessed memory (specified by the 'a' descriptor)

The file register and access memory descriptors are the same as for the byte-oriented instructions. The bit field designator 'b' selects the number of the bit affected by the operation. Table 4.2 lists the byte- and bit-oriented instructions.

4.4.3 Literal Instructions

Literal instructions may use some of the following operands:

- A literal value to be loaded into a file register (specified by 'k')
- The desired FSR register to load the literal value into (specified by 'f')
- No operand required (specified by '—')

Table 4.3 lists the literal and data/program memory operations.

4.4.4 Control Instructions

Control instructions may use some of the following operands:

- A program memory address (specified by 'n')
- The mode of the Call or Return instructions (specified by 's')
- The mode of the Table Read and Table Write instructions (specified by 'm')
- No operand required (specified by '—')

Table 4.2

PIC 18FXX2 Byte- and Bit-Oriented Instructions

Mnemonic, Operands		Description	Cycles	16-Bit Instruction Word				Status Affected	Notes
				MSb			LSb		
BYTE-ORIENTED FILE REGISTER OPERATIONS									
ADDWF	f, d, a	Add WREG and f	1	0010	01da0	ffff	ffff	C, DC, Z, OV, N	1, 2
ADDWFC	f, d, a	Add WREG and Carry bit to f	1	0010	0da	ffff	ffff	C, DC, Z, OV, N	1, 2
ANDWF	f, d, a	AND WREG with f	1	0001	01da	ffff	ffff	Z, N	1,2
CLRF	f, a	Clear f	1	0110	101a	ffff	ffff	Z	2
COMF	f, d, a	Complement f	1	0001	11da	ffff	ffff	Z, N	1, 2
CPFSEQ	f, a	Compare f with WREG, skip =	1 (2 or 3)	0110	001a	ffff	ffff	None	4
CPFSGT	f, a	Compare f with WREG, skip >	1 (2 or 3)	0110	010a	ffff	ffff	None	4
CPFSLT	f, a	Compare f with WREG, skip <	1 (2 or 3)	0110	000a	ffff	ffff	None	1, 2
DECF	f, d, a	Decrement f	1	0000	01da	ffff	ffff	C, DC, Z, OV, N	1, 2, 3, 4
DECFSZ	f, d, a	Decrement f, Skip if 0	1 (2 or 3)	0010	11da	ffff	ffff	None	1, 2, 3, 4
DCFSNZ	f, d, a	Decrement f, Skip if Not 0	1 (2 or 3)	0100	11da	ffff	ffff	None	1, 2
INCF	f, d, a	Increment f	1	0010	10da	ffff	ffff	C, DC, Z, OV, N	1, 2, 3, 4
INCFSZ	f, d, a	Increment f, Skip if 0	1 (2 or 3)	0011	11da	ffff	ffff	None	4
INFSNZ	f, d, a	Increment f, Skip if Not 0	1 (2 or 3)	0100	10da	ffff	ffff	None	1, 2
IORWF	f, d, a	Inclusive OR WREG with f	1	0001	00da	ffff	ffff	Z, N	1, 2
MOVF	f, d, a	Move f	1	0101	00da	ffff	ffff	Z, N	1
MOVFF	f_s, f_d	Move f_s (source) to 1st word	2	1100	ffff	ffff	ffff	None	
		f_d (destination) 2nd word		1111	ffff	ffff	ffff		
MOVWF	f, a	Move WREG to f	1	0110	111a	ffff	ffff	None	
MULWF	f, a	Multiply WREG with f	1	0000	001a	ffff	ffff	None	
NEGF	f, a	Negate f	1	0110	110a	ffff	ffff	C, DC, Z, OV, N	1, 2
RLCF	f, d, a	Rotate Left f through Carry	1	0011	01da	ffff	ffff	C, Z, N	
RLNCF	f, d, a	Rotate Left f (No Carry)	1	0100	01da	ffff	ffff	Z, N	1, 2
RRCF	f, d, a	Rotate Right f through Carry	1	0011	00da	ffff	ffff	C, Z, N	
RRNCF	f, d, a	Rotate Right f (No Carry)	1	0100	00da	ffff	ffff	Z, N	
SETF	f, a	Set f	1	0110	100a	ffff	ffff	None	
SUBFWB	f, d, a	Subtract f from WREG with borrow	1	0101	01da	ffff	ffff	C, DC, Z, OV, N	1, 2
SUBWF	f, d, a	Subtract WREG from f	1	0101	11da	ffff	ffff	C, DC, Z, OV, N	
SUBWFB	f, d, a	Subtract WREG from f with borrow	1	0101	10da	ffff	ffff	C, DC, Z, OV, N	1, 2
SWAPF	f, d, a	Swap nibbles in f	1	0011	10da	ffff	ffff	None	4
TSTFSZ	f, a	Test f, skip if 0	1 (2 or 3)	0110	011a	ffff	ffff	None	1, 2
XORWF	f, d, a	Exclusive OR WREG with f	1	0001	10da	ffff	ffff	Z, N	
BIT-ORIENTED FILE REGISTER OPERATIONS									
BCF	f, b, a	Bit Clear f	1	1001	bbba	ffff	ffff	None	1, 2
BSF	f, b, a	Bit Set f	1	1000	bbba	ffff	ffff	None	1, 2
BTFSC	f, b, a	Bit Test f, Skip if Clear	1 (2 or 3)	1011	bbba	ffff	ffff	None	3, 4
BTFSS	f, b, a	Bit Test f, Skip if Set	1 (2 or 3)	1010	bbba	ffff	ffff	None	3, 4
BTG	f, d, a	Bit Toggle f	1	0111	bbba	ffff	ffff	None	1, 2

Note 1: When a PORT register is modified as a function of itself (e.g., MOVF PORTB, 1, 0), the value used will be that value present on the pins themselves. For example, if the data latch is '1' for a pin configured as input and is driven low by an external device, the data will be written back with a '0'.

2: If this instruction is executed on the TMR0 register (and, where applicable, d = 1), the prescaler will be cleared if assigned.

3: If Program Counter (PC) is modified or a conditional test is true, the instruction requires two cycles. The second cycle is executed as a NOP.

4: Some instructions are 2-word instructions. The second word of these instructions will be executed as a NOP, unless the first word of the instruction retrieves the information embedded in these 16-bits. This ensures that all program memory locations have a valid instruction.

5: If the Table Write starts the write cycle to internal memory, the write will continue until terminated.

All control instructions are a single word, except for three double-word instructions. These three instructions are necessary so that all the required information can be included in 32 bits. In the second word, the 4-MSbs are 1's. If this second word is executed as an instruction (by itself), it will execute as a NOP. Table 4.4 lists the control instructions.

Table 4.3

Literal and Data/Program Memory Operations

Mnemonic, Operands		Description	Cycles	16-Bit Instruction Word MSb			LSb	Status Affected	Notes
LITERAL OPERATIONS									
ADDLW	k	Add literal and WREG	1	0000	1111	kkkk	kkkk	C, DC, Z, OV, N	
ANDLW	k	AND literal with WREG	1	0000	1011	kkkk	kkkk	Z, N	
IORLW	k	Inclusive OR literal with WREG	1	0000	1001	kkkk	kkkk	Z, N	
LFSR	f, k	Move literal (12-bit) 2nd word	2	1110	1110	00ff	kkkk	None	
		to FSRx 1st word		1111	0000	kkkk	kkkk		
MOVLB	k	Move literal to BSR<3:0>	1	0000	0001	0000	kkkk	None	
MOVLW	k	Move literal to WREG	1	0000	1110	kkkk	kkkk	None	
MULLW	k	Multiply literal with WREG	1	0000	1101	kkkk	kkkk	None	
RETLW	k	Return with literal in WREG	2	0000	1100	kkkk	kkkk	None	
SUBLW	k	Subtract WREG from literal	1	0000	1000	kkkk	kkkk	C, DC, Z, OV, N	
XORLW	k	Exclusive OR literal with WREG	1	0000	1010	kkkk	kkkk	Z, N	
DATA MEMORY ↔ PROGRAM MEMORY OPERATIONS									
TBLRD*		Table Read	2	0000	0000	0000	1000	None	
TBLRD*+		Table Read with post-increment		0000	0000	0000	1001	None	
TBLRD*-		Table Read with post-decrement		0000	0000	0000	1010	None	
TBLRD+*		Table Read with pre-increment		0000	0000	0000	1011	None	
TBLWT*		Table Write	2 (5)	0000	0000	0000	1100	None	
TBLWT*+		Table Write with post-increment		0000	0000	0000	1101	None	
TBLWT*-		Table Write with post-decrement		0000	0000	0000	1110	None	
TBLWT+*		Table Write with pre-increment		0000	0000	0000	1111	None	

Note 1: When a PORT register is modified as a function of itself (e.g., MOVF PORTB, 1, 0), the value used will be that value present on the pins themselves. For example, if the data latch is '1' for a pin configured as input and is driven low by an external device, the data will be written back with a '0'.

2: If this instruction is executed on the TMR0 register (and, where applicable, d = 1), the prescaler will be cleared if assigned.

3: If Program Counter (PC) is modified or a conditional test is true, the instruction requires two cycles. The second cycle is executed as a NOP.

4: Some instructions are 2-word instructions. The second word of these instructions will be executed as a NOP, unless the first word of the instruction retrieves the information embedded in these 16-bits. This ensures that all program memory locations have a valid instruction.

5: If the Table Write starts the write cycle to internal memory, the write will continue until terminated.

Table 4.4

Control Operations

Mnemonic, Operands		Description	Cycles	16-Bit Instruction Word				Status Affected	Notes
				MSb			LSb		
CONTROL OPERATIONS									
BC	n	Branch if Carry	1 (2)	1110	0010	nnnn	nnnn	None	
BN	n	Branch if Negative	1 (2)	1110	0110	nnnn	nnnn	None	
BNC	n	Branch if Not Carry	1 (2)	1110	0011	nnnn	nnnn	None	
BNN	n	Branch if Not Negative	1 (2)	1110	0111	nnnn	nnnn	None	
BNOV	n	Branch if Not Overflow	1 (2)	1110	0101	nnnn	nnnn	None	
BNZ	n	Branch if Not Zero	2	1110	0001	nnnn	nnnn	None	
BOV	n	Branch if Overflow	1 (2)	1110	0100	nnnn	nnnn	None	
BRA	n	Branch Unconditionally	1 (2)	1101	0nnn	nnnn	nnnn	None	
BZ	n	Branch if Zero	1 (2)	1110	0000	nnnn	nnnn	None	
CALL	n, s	Call subroutine1st word	2	1110	110s	kkkk	kkkk	None	
		2nd word		1111	kkkk	kkkk	kkkk		
CLRWDT	—	Clear Watchdog Timer	1	0000	0000	0000	0100	$\overline{\text{TO}}, \overline{\text{PD}}$	
DAW	—	Decimal Adjust WREG	1	0000	0000	0000	0111	C	
GOTO	n	Go to address1st word	2	1110	1111	kkkk	kkkk	None	
		2nd word		1111	kkkk	kkkk	kkkk		
NOP	—	No Operation	1	0000	0000	0000	0000	None	
NOP	—	No Operation	1	1111	xxxx	xxxx	xxxx	None	4
POP	—	Pop top of return stack (TOS)	1	0000	0000	0000	0110	None	
PUSH	—	Push top of return stack (TOS)	1	0000	0000	0000	0101	None	
RCALL	n	Relative Call	2	1101	1nnn	nnnn	nnnn	None	
RESET		Software device RESET	1	0000	0000	1111	1111	All	
RETFIE	s	Return from interrupt enable	2	0000	0000	0001	000s	GIE/GIEH, PEIE/GIEL	
RETLW	k	Return with literal in WREG	2	0000	1100	kkkk	kkkk	None	
RETURN	s	Return from Subroutine	2	0000	0000	0001	001s	None	
SLEEP	—	Go into Standby mode	1	0000	0000	0000	0011	$\overline{\text{TO}}, \overline{\text{PD}}$	

Note 1: When a PORT register is modified as a function of itself (e.g., MOVF PORTB, 1, 0), the value used will be that value present on the pins themselves. For example, if the data latch is '1' for a pin configured as input and is driven low by an external device, the data will be written back with a '0'.

2: If this instruction is executed on the TMR0 register (and, where applicable, d = 1), the prescaler will be cleared if assigned.

3: If Program Counter (PC) is modified or a conditional test is true, the instruction requires two cycles. The second cycle is executed as a NOP.

4: Some instructions are 2-word instructions. The second word of these instructions will be executed as a NOP, unless the first word of the instruction retrieves the information embedded in these 16-bits. This ensures that all program memory locations have a valid instruction.

5: If the Table Write starts the write cycle to internal memory, the write will continue until terminated.

Chapter 5

PIC18 Programming in C Language

5.1 C Compilers

The C programming language was developed at Bell Labs in the late 1960s and early 1970s by Dennis Ritchie. It was designed to provide many of the features of low-level languages, thus being suitable for operating system and driver development. Over the years, C has become the most widely used programming language. Several C compilers are available for PIC programming from Microchip and from other vendors.

The C compilers and versions of these compilers are specifically related to the PIC families with which they are compatible, as follows:

1. MPLAB C18 Compiler supports the PIC 18 MCUs.

2. MPLAB C30 Compiler supports PIC24 MCUs and dsPIC DSCs.

3. MPLAB C Compiler is specific for dsPIC DSCS

4. MPLAB C Compiler is specific for PIC24 MCUs.

5. MPLAB C Compiler is specific for PIC32 MCUs.

Notice that there is a C compiler that is compatible with both PIC24 and dsPIC devices, while there are also specific compilers for either the PIC24 and the dsPICs. Also note that a company named HI-TECH has developed a C compiler for the 10, 12, and 16 PICs in addition to their PIC18 MCUs compiler.

5.1.1 C versus Assembly Language

An assembler program performs a one-to-one translation of machine mnemonics into machine code. The C compiler, on the other hand, converts the C language instructions into groups of assembly language statements. In fact, a C compiler can be viewed as an assembly language expert system because its immediate product is an assembly language file. This process results in several inefficiencies:

1. C language programs are larger and occupy more memory than the equivalent ones in assembly language.

2. C language programs can never execute faster than an equivalent one in assembly language and are usually quite slower.

3. The C language does not have the full functionality of the hardware because it requires functions in libraries or the language itself to implement this functionality.

On the other hand, the advantages of using a high-level language such as C are convenience, ease, and better performance in program development. By insulating the programmer from the hardware details, a high-level language may make it easier and faster to develop an application.

The developer must be aware of these trade-offs when selecting a language. However, the MPLAB C Compilers do allow writing programs that use both C and assembly language. Mixed language programs can be based on C language applications that make use of modules containing sub-routines written in assembly language, or assembly language applications that reference sub-routines written in C. In addition, MPLAB C18 Compiler supports inline assembly code with some restrictions. This allows the programmer to include assembly language statements and code as part of a C language application.

5.1.2 MPLAB C18

MPLAB C18 C Compiler is a cross-compiler for the PC that is compatible with the Microchip PIC18 family of microcontrollers. The C18 compiler translates a text file in C language, and optionally other object and library files provided at link time, into a .hex file that can be executed by the targeted PIC microcontroller. According to Microchip, the compiler can optimize code so that routines that were developed to be used with a specific C function can be easily ported to other C functions.

Source code for the compiler is written using standard ANSI C notation. The "build" process employed by the compiler consists of compiling source code fragments into blocks, which are later linked with other blocks and placed into PIC18 memory. The compiler's "make" command makes the build process more efficient by only invoking those portions of the C source file that have changed since the last build.

The two most important associated tools of the MPLAB C18 compiler are the linker and the assembler. The compiler and the associated tools can be invoked from the command line from within the MPLAB IDE. In either case, the resulting .hex file can be used to blow a PIC and later execute in the hardware.

5.2 MPLAB C18 Installation

The installation routine for MPLAB C18 follows the conventional steps, as follows:

1. A welcome screen is deployed.

2. The user is prompted to accept the license agreement.

3. A readme screen is displayed, which includes notes regarding compatibility, bug fixes, and device support.

4. The user is prompted to approve or change the installation directory.

5. The user is prompted to select program components or approve a standard selection list.

6. Required and available disk space is displayed.

7. Installation concludes, possibly requiring a system restart.

One important prerequisite of the C18 compiler installation is that the MPLAB IDE has been previously installed in the system. One of the reasons for requiring that MPLAB C18 installation be performed last is that both packages include the MPASM and MPLINK programs and the compiler must use the ones included in its own package. This is also the case with some MPLAB C18 linker scripts.

5.2.1 MPLAB Software Components

During installation, the program prompts for the selection of software components to be included. The following are offered:

1. Executables for the compiler (MPASM) and the linker (MPLINK).

2. Assembler files including the MPASM assembler and the header files for the devices supported by MPLAB C18. These are include files with filenames in the format: p18xxxx.inc.

3. Linker script files required by the MPLINK linker. There is one file for each supported PIC18 microcontroller. Each file provides a default memory configuration for the processor and directs the linker in the allocation of code and data in the processor's memory.

4. Standard headers for the standard C library and the processor-specific libraries.

5. Standard libraries contain the standard C library, the processor-specific libraries and the start-up modules.

6. Examples and sample applications to assist users in getting started with the compiler package.

7. Library source code for the standard C library and the processor-specific libraries.

8. Preprocessor source code.

All these components, except the last one, are selected by default during the compiler installation. Inclusion of the preprocessor source code requires clicking the box attached to this item. Figure 5.1 is a screen snapshot of the component selection screen.

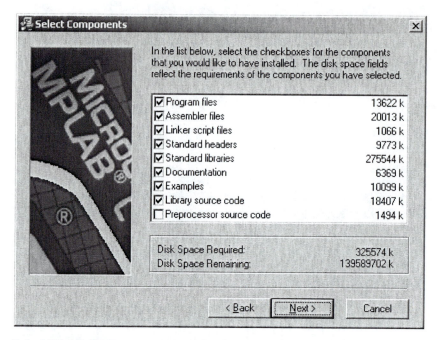

Figure 5.1 *MPLAB C18 component selection screen.*

5.2.2 Configuration Options

The configuration dialog box, displayed during installation, provides checkboxes for the following options:

1. Add MPLAB C18 to PATH environment variable. This option adds path of the MPLAB C18 executable (mcc18.exe) and the MPLINK linker executable (mplink.exe) to the beginning of the PATH environment variable. This allows MPLAB C18 and the MPLINK linker to be launched at the command prompt from any directory.

2. Add MPASM to PATH environment variable. Adds the path of the MPASM executable (mpasmwin.exe) to the beginning of the PATH environment variable. Doing this allows the MPASM assembler to be launched at the command shell prompt from any directory.

3. Add header file path to MCC_INCLUDE environment variable. Adds the path of the MPLAB C18 header file directory to the beginning of the MCC_INCLUDE environment variable. MCC_INCLUDE is a list of semicolon-delimited directories that MPLAB C18 will search for a header file if it cannot find the file in the directory list specified with the -I command-line option. Selecting this configuration option means it will not be necessary to use the -I command-line option when including a standard header file. If this variable does not exist, it is created.

4. Modify PATH and MCC_INCLUDE variables for all users. Selecting this configuration will perform the modifications to these variables as specified in the three previous options for all users.

5. Update MPLAB IDE to use this MPLAB C18. This option only appears if the MPLAB IDE is installed. Selecting this option configures the MPLAB IDE to use the newly installed MPLAB C18. This includes using the MPLAB C18 library directory as the default library path for MPLAB C18 projects in the MPLAB IDE.

6. Update MPLAB IDE to use this MPLINK linker. This option appears only if the MPLAB IDE is installed. Selecting this option configures the MPLAB IDE to use the newly installed MPLINK™ linker.

Figure 5.2 is a screen snapshot of the configuration options screen.

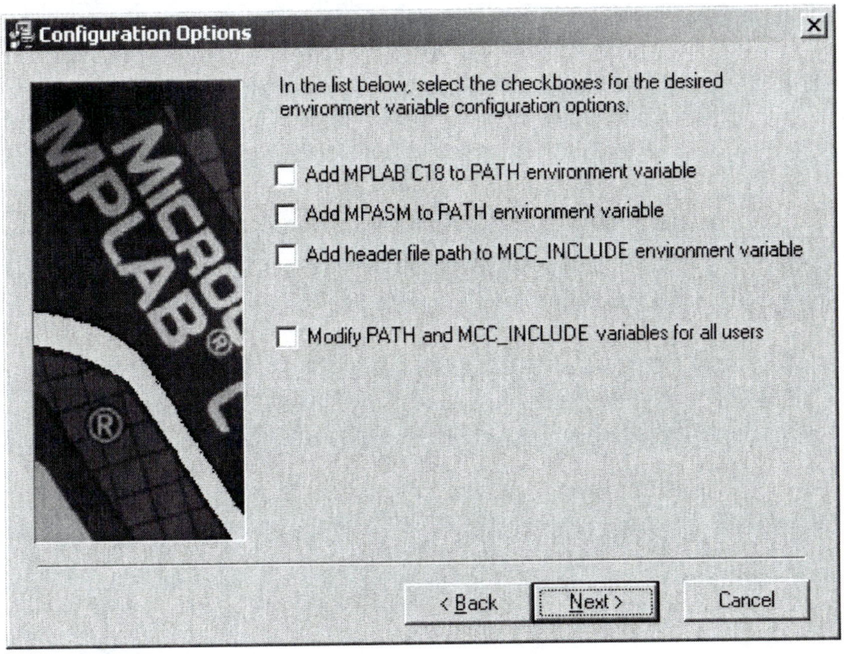

Figure 5.2 *MPLAB C18 configuration options screen.*

Notice that the last two configuration options in the previous list do not appear in the screen snapshot of Figure 5.2. This is because these options are only shown if the current user is logged on to a Windows NT or Windows 2000 computer as an administrator. This was not the case during our sample installation.

5.2.3 System Requirements

According to Microchip, the system requirements for MPLAB C18 and the MPLAB IDE are the following:

1. Intel Pentium class PC running Microsoft 32-bit Windows operating system.

2. A minimum of 250 MB hard disk space.

3. MPLAB IDE previously installed.

The default installation for MPLAB IDE may have preset selections. When installing the IDE for use with MPLAB C18, there are some IDE components that must be selected at a minimum. These are

```
MPLAB IDE Device Support

    8-bit MCUs

Microchip Applications
    MPLAB IDE

    MPLAB SIM

    MPASM Suite
```

The MPASM Suite is installed with MPLAB C18, so it does not need to be installed with MPLAB IDE.

5.2.4 Execution Flow

Flow of execution refers to the language tools used during compilation and the files that take part in this process. The flow is shown in Figure 5.3.

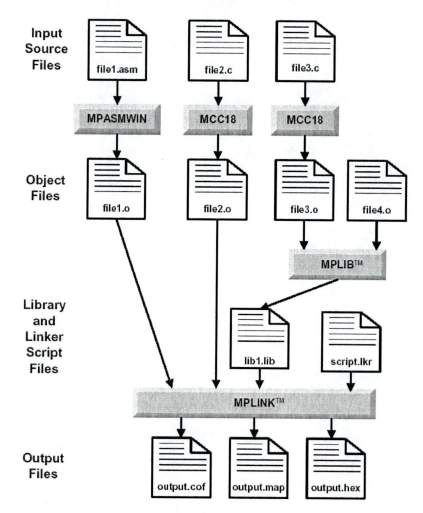

Figure 5.3 *C18 compiler flow of execution.*

In the example of Figure 5.3, a program is composed of three files: two are C language source files and one is in assembly language. These three files are the input into the development system. The first stage consists of generating the object files. The C compiler (MCC18) operates on the C language sources, and the assembler (MPASMWIN) on the assembly language file. In the following stage the linker program (MPLINK) receives as input the object files generated by the Assembler and the C compiler, the library files produced by the Library Manager application (MPLIB), and a linker script. With this input, the Linker produces the executable file (in .hex format) as well as several support files in .cof (Common Object File) and .map formats, which are used mostly in debugging.

5.3 C Compiler Project

An MPLAB IDE project is a group of related files associated with a particular application and language tool, such as MPLAB C18. Every project will have at least one source file and one linker script. In addition, a project must also identify the target hardware and define register names and other identifiers. The header files that contain this information are typically included as source files and need not be added to the project.

During the compilation process, the project produces the following output files:

- Executable code that can be loaded into the target microcontroller as firmware.

- Debugging files to help MPLAB IDE relate symbols and function names from the source files, with the executable code.

The simplest applications consist of a project with a single source file and one linker script. This is the case in many of the sample programs developed in this book.

5.3.1 Creating the Project

An MPLAB project is created either using the <Project Wizard...> or the <New...> command in the Project menu. The dialogs that follow are identical in either command; however, using the <Project Wizard...> command has the advantage of ensuring that all necessary subcommands are visited during project creation. The <Project Wizard...> command presents the following dialogs:

1. Select hardware device

2. Select a language toolsuite

3. Create the new project

4. Add existing files to the project

5. Display summary of created project

In the following sub-sections we consider each of these project creation dialogs for a program in C language and target the 18f452 microcontroller.

Select Hardware Device

After the initial welcome screen, the MPLAB IDE presents Step One of the Project Wizard consisting of selecting a particular hardware device for the project. Figure 5.4 is a screen snapshot of the device selection dialog of the project creation wizard.

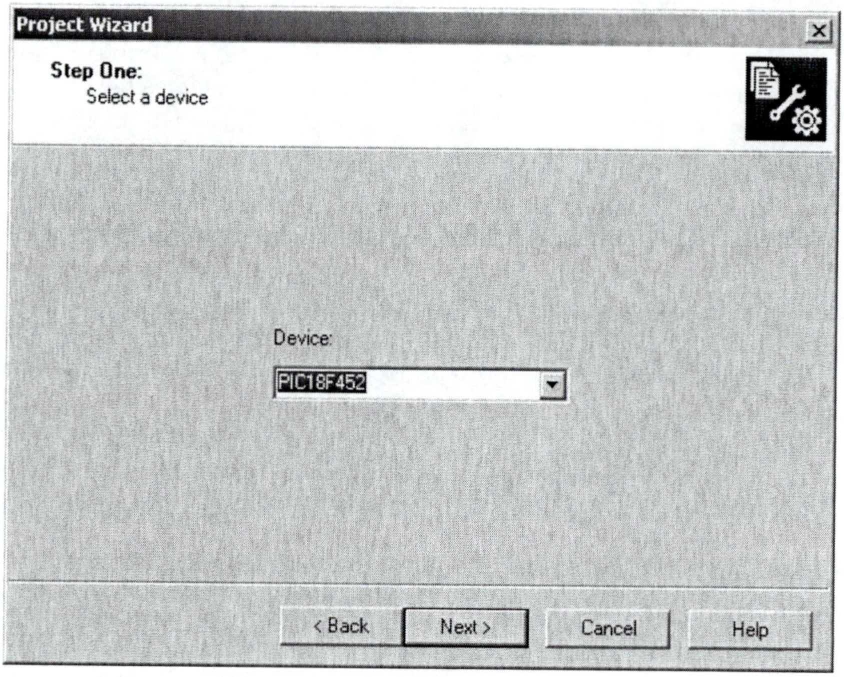

Figure 5.4 *Device selection dialog in project wizard.*

Once the device has been selected, the user presses the Next > button on the dialog screen.

Select the Language Toolsuite

The second step in the project creation wizard consists of selecting the language to be used in the project. If the compiler was installed correctly, the dialog will show Microchip C18 as the as the active language toolsuite. The toolsuite contents window will list the four programs required in project development; these are the Assembler, the Linker, the Compiler and the Librarian. Figure 5.5 is a screen snapshot of the language toolsuite selection dialog.

If the language toolsuite selection dialog does not match the one in Figure 5.5 it will be necessary to make the necessary corrections. If Microchip C18 Toolsuite is not selected in the Active Toolsuite window, you can expand the window and choose it from the drop-down list. If the Microchip C18 Toolsuite does not appear in the drop-down window, then the compiler has not been installed correctly in the development environment. The four executables required by the environment should be listed in the Toolsuite Contents window. If one or more are missing, then the compiler installation was not correct.

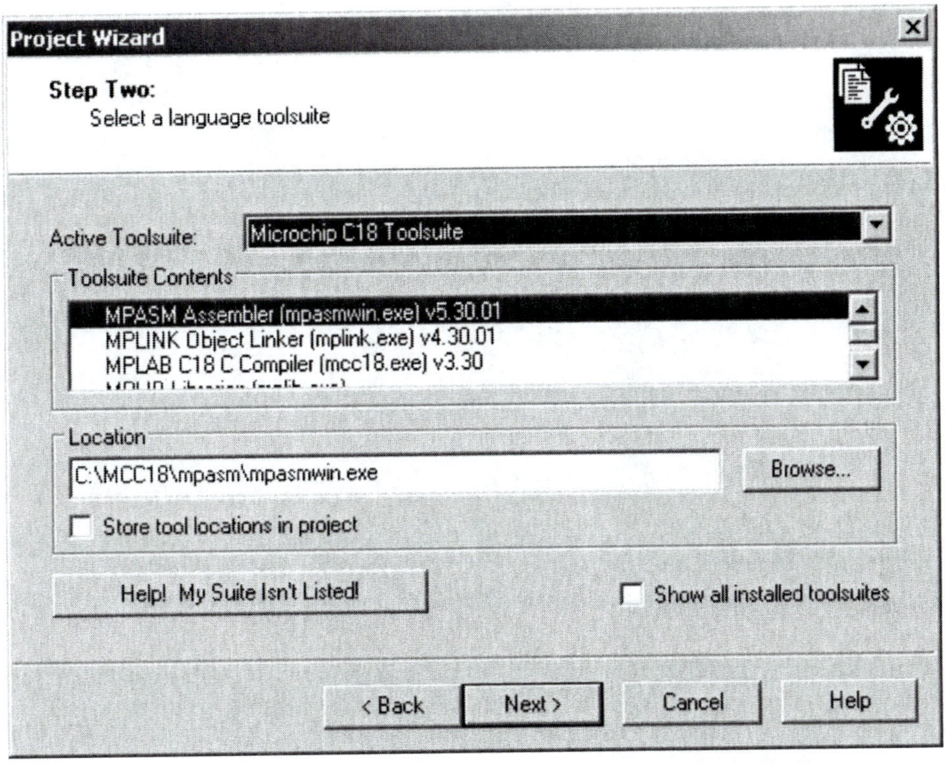

Figure 5.5 *Language toolsuite selection dialog in Project Wizard.*

The location window of the dialog box shows the location of each executable selected in the Toolsuite Contents window. The default location for each of the executables is as follows:

mpasmwin.exe at C:\MCC18\mpasm\mpasmwin.exe

mplink.exe at C:\MCC18\bin\mplink.exe

mcc18.exe at C:\MCC18\bin\mcc18.exe

mplib.exe at C:\MCC18\bin\mplib.exe

If the installation drive and directory for the C18 Toolsuite were different from the default shown in Figure 5.5, then the paths to the executables would be different. In this case, one can use the <Browse...> button to navigate to the correct path and select it into the wizard window.

Create a New Project

The third step provided by the Project Creation Wizard relates to defining the new project. The dialog box also contains a window for reconfiguring an existing project that is inactive during normal new project creation. The dialog box for the project creation step is shown in Figure 5.6.

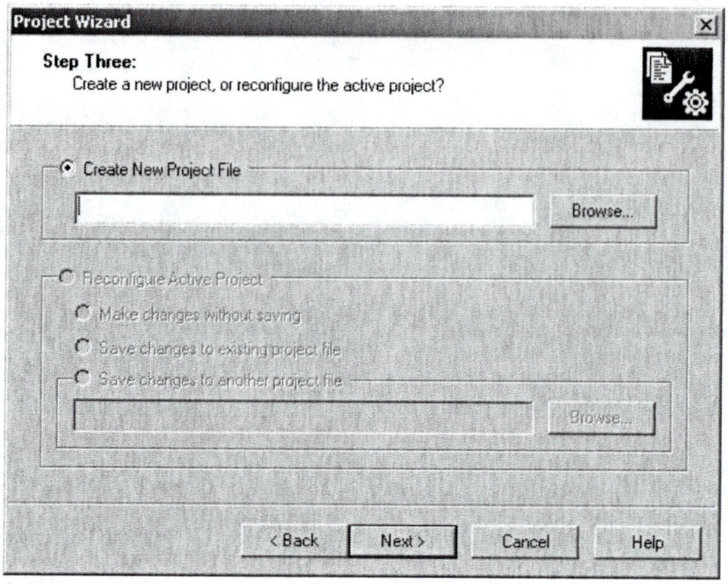

Figure 5.6   *Project creation dialog in the Project Wizard*

The dialog requires that the path for the new project be entered on the input line. Alternatively, one can use the <Browse...> button to navigate through the file system for the desired location. When using the <Browse...> option, the Save Project As dialog that follows contains an input box labeled that must be filled with the project's name. Note that the Save Project As dialog also permits creating a project folder at this time. Figure 5.7 is a screen snapshot of the Save Project As dialog.

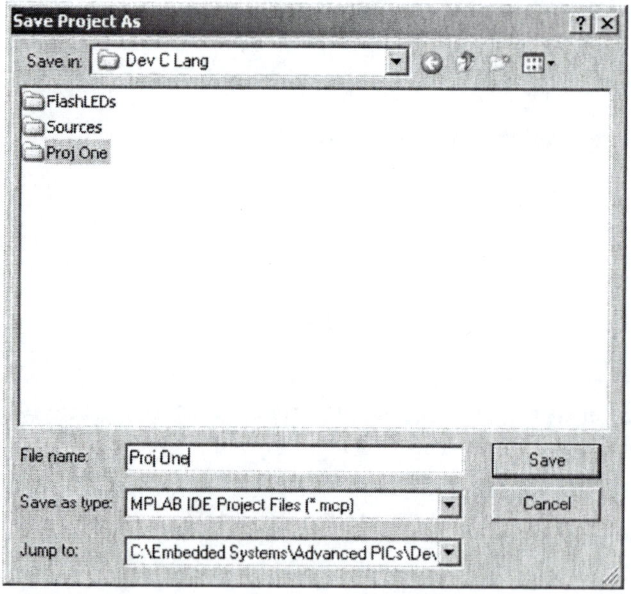

Figure 5.7   *Save Project As dialog in the Project Wizard.*

In the example in Figure 5.7, we used the <Create New Folder> button in the input box command line to create a new project folder, which we named Proj One. Then we entered the same name (Proj One) in the File Name input box. After clicking the Save button twice, we end up with a new project Folder as well as a project file of the same name. Assigning the same name to both the project folder and the project file is often a convenient simplification.

Add Files to the Project

The fourth step in the Project Creation Wizard allows adding one or more files to the new project. If the project uses a template file for the source code, it can be attached to the project at this time. Alternatively, the source file or files can be added later.

In past versions of the MPLAB IDE, it was necessary to add a device-specific linker script file to the project. With the current version of MPLAB IDE, the linker will find the appropriate file automatically. This means that adding a linker script during project creation only applies to the following cases:

1. Projects developed with MPLAB IDE versions older than 8

2. Projects using the MPLAB Assembler for PIC24 MCUs and dsPIC DSCs

3. Projects that require an edited linker script file

Also note that adding a linker script file to the project has no undesirable effects; so if in doubt, it is preferable to add the script file. The Add Files dialog is shown in Figure 5.8.

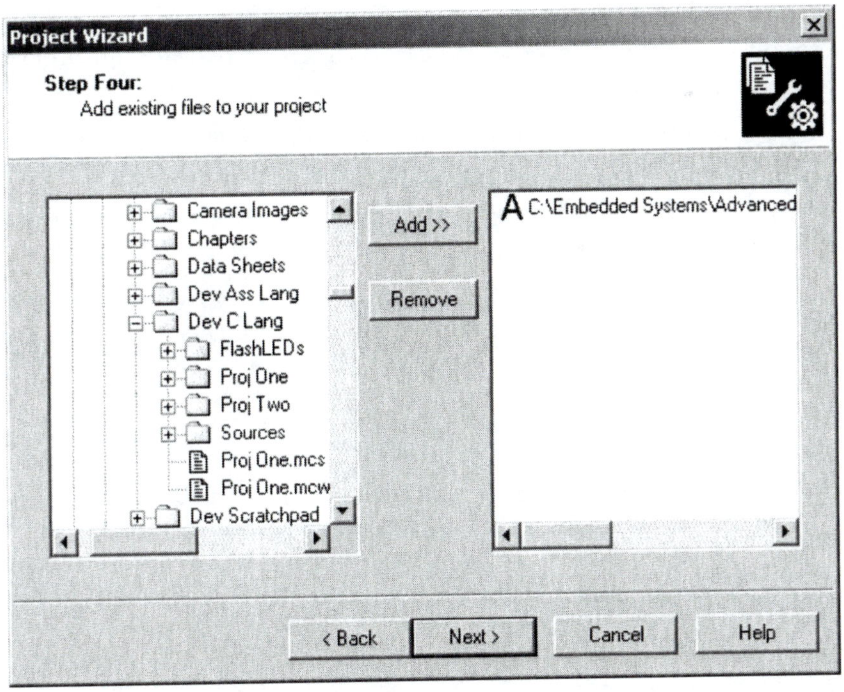

Figure 5.8 *Add files dialog in the Project Wizard.*

Once in the Add Files dialog, you can click on a file name to select that file. Alternatively, <Ctrl> + Click can be used to select more than one file. Clicking the <Add>> > button displays the filename in the right-hand window preceded by a letter icon. The letter icon can be changed by clicking on it. The following file addition modes are available:

- [A] Auto – Allow MPLAB IDE to determine whether the file's path should be relative or absolute based upon the file's location. If the project directory contains the file, the reference is a relative path. Otherwise, the reference is an absolute path.

- [U] User – Reference the file with a relative path. This is the preferred way to add files created specifically for a project.

- [S] System – Reference the file with an absolute path. There are cases when this mode is useful, but these paths often require correction when a project is moved from one system to another.

- [C] Copy – Copy a file to the project directory and add a relative path to the copied file. Optionally, edit the file name to rename the local copy of the file. This mode provides a convenient way to incorporate linker scripts and template source files into a project without introducing an absolute path.

It is possible to use the Add Files dialog box to attach a source template file to the project without overwriting the template or leaving the development environment. The following steps are required:

1. Double-click the template file to open it in the MPLAB Editor.

2. Select the Save As command from the MPLAB File Menu and enter a new name for the template file. Make sure you have navigated to the desired project directory.

3. Select the <Add Files to Project> command in the Project Menu. Select the newly named file and click the <Open> button.

4. In the Project Files display box, right-click on the filename of the original template file and select Remove.

The files and structure of a project can be seen in the MPLAB IDE Project Window. The Project Window is displayed by selecting the Project command in the View Menu. Once the Project Window is displayed, double-clicking on a source file name activates the Editor program with the selected file loaded. Figure 5.9 shows the MPLAB Project Window with a single source file.

A file can be deleted from a project by right-clicking the file name and then selecting the <Remove> command from the drop-down menu.

5.3.2 Selecting the Build Directory

A compiler operation control that is not offered by the MPLAB Project Wizard refers to the project build options. Not setting these options often has unexpected results; for example, the executable file that results from a successful build is placed in a folder not expected by the user.

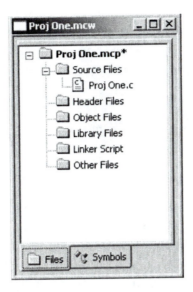

Figure 5.9 *MPLAB Project Window.*

The Build Options dialog is displayed by selecting <Build Options...> in the Project Menu and then clicking on the Project option. Of the various tabs in the Build Options For Project dialog the most often required one is the tab labeled <Directories>, shown in Figure 5.10.

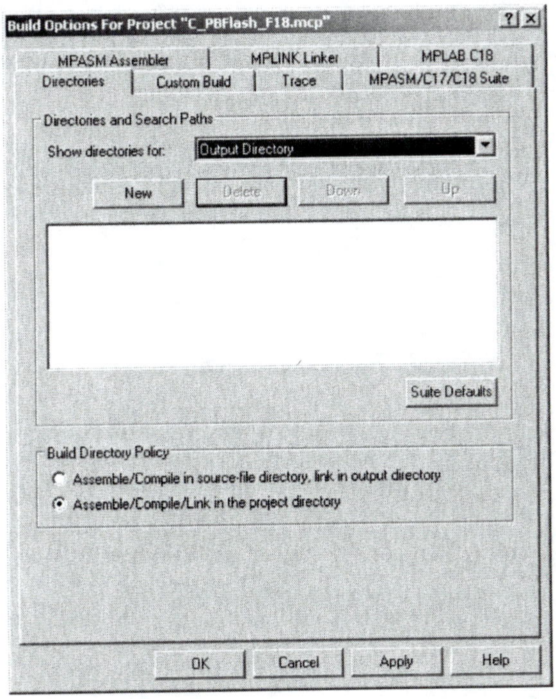

Figure 5.10 *Build Options For Project dialog.*

The Build Directory Policy, on the lower part of the dialog screen, allows selecting the destination of the files generated during assembly, compilation, and linkage. If the lower radio button is selected, then the output is all placed in the Project directory. Occasionally we may wish to place all our executables in a single directory. This can be accomplished by clicking on the upper radio button labeled Assemble/Compile in source-file directory, link in output directory.

To select or create an output directory, make sure that Output Directory is selected in the Show directories for: window. Then click the New button and the ellipses on the large screen. At this time you can navigate to any existing directory or create a new one by clicking the Make New Folder button. Once the output directory is defined, all executable generated by the linker will be placed in it.

5.4 A First Program in C

In this section and the sub-sections that follow, we present a C language version of the LedPB_F18 program developed in Chapter 4, Section 4.5.2. This presentation assumes some knowledge of the C language in the reader as implemented in the MPLAB C18 compiler and is familiar with the MPLAB IDE. Appendix A is a brief tutorial on the language to which the reader can refer.

The program's version in C language, called C_PBFlash_F18, monitors pushbutton # 1 in DemoBoardA 18F452 (or equivalent circuit). These circuits were discussed in Chapter 4, Section 4.5, and following.

- If the pushbutton is released, the four green LEDs wired to port C lines 0–3, are flashed.

- If the pushbutton is held pressed, then the four red LEDs wired to port C lines 4–7 are flashed.

The source code listing for the C_PBFlash_F18.c program is as follows:

```
/* Project name: C_PBFlash_F18
   Source files: C_PBFlash_F18.c
   Date: August 17, 2012
   Copyright 2012 by Julio Sanchez and Maria P. Canton
   Processor: PIC 18F452
   Environment:    MPLAB IDE Version 8.86
                   MPLAB C-18 Compiler

   TEST CIRCUIT: Demo Board 18F452 or circuit wired as
   follows:
   PORT        PINS        DIRECTION        DEVICE
     C          0-3         Output            Green LEDs
     C          4-7         Output            Red LEDs
     B           4          Input             Pushbutton No.
   Description:
   A demonstration program to monitor pushbutton No. 1 in
   DemoBoard 18F452 (or equivalent circuit). If the pushbutton
   is released the four green LEDs wired to port C lines 0-3
   are flashed. If the pushbutton is held pressed then the
   our red LEDs wired to port C lines 4-7 are flashed.
*/
#include <p18f452.h>
```

```
#include <delays.h>

#pragma config WDT = OFF
#pragma config OSC = HS
#pragma config LVP = OFF
#pragma config DEBUG = OFF

/* Function prototypes */
void FlashRED(void);
void FlashGREEN(void);

/*****************************************************************
                        main program
*****************************************************************/
void main(void)
{
    // Initalize direction registers
    TRISB = 0b00001000;// Port B, line 4, set for input
    TRISC = 0;              // Port C set for output
    PORTC = 0;              // Clear all port C lines

    while(1)
    {
//        if(PORTBbits.RB4)
        if(PORTB & 0b00010000)   // Alternative expression
            FlashRED();
        else
            FlashGREEN();
    }
}

/*****************************************************************
                        local functions
*****************************************************************/
void FlashRED()
{
    PORTC = 0x0f;
    Delay1KTCYx(200);
    PORTC = 0x00;
    Delay1KTCYx(200);
    return;
}
void FlashGREEN()
{
    PORTC = 0xf0;
    Delay1KTCYx(200);
    PORTC = 0x00;
    Delay1KTCYx(200);
    return;
}
```

5.4.1 Source Code Analysis

The source code listing begins with a commented header that records the program's name, files, authors, data, development environments and software, wiring of the demonstration circuit, and describes the program's operation.

Following the header is a statement line to include the processor's header file (<p18f452.h>) and the header file for the delays ry (<delays.h>). Code is as follows:

```
#include <p18f452.h>
#include <delays.h>
```

Four #pragma config statements are used to initialize the configuration bits listed in Table 4.1. The four configuration bits referred by the program turn off the watchdog timer, define a high-speed oscillator, turn off low-voltage protection, and turn off the background debugger, as follows:

```
#pragma config WDT = OFF
#pragma config OSC = HS
#pragma config LVP = OFF
#pragma config DEBUG = OFF
```

The next group of program lines are prototypes for the two local functions used by the program. By prototyping the functions, we are able to list them after the main() function. Without the prototypes, all local functions would have to be coded before main(). Adding auxiliary functions after main() seems reasonable to many programmers. Code is as follows:

```
/* Function prototypes */
void FlashRED(void);
void FlashGREEN(void);
```

main() Function

In MPLAB C18, the main() function has void return type and void parameter list. In C language, the main() function defines the program's entry point; execution starts here. The first lines in main() are to initialize the port directions as required by the program. Because the LEDs are wired to port C, its tris register (TRISC) is set for output. By the same token, because the pushbutton switch number 1 is wired to port B, line 4, this line must be initialized for input. Code is as follows:

```
// Initalize direction registers
TRISB = 0b00001000;// Port B, line 4, set for input
TRISC = 0;          // Port C set for output
PORTC = 0;          // Clear all port C lines
```

Notice that we have taken advantage of a feature of MPLAB C18 that allows declaring binary data using the 0b operator.

In C language we can create an endless loop by defining a condition that is always true. Because in C true is equivalent to the numeric value 1, we can code:

```
while(1)
{
    if(PORTBbits.RB4)
        FlashRED();
    else
        FlashGREEN();
}
```

This coding style provides a simple mechanism for implementing a routine that tests one or more input devices and proceeds accordingly. Because pushbutton switch number 1 is wired to port B line 4, code can test this bit to find out if the pushbutton is in a released or pressed state. Testing the state of a port bit is simplified by the presence of macros in the C compiler that define the state of each individual bit in the port. The if statement does this using the PORTBbits.RB4 expression. Alternatively, code can use a conventional C language bitwise AND operation on the port bits, as follows:

```
while(1)
{
    if(PORTB & 0b00010000)
        FlashRED();
    else
        FlashGREEN();
}
```

Local Functions

In either case, the previous test determines whether the pushbutton is pressed. If so, then the red LEDs must be flashed. Otherwise, the green LEDs must be flashed. The flashing is performed by two simple procedures called FlashRED() and FlashGREEN(). The FlashRED() procedure is coded as follows:

```
void FlashRED()
{
    PORTC = 0x0f;
    Delay1KTCYx(200);
    PORTC = 0x00;
    Delay1KTCYx(200);
    return;
}
```

The FlashRED() function starts by turning off the four high-order lines in Port C and turning on the four low-order lines. The hexadecimal operand 0x0f sets the port bits accordingly.

The statement that follows calls one of the delay functions in the delays General Software Library. These delay functions (whoch are revisited in Chapter 6) provide a convenient mechanism for implementing timed operations in an embedded environment. The delay functions are based on instruction cycles and are, therefore, dependent on the processor's speed. The one used in the sample program (Delay1KTCYx()) delays 1,000 machine cycles for every iteration. The number of desired iterations is entered as a function argument inside the parenthesis. In this case, 1,000 machine cycles are repeated 200 times, which implements a delay of 200,000 machine cycles. The function called FlashGREEN() proceeds in a similar manner.

Chapter 6

C Language in an Embedded Environment

6.1 MPLAB C18 System

The MPLAB C18 compiler was designed to make it easier to develop embedded applications that use the PIC microcontrollers of the PIC18XXXX family. It uses the standard C language as defined in the ANSI standard X3.159-1989 with some processor-specific extensions and documented deviations in cases of conflict with efficient support for the PICmicro MCUs. The compiler is documented to support the following features:

- Compatible with Microchip's MPLAB IDE, allowing source-level debugging with several MPLAB debugging tools and applications

- Integration with the MPLAB IDE for easy-to-use project management and source-level debugging

- Generation of relocatable object modules for enhanced code reuse

- Compatibility with object modules generated by the MPASM assembler, allowing complete freedom in mixing assembly and C programming in a single project

- Transparent read/write access to external memory

- Support for inline assembly

- Code generator engine with multi-level optimization

- Library support for PWM, SPI™, I2C™, UART, USART, string manipulation, and math

- User-level control over data and code memory allocation

Reputedly, the MPLAC compiler makes development of embedded systems applications easier and saves coding time. Appendix A is a tutorial on the fundamentals of C language for those readers who are new to C programming.

6.1.1 PIC18 Extended Mode

Some members of the PIC18 family provide support for an additional mode of operation sometimes referred to as the extended microcontroller mode. Extended mode includes memory handling facilities that improve compiler performance and changes to the instructions set in the form of modified operation of some instructions and the addition of eight new ones.

   Not all PIC18 devices support extended mode. The 18F452 which is the CPU covered in this book, does not support extended mode. In the devices that do support extended mode (such as the PIC 18F4620), the mode is controlled by the corresponding configuration bit, as follows:

```
#pragma config XINST = ON [OFF]
```

   CPUs that do not support extended mode generate a compile-time error on this directive. Using extended mode in MPLAB C18 also requires checking the extended mode checkbox in the Project>Build Options>Project and in the tab MPASM/C17/C18 Suite. Because the CPU discussed in this book does not support extended mode, the book contains no other discussion of this topic.

6.2 MPLAB C18 Libraries

A C language library is a collection of functions usually intended for a specific subject area or purpose, for example, a math library or a input/output library. A library is implemented as one or more files grouped for easy reference and convenient linking. The MPLAB applications MPASM, MPLINK, and MPLIB provide support for creating and managing libraries.

   After the installation of MPLAB C18, the library files (.lib extension) are located in the MCC18/lib directory. These files can be linked to an application directly. The sources for these library files are found in the MCC18\src directory. The subdirectory /traditional contains the files for nonextended mode operation and the directory /extended contains the files for extended mode. A third subdirectory named /pmc_common contains library support for power management control functions. Browsing through the library source files with the extensions .c, .asm, and .h is a source of valuable information for the trained programmer.

6.2.1 Start-Up Routines

The C18 compiler initializes a C application by linking-in a source file that contains what is usually called the start-up code. The source files (extension .c) for the start-up code can be found in the directories MCC18/src/tradidional/startup and MCC18/src/extended/startup. Which source is selected depends on the mode chosen for the application. In programs that run in the 18F452 chip, only the nonextended code is available.

   Three different source files are available for each mode. The ones in the traditional subdirectory are named c018.c, c018i.c, and c018iz.c The first one (c018.c) generates startup code that initializes the C language stack and jumps to the application's main() function. The second one (c018i.c) also provides code to initialized data elements (idata). This is required if global or static variables are initialized when they are defined. This second file (c018i.c) is the start-up code generated by default by the linker scripts provided with MPLAB C18. The third one (c018iz.c) initializes idata and zeros variables that are not initialized in their definitions, as is required for strict ANSI compliance.

You may explore the start-up code by selecting the MPLAB SIM debugger before building your application. This is accomplished from the IDE Debugger menu by clicking <Select Tool/MPLAB SIM>. When the program is compiled, the editor screen will now contain a tab for the start-up source used by the compiler (normally c018i.c). The entry point of the start-up code will be displayed on the editor screen, as follows:

```
#pragma code _entry_scn=0x000000
void
_entry (void)
{
_asm goto _startup _endasm

}
#pragma code _startup_scn
void
_startup (void)
{
...
```

In Appendix B we discuss the use of MPLAB SIM and show how you can step through and inspect variables in the start-up routine. MPLAB C18 provides a batch file named makestartup.bat that may be used to rebuild the start-up code and copy the generated object files to the lib directory.

6.2.2 Online Help for C18 and Libraries

The MPLAB C18 Compiler is furnished with documentation in pdf and HTML-format files (.chm extension). After compiler installation, the pdf files that document the language and the libraries are found in the directory MCC18\doc and the subdirectory MCC18\doc\periph-lib. According to the Microchip documentation, the help files for the C18 compiler and libraries should be accessible from within the IDE. Clicking <Help/Topics/Language Tools> should display the names of all available help files, including the one for the C18 language and the one for the C18 Libraries.

However, this is not always the case. Several online forums have reported that some versions of the IDE and the C18 compiler fail to install the help files correctly. One explanation for this installation error is that the MPLAB IDE was installed after the C18 compiler. Another explanation is that the user failed to check the boxes that prompt for IDE update during compiler installation. Nevertheless, even when the above-mentioned precautions are taken, the help files are sometimes not installed correctly.

One possible solution is to edit the Windows Registry and insert a new reference to the missing documentation and its location in the system. However, changing the Windows Registry should only be done by very experienced Windows users because the operation can be the cause of major system malfunctions.

A simple solution, but not without inconveniences, is to open a browser window with each of the required help files and keep these windows on the desktop. In fact, some programmers prefer to use this method even for help files available from inside the IDE because it is easier and faster to access the desired information if it is located in a separate program.

6.3 Processor-Independent Libraries

MPLAB C18 furnishes the standard C libraries by means of functions that are supported by all members of the PIC18 architecture. These functions can be divided into two groups:

- Functions in General Software Library
- Functions in Math Libraries

The General Software Library can be further divided into two groups of functions: the standard C library functions and the delay functions. The source code for the first group is found in the following directories:

- src\traditional\stdclib
- src\extended\stdclib

The source code for the delay functions is found in

- src\traditional\delays
- src\extended\delays

Here again the \traditional directories contain the functions for nonextended mode operation and the \extended directories contain those for extended mode execution.

6.3.1 General Software Library

The functions provided by the General Software Libraries are classified into the following groups:

- Character Classification Functions
- Data Conversion Functions
- Memory and String Manipulation Functions
- Delay Functions
- Reset Functions
- Character Output Functions

In the sections that follow we present a listing and a brief description of the functions in the General Software Library. The details of each of these conversion functions can be found in the MPLAB-C18-Libraries_59297f.pdf document available from the Microchip Website and in this book's software resource.

Character Classification Functions

These are standard C language library functions defined in the ANSI 1989 standard. These functions are listed and described in Table 6.1.

Table 6.1

Character Classification Fucntions

FUNCTION	DESCRIPTION
isalm	Determine if character is alphanumeric.
isalpha	Determine if character is alphabetic.
iscntrl	Determine if character is a control character.
isdigit	Determine if character is a decimal digit.
isgraph	Determine if character is a graphical character.
islower	Determine if character is a lowercase alphabetic.
isprint	Determine if character is printable.
ispunct	Determine if character is a punctuation character.
isspace	Determine if character is white space.
isupper	Determine if character is an uppercase alphabetic.
isxdigit	Determine if character is a hexadecimal digit.

Using any of these functions requires including the file ctype.h, as follows:

```
#include <ctype.h>
```

The argument passed to the functions is an unsigned character, and the return type is a Boolean unsigned char that is non-zero if the test is positive and zero otherwise. The following code fragment shows calling the isalpha function.

```
unsigned char achar = 'A';
...
if(isalpha(achar))
```

The sample program C_Char_Funcs.c in the book's software resource calls eight character classification functions on an unsigned char argument and turns on an LED on Port C if the test is positive.

Data Conversion Functions

The data conversion functions are familiar to C and C++ programmers because they are implemented in the stdlib standard C library. The implementation of these functions in MPLAB C18 is in conformance with the requirements of the ANSI C 1989 standard. Table 6.2 shows the data conversion functions.

For example, the description of the atob() function is as follows:

```
Function:     Convert a string to an 8-bit signed byte.
Include:      stdlib.h
Prototype:    signed char atob( const char * s );
Arguments:    s
              Pointer to ASCII string to be converted.
Remarks:      This function converts the ASCII string s into an 8-bit
              signed byte (-128 to 127). The input string must be in
              base 10  (decimal radix) and can begin with a character
              indicating sign ('+' or '-'). Overflow results are
```

undefined. This function is an MPLAB C18 extension to
the ANSI standard libraries.
Return Value: 8-bit signed byte for all strings in the range (-128 to
127).
File Name: atob.asm

Table 6.2

Data Conversion Functions

FUNCTION	DESCRIPTION
atob	String to an 8-bit signed byte.
atof	String into a floating point value.
atoi	String to a 16-bit signed integer.
atol	String into a long integer representation.
btoa	8-bit signed byte to a string.
itoa	16-bit signed integer to a string.
ltoa	Signed long integer to a string.
rand	Generate a pseudo-random integer.
srand	Set the pseudo-random number seed.
tolower	Character to a lowercase alphabetical ASCII character.
toupper	Character to an uppercase alphabetical ASCII character.
ultoa	Unsigned long integer to a string.

Memory and String Manipulation Functions

The functions in this group correspond with those of the ANSI C string library as defined by the 1989 standard, with some minor deviations and extensions. The functions are located in the file string.h and the implementation in the associated asm sources. Table 6.3 lists and describes the functions in the library.

Table 6.3

Memory and String Manipulation Functions

FUNCTION	DESCRIPTION
memchr memchrpgm	Search for a value in a specified memory region.
memcmp memcmppgm memcmppgm2ram memcmpram2pgm	Compare the contents of two arrays.
memcpy memcpypgm memcpypgm2ram memcpyram2pgm	Copy a buffer.
memmove memmovepgm memmovepgm2ram memmoveram2pgm	Copy a buffer. Source and destination may overlap.

(continues)

Table 6.3

Memory and String Manipulation Functions (continued)

FUNCTION	DESCRIPTION
memset memsetpgm	Initialize an array with a single repeated value.
strcat strcatpgm strcatpgm2ram strcatram2pgm	Append a copy of the source string to the end of the destination string.
strchr strchrpgm	Locate the first occurrence of a value in a string.
strcmp strcmppgm strcmppgm2ram strcmpram2pgm	Compare two strings.
strcpy strcpypgm strcpypgm2ram strcpyram2pgm	Copy a string from data or program memory into data memory.
strcspn strcspnpgm strcspnpgmram strcspnrampgm	Calculate the number of consecutive characters at the beginning of a string that are not contained in a set of characters.
strlen strlenpgm	Determine the length of a string.
strlwr strlwrpgm	Convert all uppercase characters in a string to lowercase.
strncat strncatpgm strncatpgm2ram strncatram2pgm	Append a specified number of characters from the source string to the end of the destination string.
strncmp strncmppgm strncmppgm2ram strncmpram2pgm	Compare two strings, up to a specified number of characters.
strncpy strncpypgm strncpypgm2ram strncpyram2pgm	Copy characters from the source string into the destination string, up to the specified number of characters.
strpbrk strpbrkpgm strpbrkpgmram strpbrkrampgm	Search a string for the first occurrence of a character from a set of characters.

(continues)

Table 6.3

Memory and String Manipulation Functions (continued)

FUNCTION	DESCRIPTION
strrchr strrchrpgm	Locate the last occurrence of a specified character in a string.
strspn strspnpgm strspnpgmram strspnrampgm	Calculate the number of consecutive characters at the beginning of a string that are contained in a set of characters.
strstr strstrpgm strstrpgmram strstrrampgm	Locate the first occurrence of a string inside another string.
strtok strtokpgm strtokpgmram strtokrampgm	Break a string into substrings or tokens, by inserting null characters in place of specified delimiters.
strupr struprpgm	Convert all lowercase characters in a string to uppercase.

Delay Functions

Very often microcontroller programs must include a time delay, for example; to flash an LED code must keep the port line high for a time period and low for the following time period. The shorter the time delay period the faster the LED will flash. The 18F PIC family contain sophisticated timers that can be used for this purposes; in particular, the 18F452 has four timer modules labeled Timer0 to Timer3. These modules can be programmed to implement sophisticated and powerful timing operations. Programming the timer modules is described in Chapter 9.

On the other hand, applications can produce a simple time delay without the complications of programming the timer modules by means of a do-nothing scheme. The NOP (no operation) opcode serves this purpose quite nicely because it delays one machine cycle without introducing any changes in the hardware. The following are code snippets from a do-nothing loop in the sample program C_LEDs_Flash.c:

```
// DATA VARIABLES AND CONSTANTS
unsigned int count;
#define MAX_COUNT 16000
...
    while (count <= MAX_COUNT)
    {
        count++;
    }
```

The objection to such a delay loop is that it is difficult to estimate the time delay that it will produce because it will depend on the code generated by the compiler. This means that the programmer would have to trial-and-error the value of the delay constant to find a satisfactory delay. The functions in the C18 delays library use multiples of instruction cycles to implement the delay period. Once the processor operating frequency is known, the actual time delay can be easily calculated. Table 6.4 lists the delay functions.

Table 6.4

Delay Functions

FUNCTION	DESCRIPTION
Delay1TCY	Delay one instruction cycle.
Delay10TCYx	Delay in multiples of 10 instruction cycles.
Delay100TCYx	Delay in multiples of 100 instruction cycles.
Delay1KTCYx	Delay in multiples of 1,000 instruction cycles.
Delay10KTCYx	Delay in multiples of 10,000 instruction cycles.

Using the delay function requires including the file delays.h in the program. The value of x in the name of the last four delay functions refers to the repetition parameter passed to the function, for example:

```
Delay1KTCYx(200);
```

In this case 200 multiples of 1,000 instructions (200,000 instructions) will be the delay that results from the function call. Once the instruction cycle time is known, calculating a specific delay is simple arithmetic. For example, to implement a one-second delay with a machine running at 10 MHz (10,000,000 cycles per second), we would need to delay 1,000,000 instruction cycles, which requires 1,000 calls to the Delay1KTCYx() function.

Reset Functions

The functions in the reset library can be used by code to determine the cause of a hardware Reset or wake-up and to reconfigure the processor status following the Reset. Table 6.5 lists and described the reset functions.

Table 6.5

Reset Functions

FUNCTION	DESCRIPTION
isBOR	Determine if the cause of a Reset was the Brown-out Reset circuit.
isLVD	Determine if the cause of a Reset was a low-voltage detect condition.
isMCLR	Determine if the cause of a Reset was the MCLR pin.
isPOR	Detect a Power-on Reset condition.
isWDTTO	Determine if the cause of a Reset was a Watchdog Timer time-out.
isWDTWU	Determine if the cause of a wake-up was the Watchdog Timer.
isWU	Detects if the microcontroller was just woken up from Sleep from the MCLR pin or an interrupt.
StatusReset	Set the POR and BOR bits.

For programs to use any of the Reset functions, they must include the reset.h header file.

Character Output Functions

By design, the C language is small and to keep it so, many of the input and output functions that are part of other programming languages are not part of C. Instead, these functions are implemented in libraries that are linked-in by the compiler whenever they are necessary. In C, the classic input/output library is named stdio.h, the name of which originates in standard input and output.

Although the C language stdio library includes functions for both input and output, the C18 implementation is limited to character output operations. The reason probably relates to how input is obtained directly from the hardware in an embedded environment, which typically has no keyboard or mouse devices. The character input functions that are implemented in C18 are also of limited usefulness because monitors and printers (for which these functions were designed) are not often part of an embedded system. It is not surprising that several major books on embedded C do not discuss the functions in the C18 stdio library.

C language input/output functions are based on the notion of a stream. A stream may be a file, a terminal, or any other physical device. In C18, output operations are based on the use of a destination stream, which can be a peripheral device, a memory buffer, or any other consumer of data. The destination stream is denoted by a pointer to an object of type FILE. The MPLAB C18 compiler defines two streams in the standard library:

- _H_USER output via the user-defined output function _user_putc
- _H_USART output via the library output function _usart_putc

Both streams are always open and do not require the use of functions such as fopen() and fclose(), as is the case in standard C. The global variables stdout and stderr are defined in stdio.h and have a default value of _H_USART. To change the destination to _H_USER, you will code as follows:

```
stdout = _H_USER;
```

Table 6.6 lists the character output functions in the C18 implementation.

The program named C_Printf_Demo.c in this book's software package demonstrates the use of the printf() function and how its output can be viewed using MPLAB SIM. The following code fragment from the program C_Printf_Demo shows the call to the fpring() function.

```
void main(void)
{
    printf ("Hello, World\n");

    while(1) {
    Nop();
    }
}
```

Table 6.6

Character Output Functions

FUNCTION	DESCRIPTION
fprintf	Formatted string output to a stream.
fputs	String output to a stream.
printf	Formatted string output to stdout.
putc	Character output to a stream
puts	String output to stdout.
sprintf	Formatted string output to a data memory buffer.
vfprintf	Formatted string output to a stream with the arguments for processing the format string supplied via the printf function.
vprintf	Formatted string output to stdout with the arguments for processing the format string supplied via the stdarg facility.
vsprintf	Formatted string output to a data memory buffer with the arguments for processing the format string supplied via the stdarg facility.
_usart_putc	Single character output to the USART (USART1 for devices with more than one USART).
_user_putc	Single character output in an application-defined manner.

You can test the program's output function in the MPLAB IDE by loading the C_Prinff_Demo project from the book's software resource or create your own project with the code listed previously. In order to prepare the environment, you must first select <Debugger:Select Tool:MPLAB SIM>. Then <Debugger:Settings...>. In the dialog that follows, click on the Uart1 IO tab and then the Debug Options as shown in Figure 6.1.

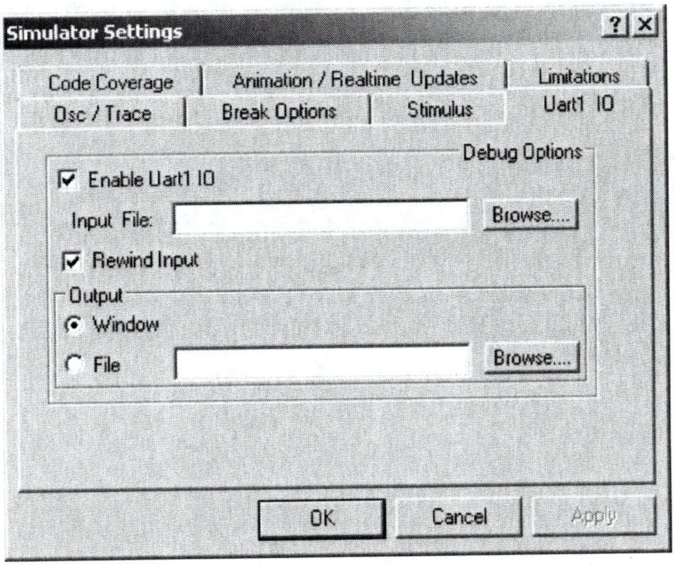

Figure 6.1 *Screen snapshot of the Simulator Settings dialog.*

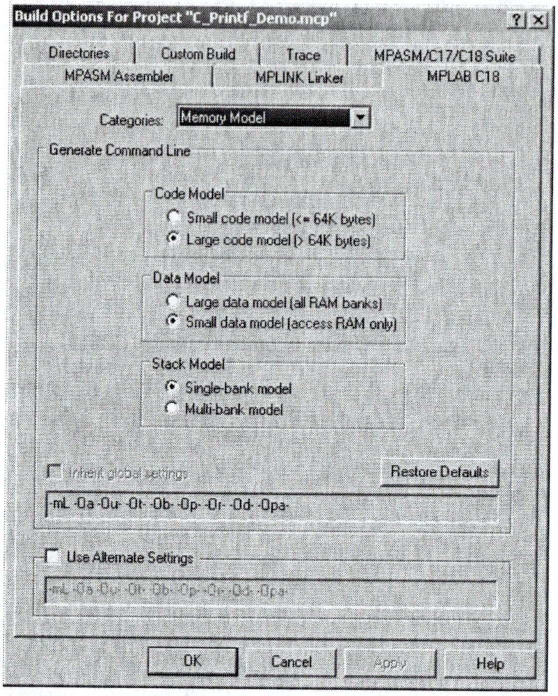

Figure 6.2 *Selecting the memory and code models.*

The next step is setting the memory model to large code and small data. This is done by selecting <Project:Build Options:Project> and then opening the MPLAB C18 window. In the Categories listbox, select Memory Model and Select the options shown in Figure 6.2. At this point, the Output window will include a tab labeled SIM Uart1. To test the program, you can now insert a breakpoint on the Nop() line in the while() loop and run a simulation. The SIM Uart1 tab in the Output window will show the "Hello, World" message, as shown in Figure 6.3.

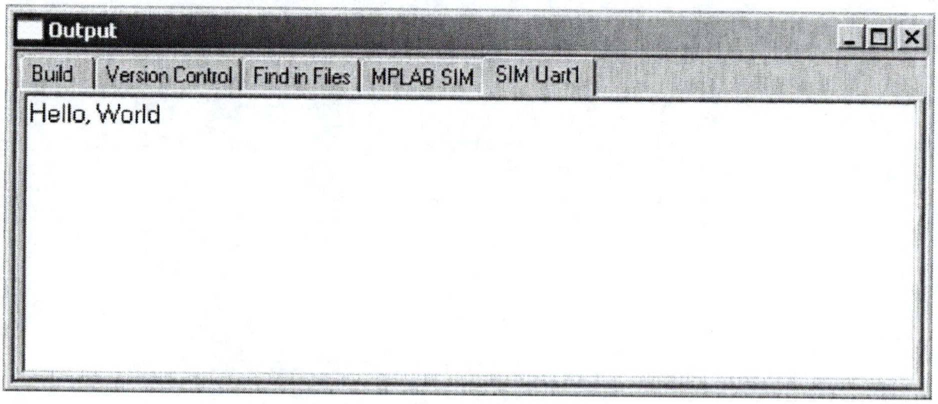

Figure 6.3 *Program output in the SIM Uart1 window.*

6.4 Processor-Specific Libraries

The processor-specific libraries that are part of MPLAB C18 contain definitions and functions that are compatible with specific devices of the PIC18 family. The processor-specific libraries include all of the peripheral routines as well as the definitions of the Special Function Register (SFR). The peripheral routines can be classified into two groups:

- Hardware Peripheral Functions
- Software Peripheral Library

The Hardware Peripheral Functions provide support for specific hardware components of the device. The source code for these functions is found in MPLAB C18 installation directory (usually named MCC18) and in the subdirectories src\traditional\pmc and src\extended\pmc.

6.4.1 Hardware Peripheral Library Functions

The following peripherals are supported by MPLAB C18:

- A/D Converter
- Input Capture
- i2C
- I/O Ports
- Microwire
- Pulse-Width Modulation (PWM)
- SPI
- Timer
- USART

These functions refer to specific hardware modules and special function registers in the PIC microcontroller. The functions allow enabling, configuring, changing the operation mode, reading from, writing to, and disabling the specific device. For example, Table 6.7 lists the functions that relate to the Anolog-to-Digital Converter module.

Table 6.7

A/D Converter Functions

FUNCTION	DESCRIPTION
BusyADC	Is A/D converter currently performing a conversion?
CloseADC	Disable the A/D converter.
ConvertADC	Start an A/D conversion.
OpenADC	Configure the A/D converter.
ReadADC	Read the results of an A/D conversion.
SetChanADC	Select A/D channel to be used.

The functions in the Hardware Peripherals Library are described and explained in the context of the specific programming topics in this book's chapters.

6.4.2 Software Peripherals Library Functions

The Software Peripherals Library contains functions that allow programming devices and components frequently found in PIC 18 circuits. The following devices are supported:

- External LCD Functions
- External CAN2510 Functions
- Software I2C Functions
- Software SPI Functions
- Software UART Functions

Some of the functions in the Software Peripherals Library complement those in the Hardware Peripherals Library, such as the I2C, SPI, and UART functions. Other functions refer to stand-alone devices that are not part of the microcontroller hardware, such as the LCD and CAN2510 functions. The functions in the Software Peripherals Library are described and explained in the context of the specific programming topics in this book's chapters.

6.4.3 Macros for Inline Assembly

The processor-specific header file also defines inline assembly macros that allow direct encoding of frequently used assembly language instructions from C code. They are provided as a convenience because any assembly language opcode can be called by inline assembly code.

For programs that use the 18F452 device, in order to access the inline assembly instructions, the following header file reference must be included in the code:

```
#include <p18f452.h>)
```

Table 6.8 lists the assembly language instruction macros.

Table 6.8

Macros for PICmicro MCU Instructions

MACRO	ACTION
Nop()	Executes a no operation (NOP)
ClrWdt()	Clears the Watchdog Timer (CLRWDT)
Sleep()	Executes a SLEEP instruction
Reset()	Executes a device reset (RESET)
Rlcf	Rotates var to the left through the carry bit
Rlncf	Rotates var to the left without going through the carry bit
Rrcf	Rotates var to the right through the carry bit
Rrncf	Rotates var to the right without going through the carry bit
Swapf	Swaps the upper and lower nibble of var

The macros Rlcf(), RIncf(), Rrcf(), Rrncf(), and Swapf() have the following proto-type:

```
Macro (var, dest, access)
```

In this case, var must be an 8-bit quantity (i.e., char) and not located on the stack. dest is a switch that determines where the result is stored. If dest = 0, the result is in the W register; if dest = 1, the result is in var. If access = 0, the access bank will be selected, overriding the BSR value. If access = 1, then the bank will be selected as per the BSR value.

6.4.4 Processor-Specific Header Files

The processor-specific header file is a C language file that contains external declarations for the special function registers and establishes structures and unions that facilitate addressing hardware devices from C. For the 18F452 device, the header file is called p18f452.h. For example, the following structure and unions define the name PORTA and various labeling for the Port A bits. It is declared as follows:

```
extern volatile near unsigned char     PORTA;
extern volatile near union {
    struct {
        unsigned RA0:1;
        unsigned RA1:1;
        unsigned RA2:1;
        unsigned RA3:1;
        unsigned RA4:1;
        unsigned RA5:1;
        unsigned RA6:1;
    } ;
    struct {
        unsigned AN0:1;
        unsigned AN1:1;
        unsigned AN2:1;
        unsigned AN3:1;
        unsigned T0CKI:1;
        unsigned SS:1;
        unsigned OSC2:1;
    } ;
    struct {
        unsigned :2;
        unsigned VREFM:1;
        unsigned VREFP:1;
        unsigned :1;
        unsigned AN4:1;
        unsigned CLKOUT:1;
    } ;
    struct {
        unsigned :5;
        unsigned LVDIN:1;
    } ;
} PORTAbits ;
```

The first statement in the declaration specifies that PORTA is a byte (unsigned char). The extern modifier is required because the variables are declared in the register definitions file. The volatile modifier tells the compiler that PORTA may not retain the values assigned to it during program execution. The near modifier specifies that the port is located in access RAM. The declaration is followed by a structure with several unions that provide alternative names for the Port A bits. For example, bit number 3 can be referenced by code as:

```
PORTAbits.RA3
PORTAbits.AN3
PORTAbits.VREFM:1
```

Some of the bit fields are padded in the third and fourth structures because they are not defined in that context. The Inline Assembly macros discussed in Section 6.4.3 are also located in the p18f452.h header file. Many other structure/union definitions are found in the processor-specific header files.

6.5 Math Libraries

The mathematical operations directly supported by the PIC18 devices are limited to addition, subtraction, and multiplication of small integers. Integer division must be provided in software, and there is no support for floating-point numbers and operations. The assembly language programmer that needs to deal with numerical calculations has indeed a chore in implementing floating-point mathematics in software. Fortunately, the C18 package includes a mathematical library (math.h) with the following support:

- 32-Bit Floating-Point Math

- The C Standard Library Math Functions

This means that developing PIC18 applications that require floating-point calculations, or even substantial integer arithmetic, practically mandate the use of C18. Developing floating-point operations in PIC assembly is a major enterprise.

The floating-point operations provided by the math library include the following primitives:

- Addition, subtraction, multiplication, and division

- Conversions between floats and integers

The floating-point implementations in C18 comply with the ANSI-IEEE 754 standard for single precision floats with two exceptions: one is regarding the handling of subnormals and the second one relates to rounding operations. Neither one is of importance to most embedded system applications. Extended and traditional modes use the same float representations, and the results of float operations are the same.

6.5.1 ANSI-IEEE 754 Binary Floating-Point Standard

The ANSI-IEEE standard for binary floating-point arithmetic was first released in 1979, and the final approval came in 1985. Because then it has defined the requirements of all floating-point by establishing the following requirements:

- All implementations must provide a consistent representation of floating-point numbers.
- All implementations must provide correctly rounded floating-point operations.
- All implementations must provide consistent treatment of exceptional situations and errors.

According to ANSI-IEEE 754 a single precision floating-point number (float variable type in C language) consists of four parts:

1. A sign
2. A significand
3. A base
4. An exponent

These components are of the form:

$$x = \pm d0.d1.d2.d3 \cdots d23 \times 2E$$

where ± is the sign of the number;, d0,d1,d2,d3, ···; d23 are the 24 significand digits; and E is the exponent to which the base 2 is raised. Each di is a digit (0 or 1). The exponent E is an integer in the range Emin to Emax, where Emin = minus 126 and Emax = 127.

Encodings

Single-precision format numbers (float tyhpe) use a 32-bit word organized as a 1-bit sign, an 8-bit biased exponent e = E + 127, and a 23-bit fraction, which is the fractional part of the significand.

The most-significant bit of the significand (d0) is not stored because it can be inferred from the value of the exponent, as follows: if the biased exponent value is 0, then d0 = 0, otherwise d0 = 1. This scheme saves one bit in the storage of the number's significand and increases the precision to 24 bits.

ANSI-IEEE 754 defines and C18 supports several numeric types:

- Normals are numbers that can be correctly represented in the format. The standard defines the smallest and largest numbers that can be represented.
- Subnormals are numbers smaller than the smallest normalized representation (2–126). C18 does not provide an operation by which a float can create a subnormal. Subnormals used in floating-point operations are automatically converted to a signed zero.
- NaNs (Not-a-Number) are used for encoding error patterns and invalid operations. Division by zero generates a NaN. Any operation with a NaN argument returns a NaN.

Rounding

The ANSI-IEEE-754 standard defines four rounding modes; however, C18 uses round-to-nearest in all rounding operations. The C18 implementation uses a modification to ANSI-IEEE 754: The threshold for rounding up is about 0.502 in C18 instead of exactly 0.5 as defined in the standard. This results in a small bias when rounding toward zero. Microchip states that this modification results in a significant savings in code space and execution time with virtually no consequences for real-world calculations.

6.5.2 Standard Math Library Functions

The operations in the C18 standard math library includes the following types of operations:

- Trigonometric functions
- Hyperbolic functions
- Exponential functions
- Conversions and number splitting
- Logarithms
- Square roots

The functions are listed and described in Table 6.9.

Table 6.9
Math Library Functions

FUNCTION	DESCRIPTION
acos	Compute the inverse cosine (arccosine).
asin	Compute the inverse sine (arcsine).
atan	Compute the inverse tangent (arctangent).
atan2	Compute the inverse tangent (arctangent) of a ratio.
ceil	Compute the ceiling (least integer).
cos	Compute the cosine.
cosh	Compute the hyperbolic cosine.
exp	Compute the exponential ex.
fabs	Compute the absolute value.
floor	Compute the floor (greatest integer).
fmod	Compute the remainder.
frexp	Split into fraction and exponent.
ieeetomchp	Convert an IEEE-754 format 32-bit floating-point value into the Microchip 32-bit floating point format.
ldexp	Load exponent - compute x * 2n.
log	Compute the natural logarithm.
log10	Compute the common (base 10) logarithm.
mchptoieee	Convert a Microchip format 32-bit floating-point value into the IEEE-754 32-bit floating-point format.
modf	Compute the modulus.
pow	Compute the exponential xy.
sin	Compute the sine.
sinh	Compute the hyperbolic sine.
sqrt	Compute the square root.
tan	Compute the tangent.
tanh	Compute the hyperbolic tangent.

6.5.3 Floating-Point Math Sample Program

The sample program C_Floats_Demo.c in the book's software package is a small demonstration of floating-point operations using the C18 math library. The mathematics are trivial: an arbitrary number is multiplied by Pi and then its square root is obtained. Meaningful code is as follows:

```
#include <p18f452.h>
#include <stdio.h>
#include <math.h>

#define PI 3.14159265359

#pragma config WDT = OFF
#pragma config OSC = HS
#pragma config LVP = OFF
#pragma config DEBUG = OFF

//*****************************************************************
//                        main program
//*****************************************************************
void main(void)
{

    float num1 = 22.334455;
    int wholePart;
    int fracPart;

    // Multiply times PI
    num1 *= PI;
    // Calculate square root of product
    num1 = sqrt(num1);

    // Find integer and fractional parts
    wholePart = num1;            // Get whole part
    fracPart = (num1 - wholePart) * 100000; // Fractional part
    fracPart = fabs(fracPart);   // Eliminate negative

    printf ("Testing floating-point math\n");
    printf ("%d.%03d\n", wholePart, fracPart);

    while(1) {
        Nop();
    }
}
```

Notice that the constant PI is created with a #define statement, thus saving variable space in the application.

In main(), the code creates three variables: one is a float type (num1) and the other two are int types (wholePart and fracPart). Product and square root are then obtained.

One difficulty in using floating-point in the MPLAB IDE is that the C18 prinf() function cannot display floating-point numbers. According to Microchip, floating-point display was not implemented because it would have added considerably to the size of the library. The integer part is obtained with a simple cast of the float operand into an int variable. The fractional part requires subtracting the integer part and then multiplying by a power of the base (10 in this case) to the desired precision. To eliminate possible negative fractions, the fabs() function is called to get the absolute value. The two resulting integer types are displayed conventionally using printf(). The result can be seen by setting a breakpoint at the end of the program in displaying the SIM Uart1 window.

6.6 C18 Language Specifics

The present section refers to the C18; we discuss the specific features of the C18 implementation that are either variations from or extensions to the ANSI standard. According to Microchip the MPLAB C18 compiler is a free-standing, optimizing ANSI C compiler specially adapted for the PIC18 family of microcontrollers. The compiler deviates from ANSI standard X3.159-1989 only where the standard conflicts with efficient PICmicro MCU support.

6.6.1 C18 Integer Data Types

The C18 compiler supports the standard integer types defined in the ANSI C standard previously mentioned. In addition no the standard types, C18 supports a 24-bit integer type named short long int (or long short int), in both signed and unsigned formats. Table 6.10 lists the specifications of the C18 integer types.

Table 6.10

C18 Integer Data Types

TYPE	SIZE	MINIMUM	MAXIMUM
char	8 bits	−128	127
signed char	8 bits	−128	127
unsigned char	8 bits	0	255
int	16 bits	−32,768	32,767
unsigned int	16 bits	0	65,535
short	16 bits	−32,768	32,767
unsigned short	16 bits	0	65,535
short long	24 bits	−8,388,608	8,388,607
unsigned short long	24 bits	0	16,777,215
long	32 bits	−2,147,483,648	2,147,483,647
unsigned long	32 bits	0	4,294,967,295

6.6.2 C18 Floating-Point Data Types

C18 supports the definition of floating-point types as float or double, however both types have identical size and limits, as shown in Table 6.11.

Table 6.11

C18 Floating-Point Size and Limits

Type	Size	Minimum Exponent	Maximum Exponent	Minimum Normalized	Maximum Normalized
float	32 bits	−126	128	$2^{-126} \approx 1.17549435e-38$	$2^{128} * (2-2^{-15}) \approx 6.805646693e+38$
double	32 bits	−126	128	$2^{-126} \approx 1.17549435e-38$	$2^{128} * (2-2^{-15}) \approx 6.805646693e+38$

6.6.3 Endianness

Endianness refers to the ordering of multiple-byte values and can be little-endian (low-byte at low memory address) or big-endian (high-byte at low memory address). C18 uses the little-endian format. For example, the hex value 0xAABBCCDD is stored in memory (low-to-high addresses) as 0xDD, 0xCC, 0xBB, and 0xAA.

6.6.4 Storage Classes

C18 supports the ANSI standard storage classes named auto, extern, register, and typedef. In addition, C18 defines a unique storage class named overlay that is available only in nonextended modes. The overlay storage class may be applied to local variables but not formal parameters, function definitions, or global variables. The overlay storage class will allocate the associated symbols into a function-specific, static overlay section. In other words, a variable of overlay type is allocated statically but reinitialized on each function entry. For example,

```
void demo(void)
{
    overlay int x = 5;
    x++;
}
```

In this case, x will be initialized to 5 upon every entry to the function named demo(), although its storage will be statically allocated. If the overlay variable is not initialized in its declaration, then its value upon function entry is undefined.

6.6.5 static Function Argument

Function parameters have the storage class auto or static. The default storage class is auto which determines that function parameters are placed in the software stack, thus making reentry possible. The auto qualifier can be overwritten by means of the static keyword. Parameters for static-type functions are allocated globally, impeding reentrancy but resulting in smaller code and enabling direct access.

6.6.6 Storage Qualifiers

C18 supports the standard ANSI C storage qualifiers const and volatile. The const qualifier is used to designate objects whose value does not change (constants). Once a variable is qualified as const, the programmer will not be allowed to change its value. The volatile qualifier, on the other hand, informs the compiler that an object can have its value changed in ways not under the control of the implementation.

In addition, the C18 compiler introduced four new qualifiers that bind to identifiers in a similar manner as const and volatile. These are named far, near, rom, and ram.

far and near Qualifiers

The far qualifier denotes a variable that is located in data memory and in a memory bank. To access a far variable a bank switching instruction is required. The near qualifier denotes that a variable is located in data memory and in access RAM.

When the far qualifier refers to a variable that is located in program memory, then the variable can be anywhere in the system's program memory space. If it refers to a pointer, then the far pointer can access up to and beyond 64K of program memory space. The near qualifier is used to denote that a variable located in program memory is found at an address less than 64K. If it refers to a pointer, then the near pointer can access only up to 64K of program memory space.

rom and ram Qualifiers

The Harvard architecture of the PIC microcontrollers uses separate program memory and data memory address buses. This means that the C18 compiler requires extensions to distinguish between data located in program memory and data located in data memory. The ANSI C standard allows for code and data to be in separate address spaces, but does not provide ways of locating data in the code space. The rom qualifier denotes that the object is located in program memory, whereas the ram qualifier denotes that the object is located in data memory. Pointers can point to either data memory (ram pointers) or program memory (rom pointers). Pointers are assumed to be ram pointers unless declared as rom. Table 6.12 shows the location of objects based on their storage qualifiers.

Table 6.12

Object Location According to Storage Qualifiers

	ROM	RAM
far	Anywhere in program memory	Anywhere in data memory (default)
near	In program memory with address less than 64K	In access memory

Chapter 7

Programming Simple Input and Output

7.1 Port-Connected I/O

In this chapter we describe very simple PIC-based circuits that consist of very few hardware components. The input and output devices are connected directly to 18F452 PIC ports. The processing routines discussed (both in assembly language and C18) consist of reading from and writing to these ports.

7.1.1 A Simple Circuit and Code

We start with a PIC 18F452 simple application and circuit to demonstrate port-based simple input and output. The circuit contains the following elements:

1. Power supply
2. Supply to the MCLR pin
3. Oscillator
4. Four red LEDs
5. Four green LEDs
6. One pushbutton switch

In addition, the circuit requires several resistors and capacitors, which can be seen in the circuit schematics. Demo board 18F452A, developed in this chapter, is compatible with the many applications discussed in the book..

7.1.2 Circuit Schematics

The test circuit can be built on a breadboard or a perfboard. Four red LEDs are wired to port C lines RC0 to RC3. Four green LEDs are wired to port C lines RC4 to RC7. The circuit can be simplified by wiring a single red LED to port C line 0 and a single green LED to port C line 7. The negative poles of the LEDs (flat pin) are connected to ground via a 330-ohm resistor. The positive pins (round pin) are wired to the corresponding line of port C. Port C is trissed for output. A pushbutton switch is wired to port B, line

RB4, and to the 5V source via a 10K resistor. This line of port B is trissed for input. The other contact of the pushbutton switch is wired to the ground line. The oscillator is a Murata Erie 20 MHz crystal resonator and is wired to ground and to PIC lines OSC1 and OSC2. The MCLR line is held high during operation by its connection to the 5V source via a 10K resistor. A pushbutton switch allows bringing MCLR low to reset the system. A single LED connected to ground serves to indicate a power-on state. The 5V source for the circuit is provided by a 9 to 12 V DC wall transformer that is regulated by a 7805 IC and the standard capacitors. The circuit schematics are shown in Figure 7.1.

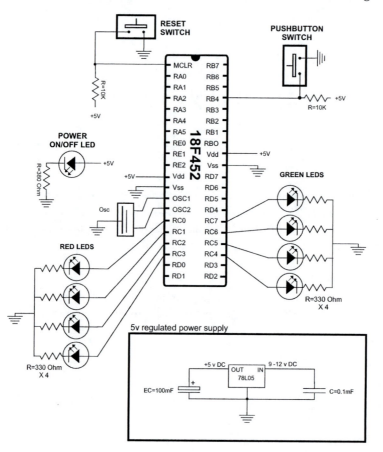

Figure 7.1 *Schematics for a LED/pushbutton circuit.*

7.1.3 Assembler Simple I/O Program

The sample program LedPB_F18.asm, in the book's on-line software package, exercises the circuit in Figure 7.1 or Demo Board 18F452A described later in this chapter. The program listing is as follows:

```
; File name: LedPB_F18.asm
; Date:         September 9, 2012
; No copyright
; Processor: PIC 18F452
;
```

```
; Port direction and wiring for this program:
; PORT        PINS      DIRECTION     DEVICE
;   C         0-3       Output        Green LEDs
;   C         4-7       Output        Red LEDs
;   B         4         Input         Pushbutton # 1
;
; Description:
; A demonstration program to monitor pushbutton # 1 in
; DemoBoard 18F452A (or equivalent circuit). If the pushbutton
; is released the four red LEDs wired to port C lines 0-3
; are flashed. If the pushbutton is held pressed then the
; four green LEDs wired to port C lines 4-7 are flashed.
;
;==========================================================
;              definition and include files
;==========================================================
     processor     18F452         ; Define processor
     #include  <p18F452.inc>
; ========================================================
;                  configuration bits
;==========================================================
     config OSC = HS        ; Assumes high-speed resonator
     config WDT = OFF       ; No watchdog timer
     config LVP = OFF       ; No low voltage protection
     config DEBUG = OFF     ; No background debugger
;
; Turn off banking error messages
     errorlevel    -302
;
;==========================================================
;                   variables in PIC RAM
; Access RAM locations from 0x00 to 0x7F
; Declare variables at 2 memory locations
j    equ       0x000             ; Counters for delay routine
k    equ       0x001


;==========================================================
;                        program
;==========================================================
     org      0      ; start at address
     goto main
; Space for interrupt handlers
;============================
;      interrupt intercept
;============================
     org  0x08      ; High-priority vector
     retfie
     org  0x18      ; Low-priority vector
     retfie
;==========================================================
;                  main program entry point
;==========================================================
main:
; Set BSR for bank 0 operations
     movlb     0              ; Bank 0
; Initialize all lines in PORT C for output
     movlw     B'00000000' ; 0 = output
     movwf     TRISC          ; Port C tris register
; Initialize line 4 in PORT B for input
     movlw     B'00010000' ; 1 = input
```

```
     movwf       TRISB            ; Port B tris register
;==============================
;    COMMAND MONITORING LOOP
;==============================
; Program loop to turn red and green LED banks on and off
flashLEDS:
     call command         ; Procedure to test switch state
; Z flag set if PB#1 pressed
     btfsc       STATUS,Z
     goto        redLEDs
     goto        greenLEDs
greenLEDs:
; Turn on lines 0 to 3 in port C. All others are off
     movlw       B'00001111'    ; LEDS 0 to 3 ON
     movwf       PORTC,0
     call        delay            ; Local delay routine
; Turn off all lines in port C to flash off LEDs
     movlw       B'00000000'    ; All LEDs OFF
     movwf       PORTC,0
     call        delay
     goto        flashLEDS

redLEDs:
; Turn on lines 7 to 4 in port C. All others are off
     movlw       B'11110000'    ; LEDS 7 to 4 ON
     movwf       PORTC,0
     call delay              ; Local delay routine
; Turn off all lines in port C to flash off LEDs
     movlw       B'00000000'    ; All LEDs OFF
     movwf       PORTC,0
     call        delay
     goto        flashLEDS


;==============================
;   test for push button 1
;==============================
command:
; The command procedure tests push button # 1 wired to port
; RB4 (active low) to see if it is pressed or released.
; If line line is low then the switch pressed. In this case the
; Z flag is set and execution returns to the caller. Otherwise
; the Z flag is cleared before executing returns
     btfsc       PORTB,4          ; Test bit 4
; The following goto executes if the the carry bit is set
; indicating that the pushbutton is pressed
     goto        pbDown           ; Jump taken if line bit set
                                  ; Indicating pushbutton is down
; At this point the pushbutton is released (line off)
; Clear zero flag in STATUS register to report back to caller
     bcf         STATUS,Z
     return
pbDown:
; Set Z flag
     bsf         STATUS,Z
     return


;===============================
;        delay subroutine
;===============================
delay:
```

```
      movlw     .200 ; w = 200 decimal
      movwf     j          ; j = w
jloop:
      movwf     k          ; k = w
kloop:
      decfsz    k,f        ; k = k-1, skip next if zero
      goto      kloop
      decfsz    j,f        ; j = j-1, skip next if zero
      goto      jloop
      return

      end       ;END OF PROGRAM
```

7.1.4 Assembler Source Code Analysis

The program skeleton is based on the assembly language coding template developed in Section 4.1.1. The program header defines the hardware and software environment and describes the application. Code then defines the processor and includes the file p18f452.inc. The minimal configuration bits for the application are set and the errorlevel -302 directive disables banking messages for the application. This is possible because the application only uses access RAM. Following, the two program variables are defined in bank 0 addresses, thus placing them in access RAM.

Processing starts at the main label where the BSR register is set to bank 0. Because the default state is to select the bank according to the state of the BSR bits, the remaining code can omit setting or clearing the "a" bit in the instructions. This scheme simplifies the coding and works as long as the BSR bits are kept in the original state.

In the following lines, port C lines are initialized for output by writing zeros to the corresponding TRIS register. Then line 4 of port B is initialized for input by writing a one to that bit position. Code is as follows:

```
main:
; Set BSR for bank 0 operations
      movlb     0          ; Bank 0
; Initialize all lines in PORT C for output
      movlw     B'00000000' ; 0 = output
      movwf     TRISC       ; port C tris register
; Initialize line 4 in PORT B for input
      movlw     B'00010000' ; 1 = input
      movwf     TRISB       ; port B tris register
```

Command Monitoring Loop

Many embedded applications applications monitor for changes in one or more input line in order to modify their action. This monitoring for a command or change of state can be interrupt-driven or it can take place in a program loop, usually called polling routines. Interrupt-driven command systems are discussed in Chapter 8.

The current program uses an endless loop to poll the state of the pushbutton switch. If it is released, then the green LEDs are flashed. If it is pressed (held down), then the red LEDs are flashed. The actual test is performed by an auxiliary procedure named command. In this example, the label command refers to a subroutine (sometimes called a procedure) that appears later in the program. The call instruction transfers control to the subroutine's label (in this case, command), which, in turn, returns execution to the line following the call by means of a return instruction.

Tests in the command procedure consist of determining the state of pushbutton # 1, wired to port RB4. This port line is wired active low. That means that the line returns zero when the pushbutton is released; in other words, the RB4 line is low when the switch is pressed. If pressed, the Z flag is set and execution returns to the caller. Otherwise, the Z flag is cleared before execution returns.

Notice that in the scheme used by this code, the processor's zero flag (Z) is used to report the condition of the switch. The bcf instruction, with the STATUS register and the Z flag as operands, is used to clear the flag. The bsf instruction is used to set the Z flag. The calling routine tests the state of the Z STATUS bit to take the corresponding conditional jump. Code is as follows:

```
btfsc    STATUS,Z
goto greenLEDs
goto redLEDs
```

This simple test-and-jump mechanism is very common in PIC programs. The btfsc instruction (mnemonics stand for bit test file register skip if clear), determines if the following line is skipped or not. Following the test, execution is directed to either one of two labels. Notice that the labels greenLEDs and redLEDs are accessed by means of goto instructions. This means that execution is transferred directly to the label and it is not a subroutine call.

Action on the LEDs

Turning on and off LEDs that are wired to a port trissed for output consist of writing ones or zeros to the corresponding port lines. Because the red LEDs are wired to the four high-order bits of port C, code writes ones to these bits to turn on the LEDs and zeros to turn them off, as follows:

```
; Turn on lines 7 to 4 in port C. All others are off
    movlw    B'11110000'   ; LEDS 7 to 4 ON
    movwf    PORTC,0
    call     delay         ; Local delay routine
; Turn off all lines in port C to flash off LEDs
    movlw    B'00000000'   ; All LEDs OFF
    movwf    PORTC,0
    call     delay
```

A Delay Routine

The routine named delay, called by the previous code fragment, is used to make the LEDs flash by remaining on or off for a fraction of a second before changing their state. There are many ways of producing timed delays in the PIC 18FXX2 devices. Later in the book you will find an entire chapter devoted to timing. The present routine is a simple wait-by-doing-nothing scheme that is simple to code and understand. Code starts by moving the literal value 200 into the work register and then initializing two counters (j and k) to this value, as follows:

```
delay:
    movlw      .200 ; w = 200 decimal
    movwf      j    ; j = w
```

The delay itself consists of two loops: an inner one (the k loop that iterates 200 times) and an outer j loop one (that also iterates 200 times), The total number of times that the inner loop executes is 40,000 times. Both loops use the decfsz instruction (mnemonic for decrement file register skip if zero) in order to decrement the counter register. When the register reaches zero, the next code line is skipped and the loop terminates, as follows:

```
    decfsz     j,f        ; j = j-1, skip next if zero
    goto jloop
    return
```

Note that it is good programming practice to use meaningful names for program labels and variables. However, simple counter variables are often given lowercase, in single-letter names reminiscent of their use in mathematics, as is the case with the variables j and k in the previous program.

7.2 C Language Simple I/O Program

The program C_LED_PB.c in this book's software package performs the same operation as the program LED_PB.asm described previously. Analysis of both code listings shows the differences between coding in either language. It is obvious that the C language version is shorter and easier to code and understand. Following is the program listing:

```
// Project name: C_LED_PB
// Source files: C_LED_PB.c
// Date: September 9, 2012
// Processor: PIC 18F452
// Environment: MPLAB IDE Version 8.86
//              MPLAB C-18 Compiler
//
// TEST CIRCUIT: Demo Board 18F452 or circuit wired as
// follows:
// PORT      PINS      DIRECTION      DEVICE
//  C        0-3       Output         Green LEDs
//  C        4-7       Output         Red LEDs
//  B        4         Input          Pushbutton No. 1
//
//  Note: Pushbutton # 1 switch is wired active low
//
// Description:
// A demonstration program to monitor pushbutton No. 1 in
// DemoBoard 18F452A (or equivalent circuit). If the pushbutton
// is held pressed then the four red LEDs wired to port C
// lines 0-3 are flashed.


#include <p18f452.h>
#include <delays.h>

#pragma config WDT = OFF
```

```
#pragma config OSC = HS
#pragma config LVP = OFF
#pragma config DEBUG = OFF

/* Function prototypes */
void FlashRED(void);
void FlashGREEN(void);

//*******************************************************************
//                         main program
//*******************************************************************
void main(void)
{
// Initialize direction registers
    TRISB = 0b00001000; // Port B, line 4, set for input
//                  |
//                  |_____ Pushbutton # 1
    TRISC = 0;              // Port C set for output
    PORTC = 0;              // Clear all port C lines

    while(1)
    {
        if(!PORTBbits.RB4)
            FlashRED();
        else
            FlashGREEN();
    }
}
//*******************************************************************
//                        local functions
//*******************************************************************

void FlashRED()
{
    PORTC = 0x0f;
    Delay1KTCYx(200);
    PORTC = 0x00;
    Delay1KTCYx(200);
    return;
}
void FlashGREEN()
{
    PORTC = 0xf0;
    Delay1KTCYx(200);
    PORTC = 0x00;
    Delay1KTCYx(200);
    return;
}
```

7.2.1 C Source Code Analysis

The C language coding template is similar to the one developed for assembly language in Section 4.1.1. The program header defines the development dates, the hardware and software environment, and describes the application. Code then selects the processor and includes the file delays.h, which is used in implementing timing using the library functions described in Section 6.3.1. The minimal configuration bits for the application are set using the C language #pragma config directive with the defined operands. The prototypes for the two local functions are listed.

The next group of program lines are prototypes for the two local functions used by the program. By prototyping the functions we are able to list them after the main() function. Without the prototypes, all local functions would have to be coded before main(). Code is as follows:

```
/* Function prototypes */
void FlashRED(void);
void FlashGREEN(void);
```

main() Function

The in MPLAB C18 the main() function has void return type and void parameter list. The C language main() function defines the program's entry point; execution starts here. The first lines in main() are to initialize the port directions as required by the program. Because the LEDs are wired to port C, its tris register (TRISC) is set for output. By the same token, because the pushbutton switch number 1 is wired to port B, line 4, this line must be initialized for input. Code is as follows:

```
// Initalize direction registers
TRISB = 0b00001000;// Port B, line 4, set for input
TRISC = 0;               // Port C set for output
PORTC = 0;               // Clear all port C lines
```

Notice that we have taken advantage of a feature of MPLAB C18 that allows declaring binary data using the 0b operator.

In C language, we can create an endless loop by defining a condition that is always ture. Because in C true is equivalent to the numeric value 1, we can code:

```
while(1)
{
    if(PORTBbits.RB4)
        FlashRED();
    else
        FlashGREEN();
}
```

This coding style provides a simple mechanism for implementing a routine that tests one or more input devices and proceeds accordingly. Because pushbutton switch number 1 is wired to port B line 4, code can test this bit to find out if the pushbutton is in a released or pressed state. Testing the state of a port bit is simplified by the presence of macros in the C compiler that define the state of each individual bit in the port. The if statement does this using the PORTBbits.RB4 expression. Alternatively, code can use a conventional C language bitwise AND operation on the port bits, as follows:

```
while(1)
{
    if(PORTB & 0b00001000)
        FlashRED();
    else
        FlashGREEN();
}
```

In either case, the test determines that the pushbutton is pressed. If so, then the red LEDs are flashed. Otherwise, the green LEDs are flashed. The flashing is performed by two simple procedures named FlashRED() and FlashGREEN(). The FlashRED() procedure is coded as follows:

```
void FlashRED()
{
    PORTC = 0x0f;
    Delay1KTCYx(200);
    PORTC = 0x00;
    Delay1KTCYx(200);
    return;
}
```

The FlashRED() function starts by turning off the four high-order lines in port C and turning on the four low-order lines. The hexadecimal operand 0x0f sets the port bits accordingly.

The statement that follows calls one of the delay functions in the delay's General Software Library. These delay functions (which were visited in Chapter 6) provide a convenient mechanism for implementing timed operations in an embedded environment. The delay functions are based on instruction cycles and are, therefore, dependent on the processor's speed. The one used in the sample program (Delay1KTCYx()) delays 1000 machine cycles for every iteration. The number of desired iterations is entered inside the function's parentheses. In this case, 1,000 machine cycles are repeated 200 times, which implements a delay of 200,000 machine cycles. The function named FlashGREEN() proceeds in a similar manner.

In main(), the tris registers for ports B and C are initialized and the port C latches are cleared. The while() statement is an endless loop that uses an if statement with the expression PORTBbits.RB4 to identify the corresponding port bit.

7.3 Seven-Segment LED Programming

A seven -segment display can be connected to output ports on the PIC and used to display numbers and some digits. The circuit in Figure 7.2 shows the wiring used in the sample programs developed in this chapter. This is also the same wiring as that in Demo Board 18F452A.

As the name indicates, the seven-segment display has seven linear LEDs that allow forming all the decimal and hex digits and some symbols and letters. Once the mapping of the individual bars of the display to the PIC ports has been established, digits and letters can be shown by selecting which port lines are set and which are not. For example, in the seven-segment LED of Figure 7.2, the digit 2 can be displayed by setting segments a, b, g, e, and d. In this particular wiring, these segments correspond to port C lines 0, 1, 6, 4, and 5.

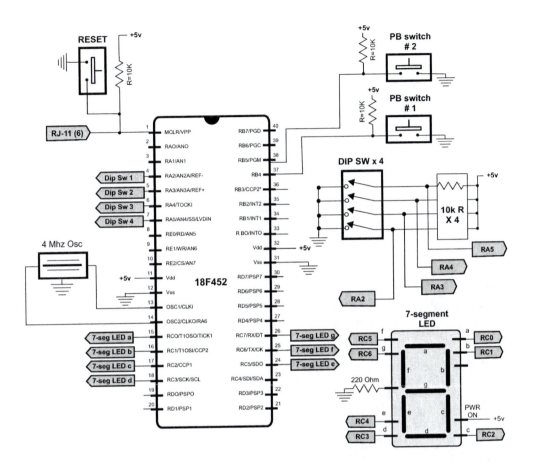

Figure 7.2   *Seven-segment LED and DIP switch circuit.*

7.3.1 Computed Goto

In assembly language conversion of the decimal digits and some letters to port display codes can be achieved by means of a lookup table using a mechanism sometimes called a "computed Goto." The processing depends on three special features of PIC18 assembly language:

- The program counter (PC) file registers (labeled PCU, PCH, and PCL) hold the address in memory of the current instruction (see Figure 2.5 and Section 2.1.4). Because each PIC18 instruction takes up two bytes, (except for those that modify the PC), one can jump to consecutive entries in a table by adding two to the value in the program counter.

- The addwf instruction can be used to add a value in the w register to the PCL register.

- The retlw instruction returns to the caller a literal value stored in the w register. In the case of the retlw, the literal value is the instruction operand.

If the lookup table is located at a subroutine called getcode, then the processing can be implemented as follows:

```
getcode:
        addwf   PCL,f    ; Add value in w register to PCL
        retlw   0x3f ; code for number 0
        retlw   0x06 ; code for number 1
        retlw   0x5b ; code for number 2
        ...
        retlw   0x6f ; code for number 9
```

The calling routine places in the w register the numeric value whose code is desired, and then calls the table lookup as follows:

```
        movlw   .6   ; Code for number 3 (2 times offset in table)
        call    getcode
        movwf   PORTC   ; Display 3 in 7-segment display
```

This computer goto scheme is very popular in programming PIC16 and earlier devices, but it is not without problems. We have seen that code uses the instruction:

```
        addwf   PCL,F
```

Code assumes that twice the offset into the table is stored in the W register at the time the instruction executes. Keep in mind that the PCL register stores the low-order byte of the instruction address. So what happens if the table is located at address 0x4f0 and the value in the W register is 0x40? Adding 0x40 to 0x4f0 results in 0x530. In this case, the new address requires changing the PCH register because the add operation straddled a code page boundary. However, the addwf opcode does not update PCH so the resulting address (0x430) is not the desired one and the computed goto will fail.

Several more-or-less-complicated methods of avoiding this flaw have been published and can be found in the online literature. It is possible to write code that anticipates if a page boundary will be crossed by estimating the table size and the initial PCL value. If so, the high part of the address can be adjusted. A simpler solution is to place the table at a known location in memory where the PCL value is sufficiently low to accommodate the worst case. This method is used by the program described in the following subsection.

7.3.2 Assembler Seven-Segment LED Program

The sample program DIPs_to_7Seg.asm in this book's software resource monitors the state of four toggle (DIP) switches in port A lines 2 to 5 and displays the selected hexadecimal digit on the seven-segment LED wired to port C.

Access Bank Operation

The main() function in the program DIPs_to_7Seg.asm begins as follows:

```
;===============================================================
;                       main program entry point
;===============================================================
main:
; Set BSR for bank 0 operations
      movlb   0              ; Bank 0
```

In this fragment, code uses the movlb instruction to move a literal value to the low nibble in the Bank Select Register. The result is that the program will use the access bank (BSR = 0) in all instructions whether the access bank bit is set to use the BSR register (value = 1) or to the access bank (value = 0). This simplifies memory addressing for applications that do not require more than 128 bytes of data space by forcing all operations to take place in the access bank. For example, a program defines a data element at physical address 0x20 (bank 0) as follows:

```
TEMP equ  0x020;
```

Code then proceeds to clear the BSR bit

```
      movlb    0
```

Hereafter, all instructions refer to the access bank, whether the access bank bit (bit a) is set to A (value = 0) or to B (value = 1). The result is that banking has been effectively eliminated and the program executes in a flat data space that extends from 0x0 to 0x7f. The simplification is valid as long as data is not placed outside the limits of the access bank. The following code lines show possible variations:

```
      movf      TEMP,W,B  ; Move from TEMP to W using BSR
                          ; Because BSR = 0 access bank is used
```

or

```
      movf      TEMP,W,A  ; Move from TEMP to w in access bank
```

or

```
      movf      TEMP,W    ; Same action with no BSR bit
                          ; Access bank is assumed when BSR
                          ; bit is bit omitted.
```

In the remaining part of the program the a bit (bank select bit) is omitted in the source.

Port A for Digital Operation

The following code lines relate to the fact that the circuit used by this application (see Figure 7.2) is wired so that the four lines of the DIP switch are connected to lines 2 to 5 in port A, and that port A defaults to analog operation. Conversion from analog to digital is ensured as follows:

```
; Init Port A for digital operation
    clrf    PORTA,0
    clrf    LATA,0
; ADCON1 is the configuration register for the A/D
; functions in Port A. A value of 0b011x sets all
; lines for digital operation
    movlw   B'00000110'  ; Digital mode
    movwf   ADCON1,0
```

At this point, port A lines operate as digital input sources and will correctly report action on the DIP switch. The code that follows sets port A lines 2 to 5 for input and port C lines for output.

```
; Port A. Set lines 2 to 5 for input
    movlw     B'00111100'   ; w = 00111100 binary
    movwf     TRISA,0       ; port A (lines 2 to 5) to input
; Initialize all lines in PORT C for output
    movlw     B'00000000'   ; 0 = output
    movwf     TRISC,0       ; Port C tris register
```

DIP Switch Processing

The state of the DIP switch on Port A is obtained by reading the port value. In this sample program code uses a local variable named TEMP in order to avoid writing to Port A in the required manipulations. In addition, the DIP switch device is wired active low, therefore the bits must be inverted in order to reflect their physical state. Also, the bits in port A wired to the DIP switch are 2 to 5; therefore the unused bits must be masked out and the remaining ones shifted right two positions. The bitwise manipulations performed by the code are as follows:

```
;================================
;    DIP switch processing
;================================
DIPState:
; Read Port A and move to TEMP register to avoid read/math
; operations on Port A
    movf    PORTA,W
    movwf   TEMP
; Because board is wired active low then all switch bits
; must be negated.  This is done by XORing with 1-bits
    movlw   b'11111111'
    xorwf   TEMP,1  ; Invert all bits
; Mask off all unused bits
    movlw   b'00111100'
    andwf   TEMP,1
; Rotate port value right, twice
    rrncf   TEMP,1
    rrncf   TEMP,1
    .
    .
    .
```

At this point, the local variable named TEMP holds the switch value in the range 0x0 to 0xf.

Seven-Segment Code with Computed Goto

The program uses a computed goto (see Section 7.3.1) in order to obtain the seven-segment display code for the numeric value entered in the seven-segment LED. In order to avoid the addressing problems mentioned in Section 7.3.1, the table containing the seven-segment code is placed in a memory location where a code page boundary will not be exceeded during table access. Because the table consists of eighteen entries and each one occupies 2 bytes in program memory, a space of 36 bytes must be available in the code page where the table is located. To ensure this, the table is placed as follows:

```
          .
          .
          .
     org       0x018      ; Low-priority vector
     retfie
;=================================
;  Table to returns 7-segment
;          codes
;=================================
     org       $+2
;
codeTable:
     addwf     PCL,F           ; PCL is program counter latch
     retlw     0x3f ; 0 code
     retlw     0x06 ; 1
     retlw     0x5b ; 2
     retlw     0x4f ; 3
     retlw     0x66 ; 4
     retlw     0x6d ; 5
     retlw     0x7d ; 6
     retlw     0x07 ; 7
     retlw     0x7f ; 8
     retlw     0x6f ; 9
     retlw     0x77 ; A
     retlw     0x7c ; B
     retlw     0x39 ; C
     retlw     0x5b ; D
     retlw     0x79 ; E
     retlw     0x71 ; F
     retlw     0x00 ; Padding
```

The org statement for the table leaves 2 bytes for the low-priority interrupt vector and places the table at address 0x1A, which corresponds to a decimal offset of 26 bytes into the code page. Because 26 plus 36 (length of table) equals 62 and the code page is 256 bytes, we are sure that there is sufficient space for the table in the code page and that page boundary problems are avoided. Continuing from the previous code fragment, the code for accessing the table is as follows:

```
          .
          .
          .
; At this point the TEMP register contains a 4-bit value
; in the range 0 to 0xf. In PIC18 devices this value must
; be doubled to obtain offset into table since the program
; counter increments by 2 to access sequential instructions
     movf    TEMP,W      ; Offset to W
```

```
    addwf    TEMP          ; Add to TEMP
; Use value in TEMP to obtain Seven-Segment display code
    movf     TEMP,W        ; TEMP to W
    call     codeTable
    movwf    PORTC            ; Display switch bits
    goto     DIPState            ; Loop end
```

7.3.3 Assembler Table Lookup Sample Program

The sample program DIPs_to_7Seg_Tbl.asm in this book's software resource monitors the state of four toggle (DIP) switches in Port A lines 2 to 5 and displays the selected hexadecimal digit on the seven-segment LED wired to Port C. This program uses a read operation of a table in program memory instead of the computed goto method of the previous program. The table access technique has several advantages over the computed goto:

1. The table size is only limited by the amount of available RAM.

2. Each table entry takes up 8 instead of 16 bits.

3. The table can be located anywhere in the device's code memory space.

The sample program initialization and initial processing is identical to the program DIPs_to_7Seg.asm previously discussed. The table of seven-segment codes is defined as follows:

```
;=================================
;   Table of 7-segment codes
;=================================
Table:
    db   0x3f, 0x06, 0x5b, 0x4f, 0x66, 0x6d, 0x7d, 0x07
    db   0x7f, 0x6f, 0x77, 0x7c, 0x39, 0x5b, 0x79, 0x71
```

The use of commas with the db directive has the effect of allocating single bytes for the data. Had every data byte been defined by means of an individual db directive, the assembler would have allocated two bytes per entry.

The processing operations listed below assume that the TEMP register already holds the desired offset into the table, not double the offset as in the previous program. The next step in the code is to set the table pointer to the table address. The 21-bit table pointer is contained in three different registers named TBLPTRU <4:0>, TBLPTRH <7:0> and TBLPTRL <7:0>. The address range of the 21-bit pointer is 2 Mbytes. The example code uses the LOW, HIGH, and UPPER operators to obtain the three components of the address, as follows:

```
    movlw    LOW Table    ; Get low address of Table
    movwf    TBLPTRL      ; Store in table pointer low register
    movlw    HIGH Table   ; Get high byte
    movwf    TBLPTRH      ; Store it
    movlw    UPPER Table  ; Get upper byte
    movwf    TBLPTRU      ; Store it
    movf     TEMP,W       ; index to W
```

Now the offset of the desired table entry (stored in the variable named TEMP) must be added to the table pointer. Code must take into account the possibility of overflowing the low and high pointer bytes. Processing is as follows:

```
; Add offset to table pointer accomodating possible overflow
    addwf   TBLPTRL,f    ; Add index to table pointer low
    btfss   STATUS,C     ; Is there a carry?
    goto    readTbl      ; Go if no carry
    incf    TBLPTRH,F    ; Add one to high pointer
    btfss   STATUS,C     ; Test carry again
    goto    readTbl
    incf    TBLPTRU,F    ; Add one to upper pointer
readTbl:
```

Reading the table value using the TBLPTR register requires using the special table read instruction tblrd. The instruction can be formatted using four different operands, as shown in Table 7.1

Table 7.1

Formats for the tblrd Instructions

INSTRUCTION	ACTION
tblrd *	Read program memory into TABLAT using TBLPTR. No change to TBLPTR.
tblrd *+	Read program memory into TABLAT using TBLPTR. TBLPTR incremented after read operation.
tblrd *-	Read program memory into TABLAT using TBLPTR. TBLPTR decremented after read operation.
tblrd +*	Read program memory into TABLAT using TBLPTR. TBLPTR incremented before read operation.

Note that all versions of the tblrd instruction read the table entry into the table latch (TABLAT) register. The last three variations are reminiscent of C language pointer arithmetic and serve to update the pointer when sequential reads are performed. Once the table entry has been read into TABLAT, this register can be moved to the output port directly (movwff) or through the W register, as shown below.

```
readTbl:
    tblrd   *            ; Read byte from table (into TABLAT)
    movf    TABLAT,W     ; Move TABLAT to W
    movwf   PORTC        ; Display switch bits
    goto    DIPState
```

7.4 C Language Seven-Segment LED Programs

In the sections that follow we discuss the C language code for programs that read the state of four DIP switches and display the corresponding hex code in a seven-segment LED. These are high-lelvel versions of the programs DIPs_to_7Seg.asm and DIPs_to_7Seg_Tbl.asm presented previously.

7.4.1 Code Selection by Switch Construct

The sample program C_DIPs_to_7Seg.c in this book's software package reads the state of four DIP switches on Port A lines 2 to 5 and displays the corresponding hexadecimal code on the seven-segment LED wired to Port C lines 0 to 6. The sample program uses a swuitch construct to obtain the seven-segment hex code. After initializing Port A for digital input and Port C for output, the program proceeds as follows:

```
unsigned char DIPs = 0;
unsigned char digitCode;

while(1)
{
    // Read DIP switches and shift left
    DIPs = (PORTA >> 2);
    // DIPs are active low. Invert bits
    DIPs ^= 0xff;        // All bits XORed
    // Mask off high order nibble
    DIPs &= 0x0f;

    switch (DIPs)
    {
        case 0x0:
            digitCode = 0x3f;
            break;
        case 0x1:
            digitCode = 0x06;
            break;
        case 0x2:
            digitCode = 0x5b;
            break;
        case 0x3:
            digitCode = 0x4f;
            break;
            .
            .
            .
        case 0x0f:
            digitCode = 0x71;
            break;
        default:
            digitCode = 0x00;
            break;
    }
// Display digit code
    PORTC = digitCode;
    }
}
```

The C language processing requires no further comment.

7.4.2 Code Selection by Table Lookup

The sample program C_DIPs_to_7Seg_Tbl.c also reads the state of four DIP switches and displays the corresponding hex digit on a seven-segment LED. In this case, the program finds the corresponding hex code by looking up in an array-based table. The resulting source code and program is more compact and efficient.

The array of a hex codes can be placed in program memory or in data memory. Usually the programmer chooses whatever memory space is more abundant in the program. The rom and ram qualifiers that are part of the C18 implementation (see Section 6.6.6) allow placing the table in data or program memory. If in program memory (rom qualifier) then the declaration must be global because auto data cannot be placed in rom. The sample program defines the global table of seven-segment codes as follows:

```
rom unsigned char codeTable[]={0x3f, 0x06, 0x5b, 0x4f,
                               0x66, 0x6d, 0x7d, 0x07,
                               0x7f, 0x6f, 0x77, 0x7c,
                               0x39, 0x5b, 0x79, 0x71};


/****************************************************************
                      main program
****************************************************************/

void main(void)
{
```

Once the port and tris registers are initialized code can access the table as follows:

```
while(1)
    {
        // Read DIP switches and shift left
        DIPs = (PORTA >> 2);
        // DIPs are active low. Invert bits
        DIPs ^= 0xff;       // All bits XORed
        // Mask off high order nibble
        DIPs &= 0x0f;
        // Look up in array table and display code
        PORTC = codeTable[DIPs];
    }
```

7.5 A Demonstration Board

Demonstration (or demo) boards are a useful tool in mastering PIC programming. Many are available commercially and, like programmers, you will find a cottage industry of PIC demo boards on the Internet. Constructing your own demo boards and circuits is not a difficult task and is a valuable learning experience. Alternatively, components can be placed on a breadboard or wire-wrapped onto a special circuit board. Printed circuit boards can be home-made or ordered through the Internet. Appendix C contains instructions on how to build your own PCBs. Figure 7.3 shows a 18F452 demo board with the following elements:

```
bank of eight LEDs: 4 green and 4 red
seven-segment LED
buzzer
Liquid Crystal Display
Two pushbutton switches
DIP switch with four toggle arms
5K potentiometer
LM35 temperature sensor
```

```
NJU6355 real-time clock
Piezo buzzer
Reset switch
RJ-11 connector to external debugger
78L05 power supply
```

Figure 7.3 shows the wiring diagram for the Demo Board 18F452-A.

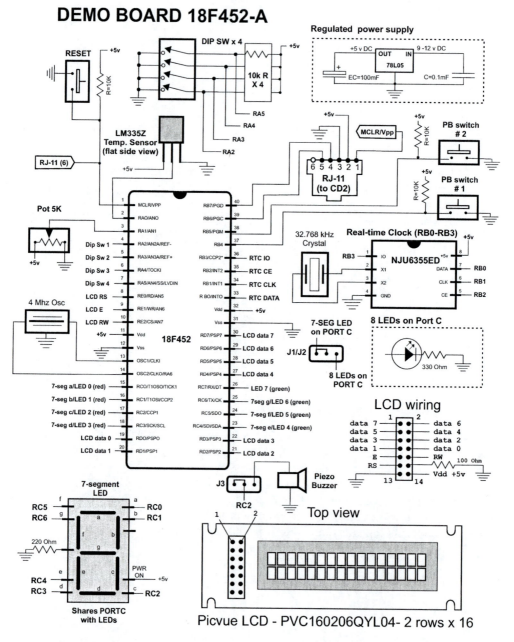

Figure 7.3 *Wiring diagram for Demo Board 18F452-A.*

7.6.1 Power Supply

Most PIC-based circuits require a +5V power source. One possible source of power is one or more batteries. There is an large selection of battery types, sizes, and qualities. The most common one for use in experimental circuits are listed in Table 7.2.

Table 7.2
Common Dry Cell Alkaline Battery Types

DESIGNATION	VOLTS	LENGTH	DIAMETER
D	1.5	61.5 mm	34.2 mm
C	1.5	50 mm	26.2 mm
AA	1.5	50 mm	14.2 mm
AAA	1.5	44.5 mm	10.5 mm
AAAA	1.5	42.5 mm	8.3 mm

All the batteries in Table 7.2 produce 1.5 volts. This means that for a PIC with a supply voltage from 2 to 6 volts, two to four batteries will be adequate. Note that in selecting the battery power source for a PIC-based circuit, other elements besides the microcontroller itself must be considered, such as the oscillator. Holders for several interconnected batteries are available at electronic supply sources.

Alternatively, the power supply can be a transformer with 120VAC input and 3 to 12VDC. These are usually called AC/DC adapters. The most useful types for the experimenter are the ones with an ON/OFF switch and several selectable output voltages. Color-coded alligator clips at the output wires are also a convenience.

Voltage Regulator

A useful device for a typical PIC-based power source is a voltage regulator IC. The 7805 voltage regulator is ubiquitous in most PIC-based boards with AC/DC adapter sources. The IC is a three-pin device whose purpose is to ensure a stable voltage source not to exceed the device rating. The 7805 is rated for 5V and will produce this output from any input source in the range 8 to 35V. Because the excess voltage is dissipated as heat, the 7805 is equipped with a metallic plate intended for attaching a heat sink. The heat sink is not required in a typical PIC application but it is a good idea to maintain the supply voltage closer to the device minimum rather than its maximum. The voltage regulator circuit also requires two capacitors: one electrolytic and the other one not. Figure 7.4 shows a power source circuit that uses the 7805 regulator.

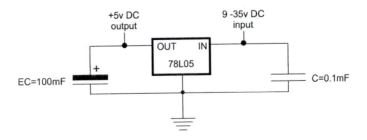

Figure 7.4 *Voltage stabilizer circuit.*

Chapter 8

Interrupts

8.1 Interrupt Mechanism

An interrupt is an asynchronous signal for processor attention that can originate in hardware or in software. The interrupt mechanism provides a way to avoid wasting processor time by avoiding ineffective polling routines in closed loops. Interrupts allow the processor to continue its work until the event that triggers the interrupt takes place. It also ensures that the CPU will receive a signal whenever an event occurs that requires its attention.

Interrupts are useful in many programming situations; for example,

- Preventing the CPU from being tied up while waiting for a process to begin or terminate. One common use for an interrupt is to notify the processor that a data transfer can take place.

- Responding to a hardware condition such as the pressing of a switch, the tripping of a lever, or action on a sensor.

- Responding to time-critical events such as an action that must take place immediately on a power failure condition.

- Providing an exit from a routine or an application on the occurrence of an error condition.

- Keeping track of time and updating time-keeping registers.

- Task switching in operating systems or multitasking environments.

8.2 PIC18 Interrupt System

The PIC 18 interrupt system consists of two vectors. The high-priority interrupt vector is located at address 0x08 and the low-priority vector is at 0x18. A high-priority event overrides a low-priority event in progress. Each interrupt source can be assigned a high-priority or a low-priority level.

The following code fragment used in the assembler programs previously discussed shows how the template provides links for the low-priority and high-priority interrupt vectors.

```
;=============================================================
;                              program
;=============================================================
     org      0        ; start at address
     goto main
; Space for interrupt handlers
;=============================
;      interrupt intercepts
;=============================
     org      0x008       ; High-priority vector
     retfie
     org      0x018       ; Low-priority vector
     retfie


;=============================================================
;                      main program entry point
;=============================================================
main:
     ...
```

8.2.1 Hardware Sources

In the 18F PIC family, interrupts can originate in the following hardware sources:

- External interrupt from the INT, INT1, and INT2 pins
- Interrupt on change on RB7:RB4 pins
- Timer overflow on TMR0, TMR1, TMR2, AND TMR3
- USART interrupts
- SSP interrupt
- C bus collision interrupt
- A/D conversion complete
- CCP interrupt
- LVD interrupt
- Parallel Slave Port
- CAN interrupts

The first three interrupt sources are covered in the present chapter. The remaining ones are discussed in the context of the specific modules.

8.2.2 Interrupt Control and Status Registers

Several SFRs are devoted to controlling interrupts and recoding their status. These are the following registers:

INTCON, INTCON2, INTCON3, PIR1, PIR2, PIE1, PIE2, IPR1, and IPR2.

bit 7							bit 0
GIE/ GIEH	PEIE/ GIEL	TMROIE	INTOIE	RBIE	TMROIF	INTOIF	RBIF

bit 7   **GIE/GIEH:** Global Interrupt Enable bit
      When IPEN = 0:
          1 = Enables all unmasked interrupts
          0 = Disables all interrupts
      When IPEN = 1:
          1 = Enables all high priority interrupts
          0 = Disables all interrupts

bit 6   **PEIE/GIEL:** Peripheral Interrupt Enable bit
      When IPEN = 0:
          1 = Enables all unmasked peripheral interrupts
          0 = Disables all peripheral interrupts
      When IPEN = 1:
          1 = Enables all low priority peripheral interrupts
          0 = Disables all low priority peripheral interrupts

bit 5   **TMROIE:** TMR0 Overflow Interrupt Enable bit
          1 = Enables the TMR0 overflow interrupt
          0 = Disables the TMR0 overflow interrupt

bit 4   **INTOIE:** INT0 External Interrupt Enable bit
          1 = Enables the INT0 external interrupt
          0 = Disables the INT0 external interrupt

bit 3   **RBIE:** RB Port Change Interrupt Enable bit
          1 = Enables the RB port change interrupt
          0 = Disables the RB port change interrupt

bit 2   **TMROIF:** TMR0 Overflow Interrupt Flag bit
          1 = TMR0 register has overflowed
            (must be cleared in software)
          0 = TMR0 register did not overflow

bit 1   **INTOIF:** INT0 External Interrupt Flag bit
          1 = The INT0 external interrupt occurred
            (must be cleared in software)
          0 = The INT0 external interrupt did not occur

bit 0   **RBIF:** RB Port Change Interrupt Flag bit
          1 = At least one of the RB7:RB4 pins changed state
            (must be cleared in software)
          0 = None of the RB7:RB4 pins have changed state
          Note: A mismatch condition will continue to set
              this bit. Reading PORTB will end the
              mismatch condition and allow the bit to
              be cleared.

Figure 8.1 *INTCON register bitmap.*

The INTCON register contains a bit labeled GIE that serves to enable and disable global interrupts. If this bit is set, all interrupts are enabled. Some specific members of the PIC 18F family have additional INTCON, PIR, PIE, and IPR registers to support their hardware.

INTCON Registers

The INTCON registers contain various bits that relate to interrupts and that allow enabling and disabling, establishing priorities, and determining interrupt status. Figure 8.1 is a descriptive bitmap of the INTCON register. Note that the interrupt flag bits are set whenever an interrupt condition takes place, regardless of the state of its corresponding enable bit or the global enable bit. User software must ensure the appropriate interrupt flag bits are clear prior to enabling an interrupt. This feature allows for software polling.

bit 7   bit 0

$\overline{REPU}$	INTEDGO	INTEDG1	INTEDG2	-	TMROIP	-	RBIP

bit 7   **RBPU:** PORTB Pull-up Enable bit
         1 = All PORTB pull-ups are disabled
         0 = PORTB pull-ups are enabled by individual
             port latch values
bit 6   **INTEDG0:** External Interrupt0 Edge Select bit
         1 = Interrupt on rising edge
         0 = Interrupt on falling edge
bit 5   **INTEDG1:** External Interrupt1 Edge Select bit
         1 = Interrupt on rising edge
         0 = Interrupt on falling edge
bit 4   **INTEDG2:** External Interrupt2 Edge Select bit
         1 = Interrupt on rising edge
         0 = Interrupt on falling edge
bit 3   **Unimplemented:** Read as '1'
bit 2   **TMR0IP:** TMR0 Overflow Interrupt Priority bit
         1 = TMR0 Overflow Interrupt is high priority
         0 = TMR0 Overflow Interrupt is low priority
bit 1   **Unimplemented:** Read as '1'
bit 0   **RBIP:** RB Port Change Interrupt Priority bit
         1 = RB Port Change Interrupt is high priority
         0 = RB Port Change Interrupt is low priority

Figure 8.2 *INTCON2 register bitmap.*

The IPEN bit referred to in Figure 8.1 is located in the RCON register. This bit allows enabling and disabling priority levels of the interrupt system. If the bit is set, priority levels are enabled. Otherwise, priority levels are disabled. The INTCON2 and INTCON3 registers in the 18F452 device perform additional interrupt control, priorities, and status functions. Figure 8.2 is a descriptive bitmap of the INTCON2 register. Figure 8.3 is a bitmap of the INTCON3 register.

bit 7   bit 0

INT2IP	INT1IP	-	INT2IE	INT1IE	-	INT2IF	INT1IF

bit 7   **INT2IP:** INT2 External Interrupt Priority bit
         1 = INT2 External Interrupt is a high priority event
         0 = INT2 External Interrupt is a low priority event
bit 6   **INT1IP:** INT1 External Interrupt Priority bit
         1 = INT1 External Interrupt is a high priority event
         0 = INT1 External Interrupt is a low priority event
bit 5   **Unimplemented:** Read as '0'
bit 4   **INT2IE:** INT2 External Interrupt Enable bit
         1 = Enables the INT2 external interrupt
         0 = Disables the INT2 external interrupt
bit 3   **INT1IE:** INT1 External Interrupt Enable bit
         1 = Enables the INT1 external interrupt
         0 = Disables the INT1 external interrupt
bit 2   **Unimplemented:** Read as '0'
bit 1   **INT2IF:** INT2 External Interrupt Flag bit
         1 = The INT2 external interrupt occurred
             (must be cleared in software)
         0 = The INT2 external interrupt did not occur
bit 0   **INT1IF:** INT1 External Interrupt Flag bit
         1 = The INT1 external interrupt occurred
             (must be cleared in software)
         0 = The INT1 external interrupt did not occur

Figure 8.3 *INTCON3 register bitmap.*

The PIE Registers

The registers generically referred to as PIE (Peripheral Interrupts Enable) provide information and control over the interrupts related to peripheral devices. The number of PIE registers is device dependent and so is their function. In the 18F452 there are two PIE registers labeled PIE1 and PIE2. Figure 8.4 is a descriptive bitmap of the PIE1 register. Figure 8.5 shows the PIE2 register.

bit 7   bit 0

PSPIE	ADIE	RCIE	TXIE	SSPIE	CCP1IE	TMR2IE	TMR1IE

```
bit 7      PSPIE(1): Parallel Slave Port Read/Write Interrupt Enable bit
               1 = Enables the PSP read/write interrupt
               0 = Disables the PSP read/write interrupt
bit 6      ADIE: A/D Converter Interrupt Enable bit
               1 = Enables the A/D interrupt
               0 = Disables the A/D interrupt
bit 5      RCIE: USART Receive Interrupt Enable bit
               1 = Enables the USART receive interrupt
               0 = Disables the USART receive interrupt
bit 4      TXIE: USART Transmit Interrupt Enable bit
               1 = Enables the USART transmit interrupt
               0 = Disables the USART transmit interrupt
bit 3      SSPIE: Master Synchronous Serial Port Interrupt Enable bit
               1 = Enables the MSSP interrupt
               0 = Disables the MSSP interrupt
bit 2      CCP1IE: CCP1 Interrupt Enable bit
               1 = Enables the CCP1 interrupt
               0 = Disables the CCP1 interrupt
bit 1      TMR2IE: TMR2 to PR2 Match Interrupt Enable bit
               1 = Enables the TMR2 to PR2 match interrupt
               0 = Disables the TMR2 to PR2 match interrupt
bit 0      TMR1IE: TMR1 Overflow Interrupt Enable bit
               1 = Enables the TMR1 overflow interrupt
               0 = Disables the TMR1 overflow interrupt
```

Figure 8.4 *PIE1 register bitmap*

bit 7   bit 0

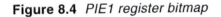

—	—	—	EEIE	BCLIE	LVDIE	TMR3IE	CCP2IE

```
bit 7-5    Unimplemented: Read as '0'
bit 4      EEIE: Data EEPROM/FLASH Write Operation Interrupt Enable bit
               1 = Enabled
               0 = Disabled
bit 3      BCLIE: Bus Collision Interrupt Enable bit
               1 = Enabled
               0 = Disabled
bit 2      LVDIE: Low Voltage Detect Interrupt Enable bit
               1 = Enabled
               0 = Disabled
bit 1      TMR3IE: TMR3 Overflow Interrupt Enable bit
               1 = Enables the TMR3 overflow interrupt
               0 = Disables the TMR3 overflow interrupt
bit 0      CCP2IE: CCP2 Interrupt Enable bit
               1 = Enables the CCP2 interrupt
               0 = Disables the CCP2 interrupt
```

Figure 8.5 *PIE2 register bitmap.*

Note that if the device has a PIE register and IPEN = 0, the PEIE bit must be set to enable the specific peripheral interrupt.

PIR Registers

The Peripheral Interrupt Request Registers (labeled PIR) contain the individual flag bits for the peripheral interrupts. In the 18F452 device there are two registers: PIR1 and PIR2. Figure 8.6 is a descriptive bitmap of the PIR1 register, Figure 8.7 of the PIR2 register.

bit 7   bit 0

PSPIF	ADIF	RCIF	TXIF	SSPIF	CCP1IF	TMR2IF	TMR1IF

```
bit 7     PSPIF(1): Parallel Slave Port Read/Write Interrupt Flag bit
              1 = A read or a write operation has taken place
                  (must be cleared in software)
              0 = No read or write has occurred
bit 6     ADIF: A/D Converter Interrupt Flag bit
              1 = An A/D conversion completed
                  (must be cleared in software)
              0 = The A/D conversion is not complete
bit 5     RCIF: USART Receive Interrupt Flag bit
              1 = The USART receive buffer, RCREG, is full
                  (cleared when RCREG is read
              0 = The USART receive buffer is empty
bit 4     TXIF: USART Transmit Interrupt Flag bit
              1 = The USART transmit buffer, TXREG, is empty
                  (cleared when TXREG is written)
              0 = The USART transmit buffer is full
bit 3     SSPIF: Master Synchronous Serial Port Interrupt Flag bit
              1 = The transmission/reception is complete
                  (must be cleared in software)
              0 = Waiting to transmit/receive
bit 2     CCP1IF: CCP1 Interrupt Flag bit
          Capture mode:
              1 = A TMR1 register capture occurred
                  (must be cleared in software)
              0 = No TMR1 register capture occurred
          Compare mode:
              1 = A TMR1 register compare match occurred
                  (must be cleared in software)
              0 = No TMR1 register compare match occurred
          PWM mode:
              Not used in this mode
bit 1     TMR2IF: TMR2 to PR2 Match Interrupt Flag bit
              1 = TMR2 to PR2 match occurred (must be cleared in software)
              0 = No TMR2 to PR2 match occurred
bit 0     TMR1IF: TMR1 Overflow Interrupt Flag bit
              1 = TMR1 register overflowed (must be cleared in software)
              0 = MR1 register did not overflow
```

Figure 8.6 *PIR1 register bitmap.*

IPR Registers

The Peripheral Interrupt Priority Registers (IPR) contain the individual bits for setting the priority of the various peripheral interrupts. In the 18F452, there are two IPR registers, labeled IPR1 and IPR2, respectively. For the priority bits to take effect the IPEN bit in the RCON register must be set. Figure 8.8 is a descriptive bitmap of the IPR1 register and Figure 8.9 of the IPR2 register.

bit 7   bit 0

-	-	-	EEIF	BCLIF	LVDIF	TMR3IF	CCP2IF

bit 7-5   **Unimplemented:** Read as '0'
bit 4   **EEIF:** Data EEPROM/FLASH Write Operation Interrupt Flag bit
           1 = The Write operation is complete
               (must be cleared in software)
           0 = The Write operation is not complete, or has not
               been started
bit 3   **BCLIF:** Bus Collision Interrupt Flag bit
           1 = A bus collision occurred
               (must be cleared in software)
           0 = No bus collision occurred
bit 2   **LVDIF:** Low Voltage Detect Interrupt Flag bit
           1 = A low voltage condition occurred
               (must be cleared in software)
           0 = The device voltage is above the Low Voltage Detect
               trip point
bit 1   **TMR3IF:** TMR3 Overflow Interrupt Flag bit
           1 = TMR3 register overflowed
               (must be cleared in software)
           0 = TMR3 register did not overflow
bit 0   **CCP2IF:** CCPx Interrupt Flag bit
           Capture mode:
               1 = A TMR1 register capture occurred
                   (must be cleared in software)
               0 = No TMR1 register capture occurred
           Compare mode:
               1 = A TMR1 register compare match occurred
                   (must be cleared in software)
               0 = No TMR1 register compare match occurred
           PWM mode:
               Not used in this mode

Figure 8.7 *PIR2 register bitmap.*

bit 7   bit 0

PSPIP	ADIP	RCIP	TXIP	SSPIP	CCP1IP	TMR2IP	TMR1IP

bit 7   **PSPIP(1):** Parallel Slave Port Read/Write Interrupt Priority bit
           1 = High priority
           0 = Low priority
bit 6   **ADIP:** A/D Converter Interrupt Priority bit
           1 = High priority
           0 = Low priority
bit 5   **RCIP:** USART Receive Interrupt Priority bit
           1 = High priority
           0 = Low priority
bit 4   **TXIP:** USART Transmit Interrupt Priority bit
           1 = High priority
           0 = Low priority
bit 3   **SSPIP:** Master Synchronous Serial Port Interrupt Priority bit
           1 = High priority
           0 = Low priority
bit 2   **CCP1IP:** CCP1 Interrupt Priority bit
           1 = High priority
           0 = Low priority
bit 1   **TMR2IP:** TMR2 to PR2 Match Interrupt Priority bit
           1 = High priority
           0 = Low priority
bit 0   **TMR1IP:** TMR1 Overflow Interrupt Priority bit
           1 = High priority
           0 = Low priority

Figure 8.8 *IPR1 register bitmap.*

bit 7   **bit 0**

–	–	–	EEIP	BCLIP	LVDIP	TMR3IP	CCP2IP

bit 7-5   **Unimplemented:** Read as '0'
bit 4   **EEIP:** Data EEPROM/FLASH Write Operation Interrupt Priority bit
               1 = High priority
               0 = Low priority
bit 3   **BCLIP:** Bus Collision Interrupt Priority bit
               1 = High priority
               0 = Low priority
bit 2   **LVDIP:** Low Voltage Detect Interrupt Priority bit
               1 = High priority
               0 = Low priority
bit 1   **TMR3IP:** TMR3 Overflow Interrupt Priority bit
               1 = High priority
               0 = Low priority
bit 0   **CCP2IP:** CCP2 Interrupt Priority bit
               1 = High priority
               0 = Low priority

Figure 8.9 *IPR2 register bitmap.*

8.2.3 Interrupt Priorities

Any interrupt in the 18F family can be assigned a priority level by clearing or setting the corresponding interrupt priority bit in the its IPR register or the corresponding INTCON register. The interrupt priority bits are set on a device reset; in other words, all interrupts are assigned high-priority at reset. The IPEN bit in the RCON register enables priority levels for interrupts. If clear, all priorities are high.

High-Priority Interrupts

A global interrupt enable bit labled GIE/GIEH and located in the INTCON register is set to enable all unmasked interrupts or cleared to disable them. When the GIE/GIEH bit is enabled, and the priority is high, and the interrupt's flag bit and enable bit are set, then the interrupt will take place and execution continues at its vector address.

Individual interrupts can be disabled through their corresponding enable/disable bits in the various registers previously listed. However, the individual interrupt flag bits are set regardless of the status of the GIE/GIEH bit. The GIE/GIEH bit is cleared on reset.

When the system responds to a high-priority interrupt, the GIE/GIEH bit is automatically cleared to disable any further interrupts, the return address is pushed onto the stack, and the PC is loaded with the address of the interrupt vector. In the interrupt service routine the source of the interrupt can be determined by testing interrupt flag bits. To avoid recursive interrupts, these flag bits must be cleared before reenabling interrupts. Most flag bits are required to be cleared by the application software, although some are automatically cleared by the hardware.

The "return from interrupt" instruction, retfie, terminates the interrupt routine and sets the GIE/GIEH bit. This action re-enables the high-priority interrupts.

Low-Priority Interrupts

Low-priority interrupts are defined by having zero in their interrupt priority register IPRx. The IPEN bit must be set in order to enable a low-priority interrupt. When the IPEN is set, the PEIE/GIEL bit in the INTCON register is not used to enable peripheral interrupts. Its new function is to globally enable and disable low-priority interrupts only. When the service routine for a low-priority interrupt executes, the PEIE/GIEL bit is automatically cleared in hardware in order to disable any further low-priority interrupts. The return address is pushed onto the stack and the PC is loaded with 0x00018 instead of 0x00008. All low-priority interrupts vector at address 0x00018.

In the interrupt service routine, the source of the low-priority interrupt can be determined by testing the low-priority interrupt flag bits. The interrupt flag bit(s) must be cleared before reenabling interrupts to avoid recursive interrupts. Most flag bits are required to be cleared by the application software although some are automatically cleared by the hardware. On terminating a low-priority interrupt, the retfie instruction resets the PEIE/GIEL bit. Notice that the GIE/GIEH bit's function has not changed in the low-priority interrupts because it still enables/disables all interrupts; however, it is only cleared by hardware when servicing a high-priority interrupt.

An Interrupt Interrupting Another One

If a high-priority interrupt takes place while a low-priority interrupt is in progress, the low-priority interrupt will be interrupted regardless of the state of the PEIE/GIEL bit. This is due to the fact that the PEIE/GIEL bit is used to disable/enable low-priority interrupts only. In this case, the GIE/GIEH bit is cleared by hardware to disable any further high- and low-priority interrupts, the return address is pushed onto the stack, and the PC is loaded with 0x00008, which is the high-priority interrupt vector. In the interrupt high-priority service routine, the source of the interrupt can be determined by testing the interrupt flag bits. The interrupt flag bit(s) must be cleared in software before reenabling interrupts to avoid recursive interrupts. Keep in mind that the GIEH bit, when cleared, will disable all interrupts regardless of priority. On the other hand, a low-priority interrupt cannot interrupt a high-priority ISR. In this case the low-priority interrupt will be serviced after all high-priority interrupts have terminated.

If a high-priority and a low-priority interrupt take place simultaneously, the high-priority interrupt service routine is always serviced first. In this case, the GIE/GIEH bit is cleared by the hardware and the device vectors to location 0x00008, which is the high-priority vector. In all cases, after the interrupt is serviced, the corresponding interrupt flag should be cleared to avoid a recursive interrupt. The retfie instruction on a high-priority interrupt handler resets the GIE/GIEH bit, and if no other high-priority interrupts are pending, the low-priority interrupt is serviced.

8.2.4 Context Saving Operations

The PIC 18F devices provide a "fast context saving" option that is coded as follows:

```
retfie    0x01
```

The special operand creates a shadow register that stores the values in the WREG, BSR and STATUS register. The shadow registers are only one level deep and are not readable by software. They are loaded with the current value of their corresponding register when the processor vectors for a high-priority interrupt. The values in the shadow registers are restored into the actual register when the special instruction (retfie 0x01) is encountered.

Fast context saving can only be used if the high- and low-priority interrupts are enabled. Any interrupt, high or low-priority, pushes values into the shadow registers. Because both low- and high-priority interrupts must be enabled, the shadow registers cannot be used reliably for low-priority interrupts. The reason is that a high-priority interrupt event will overwrite the shadow registers if a low-priority interrupt is in progress.

Context Saving during Low-Priority Interrupts

Low priority interrupts may also use the fast saving option described in the previous section; however, if both high- and low-priority interrupts are active, then the fast save option cannot be used with the low-priority interrupt because a high-priority event will overwrite the shadow registers. In this case, the low-priority handler can save and restore the key registers manually on the stack, as described later in this chapter.

The following code fragment shows the elements of an interrupt service routine to handle both low- and high-priority interrupts.

```
; Access RAM locations from 0x00 to 0x7F
W_high          equ  0x000        ; Temporary registers
BSR__high       equ  0x001
STATUS_high     equ  0x002
W_low           equ  0x003
BSR__low        equ  0x004
STATUS_low      equ  0x002


.    org      0
     goto     main
;*************************************
;   high-priority vector
;*************************************
     org      0x08                ; High priority vector
     movwf         W_high
     movff         BSR, BSR_high
     movff         STATUS, STATUS_high
;*************************************
;   code for high-priority ISR here
;       or jump to ISR routine
;*************************************
     movff         BSR_high, BSR
     movf     W_high, W
     movff         STATUS_high, STATUS
     retfie   0x00
;*************************************
;   low-priority vector
;*************************************
     org      0x18                ; Low priority vector
     movwf         W_low
     movff         BSR, BSR_low
```

```
    movff           STATUS, STATUS_low
;
;*****************************************
;   code for low-priority ISR here
;      or jump to ISR routine
;*****************************************
;
    movff           BSR_low, BSR
    movf    W_low, W
    movff           STATUS_low, STATUS
    retfie  0x00


;========================================
;         main program entry point
;========================================
main:
```

8.3 Port B Interrupts

Two types of interrupts are related to Port B:

1. Port B External Interrupts labeled RB0, RB1, and RB2.

2. Port B Interrupt On Change tied to Port B lines 4 to 7.

We have developed a simple circuit that allows testing both types of Port B interrupts. The circuit, which can be implemented as a demo board, can be seen in Figure 8.10.

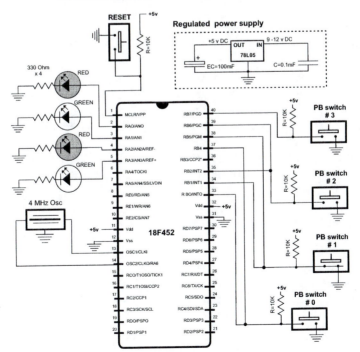

Figure 8.10 *Port B interrupt testing circuit.*

The four pushbutton switches on the interrupts demo board in Figure 8.10 are wired to Port B lines. The switches labeled 0, 1, and 2 are wired to Port B lines 0 and 4, 1 and 5, and 2 and 6, respectively. Switch number 3 is wired to Port B line 7. Circuit output is on Port A lines 0 to 3 which are wired to LEDs.

8.3.1 Port B External Interrupt

This external interrupt is triggered by either the rising or falling edge of the signal on port B, lines 0, 1, and 2. The interrupts are labeled INT0, INT1, and INT2, respectively. Whether the interrupt takes place on the rising or the falling edge of the signal depends the setting of the INTEDG0, INTEDG1, and INTEDG2 bits of the INTCON2 register (see Figure 8.2). When a valid edge appears on the RBx/INTx pin, the corresponding flag bit INTxF is set. This interrupt can be disabled by clearing the corresponding enable bit INTxE. Flag bit INTxF must be cleared in software in the Interrupt Service Routine before reenabling the interrupt.

The external interrupts (INT0, INT1, and INT2) can wakeup the processor from SLEEP. This happens if bit INTxE is set prior to going into SLEEP mode. If the global interrupt enable bit GIE is set, the processor will branch to the interrupt vector following wake-up. Interrupt priority for INT1 and INT2 is determined by the value contained in the interrupt priority bits INT1IP and INT2IP in the INTCON3 register. The INT0 interrupt is always given high-priority.

The Port B interrupts are useful in detecting and responding to external events, for example, in measuring the frequency of a signal or in responding to a change in the state of a hardware device. A simple application of this interrupt would be a circuit containing an emergency switch that can be pressed by the user. One possible approach is to check the state of the switch by continuously polling the port to which it is wired. But in a complex program, it may be difficult to make sure that the switch polling routine is called with sufficient frequency so that an emergency event is detected immediately, or that the switch is not released before it is polled. A more effective solution is to connect the emergency switch to an interrupt line in Port B and set up the Port B external interrupt source. With this scheme, whenever the emergency switch is activated, the program immediately responds via the interrupt mechanism. Furthermore, once the interrupt code has been developed and debugged, it will continue to function correctly no matter what changes are made to the rest of the program.

8.3.2 INT0 Interrupt Demo Program

The program named RB0Int_Demo.asm in the book's software package demonstrates the INT0 interrupt. The program uses the circuit shown in Figure 8.10. A pushbutton switch is connected to port RB0. The pushbutton toggles a LED on port A, line 0. Another LED on port A, line 1, flashes on and off at 1/2 second intervals.

cblock Directive

The program uses three variables in time delay operations. These variables are defined in a single block using the cblock directive, as follows:

```
; Access RAM locations from 0x00 to 0x7F
    cblock      0x000           ; Start of block
    j                           ; counter j
    k                           ; counter k
    count
    endc
```

The cblock directive defines a list of named sequential symbols. Its purpose is to assign sequential addresses to several labels. The list of variable names ends with the endc directive. The expression following the cblock keyword indicates the address for the first name in the block. If there is a previous cblock and no address is found, the block will be assigned an address one higher than the last entry in the previous cblock. If no address is assigned to the first cblock, it will be assigned a value of zero. The cblock directive cannot be used for relocatable code. A cblock is often used to replace several equ directives.

An optional increment keyword can be used after each label in the block; for example,

```
cblock     0x000            ; Start of block
    val1  :2
    val2  :4
    count
    endc
```

In this case, the name val2 is allocated 2 bytes from val1 and count is 4 bytes from val2. Multiple names may be given on the same line by separating them with commas. Two names can be defined at the same address by giving the first one an increment of zero.

Vectoring the Interrupt

Our programming template contains code to vector the high- and low-priority interrupts. The first one originates at 0x08, and the low-priority interrupt at 0x18. An interrupt handler can be located at these addresses but the high-priority handler will have to reside from 0x08 to 0x18, which allows a total of 16 instructions. Because the program's entry point (main label) can be anywhere in its code space, there is no space restriction for coding the low-priority interrupt at the 0x18 vector.

A more common and more reasonable approach is to locate the handler elsewhere in the program's code and provide a jump (goto instruction) at vector address, as in the following code fragment from the sample program RB0Int_Demo.asm.

```
;===============================================================
;                          program
;===============================================================
    org     0       ; start at address
    goto    main
; Space for interrupt handlers
;============================
;      interrupt intercept
;============================
    org     0x008       ; High-priority vector
    goto    IntServ
;
```

```
    org     0x018      ; Low-priority vector
    retfie
;===========================================================
;                       main program entry point
;===========================================================
main:
    ...
```

In this case the high-priority handler is located at the label named IntServ. There is no low-priority interrupt so the low-priority vector is left unimplemented.

Initialization

The program's circuit has four LEDs on Port A lines 0 to 3 and monitors a pushbutton switch on Port B, line 0 (INT0 line). Code must initialize the hardware accordingly. Code is as follows:

```
main:
; Set BSR for bank 0 operations
    movlb   0              ; Bank 0
; Init Port A for digital operation
    clrf    PORTA,0
    clrf    LATA,0
; ADCON1 is the configuration register for the A/D
; functions in Port A. A value of 0b011x sets all
; lines for digital operation
    movlw   B'00000110'    ; Digital mode
    movwf   ADCON1,0
; Port A lines to output
    movlw   B'00000000'    ; w = 0
    movwf   TRISA,0        ; port A to output
; Initialize all lines in PORT B for input
    movlw   B'11111111'    ; 1 = input
    movwf   TRISB,0        ; Port B tris register
```

Because Port A on the 18F452 is by default an analog port, software must reconfigure it for digital operation. This requires setting the corresponding code in the ADCON1 register. Then, Port A lines are trissed for output and Port B for input. In this case, it does not matter if the unused lines are trissed either way.

Set up INT0

The interrupt-related registers must then be set so that the INT0 interrupt is recognized. This requires setting the interrupt priority bit in the RCON register, enabling INT0 and high-priority interrupts in the INTCON register, and selecting falling edge operation so that the pushbutton switch generates the interrupt when it is pressed. Keep in mind that the switch is wired active low so that the edge falls when it is pressed. Code is as follows:

```
;===============================
; Set up interupt on Port B
;===============================
; Set interrupt priority bit in RCON register
    bsf     RCON,IPEN      ; Set bit
    bcf     INTCON,INT0IF  ; Clear TMR0IF flag
; INTCON register initialized as follows:
; (IPEN bit is set)
```

```
;               |------------ enable high-priority interrupts
;               |  |--------- enable INT0
    movlw   b'10010000
    movwf   INTCON
; Set INTCON2 for falling edge operation
; (button is active low)
    bcf     INTCON2,INTEDG0
```

At this point in the program, a high-priority interrupt will take place whenever pushbutton number 0 on Port B, line 0, is pressed.

Program Foreground

In order to demonstrate interrupt action, the program executes in an endless loop that flashes the LED wired to Port B, line 1. To turn the Port B bit on and off, the program uses an XOR bitwise operation because xoring with a one bit inverts the corresponding bit in the other operand. The delay subroutine is a simple do-nothing loop that is called three times consecutively. Code is as follows:

```
;=============================
;           flash LED
;=============================
; Program flashes LED wired to port B, line 1
lights:
    movlw   b'00000010'     ; Mask with bit 1 set
    xorwf   PORTA,F         ; Complement bit 1
    call    long_delay      ; Local delay routine
    call    long_delay
    call    long_delay
    goto    lights
;=============================
;   long delay sub-routine
;=============================
long_delay
    movlw   D'200'      ; w = 200 decimal
    movwf   j           ; j = w
jloop:
    movwf   k           ; k = w
kloop:
    decfsz  k,f         ; k = k-1, skip next if zero
    goto    kloop
    decfsz  j,f         ; j = j-1, skip next if zero
    goto    jloop
    return
```

Interrupt Service Routine

The interrupt service routine is located at the label IntServ. Code is considerably simplified because we are using the fast context saving option and the critical registers are saved automatically on the stack.

The first step in the ISR is to make sure that the interrupt originated in the INT0 pin. This is done by testing the INT0IF flag in the INTCON register. If the flag is not set, execution jumps to the exit routine and processing does not take place. If it is an INT0 interrupt, then the INT0IF flag is cleared by code.

The next step is to make sure that the interrupt occurred on the falling edge of the signal, that is, when the button was pressed. This is accomplished by making sure that the bit mapped to the switch register is clear because it is wired active low. If not so, then a quick exit from the ISR takes place and execution is aborted.

Switch Debouncing

Contact bounce is a fact in electrical switches. The switch elements are metal surfaces that are forced into contact by an actuator. The striking action of the contacts causes a rapidly pulsating electrical current instead of a clean transition. This is due to momentum and elasticity as well as parasitic inductance and capacitance in the circuit. The result is a series of sinusoidal oscillations.

Switch bounce often causes problems in circuits that are not designed to cope with oscillating voltages, particularly in digital devices. Several methods of hardware switch debouncing have been developed based on hysteresis. Switches can also be debounced in software by adding sufficient delay before reading the switch so as to prevent the bounce from being detected.

The sample program RB0Int_Demo.asm presently under discussion uses a simple time delay loop that ensures that a number of samplings of the switch at the desired level are received before code assumes that a specific switch action has taken place. In this implementation, a counter is initialized to a value of 10 and the port read operation is repeated as many times. If during any read iteration the switch is detected to be in the opposite state, execution of the service routine is aborted and the switch action is assumed to have been a bounce. Code is as follows:

```
        movlw       D'10'       ; Number of repetitions
        movwf       count       ; To counter
wait:
; Check to see that port B bit 0 is still 0
; If not, wait until it changes
        btfsc   PORTB,0         ; Is bit set?
        goto    exitISR         ; Go if bit not 0
;
; At this point RB0 bit is clear
        decfsz  count,f         ; Count this iteration
        goto    wait         ; Continue if not zero
```

Interrupt Action

In this simple demonstration, the action taken in the interrupt service routine consists of toggling on and off the LED wired to port A, line 0. Here again, we use the action of the bitwise XOR operation to turn the port bit to its opposite state, as follows:

```
; Interrupt action consists of toggling bit 0 of
; port A to turn LED on and off
      movlw    b'00000001'           ; Xoring with a 1-bit produces
                                     ; the complement
      xorwf    PORTA,f               ; Complement bit 2, port B .
```

The final action of the service routine is to use the fast exit option of the retfie instruction, as follows:

```
retfie   0x01          ; Fast return
```

8.3.3 Port B Line Change Interrupt

Another interrupt related to the port B register is determined by a change in value in any of port B lines 4 to 7. When this interrupt is enabled, any change in the status of any of the four port B pins (RB7, RB6, RB5, and RB4) can trigger an interrupt. The interrupt can be set up to take place when the status changes from logic one to logic zero, or vice-versa.

There are features of the line change interrupt that limit its usefulness. One of them is that for this interrupt to take place, all four port B pins 4 to 7 must be defined as input. This means that an application that wishes to use only one of the Port B lines 4 to 7 as input, it must also initialize for input the other three. Another limitation is that there is no control over which of the four lines generates the interrupt. This means that the determination of which line generated the interrupt must be made inside the handler because the interrupt sources cannot be enabled or disabled individually.

In spite of these limitation, the port B line change interrupt finds use in monitoring up to four different interrupt sources, typically originating in hardware devices. When the interrupt is enabled, the current state of the port B lines is constantly compared to the old values. If there is a change in state in any of the four lines, the interrupt is generated.

Implementation of the line change interrupt is not without complications. The circuit and software designer must take into account the characteristics of the external signal because only then can code be developed that will correctly handle the various possible sources. Two pieces of information that are necessary in this case are

1 .The signal's rising edge and falling edges

2. The pulse width of the interrupt trigger

Knowledge of the signals' rising and falling edges is necessary to ensure that the service routine is only entered for the desired edge. For example, if the device is an active-low pushbutton switch, an interrupt will typically be desired on the signal's falling edge, that is, when it goes from high to low.

Knowledge of the signal's width is necessary in order to determine the processing required by the service routine. If the triggering signal has a small pulse width compared to the time of execution of the interrupt handler, then the interrupt line will have returned to the inactive state before the service routine completes and a possible false interrupt on the signal's falling edge is not possible. On the other hand, if the pulse width of the interrupt signal is large and the service routine completes before the signal returns to the inactive state, then the signal's falling edge can trigger a false interrupt. Figure 8.11 shows both situations.

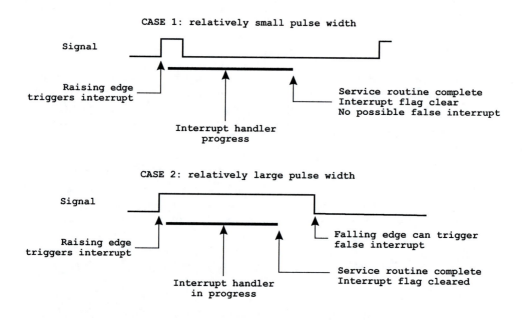

Figure 8.11 *Signal pulse width and interrupt latency.*

In Figure 8.11 the period between the edge that triggers the interrupt and the termination of the interrupt handler is sometimes called the "mismatch period." The mismatch period ends when the service routine completes and the corresponding interrupt is reenabled. If this happens after the interrupt signal is reset, no possible false interrupt can take place and no special provision is required in the handler. In fact, the interrupt handler will run correctly as long as the service routine takes longer to execute than the interrupt frequency. However, if the handler terminates before the signal returns to its original state, then the handler must make special provisions to handle a possible false interrupt. In order to do this the handler must first determine if the interrupt took place on the rising or the falling signal edge, which can be done by examining the corresponding port B line. For example, if the interrupt is to take place on the rising edge only, and the line is low, then it can be ignored because it took place on the falling edge.

When an interrupt can take place on either the rising or the falling edge of the triggering signal, the interrupt source must have a minimum pulse width in order to ensure that both edges are detected. In this case the minimum pulse width is the maximum time from the edge that triggered the interrupt to the moment when the interrupt flag is cleared. Otherwise, the interrupt will be lost because the interrupt mechanism is disabled at the time it takes place.

Reentrant Interrupts

The preceding discussion leads to the possibility of an interrupt taking place while the service routine of a previous interrupt is still in progress. These are called "reentrant" or "nested interrupts." Several events must take place in order to make possible reentrant interrupts. One of them is that interrupts are reenabled before the handler terminates. In addition, the service routine must be able to create different instances of the variables in use, usually allocated in the stack. The PIC interrupt mechanism itself forces the conclusion that reentrant interrupts are not recommended in PIC programs.

Multiple External Interrupts

One of the practical applications of the port B line change interrupt is in handling several different interrupt sources. For example, a circuit containing four pushbutton switches that activate four different circuit responses. If the switches are wired to the corresponding pins in port B (RB4 to RB7) and the line change interrupt is enabled, then the interrupt will take place when any one of the four switches changes level, that is, when any one of the interrupt lines go from high to low or from low to high. The interrupt handler software can easily determine which of the switches changed state and if the change took place on the signal's rising or falling edge. A propos software routines will then handle each case. Later in this chapter we develop a sample program that uses the port B line change interrupt to respond to action on four pushbutton switches.

8.3.4 Port B Line Change Interrupt Demo Program

The sample program RB4_to_RB7Int_Demo.asm in this book's software resource, is a simple demonstration of the Port B line change interrupt. The program uses the same circuit shown in Figure 8.10, which has pushbutton switches connected to Port B lines RB4 to RB7 and LEDs on Port A lines RA0 to RA3. The interrupt handler checks for action on Port B lines 4 and 7 in order to toggle the state of an LED. If the pushbutton switch wired to port RB4 is the one generating the interrupt, then the state of the LED on Port A line 1 is toggled. If the interrupt originated in the pushbutton switch wired to Port B, line 7, then the state of the LED wired to port A, line 0, is toggled. Interrupts generated by action on Port B, lines 5 and 6, are ignored by the code.

In the sample program RB4_to_RB7Int_Demo.asm, we have not used the fast context saving option described for the previous sample program. Our purpose has been to show how the register context can be saved manually although, in this application, the fast context save would have been possible and would have simplified the code. In the comments that follow, we focus on the features of the program RB4_to_RB7Int_Demo.asm that are different from the program RB0Int_Demo.asm previously discussed. For the unexplained features, the reader should refer to the preceding sample program.

Setting Up the Line Change Interrupt

The hardware initialization for Port A and Port B is the same for both programs. The code for setting up the Port B line change interrupt starts by clearing the IPEN bit in the RCON register. This bit enables or disables the interrupt priority mechanism of the PIC 18 family. Disabling interrupt priority simulates a system compatible with the mid-range PIC family and is sometimes called the "compatibility mode." In this case, all interrupts are vectored to address 0x08 as they occur.

Code must then proceed to enable interrupts and to enable the RB4 to RB7 line change interrupt. This is accomplished by setting bits 3 and 7 in the INTCON register. The falling edge operation is set by clearing the INTEDG0 bit in the INTCON2 register. The RBIF flag in the INTCON register must also be cleared to allow the interrupt to take place. Code is as follows:

```
;==============================
; Set up line change interrupt
;==============================
; Disable interrupt priority levels in the RCON register
; setting up the mid range compatibility mode
    bcf      RCON,IPEN       ; Clear bit
    bcf      INTCON,RBIF     ; Clear RB4-7 change flag
; INTCON register initialized as follows:
; (IPEN bit is clear)
;                  |------------ enable unmasked interrupts
;                  |   |-------- enable RB4-7 interrupt
    movlw    b'10001000'
    movwf    INTCON
; Set INTCON2 for falling edge operation
; (button is active low)
    bcf      INTCON2,INTEDG0
```

The program loop that follows the interrupt setup code is a do-nothing routine because the program performs no other actions outside of the interrupt handler.

Interrupt Service Routine

The interrupt service routine begins by ensuring that the cause of the interrupt was the Port B line change. This is accomplished by testing the RBIF bit in the INTCON register. The bit must be set if this is a Port B line change event. If not, a quick exit from the handler takes place. Code is as follows:

```
IntServ:
; First test: make sure source is an RB4-7 interrupt
    btfss    INTCON,RBIF        ; RBIF flag is interrupt
    goto     notRBIF            ; Go if not RBIF origin
```

The next step is saving the context registers because we have opted not to use the fast context save option in this program. For many applications, the context that must be saved in the handler is limited to the w and the STATUS registers. If the handler uses any other register or variable that is shared with the main code, then it must also be saved. For example, suppose a handler that accesses a memory area different from the one used by the main program. Because the BSR (bank select) register will be changed in the handler it must also be saved and restored by the ISR. This is not the case in the present sample code so only the w and the STATUS registers are preserved.

Saving the W and the STATUS registers requires using register variables, but the process requires special care. Saving the W register is simple enough: its value at the start of the Service Routine is stored in a local variable from which it is restored at termination. Saving the STATUS register cannot be done with MOVF instruction because this instruction itself changes the zero flag. The solution is to use the swapf instruction, which does not affect any of the flags. Of course, swapf inverts the nibbles in the operand, so it must be repeated in order to restore the original state. The following code fragment assumes that file register variables named old_w and old_status were previously created.

```
save_cntx:
      movwf    old_w            ; Save w register
      swapf    STATUS,w  ; STATUS to w
      movwf    old_status       ; Save STATUS
;
; Interrupt handler operations go here
;
      swapf    old_status,w  ; Saved status to w
      movfw    STATUS    ; To STATUS register
; At this point all operations that change the
; STATUS register must be avoided, but swapf does not.
      swapf    old_w,f   ; Swap file register in itself
      swapf    old_w,w   ; re-swap back to w
```

Once W and the STATUS register have been saved in variables, the ISR must determine which Port B line originated the interrupt. In this example Port B line 4 requires one action, Port B line 7 a different one, and lines 4 and 5 are ignored. Because the interrupt mechanism in the PIC 18F family does not provide a way of knowing which Port B line generated the interrupt, code must keep record of the Port B value during the previous interrupt and test if it has changed in the previous intercept. The sample program uses two variables for this purpose: a variable named bitsB47 holds the value in Port B during the previous interrupt, and the variable named temp to save the current Port B value.

Processing consists of XORing the value in Port B during the previous iteration of the interrupt with the Port B present value. The XOR operation only results in a 1 bit if the two operands have opposite values. So any 1 bit in the result indicates a bit that has changed from the previous interrupt iteration. Code is as follows:

```
; The interrupt action takes place when any of port B bits
; 4 and 7 have changed status.
      movf    PORTB,w          ; Read port B bits
      movwf   temp      ; Save reading
      xorwf   bitsB47,f  ; Xor with old bits, result in f
; Test each meaningful bit (4 and 7 in this example)
      btfsc   bitsB47,4  ; Test bit 4
      goto    bit4Chng         ; Routine for changed bit 4
; At this point bit 4 did not change
      btfsc   bitsB47,7  ; Test bit 7
      goto    bit7Chng         ; Routine for changed bit 7
; Invalid port line change. Exit
      goto    pbRelease
```

The last line in the previous code snippet ensures that any action on the other Port B lines is ignored. If either bit 4 or bit 7 of Port B has changed, then the interrupt handler toggles the state of the corresponding LED on Port A. Also at this time, code checks that the signal took place on the falling edge and ignores the interrupt if it did not. Toggling the LED on Port A is accomplished by XORing with a one bit in the corresponding mask. For change in Port B, bit 4, code is as follows:

```
;========================
; bit 4 change routine
;========================
; Check for signal falling edge, ignore if not
bit4Chng:
      btfsc    PORTB,4          ; Is bit 4 high
      goto     pbRelease        ; Bit is high. Ignore
; Toggling bit 1 of port A turns LED on and off
      movlw    b'00000010'      ; Xoring with a 1-bit produces
                                ; the complement
      xorwf    PORTA,f          ; Complement bit 1, port A
      goto     pbRelease
```

The code provides two exits for the interrupt handler. One for the case in which the W and STATUS registers were saved in variables, and another exit for when the interrupt did not originate in a Port B line change. In the first case, the current value of Port B is saved for the next iteration in the variable bitsB47. Also, the variables old_w and old_STATUS that were used to preserve the value of the W and STATUS registers at the start of the interrupt, are now used to restore these two registers. swapf instructions are used to avoid changing the Z bit in the STATUS register. Code is as follows:;========================

```
;           exit ISR
;========================
exitISR:
; Store new value of port B
      movf temp,w           ; This port B value to w
      movwf    bitsB47       ; Store
; Restore context
      swapf    old_STATUS,w  ; Saved STATUS to w
      movwf    STATUS        ; To STATUS register
      swapf    old_w,f       ; Swap file register in itself
      swapf    old_w,w       ; re-swap back to w
; Reset,interrupt
notRBIF:
      bcf  INTCON,RBIF       ; Clear INTCON bit 0
      retfie
```

8.4 Sleep Mode and Interrupts

The PIC microcontroller sleep mode provides a useful mechanism for saving power that is particularly useful in battery-operated devices.

The sleep mode is activated by executing the sleep instruction, which suspends all normal operations and switches off the device oscillator so no clock cycles take place. The sleep instruction takes no operands. The sleep mode is suitable for applications that are not required to run continuously. For example, a device that records temperature at daybreak can be designed so that a light-sensitive switch generates an interrupt that turns on the device each morning. Once the data is recorded, the device goes into the sleep mode until the next daybreak.

The following actions and states take place during sleep mode:

- Watchdog Timer is cleared but keeps running
- The PD bit in the RCON register is cleared and the TO bit is set
- The oscillator driver is turned off
- The I/O ports maintain the status during sleep mode

To ensure the lowest possible current consumption during sleep, the following precautions should be taken:

- All I/O pins should be either at VDD or VSS.
- No external circuitry should draw current from an I/O pin.
- Modules that are specified to have a delta sleep current should be disabled.
- I/O pins that are hi-impedance inputs should be pulled high or low externally.
- The contribution from on-chip pull-ups on PORTB should be considered.
- The MCLR pin must be at a valid high level.

Some features of the device consume a delta current. These are enabled/disabled by configuration bits. These include the Watchdog Timer (WDT), LVD, and Brown-out Reset (BOR) circuitry modules.

8.4.1 Wake Up from SLEEP

Several actions wake the controller from SLEEP:

- The WDT times out
- A RESET
- Interrupts from peripherals or external sources

In addition, the sleep mode terminates by one of the following events:

1. Any device RESET such as MCLR pin

2. Watchdog Timer Wake-up (if WDT was enabled)

3. Any peripheral module that can set its interrupt flag while in SLEEP, such as

   An external INT pin

   A port pin

   Comparators

   A/D converters

   Timer1 and Timer 3

   LVD

   MSSP

   Capture and Compare

   PSP read or write

   CCP1 and CCP2

Addressable USART

PORTB Interrupt on Change

External Interrupts

Parallel Slave Port

Voltage Reference (bandgap)

WDT

Action on the MCLR pin will reset the device upon wake-up. The second and third events on the previous list will wake the device but do not reset and program execution is resumed. The TO and PD bits in the RCON register can be used to determine the cause of device RESET. The PD bit is set on power-up and is cleared when SLEEP is invoked. The TO bit is cleared if WDT time-out occurred (and caused a wake-up).

For the wake-up through an interrupt event to take place, the corresponding interrupt enable bit must be set. Wake-up takes place regardless of the state of the GIE bit; but if the GIE bit is clear (disabled), the device continues execution at the instruction after the SLEEP instruction. If the GIE bit is set (enabled), the device executes the instruction after the SLEEP instruction and then branches to the interrupt address. In cases where the execution of the instruction following SLEEP is not desirable, the user should insert a nop opcode after the sleep instruction.

8.4.2 Sleep_Demo Program

The program named Sleep_Demo in this book's online software package is a trivial demonstration of using the RB0 interrupt to wake up the processor from the sleep mode. The program can be tested using the circuit in Figure 8.10. Sleep_Demo flashes the LED on Port A, line 1, at one-half-second intervals during ten cycles and then goes into the sleep mode. Pressing the pushbutton switch on line RB0 generates an interrupt that wakes the processor from the sleep mode. Much of the initialization and interrupt processing is the same as program RB0Int_Demo.asm developed previously in this chapter. The following code fragment shows the coding of the main loop in the program.

```
;===============================
; flash LED on and off 10 times
;===============================
wakeUp:
; Program flashes LED wired to port A, line 1,
; 10 cycles before entering the sleep state
    movlw    D'20'          ; Number of iterations
    movwf    count2         ; To counter
lights:
    movlw    b'00000010'    ; Mask with bit 1 set
    xorwf    PORTA,f        ; Complement bit 1
    call     long_delay
    call     long_delay
    call     long_delay
    decfs    zcount2,f ; Decrement counter
    goto lights
; 20 iterations have taken place
```

```
        sleep
        nop                     ; Recommended!
        goto wakeUp             ; Resume execution
;=============================
;   long delay sub-routine
;=============================
long_delay
        movlw    D'200'          ; w = 200 decimal
        movwf    j               ; j = w
jloop:
        movwf    k               ; k = w
kloop:
        decfsz   k,f             ; k = k-1, skip next if zero
        goto     kloop
        decfsz   j,f             ; j = j-1, skip next if zero
        goto     jloop
        return
```

In the Sleep_Demo program the Interrupt Service Routine does nothing because the occurrence of an interrupt automatically wakes up the processor from the sleep state. Its coding is as follows:

```
;=========================================================
;                Interrupt Service Routine
;=========================================================
; Service routine receives control when there is
; action on pushbutton switch wired to port B, line 0
IntServ:
; Clear the INT0 External Interrupt flag
    bcf     INTCON,INT0IF
    retfie  0x01                ; Fast return
```

8.5 Interrupt Programming in C Language

The fact that interrupts are low-level program elements makes them easier to understand and operate in a low-level language (assembler) than in a high-level one (C18). Nevertheless, the C18 compiler does provide functionality for interrupts, and the necessary manipulations are not difficult to implement in code. The following #pragma directives are used in implementing interrupts:

    1 #pragma interruptlow fname

    2. #pragma interrupt fname

    3. #pragma code

The interruptlow pragma declares a C18 function that will be a low-priority interrupt service routine. This routine will be placed at the low-priority interrupt vector at address 0x018. The interrupt pragma declares the interrupt to be a high-priority service routine located at the high-priority vector at address 0x08.

8.5.1 Interrupt Action

An interrupt suspends the execution of a running application, saves the current context, and transfers control to a service routine. Once the service routine has concluded its actions, the previous context is restored and execution of the application resumes at the location where the interrupt took place. The minimal registers that are saved as context and restored at the conclusion of the interrupt are WREG, BSR, and STATUS. A high-priority interrupt uses the shadow registers to save and restore the minimal context automatically, while a low-priority interrupt requires program action to save the context in the software stack. As in the low-level routines, a high-priority interrupt terminates with a fast "return from interrupt," while a low-priority interrupt terminates with a normal "return from interrupt."

Context in the Stack

When the context is placed in the software stack, the C compiler must use two MOVFF instructions for each byte of context preserved, except for saving WREG, which requires a MOVWF and a MOVF instruction. This means that in order to preserve the minimal context registers (WREG, BSR, and STATUS) during a low-priority interrupt, an additional ten words of storage are necessary beyond the requirements of a high-priority interrupt.

In C language it is possible to add a save clause to the pragma statement that defines the low- or high-priority interrupt. This clause informs the C compiler that it must generate the necessary code for saving additional registers in the software stack and restoring them when the ISR terminates. For example,

```
#pragma interruptlow MyISR save = PORTA, PORTC
```

This statement directs execution to a service routine named MyISR. The interrupt mechanism automatically saves the context registers WREG, BSR, and STATUS in the software stack. In this case, the save clause also directs the compiler to save the registers listed after the = sign, in this case PORTA and PORTC.

Interrupt Data

Interrupt service routines created by the C18 compiler use a temporary data section that is distinct from that used by normal C functions. During the evaluation of expressions in the interrupt service routine, data is allocated in a special section and is not overlaid with the temporary locations of other functions, including other interrupts.

The interrupt-creating pragma allows naming the special data section. For example,

```
void myInt(void);
...
#pragma interrupt myInt

void myInt(void)
{
// Interrupt handler code here
}
```

In this case, the temporary variables for interrupt service routine name myInt will be placed in the udata section myInt_tmp. If no name is provided, then the temporary variables are created in a udata section named fname_tmp.

8.5.2 Interrupt Programming in C18

In order to implement an interrupt handler and vector execution to the appropriate address, the following operations are available:

1. Prototyping the ISR to make possible locating its code anywhere in the program
2. Defining the interrupt vector or vectors
3. Using inline assembly at the vector address to provide a jump to the handler
4. Restoring the compiler's code locating privilege
5. Defining the handler name and optionally saving additional data
6. Coding the interrupt handler

Not all of these operations are required in every handler. For example, if the handler can be located before it is referenced then its prototyping (Step 1) will not be necessary. However, the possibility of locating the handler inline may be limited by its byte size. This is true with the high-priority handler located at vector address 0x08. Because the vector address for the low-priority handler must be at 0x18 there is a space of 16 bytes between both vectors. If we were to code a high-priority handler that was to be placed inline at address 0x08, and if this handler exceeded the 16-byte limit, then its code will interfere with the operation of a low-priority handler at vector address 0x18. To avoid this possibility, it is safer practice to place a jump to the ISR at the vector address and locate the service routine elsewhere in the program.

The following code fragment, from the program C_RB4-7LowInt_Demo developed later in this chapter, shows the coding and location of the operations listed above:

```
        .
        .
        .
// Prototyping the ISR (Step 1)
void low_ISR(void);

// Locating the interrupt vector (Step 2)
#pragma code low_vector = 0x18

// Implementing jump to the handler (Step 3)
void low_interrupt(void)
{
    _asm
    goto            low_ISR
    _endasm
}

// Restoring code addressing to C18 (Step 4)
#pragma code

// Defining the handler (Step 5)
#pragma interruptlow low_ISR save = PROD

// Coding the interrupt handler (Step 6)
void low_ISR(void)
{
```

```
    int counter;
    unsigned char switches;

    switches = PORTB;
    PORTA = (switches >> 4);
    // Short delay to stabilize LEDs
    for(counter = 0; counter < 2000; counter++) {
        Nop();
        Nop();
    }
    INTCONbits.RBIF = 0;    // Clear flag
}

/*************************************************************
                        main program
*************************************************************/
void main(void)
    .
    .
    .
```

In this code fragment, the interrupt handler is named low_ISR while the jump to this handler is located at vector address 0x18 and the C18 procedure is named low_interrupt. Note that these names are conventional and could be changed for any other identifier. The code listing shows two instances of the #pragma code statement: the first one ensures that the jump to the handler is located at the correct vector address (0x18). The second instance of the #pragma code statement follows the goto opcode and returns addressing to the C18 compiler.

The #pragma statement in Step 5 names the ISR and provides a safe clause (which is actually redundant in this case). The identifier interruptlow in the #pragma statement is required when defining the low-priority interrupt. The keyword for the high-priority interrupt is interrupt. In programs that use both the high- and low-priority interrupts, several #pragma code statements will be necessary in order to make sure that the vectors at 0x08 and 0x18 are preserved and that the corresponding jumps to the handlers are located at these vectors.

Sleep Mode and RB0 Interrupt Demo Program

The program named C_RB0HighInt_Demo in this book's software resource is a simple demonstration of using the RB0 interrupt to wake up the processor from the Sleep mode. The program is a C18 implementation of the Sleep_Demo program developed in assembly language earlier in this chapter. The reader should refer to Section 8.4.2 for information on the Sleep mode. The C18 version uses the Sleep() macro to implement the sleep opcode. The actual processing is trivial because the RB0 interrupt wakes up the processor from sleep without any other manipulations.

Defining the high-level handler and the ISR follows the same six steps described in Section 8.5.2 for the low-level handler. The only variations are the identifiers and the location of the high-level vector at address 0x08. The program flashes LEDs on port A, lines 0 to 3, on and off at half-second intervals, for 10 cycles. On the tenth cycle, the program enters the SLEEP state and LED flashing stops. Pressing the pushbutton switch on line RB0 generates an interrupt that ends the SLEEP and repeats the flashing cycle.

The Port B external interrupt was discussed in Section 8.3.1. The sample program RB0Int_Demo.asm developed in Section 8.3.2 is the equivalent code in assembly language. The C18 program sets up the interrupt by manipulating the same registers as the assembly language version. C language code is as follows:

```
void main(void)
{
    unsigned char flashes = 0;

    // Init Port A for digital operation
    PORTA = 0;          // Clear port
    LATA = 0;        // and latch register
    // ADCON1 is the configuration register for the A/D
    // functions in Port A. A value of 0b011x sets all
    // lines for digital operation
    ADCON1 = 0b00000110;// Code for digital mode
    // Initialize direction registers
    TRISA = 0x00;        // Port A lines for output
    TRISB = 0xff;        // Port B lines for input
    PORTB = 0x0;         // Clear all Port B lines
    PORTA = 0x0;         // and Port A
    // Setup RB0 interrupt
    RCONbits.IPEN = 1; // Set interrupt priority bit
    INTCONbits.INT0IF = 0;  // Clear flag
    // Initialize INTCON register for high-priority
    // and INT0
    INTCONbits.GIEH = 1;     // High priority enabled
    INTCONbits.INT0IE = 1;  // INT0 active
    // Set INTCON2 for falling edge of switch
    INTCON2bits.INTEDG0 = 0;
    .
    .
    .
```

Once the interrupt is initialized, the program enters an endless loop that flashes the LEDs on Port A and then goes into the sleep mode, as follows:

```
    // Endless loop to flash LEDs 10 times then go
    // into sleep mode
    while(1)
    {
        for(flashes = 0; flashes < 10; flashes++)
        {
            FlashLEDs();
        }
        Sleep(); // Macro for sleep opcode
    }
```

The interrupt handler receives control whenever the pushbutton connected to line 0 in Port B is pressed. The handler is coded as follows:

```
// Interrupt wakes up from sleep automatically.
// No action is necessary except resetting the
// INT0 flag
void high_ISR(void)
{
    INTCONbits.INT0IF = 0;  // Clear flag
}
```

Port B Interrupt on Change Demo Program

The second C18 sample program demonstrates the Port B Interrupt on Change. LEDs on port A, lines 0 to 3 reflect the status of the Port B lines 4 to 7 by turning off the corresponding LED when the switch is down. The program uses the low-priority interrupt vector and saves the context in the software stack Here again, the C18 program is a version of the assembler program RB4_to_RB7Int_Demo developed in Section 8.3.4.

The interrupt on change is implemented by following the same six steps described in Section 8.5.2. Code is as follows:

```
// Prototyping the ISR (Step 1)
void low_ISR(void);

// Locating the interrupt vector (Step 2)
#pragma code low_vector = 0x18

// Implementing jump to the handler (Step 3)
void low_interrupt(void)
{
    _asm
    goto        low_ISR
    _endasm
}

// Restoring code addressing to C18 (Step 4)
#pragma code

// Defining the handler (Step 5)
#pragma interruptlow low_ISR save = PROD

// Coding the interrupt handler (Step 6)
void low_ISR(void)
{
    int counter;
    unsigned char switches;

    switches = PORTB;
    PORTA = (switches >> 4);
    // Short delay to stabilize LEDs
    for(counter = 0; counter < 2000; counter++) {
        Nop();
        Nop();
    }
    INTCONbits.RBIF = 0;     // Clear flag
}
```

Code for activating the IOC is based on manipulating the hardware interrupt registers much like in the assembly language version. The one variation we introduced in the code is to disable the IOC interrupt until the system is completed. This is accomplished by setting the interrupt flag. Code is as follows:

```
// Turn off IOC while seting up system
    INTCONbits.RBIF = 1;    // Interrupt off
    // Initialize INTCON register
    INTCONbits.GIE = 1;// Enable unmasked interrupts
    INTCONbits.GIEL = 1;    // Low priority interrupts
    RCONbits.IPEN = 1; // Turn on priority system
```

```
    INTCONbits.RBIE = 1;      // Port B IOC enabled
    INTCON2bits.RBIP = 0;     // Int on change low-priority
    // Set INTCON2 for falling edge of switch
    INTCON2bits.INTEDG0 = 0;
    // Turn on all 4 LEDS in Port A
    PORTA = 0x0f;
    // Turn on IOC
    INTCONbits.RBIF = 0;      // Interrupt on
    // Main program does nothing
    while(1)
    {
        Nop();
    }Other Interrupt Demo Programs
```

As we discuss the modules and peripheral devices of the 18F452 device, we will develop demonstration programs that exercise the corresponding interrupts.

Chapter 9

Delays, Counters, and Timers

9.1 PIC18 Family Timers

Microcontroller timers in general belong to one of two groups:

1. Delay timers used during system power-up, reset, and watchdog operations
2. Timer-counters used in implementing and measuring time periods and waveforms

9.2 Delay Timers

PIC18 microcontrollers have hardware resources that provide a delay period during reset operations. Reset operations were discussed in Section 2.4 and following. The timers associated with the reset action are

1. Power-Up Timer (PWRT)
2. Oscillator Start-Up Timer
3. Phase Lock Loop (PLL) timer
4. Watchdog timer

9.2.1 Power-Up Timer (PWRT)

The Power-up Timer provides a fixed time-out period from Power-On Reset (POR). The timer operates on an internal RC oscillator that keeps the chip in the RESET state. This delay allows the Vdd signal to rise to an acceptable level. The nominal delay period is documented to take 72 ms but it is said to vary from chip to chip, due to differences in the Vdd and changes in temperature. A configuration bit is provided to enable/disable the PWRT, as follows:

```
PWRT = ON          Power-up timer enabled
PWRT = OFF         Power-up timer disabled
```

The default state is disabled but there is no reason why the PWRT timer should not be enabled for most applications.

9.2.2 Oscillator Start-Up Timer (OST)

The Oscillator Start-Up Timer (OST) ensures that 1,024 oscillator cycles take place after the PWRT delay is over and before the RESET stage ends. This delay ensures that the crystal oscillator or resonator has started and is stable on power-up. The oscillator time-out is invoked for the following oscillator options: XT, LP, and HS. The delay takes place on Power-on Reset, Brown-out Reset, or wake-up from SLEEP. It also takes place on the transition from Timer1 input clock as the system clock to the oscillator. The OST is disabled for all resets and wake-ups in RC and EC oscillator options.

The OST function counts oscillator pulses on the OSC1/CLKIN pin. The counter starts incrementing after the amplitude of the signal reaches the oscillator input thresholds. This initial delay allows the crystal oscillator or resonator to stabilize before the device exits the OST delay. The length of the time-out is a function of the crystal/resonator frequency. For low-frequency crystals, this start-up time can become quite long. That is because the time it takes the low-frequency oscillator to start oscillating is longer than the power-up timer's delay.

The time from when the power-up timer times out to when the oscillator starts to oscillate is referred to as dead time. There is no minimum or maximum time for this dead time because it is dependent on the time required for the oscillator circuitry to have "good" oscillations.

9.2.3 Phase Locked Loop (PLL)

The Phase Locked Loop (PLL) circuit is a programmable option that allows multiplying by 4 the frequency of the crystal oscillator signal. Selecting the PLL option results in an input clock frequency of 10 MHz of the internal clock being multiplied to 40 MHz.

The PLL can only be enabled when the oscillator configuration bits are programmed for HS mode. In all other modes, the PLL option is disabled and the system clock will come directly from the OSC1 pin. The configuration bit for HS and PLL are selected with the following statement:

```
#pragma config OSC = HSPLL
```

When the Phase Locked Loop Oscillator Mode is selected, the time-out sequence following a Power-on Reset is different from the other oscillator modes. In this case, a portion of the Power-up Timer is used to provide a fixed time-out that is sufficient for the PLL to lock to the main oscillator frequency. This PLL lock time-out (TPLL) is typically 2 ms and follows the oscillator start-up time-out (OST).

Power-Up Delay Summary

Two timers are used in controlling the power-up delays: the Power-up Delay Timer (PWRT) and the Oscillator Start-up Timer (OST). This duplication ensures that no external reset circuitry is required for most applications. Their joint action guarantees that the device is kept in RESET until both, the device power supply and the clock, are stable.

When the PLL is enabled (HSPLL oscillator mode), the Power-up Timer (PWRT) is used to keep the device in RESET for an extra nominal delay. This additional delay ensures that the PLL is locked to the crystal frequency.

9.2.4 Watchdog Timer

The Watchdog Timer was discussed in Section 2.1.5. In summary, the Watchdog Timer is an independent timer with its own clock source. Its purpose is to provide a mechanism by which the processor can recover from a software error that impedes program continuation, such as an endless loop. The WatchDog Timer is not designed to recover from hardware faults, such as a brown-out.

The hardware of the Watchdog Timer is independent of the PIC's internal clock. Its time-out period can range from approximately 18 milliseconds to 2.3 seconds, depending on whether the prescaler is used. According to Microchip, the Watchdog Timer is not very accurate and in the worst case scenario, the time-out period can extend to several seconds. When the WDT times out, the TO flag in the STATUS register is cleared and the program counter is reset to 0x000 so that the program restarts. Applications can prevent the reset by issuing the clrwdt instruction before the time-out period ends. When clrwdt executes, the WDT time-out period restarts.

The clrwdt and sleep instructions clear the WDT and the postscaler (if assigned to the WDT) and prevent it from timing out and generating a device RESET condition. The WDT has a postscaler field that can extend the WDT Reset period. The postscaler is selected by the value written to three bits in the CONFIG2H register during device programming. When a clrwdr instruction is executed and the postscaler is assigned to the WDT, the postscaler count will be cleared, but the postscaler assignment is not changed.

Watchdog Timer Uses

Not much information is available regarding the practical uses of the watchdog timer in any of the PIC microcontrollers, but it is clear that there is more to it than just restarting the counter with the clrwdt instruction. The timer is supposedly designed to detect software errors that can hang up a program, but how detects these errors and which conditions trigger the WDT operation is not clear from the information currently available. For example, an application that contains a long delay loop may find that the Watchdog Timer forces an untimely break out of the loop. The Watchdog Timer provides a powerful error-recovery mechanism but its use requires careful consideration of program conditions that could make the timer malfunction.

9.3 Hardware Timer-Counters

The PIC 18 family of microcontrollers has facilities and devices for controlling and manipulating time lapses in a program. These are most frequently required in timing, measuring, and counting operations. It is difficult to imagine an embedded application of any complexity that does not require some form of counting or timing. In some of the programs previously developed, we have provided a timed delay using a do-nothing loop that wastes a series of machine cycles. In the sections that follow, we investigate and expand the theory and use of delay loops and explore the use of built-in timing and counting circuits on the PIC 18F devices. The following are possible applications of the timing hardware:

1. Measuring and comparing the arrival time of an event
2. Generating a periodic interrupt
3. Measuring period and pulse width
4. Measuring the frequency and duty cycle of periodic signals
5. Generating specific waveforms
6. Establishing a time reference for an event
7. Counting events

The most frequently used modules in timing operations are the four (or five) hardware timers of the PIC 18F family, labeled Timer0 to Timer 3. and the Capture Compare and PWM module (CCP). Timer0, Timer1, and Timer3 are 8- or 16-bit timers while Timer2 is an 8-bit timer. 16-bit timers have internal registers that operate in the range 0 to 0xffff (0 to 65,535). 8-bit timers operate in the range 0 to 0xff (0 to 255). Timer2 and Timer4 use the system's internal clock as their clock source, while the other timers can also use an external clock signal.

In the remaider of the chapter we discuss the four timer modules available in the 18F452.

9.4 Timer0 Module

The basic timer facility on the PIC 18F family is known as the Timer0 module. It is described as a free-running timer, as a timer/counter, or simply as TMR0. Timer0 can be configured as an 8- or 16-bit device. It can be made to run off the internal timer or off an external one on the TOCKI pin. Its principal features are as follows:

1 Software selectable as an 8-bit or 16-bit timer/counter
2 Readable and writable
3. 8-bit software programmable prescaler
4. Clock source can be external or internal
5. Interrupt-on-overflow from 0xff to 0x00 in 8-bit or on 0xffff in 16-bit mode
6. Edge select for external clock

Control or Timer0 is mainly through the T0CON register shown in Figure 9.1.

bit 7   bit 0

TMR0ON	T08BIT	T0CS	T0SE	PSA	T0PS2	T0PS1	T0PS0

bit 7   **TMR0ON:** Timer0 On/Off Control bit
         1 = Enables Timer0
         0 = Stops Timer0
bit 6   **T08BIT:** Timer0 8-bit/16-bit Control bit
         1 = Timer0 is configured as an 8-bit timer/counter
         0 = Timer0 is configured as a 16-bit timer/counter
bit 5   **T0CS:** Timer0 Clock Source Select bit
         1 = Transition on T0CKI pin
         0 = Internal instruction cycle clock (CLKO)
bit 4   **T0SE:** Timer0 Source Edge Select bit
         1 = Increment on high-to-low transition on T0CKI pin
         0 = Increment on low-to-high transition on T0CKI pin
bit 3   **PSA:** Timer0 Prescaler Assignment bit
         1 = TImer0 prescaler is NOT assigned. Timer0 clock
             input bypasses prescaler.
         0 = Timer0 prescaler is assigned. Timer0 clock input
             comes from prescaler output.
bit 2-0 **T0PS2:T0PS0:** Timer0 Prescaler Select bits
         111 = 1:256 prescale value
         110 = 1:128 prescale value
         101 = 1:64 prescale value
         100 = 1:32 prescale value
         011 = 1:16 prescale value
         010 = 1:8 prescale value
         001 = 1:4 prescale value
         000 = 1:2 prescale value

Figure 9.1 *Timer0 control register (T0CON) bitmap.*

All bits in the T0CON register are readable and writeable.

The Timer0 module is the first peripheral device discussed in this book. Peripheral devices add specific functionality to the microcontroller. Learning to program the Timer0 module serves as an introduction to programming PIC 18F peripherals, of which there is a long list. Figure 9.2 is a block diagram of the Timer0 module in 8-bit mode.

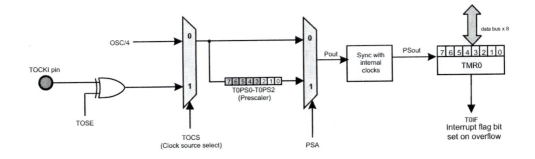

Figure 9.2 *Timer0 module bitmap.*

9.4.1 Timer0 Architecture

Several setup operations are required in programming the Timer0 module in its various modes. The various steps required for setting up the Timer0 module are as follows:

1. Enable the timer.

2. Select the 8- or 16-bit mode.

3. Select the internal or external clock source.

4. If the timer is used as a counter, then code must select whether the increment takes place on the falling or the rising edge of the signal.

5. Turn the prescaler function on or off.

6. If the prescaler is on, then select the prescaler value.

16-bit Mode Operation

Normally, applications will not change the timer mode once it has been selected. Nevertheless, code can change between 8- and 16-bit modes by carefully considering when interrupts are generated during counter rollover. Rules are as follows:

1. When Timer0 is changed from 8- to 16-bit mode on the same counting cycle as the rollover condition occurs, then no interrupt is generated.

2. When Timer0 is changed from 16- to 8-bit mode on the same counting cycle as the rollover condition occurs, then an interrupt is generated.

The high byte of the Timer0 counter (TMR0H) is not directly readable or writable by code. In fact, TMR0H is not the high byte of the timer/counter, but a buffered version of this byte. TMR0H is updated with the contents of the high byte of Timer0 during a read of the timer low byte (TMR0L). This design allows code to read all 16 bits of Timer0 without worrying that a rollover condition did not occur between the read of the high and low bytes. Code simply reads the low byte of Timer0, followed by a read of TMR0H, which contains the value in the high byte at the time that the low byte was read.

Writing to the high byte of Timer0 must take place through the TMR0H buffer register. In this case, Timer0 high byte is updated with the contents of TMR0H when a write occurs to TMR0L. This design allows code to update all 16 bits of Timer0 (high and low bytes) at the same time. When performing a write of TMR0, the carry is held off during the write of the TMR0L register. Writes to the TMR0H register only modify the holding latch, not the timer. The operation requires the following steps:

1. Load the TMR0H register.

2. Write to the TMR0L register.

Some instructions (bsf and bcf) are used to read the contents of a register, make changes to that content, and write the result back to the register. This sequence is known as a read-modify-write. With regard to the TMR0L register, the read cycle of the read-modify-write operation does not update the TMR0H register; therefore the TMR0H buffer remains unchanged. When the write cycle takes place, then the contents of TMR0H are placed in the high bytes of the Timer0 register.

The sample program Timer0_Delay.asm developed later in this chapter shows the setup and operation of Timer0 in 16-bit mode as well as reading and writing to the TMR0H and TMR0L registers. Many reports of bugs in the Timer0 16-bit mode found on the Internet are due to programs that have not followed the correct read/write sequence when accessing the Timer0 high byte.

Timer and Counter Modes

Timer0 can operate in a timer or a counter mode. The timer mode is selected by clearing the T0CS bit in the T0CON register. Without a prescaler, in timer mode the Timer0 module increments on every instruction cycle. If the TMR0 register is written, the increment is inhibited for the following two instruction cycles. Code can work around this by writing an adjusted value to the TMR0 register.

Counter mode is selected by setting the T0CS bit (T0CON register). In counter mode, Timer0 increments either on every rising or falling edge of the T0CKI pin. The edge is determined by the Timer0 Source Edge Select bit T0SE in the T0CON register. Clearing the T0SE bit selects the rising edge of the signal.

Timer0 Interrupt

When the interrupt flag bit is set, Timer0 generates an interrupt when the TMR0 register overflow. In the 8-bit mode, this takes place when the count goes from 0xff to 0x00. In the 16-bit mode, the interrupt is generated when the counter goes from 0xffff to 0x0000.

This interrupt overflow sets the TMR0IF bit in the INTCON register. The interrupt can be disabled by clearing the TMR0IE bit in the INTCON register. The TMR0IF flag bit must be cleared in software in the interrupt service routine. The TMR0 interrupt cannot awaken the processor from SLEEP, as the timer is shut off during SLEEP.

External Clock Source

When an external clock signal is selected, the Timer0 hardware must ensure that the clock signal can be synchronized with the internal clock.

When no prescaler is used, the external clock input is used instead of the prescaler output. When a prescaler is used, the external clock input is divided by the prescaler so that the prescaler output is symmetrical. For the external clock to meet the sampling requirement, the ripple-counter must be taken into account. Therefore, it is necessary for T0CKI to have a period of at least 4TSCLK (and a small RC delay) divided by the prescaler value. The only requirement on T0CKI high and low time is that they do not violate the minimum pulse width requirement. Because the prescaler output is synchronized with the internal clock, there is a small delay from the time the external clock edge occurs to the time the Timer0 module is actually incremented. The actual magnitude of this delay can be obtained from the devices' data sheets.

Timer0 Prescaler

Timer0 contains a prescaler that allows controlling the timer's rate by acting as a cycle divider. The PSA bit in the T0CON register (Prescaler Assignment Bit) allows turning the prescaler on and off.

Past errors in some PIC18 data sheets have created confusion regarding the action of the PSA bit. For example, the 18F Family Reference Manual (DS39513A) states on page 13-7 that "Setting the PSA bit will enable the prescaler." In that same document, the T0CON register bitmap shows that it is a value of 0 in the PSA bit that assigns the prescaler to Timer0. Actually, this is the case. Regarding Timer0, the PSA bit is active low, so a value of 0 turns on the prescaler while a value of 1 turns off the prescaler assignment.

The rate of the prescaler is determined by bits 0:2 in the T0CON register as shown in Figure 9.1. The 3-bit field allows selecting eight different prescaler rates: a value of 0x7 enables a 1:256 prescaler value while a value of 0x0 selects a prescaler rate of 1:2. The prescaler select bits are readable and writable but the prescaler count cannot be read or written. All instructions that write to the Timer0 register, such as clrf TMR0, bsf TMR0,x, movwf TMR0, and others) will clear the prescaler count if the prescaler has been enabled. However, writes to TMR0H do not clear the prescaler count because writing to the latch does not change the contents of Timer0. The prescaler is cleared by writing to TMR0L.

9.4.2 Timer0 as a Delay Timer

One of the simplest and most useful applications of the Timer0 module is as a simple delay timer. Two common techniques are available:

1. Polling the value in the timer counter register to detect when the counter rolls over

2. Enabling an interrupt that takes place when the counter rolls over

We begin by investigating the first case, that is, Timer0 registers are polled to implement a delay loop. Applications in which the Timer0 register is polled directly are said to use a free running timer. There are two advantage in free running timers over conventional delay loops:

1. The prescaler provides a way of slowing down the count.

2. The delay is independent of the number of machine cycles in the loop body.

These factors determine that, in most cases, it is easier to implement an accurate time delay using the Timer0 module than by counting instruction cycles.

Calculating the time taken by each counter iteration consists of dividing the clock speed by 4. For example, a 18F452 PIC running on a 4 MHz oscillator clock, increments the counter every 1 MHz. If the prescaler is not used, the counter register is incremented at a rate of 1 µs. or 1,000,000 times per second. If the prescaler is set to the maximum divisor value (256), then each increment of the timer takes place at a rate of 1,000,000/256 µs, which is approximately 3.906 ms. This is the slowest possible rate of the timer in a machine running at 4 MHz. It is often necessary to employ supplementary counters in order to achieve larger delays.

Recall that the timer register (TMR0) is both readable and writable. This makes possible several timing techniques; for example, code can set the timer register to an initial value and then count up until a predetermined limit is reached. Suppose that we define that the difference between the limit and the initial value is 100; then the routine will count 100 times the timer rate per beat.

As another example, consider a routine in 8-bit mode that allows the timer to start from zero and count up unrestricted. In this case, when the count reaches the maximum value (0xff), the routine would have introduced a delay of 256 times the timer beat rate. Now consider the case in which the maximum value (256) was used in the prescaler and the timer ran at a rate of 1,000,000 beats per second. This means that each timer beat will take place at a rate of 1,000,000/256, or approximately 3,906 timer beats per second. If now we develop a routine that delays execution until the maximum value has been reached in the counter register, then the delay can be calculated by dividing the number of beats per second (3,906) by the number of counts in the delay loop. In this case, 3,906/256 results in a delay of approximately 15.26 iterations of the delay routine per second.

A general formula for calculating the number of timer beats per second is as follows:

$$T = \frac{C}{4PR}$$

where T is the number of clock beats per second, C is the system clock speed in Hz, P is the value stored in the prescaler, and R is the number of iteration, counted in the tmr0 register. The range of both P and R in this formula is from 1 to 256. Also notice that the reciprocal of T (1/T) gives the time delay, in seconds, per iteration of the delay routine.

Long Delay Loops

In the previous section we saw that even when using the largest possible prescaler and counting the maximum number of timer beats, the longest timer delay that can be obtained in a 4-MHz system is approximately 1/15th second. Consequently, applications that measure time in seconds or in minutes must find ways of keeping count of large number of repetitions of the timer beat.

In implementing counters for larger delays we must be careful not to introduce round-off errors. In the previous example, a timer cycles at the rate of 15.26 times per second. The closest integer to 15.25 is 15, so if we now set up a seconds counter that counts 15 iterations, the counter would introduce an error of approximately 2%. Considering that each iteration of the timer contains 256 individual beats, there are 3,906.25 individual timer beats per second at the maximum pre-scaled rate.

This means that if we were to implement a counter to keep track of individual pre-scaled beats, instead of timer iterations, the count would proceed from 0 to 3,906 instead of from 0 to 15. Approximating 3,906.25 by the closest integer (3,906) introduces a much smaller round-off error than approximating 15.26 with 15. In this same example, we could eliminate the prescaler so that the timer beats at the clock rate, that is, at 1,000,000 beats per second. In this option a counter that counts from 0 to 1,000,000 would have no intrinsic error due to round-off. Which solution is more adequate depends on the accuracy required by the application and the acceptable complexity of the code.

Delay Accuracy Issues

The actual implementation of a delay routine based on multi-byte counters presents some difficulties. If the timer register (TMR0)is used to keep track of timer beats, then detecting the end of the count presents a subtle problem. Our program could detect timer overflow reading the tmr0 and testing the zero flag in the status register. Because the movf instruction affects the zero flag, one could be tempted to code:

```
wait:
    movf      tmr0,w    ; Timer value into w
    btfss          status,z  ; Was it zero?
    goto      wait
; If this point is reached tmr0 has overflowed
```

But there is a problem: the timer ticks as each instruction executes. Because the goto instruction takes two machine cycles, it is possible that the timer overflows while the goto instruction is in progress; therefore the overflow condition would not be detected. One possible solution found in the Microchip documentation is to check for less than a nominal value by testing the carry flag, as follows:

```
wait1:
    movlw         0x03      ; 3 to w
    subwf         TMR0,w    ; Subtract w - tmr0
    btfsc         status,c  ; Test carry
    goto      wait1
```

One adjustment that is sometimes necessary in free running timers results from the fact that when the TMR0 register is written, the count is inhibited for the following two instruction cycles. Software can usually compensate for the skip by writing an adjusted value to the timer register. If the prescaler is assigned to timer0, then a write operation to the timer register determines that the timer will not increment for four clock cycles.

Black–Ammerman Method

A more elegant and accurate solution has been described by Roman Black in a Web article titled Zero-error One Second Timer. Black credits Bob Ammerman with the suggestion of using Bresenham's algorithm for creating accurate PIC timer periods. In the Black–Ammerman method, the counter works in the background, either by polling or interrupt-driven. In either case, the timer count value is stored in a 3-byte register which is decremented by the software.

In their interrupt-driven version, TMR0 generates an interrupt whenever the counter register overflows, that is, every 256th timer beat (assuming no prescaler). The interrupt handler routine decrements the mid-order register that holds the 3-byte timer count. This is appropriate because every unit in the mid-order register represents 256 units of the low-order counter, which in this case is the tmr0 register. If the mid-order register underflows when decremented, then the high-order one is decremented. If the high-order one underflows, then the count has reached zero and the delay ends. Because the counter is interrupt-driven, the processor can continue to do other work in the foreground.

An even more ingenious option proposed by Black is a background counter that does not rely on interrupts. This is accomplished by introducing a 1:2 delay in the timer by means of the prescaler. Because now the timer beats at one-half the instruction rate, 128 timer cycles will be required for one complete iteration at the full instruction rate. By testing the high-order bit of the timer counter, the routine can detect when the count has reached 128. At that time, the mid-range and high-range counter variables are updated (as in the non-interrupt version of the software described in the previous paragraph). The high-order bit of the timer is then cleared, but the low-order bits are not changed. This allows the timer counter not to lose step in the count, which remains valid until the next time the high-order bit is again set. During the period between the updating of the 3-byte counter and the next polling of the timer register, the program can continue to perform other tasks.

Delays with 16-Bit Timer0

In many cases the complications mentioned in the previous sections can be avoided by running Timer0 in the 16-bit mode. For example, if the maximum delay that can be obtained in 8-bit mode, given a machine running at 4MHz, is 1/15th second (0.0666 second), then switching to 16-bit mode makes the maximum delay of approximately 17 seconds.

9.4.3 Counter and Timer Programming

Software routines that use the Timer0 module range in complexity from simple, approximate delay loops to configurable, interrupt-driven counters that must meet very high timing accuracy requirements. When the time period to be measured does not exceed the one that can be obtained with the prescaler and the timer register count, then the coding is straightforward and the processing is uncomplicated. However, if this is not the case, the following elements should be examined before attempting to design and code a Timer0-based routine:

1. What is the required accuracy of the timer delay?

2. Does the program suspend execution while the delay is in progress, or does the application continue executing in the foreground?

3. Can the timer be interrupt-driven or must it be polled?

4. Will the delay be the same on all calls to the timer routine, or must the routine provide delays of different magnitudes?

5. How long must the delay last?

In this section we explore several timer and counter routines of different complexity and requirements. The first one uses the Timer0 module as a counter. Later we develop a simple delay loop that uses the timer0 register instead of the do-nothing instruction count covered previously. We conclude with an interrupt-driven timer routine that can be changed to implement different delays.

Programming a Counter

The 18F452 PIC can be programmed so that port RA4/TOCKI is used to count events or pulses by initializing the Timer0 module as a counter. When interrupts are not used, the process requires the following preparatory steps:

1. Port A, line 4, (RA4/TOCKI) is defined for input.

2. The Timer0 register (TMR0) is cleared.

3. The Watchdog Timer internal register is cleared by means of the clrwdt instruction.

4. The T0CON register bits PSA and PSO:PS2 are initialized if the prescaler is to be used.

5. The T0CON register bit TOSE is set so as to increment the count on the high-to-low transition of the port pin if the port source is active low. Otherwise the bit is cleared.

6. The T0CON register bit TOCS is set to select action on the RA4/TOCKI pin.

Once the timer is set up as a counter, any pulse received on the RA4/TOCKI pin that meets the restrictions mentioned earlier is counted in the TMR0L and TMR0H registers. If Timer0 is set in the 8-bit mode, then the TMR0H register is not used. Software can read and write to the Timer0 registers in order to obtain or change the event count. If the timer interrupt is enabled when the timer is defined as a counter, then an interrupt takes place every time the counter overflows, that is, when the count cycles from 0xff to 0x00 or from 0xffff to 0x0000 according to the active mode.

Timer0_as_Counter.asm Program

The program named Timer0_as_Counter.asm, listed later in this chapter and contained in this book's online software package, uses the circuits mentioned in the previous paragraph to demonstrate the programming of the Timer0 module in the counter mode. The program detects and counts action on DIP switch #3, wired to port RA4/TOCKI. The value of the count in hex digits in the range 0x00 to 0x0f is displayed in the seven-segment LED connected to Port B.

The location and use of the code table were discussed in Sections 7.3 and 7.4. The main() function starts by selecting bank 0, initializing Port A for digital operations, and trissing Port A for input and Port C for output as in several preceding programs. Code first clears the watchdog time and the TMR0L register, and then proceeds as follows:

```
;==================================
; Check value in TMR0L and display
;==================================
; Every closing of DIP switch # 3 (connected to line
```

```
; RA4/TOCKI) adds one to the value in the TMR0L register.
; Loop checks this value, adjusts to the range 0 to 15
; and displays the result in the seven-segment LED on
; port B
checkTmr0:
        movf      TMR0L,w       ; Timer register to w
; Eliminate four high order bits
        andlw     b'00001111'   ; Mask off high bits
; At this point the w register contains a 4-bit value
; in the range 0 to 0xf. Use this value (in w) to
; obtain seven-segment display code
        call      codeTable
        movwf     PORTC         ; Display switch bits
        goto      checkTmr0
```

Notice that the program provides no way of detecting when the count exceeds the displayable range. This means that no display update takes place as the timer cycles from binary 00001111 to binary 11111111.

A Timer/Counter Test Circuit

Either the circuit in Figure 7.2 or the or Demo Board 18F452-A (in Figure 7.3) can be used to demonstrate the Timer0_as_Counter program. Both circuits have a seven-segment LED wired to lines RC0:RC6 and a DIP switch wired to Port A, line 4, which is the RA4/T0CKI line. By selecting the counter mode of Timer0, any action on the T0CKI line will be reflected in the TMR0x registers.

Timer0 _Delay.asm Program

One of the simplest uses of the Timer0 module is to implement a delay loop. In this case the Timer0 module is initialized to use the internal clock by clearing the TOCS bit of the T0CON register. If the prescaler is to be used, the PSA bit is cleared and the desired pre-scaling is selected by means of bits 2:0 of the T0CON register. Either the circuit in Figure 7.2 or the or Demo Board 18F452-A (in Figure 7.3) can be used to demonstrate a simple application that uses Timer0 as a delay timer. Both circuits have eight LEDs wired to lines RC0:RC7.

The program named Timer0_Delay.asm, listed later in this chapter and contained in this book's online software package, uses a timer-based delay loop to flash in sequence eight LEDs that display the binary values from 0x00 to 0xff. The delay routine executes in the foreground, so that processing is suspended while the count is in progress. The program executes in the 16-bit mode so the code can demonstrate the issues related to reading and writing to the 16-bit registers TMR0L and TMR0H. These issues were discussed in Section 9.6.1.

Setting up Timer0 as a delay counter requires selecting the required bits in the T0CON register. The following code fragment shows the program's initialization routine to set up the timer.

```
;==============================
;  setup Timer0 as delay timer
;==============================
        clrf      TMR0H         ; Clear high latch
        clrf      TMR0L         ; Write both bytes
        clrwdt                  ; Clear watchdog timer
```

```
; Setup the T0CON register
;                  |------------- On/Off control
;                  |               1 = Timer0 enabled
;                  ||------------ 8/16 bit mode select
;                  ||              0 = 16-bit mode
;                  |||----------- Clock source
;                  |||             0 = Internal clock
;                  ||||---------- Source edge select
;                  ||||            1 = high-to-low
;                  |||||--------- Prescaler assignment
;                  |||||           1 = prescaler not assigned
;                  |||||||| ----- No prescaler
;                  ||||||||
      movlw    b'10011000'
      movwf    T0CON
```

The previous code snippet starts by clearing both counter registers. This requires first clearing the buffer register TMR0H and then the low-byte register TMR0L. This last write operation updates both the high and the low byte of the timer simultaneously. The bits selected in the T0CON register enable the timer, select the 16-bit mode, enable the clock source as the internal clock, activate the signal edge in high-to-low mode, while the prescaler is left unassigned.

The program then proceeds to an endless loop that increments the value in the Port C register by one. Because Port C is wired to eight LEDs on the demo circuit, the display shows the binary value in the port. The routine calls a procedure that implements a delay in a do-nothing loop that uses Timer0 overflow. Code is as follows:

```
;=================================
;      endless loop calling
;        delay routiney
;=================================
; Display Port C count on LEDs
showLEDs:
      incf     PORTC,f      ; Add one to register
      call     tmr0_delay   ; Delay routine
      goto     showLEDs
;=================================
;      Timer0 delay routine
;=================================
tmr0_delay:
cycle:
      movf     TMR0L,w      ; Read low byte to latch
                            ; high byte
      movf     TMR0H,w      ; Now read high byte
      sublw    0xff         ; Subtract maximum count
      btfss    STATUS,Z     ; Test zero flag
      goto     cycle
; Reset counter
      clrf     TMR0H        ; Clear high byte buffer
      clrf     TMR0L        ; Write both low and high
      return
      end
```

The delay routine is the procedure named tmr0_delay. To make the code more readable, we have added a second label named cycle at this same address. The code reads the high byte of the timer, then the low one (this updates both bytes.) The value 0xff is then subtracted from the high byte. The subtraction returns zero and sets the zero flag if the value in TMR0H is also 0xff. If the test is true, then the goto cycle instruction is skipped, both timer registers are cleared, and execution returns to the caller. If the test is false, then the timer register test loop repeats. In a 4-MHz test circuit the entire cycle takes approximately 15 seconds.

A Variable Time-Lapse Routine

A variable time-lapse routine can be designed so that it can be adjusted to produce delays within a certain time range. Such a procedure would be a useful tool in a programmer's library. In previous sections we have developed delay routines that do so by counting timer pulses. This same idea can be used to develop a routine that can be adjusted so as to produce accurate delays within a certain range.

The routine itself can be implemented to varying degrees of sophistication regarding the control parameters. One implementation could receive the desired time lapse as parameters passed by the caller. Another option would be a procedure that reads the desired time lapse from program constants. In the program named Timer0_VarDelay.asm listed later in this chapter and contained in this book's software, we develop a procedure in which the desired time delay is loaded from three constants defined in the source. These constants contain the values that are loaded into local variables as they represent the desired wait period in machine cycles. Using machine cycles instead of time units (such as microseconds or milliseconds) the procedure becomes easily adaptable to devices running at different clock speeds. Because each PIC instruction requires four clock cycles, the device's clock speed in Hz is divided by four in order to determine the number of machine cycles per time unit.

For example, a processor equipped with a 4-MHz clock executes at a rate of 4,000,000/4 machine cycles per second, that is, 1,000,000 instruction cycles per second. To produce a one-quarter second delay requires a wait period of 1,000,000/4 or 250,000 instruction cycles. By the same token, an 18f452 running at 8 MHz executes 2,000,000 instructions per second. In this case, a one-quarter second delay would require waiting 500,000 instruction cycles.

Timer0_VarDelay.asm Program

The program titled Timer0_VarDelay.asm, listed later in this chapter and contained in the book's software package, uses timer0 to produce a variable-lapse delay. As previously described, the delay is calculated based on the number of machine cycles necessary for the desired wait period. The program uses a variation of the Black-Ammerman method described earlier in this chapter. Code requires a prescaler of 1:2 so that each timer iteration takes place at one-half the clock rate. This scheme simplifies using the Timer0 beat as an iteration counter. After initializing Port C for output, the program sets or clears the T0CON register bits as follows:

```
;=================================
;   setup Timer0 as counter
```

```
;          8-bit mode
;================================
; Prescaler is assigned to Timer0 and initialzed
; to 2:1 rate
; Setup the T0CON register
;                 |------------- On/Off control
;                 |                 1 = Timer0 enabled
;                 ||------------ 8/16 bit mode select
;                 ||                1 = 8-bit mode
;                 |||----------- Clock source
;                 |||               0 = internal clock
;                 ||||---------- Source edge select
;                 ||||              1 = high-to-low
;                 |||||--------- Prescaler assignment
;                 |||||             0 = prescaler assigned
;                 ||||||||------- Prescaler select
;                 ||||||||           1:2 rate
      movlw      b'11010000'
      movwf      T0CON
; Clear registers
      clrf       TMR0L
      clrwdt               ; Clear watchdog timer
```

The constants that define the time lapse period are entered in #define statements so they can be easily edited to accommodate other delays and processor speeds. In a 4-MHz system, a delay of one-half second requires 500,000 timer cycles, while a delay of one-tenth second requires a count of 10,000. Because this delay value must be entered in three 1-byte variables, the value is converted to hexadecimal so it can be installed in three constants; for example,

```
1,000,000 = 0x0f4240 = one second at 4MHz
  500,000 = 0x07a120 = one-half second at 4MHz
   10,000 = 0x002710 = one-tenth second at 4MHz
```

For example, values for one-half second are installed in constants as follows:

```
500,000 = 0x07 0xa1 0x20
          ---- ---- ----
           |    |    |___ lowCnt
           |    |_____ midCnt
           |_____ highCnt
```

Code can read these constants and move them to local variables at the beginning of each timer cycle, as in the following code fragment

```
;================================
;   set register variables
;================================
; Procedure to initialize local variables for a
; delay period defined in local constants highCnt,
; midCnt, and lowCnt.
setVars:
      movlw      highCnt      ; From constants
      movwf      countH
      movlw      midCnt
      movwf      countM
```

```
      movlw     lowCnt
      movwf     countL
      return
```

The actual delay routine is a variation of the Black-Ammerman method described in Section 9.4.2. In this case, a background counter is made possible by introducing a 1:2 timer delay by means of the prescaler. This delay makes the timer beat run at one-half the instruction rate, that is, 128 timer cycles represent one complete timer cycle. By testing the high-order bit of the timer counter (TMR0L in 8-bit mode), the routine easily detects when the count has reached 128. At that time, the mid-range and high-range counter variables are updated by decrementing the counters, thus taking care of possible overflows. When the housekeeping has concluded, the high-order bit of the timer is cleared, but the low-order bits are not changed. Because the count is kept in the low-order bits during housekeeping operations, the timer counter does not lose step, which remains valid until the next time the high-order bit is again set.

In implementing this scheme, the TMR0L register provides the low-order level of the count. Because the counter counts up from zero, code must pre-install a value in the counter register that represents one-half the number of timer iterations (prescaler is in 1:2 mode) required to reach a count of 128. For example, if the value in the low counter variable is 140, then

```
      140/2 = 70
      128 - 70 = 58
```

Because the timer starts counting up from 58, when the count reaches 128, 140 timer beats would have elapsed. The formula for calculating the value to pre-install in the low-level counter is as follows:

```
      Value in TMR0L = 128 - (x/2)
```

where x is the number of iterations in the low-level counter variable.

Code is as follows:

```
;===================================
;   variable-lapse delay procedure
;         using Timer0
;===================================
; ON ENTRY:
;         Variables countL, countM, and countH hold
;         the low-, middle-, and high-order bytes
;         of the delay period, in timer units
TM0delay:
; Formula:
;            Value in TMR0L = 128 - (x/2)
; where x is the number of iterations in the low-level
; counter variable
; First calculate xx/2 by bit shifting
    rrncf   countL,f    ; Divide by 2
; now subtract 128 - (x/2)
    movlw   d'128'
```

```
; Clear the borrow bit (mapped to Carry bit)
    bcf       STATUS,C
    subfwb    countL,w
; Now w has adjusted result. Store in TMR0L
    movwf     TMR0L
; Routine tests timer overflow by testing bit 7 of
; the TMR0L register.
cycle:
    btfss     TMR0L,7      ; Is bit 7 set?
    goto      cycle        ; Wait if not set
; At this point TMR0 bit 7 is set
; Clear the bit
    bcf       TMR0L,7         ; All other bits are preserved
; Subtract 256 from beat counter by decrementing the
; mid-order byte
    decfsz    countM,f
    goto      cycle        ; Continue if mid-byte not zero
; At this point the mid-order byte has overflowed.
; High-order byte must be decremented.
    decfsz    countH,f
    goto      cycle
; At this point the time cycle has elapsed
    return
```

Interrupt-Driven Timer

Interrupt-driven timers and counters have the advantage over polled routines that the time lapse counting takes place in the background, which makes it possible for an application to continue to do other work in the foreground. Developing a timer routine that is interrupt-driven presents no major programming challenges. The initialization consists of configuring the OPTION and the INTCON register bits for the task at hand. In the particular case of an interrupt-driven timer, the following are necessary:

1. The external interrupt flag (INTF in the INTCON Register) must be initially cleared.

2. Global interrupts must be enabled by setting the GIE bit in the INTCON register.

3. The timer0 overflow interrupt must be enabled by setting the T0IE bit in the INTCON register.

In the present example program, named Timer0_VarInt, the prescaler is not used with the timer, so the initialization code sets the PSA bit in the OPTION register in order to have the prescaler assigned to the Watchdog Timer. The following code fragment is from the Timer0_VarInt program:

```
main:
; Set BSR for bank 0 operations
    movlb     0              ; Bank 0
; Initialize all lines in PORT C for output
    movlw     B'00000000'    ; 0 = output
    movwf     TRISC          ; Port C tris register
    movwf     PORTC
;==============================
;  setup Timer0 as counter
;       8-bit mode
;==============================
```

```
        bcf      INTCON,TMR0IE
; Setup the T0CON register
;                 |------------- On/Off control
;                 |              1 = Timer0 enabled
;                 ||------------ 8/16 bit mode select
;                 ||             1 = 8-bit mode
;                 |||----------- Clock source
;                 |||            0 = internal clock
;                 ||||---------- Source edge select
;                 ||||           1 = high-to-low
;                 |||||--------- Prescaler assignment
;                 |||||          1 = prescaler not assigned
;                 ||||||||| ----- Prescaler select
;                 ||||||||        1:2 rate
        movlw    b'11011000'
        movwf    T0CON
; Clear registers
        clrf     TMR0L
        clrwdt                  ; Clear watchdog timer
;===============================
; Set up for Timer0 interupt
;===============================
; Disable interrupt priority levels in the RCON register
; setting up the mid range compatibility mode
        bsf      RCON,IPEN       ; Enable interrupt priorities
; INTCON register initialized as follows:
; (IPEN bit is clear)
;                 |------------ high-priority interrupts
;                 ||----------- low-priority peripheral
;                 |||---------- timer0 overflow interrupt
;                 ||||--------- external interrupt
;                 |||||-------- port change interrupt
;                 ||||||------- overflow interrupt flag
;                 |||||||------ external interrupt flag
;                 ||||||||----- RB4:RB7 interrupt flag
        movlw    b'10100000'
        movwf    INTCON
; Set INTCON2 for falling edge operation
        bcf      INTCON2,INTEDG0
; Re-enable timer0 interrupt
        bsf      INTCON,TMR0IE   ; Activate Timer0 interrupt
        bcf      INTCON,TMR0IF   ; Clear interrupt f
```

As in the program Timer0_VarDelay developed previously in this chapter, the timer operates by decrementing a 3-byte counter that holds the number of timer beats required for the programmed delay. In the case of the Timer0_VarInt program, the routine that initializes the register variables for a one-half second delay also makes the adjustment so that the initial value loaded into the tmr0 register is correctly adjusted. The code is as follows:

```
;===============================
;   set register variables for
;       one-half second delay
;===============================
; Procedure to initialize local variables for a
; delay of one-half second on a 16F84 at 4 MHz.
; Timer is setup for a 500,000 clock beats as
; follows: 500,000 = 0x07 0xa1 0x20
```

```
; 500,000 = 0x07 0xa1 0x20
;               ---- ---- ----
;               |    |    |___ countL)
;               |    |_____ countM
;               |_____ countH
onehalfSec:
     movlw      0x07
     movwf      countH
     movlw      0xa1
     movwf      countM
     movlw      0x20
     movwf      countL
; The tmr0 register provides the low-order level of
; the count. Because the counter counts up from zero,
; in order to ensure that the initial low-level delay
; count is correct, the value 256 - xx must be calculated
; where xx is the value in the original countL variable.
     movf       countL,w ; w holds low-order byte
     sublw      d'256'
; Now w has adjusted result. Store in tmr0
     movwft     mr0
     return
```

The interrupt service routine in the Timer0_VarInt program receives control when the tmr0 register overflows, that is, when the count goes from 0xff to 0x00. The service routine then proceeds to decrement the mid-range counter register and adjust, if necessary, the high-order counter. If the count goes to zero, the handler toggles the LED on port B, line 0, and re-initializes the counter variables by calling the onehalfSec procedure described previously. The interrupt handler is coded as follows:

```
;==========================================================
;            Interrupt Service Routine
;==========================================================
; Service routine receives control when the timer
; register tmr0 overflows, that is, when 256 timer beats
; have ellapsed
IntServ:
; First test if source is a timer0 interrupt
     btfss      INTCON,toif   ; TOIF is timer0 interrupt
     goto       notTOIF       ; Go if not RB0 origin
; If so clear the timer interrupt flag so that count continues
     bcf        INTCON,toif   ; Clear interrupt flag
; Save context
     movwf      old_w         ; Save w register
     swapf      STATUS,w  ; STATUS to w
     movwf      old_status    ; Save STATUS
;=========================
;    interrupt action
;=========================
; Subtract 256 from beat counter by decrementing the
; mid-order byte
     decfsz     countM,f
     goto       exitISR       ; Continue if mid-byte not zero
; At this point the mid-order byte has overflowed.
; High-order byte must be decremented.
     decfsz     countH,f
     goto       exitISR
```

```
; At this point count has expired so the programmed time
; has ellapsed. Service routine turns the LED on line 0,
; port B on and off at every conclusion of the count.
; This is done by xoring a mask with a one-bit at the
; port B line 0 position
    movlw     b'00000001'        ; Xoring with a 1-bit produces
                                 ; the complement
    xorwf     portb,f            ; Complement bit 2, port B
; Reset one-half second counter
    call      onehalfSec
;==========================
;        exit ISR
;==========================
exitISR:
; Restore context
    swapf     old_status,w  ; Saved status to w
    movfw     STATUS     ; To STATUS register
    swapf     old_w,f    ; Swap file register in itself
    swapf     old_w,w    ; re-swap back to w
; Return from interrupt
notTOIF:
    retfie
```

Notice that one of the initial operations of the service routine is to clear the TOIF bit in the INTCON register. This action reenables the timer interrupt and prevents counting cycles from being lost. Because the interrupt is generated every 256 beats of the timer, there is no risk that by enabling the timer interrupt flag a reentrant interrupt will take place.

The interrupt-based timer program named Timer0_VarInt can be tested either on the circuit in Figure 7.2 or the or Demo Board 18F452-A (in Figure 7.3).

9.5 Other Timer Modules

The PIC 18F family of microcontrollers provide either three or four timer modules in addition to Timer0. These are designated Timer1, Timer2, Timer3, and Timer4 modules. The programming and application of these other timer modules are similar to that of the Timer0 module previously described. The main difference between Timer0 and the other three timer modules relate to the available clock sources and special features that provide interaction with other hardware modules. The Timer4 module is only available in some specific devices of the PIC 18F family.

9.5.1 Timer1 Module

The Timer1 module is a 16-bit device that can perform timing and counting operations. It contains two 8-bit registers labeled TMR1H and TMR1L. Both registers are readable and writable. The register pair increments from 0000H to FFFFH and then rolls over back to 0000H. Timer1 can be enabled to generate an interrupt on overflow of the timer registers. In this case, the interrupt is reflected in the interrupt flag bit TMR1IF. The interrupt is enabled by setting the setting the TMR1IE interrupt enable bit.

Timer1 can operate in one of three modes:

1. As a synchronous timer

2. As a synchronous counter

3. As an asynchronous counter

The operating mode is selected by clock select bit, TMR1CS (T1CON register), and the synchronization bit, T1SYNC. In the timer mode, Timer1 increments every instruction cycle. In the counter modes, it increments on every rising edge of the external clock input pin T1OSI. Timer1 is turned on and off by means of the TMR1ON control bit in the T1CON register.

Timer1 has an internal reset input that can be generated by a CCP module as well as the capability of operating off an external crystal. When the Timer1 oscillator is enabled (T1OSCEN is set), the T1OSI and T1OSO pins become inputs and their corresponding TRIS values are ignored. Figure 9.3 shows the bitmap of the Timer1 control register (T1CON.)

```
 bit 7                                                               bit 0
```

RD16	–	T1CKPS1	T1CKPS2	T1OSCEN	T1SYNC	TMR1CS	TMR1ON

```
bit 7     Rd16: 16-bit Read/Write Mode Enable bit
              1 = Enables register Read/Write in one 16-bit operation
              0 = Enables register Read/Write in two 8-bit operations
bit 6     Unimplemented:
bit 5:4   T1CKPS1:T1CKPS0: Timer1 Input Clock Prescale Select bits
              11 = 1:8 Prescale value
              10 = 1:4 Prescale value
              01 = 1:2 Prescale value
              00 = 1:1 Prescale value
bit 3     T1OSCEN: Timer1 Oscillator Enable bit
              1 = Timer1 oscillator is enabled
              0 = Timer1 oscillator is turned off
bit 2     T1SYNC: Timer1 External Clock Input Synchronization Select
              When TMR1CS = 1:
              1 = Do not synchronize external clock input
              0 = Synchronize external clock input
              When TMR1CS = 0
              Timer1 uses the internal clock
bit 1     TMR1CS: Timer1 Clock Source Select bit
              1 = External clock from pin T1OSO/T13CKI (rising edge)
              0 = Internal clock at FOSC/4
bit 0     TMR1ON: Timer1 On bit
              1 = Timer1 enabled
              0 = Timer1 stopped
```

Figure 9.3 *Timer1 control register (T1CON) bitmap.*

Timer1 in Timer Mode

Timer1 is set in timer mode by clearing the TMR1CS (T1CON register) bit (see Figure 9.3). In the timer mode the input clock to the timer is the processor's main clock at FOSC/4. In this mode, the synchronize control bit, T1SYNC (T1CON register), has no effect because the internal clock is always synchronized.

Timer1 in Synchronized Counter Mode

Synchronized counter mode is selected by setting the TMR1CS bit (see Figure 9.3). In this mode, the timer increments on every rising edge of input signal on the T1OSI pin (when the Timer1 oscillator enable bit (T1OSCEN) is set) or the T1OSO/T13CKI pin (when the T1OSCEN bit is cleared.) If the T1SYNC bit is cleared, then the external clock input is synchronized with internal phase clocks. During SLEEP mode, Timer1 will not increment even if the external clock is present, because the synchronization circuit is shut off. The prescaler, however, will continue to increment.

External Clock Input Timing in Synchronized Mode

When Timer1 is set to use an external clock input in synchronized counter mode, it must meet the following requirements:

1. There is a delay in the actual incrementing of TMR1 after synchronization.

2. When the prescaler is 1:1, the external clock input is the same as the prescaler output.

3. The synchronization of T1CKI with the internal phase clocks is accomplished by sampling the prescaler output on alternating TSCLK clocks of the internal phase clocks. Therefore, it is necessary for the T1CKI pin to be high for at least 2TSCLK (and a small RC delay) and low for at least 2TSCLK (and a small RC delay).

4. When a prescaler other than 1:1 is used, the external clock input is divided by the asynchronous prescaler so that the prescaler output is symmetrical.

5. In order for the external clock to meet the sampling requirement, the prescaler counter must be taken into account. Therefore, it is necessary for the T1CKI pin to have a period of at least 4TSCLK (and a small RC delay) divided by the prescaler value.

6. Finally, the T1CKI pin high and low times cannot violate the minimum pulse width requirements.

Timer1 Read and Write Operations

Timer1 read and write modes allow the 16-bit timer register to be read/written as two 8-bit registers or as one 16-bit register. The mode is selected by means of the RD16 bit. When the RD16 control bit (T1CON register) is set (see Figure 9.3), the address for TMR1H is mapped to a buffer register for the high byte of Timer1. This determines that a read from TMR1L will load the contents of the high byte of Timer1 into the Timer1 high byte buffer. This scheme makes it possible to accurately read all 16 bits of Timer1 without having to determine if a rollover took place when a read of the high byte was followed by a read of the low byte.

16-bit Mode Timer1 Write

As is the case with a read operation, a write to the high byte of Timer1 must also take place through the TMR1H buffer register. Therefore Timer1 high byte is updated with the contents of TMR1H when a write occurs to TMR1L. This allows writing all 16 bits to both the high and low bytes of Timer1 in a single operation. Figure 9.4 shows the architecture of the Timer1 when configured for 16-bit Read/Write mode.

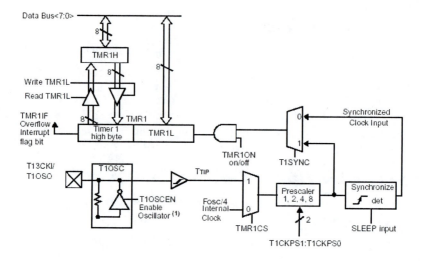

Figure 9.4 *Timer1 block diagram.*

Notice in Figure 9.4 that the high byte of Timer1 is not directly readable or writable in the 16-bit mode. Instead, all reads and writes take place through the Timer1 high byte buffer register. Also notice that writes to TMR1H do not clear the Timer1 prescaler.

16-bit Read-Modify-Write

Read-modify-write instructions, such as BSF and BCF, read the contents of a register, make the appropriate changes, and then place the result back into the same register. When Timer1 is configured in 16-bit mode, the read portion of a read-modify-write instruction of TMR1L will not update the contents of the TMR1H buffer. The TMR1H buffer will remain unchanged. When the write of TMR1L portion of the instruction takes place, the contents of TMR1H are placed into the high byte of Timer1.

Reading and Writing Timer1 in Two 8-bit Operations

When Timer1 is in Asynchronous Counter Mode for 16-bit operations (RD16 = 1), the hardware ensures a valid read of TMR1H or TMR1L. However, reading the 16-bit timer in two 8-bit values (RD16 = 0) poses the problem of a possible timer overflow between the reads. For write operations, the program can stop the timer and write the desired values. Turning off the timer prevents a write contention that could occur when writing to the timer registers while the register is incrementing. On the other hand, reading may produce an unpredictable value in the timer register and requires special care in some cases. This happens because two separate reads are required to read the entire 16-bits.

The following code fragment shows a routine to read the 16-bit timer value without experiencing the timer overflow issues previously mentioned. This scheme is useful if the timer cannot be stopped.

```
; Reading a 16-bit timer
; Code assumes the variables named tmph and tmpl
; All interrupts are disabled
    movf    TMR1H,w  ; Read high byte
```

```
        movwf    tmph
        movf     TMR1L,w    ; Read low byte
        movwf    tmpl
        movf     TMR1H,w    ; Read high byte
        subwf    tmph,w     ; Subtract 1st read and 2nd read
        btfsc    STATUS,z   ; is result = 0 ?
        goto     CONTINUE   ; good 16-bit read
; TMR1L may have rolled over between the read of the high
; and low bytes.  Reading the high and low bytes now will
; read a good value.
        movf     TMR1H,w    ; Read high byte
        movwf    TMPH
        movf     TMR1L, w   ; Read low byte
        movwf    TMPL
CONTINUE:
; Code continues at this label
```

Writing a 16-bit value to the 16-bit TMR1 register is straightforward. First the TMR1L register is cleared to ensure that there are many Timer1 clock/oscillator cycles before there is a rollover into the TMR1H register. The TMR1H register is then loaded, and then the TMR1L register, as shown in the following code fragment.

```
; Writing a 16-bit timer
; All interrupts are disabled
; Code assumes the variables names hi_byte and low_byte
    clrf     TMR1L      ; Clearing the low byte
                        ; to ensure no rollover
                        ; into TMR1H
    movlw    hi_byte    ; Value to load into tmr1h
    movwf    TMR1H,f    ; Write high byte
    movlw    lo_byte    ; value to load into TMR1L
    movwf    TMR1L,f    ; Write low byte
; re-enable the interrupt (if required)
; Code continues here
```

9.5.2 Timer2 Module

The Timer2 is an 8-bit timer with a prescaler, a postscaler, and a period register. By using the prescaler and postscaler at their maximum settings it is possible to obtain a time period equal to the one of a 16-bit timer. Timer2 is designed to be used as the time-base for the PWM module. Figure 9.5 IS a block diagram of Timer2.

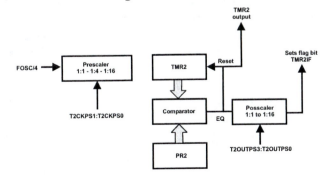

Figure 9.5 *Timer2 block diagram.*

In Figure 9.5 note that the postscaler counts the number of times that the TMR2 register matched the PR2 register. This can be useful in reducing the overhead of the interrupt service routine on the CPU performance. Figure 9.6 is a bitmap of the T2CON register.

```
 bit 7                                                                    bit 0
```

| - | TOUTPS3 | TOUTPS2 | TOUTPS1 | TOUTPS0 | TMR2ON | T2CKPS1 | T2CKPS0 |

```
bit 7      Unimplemented (reads as 0)

bit 6-3    TOUTPS3:TOUTPS0: Postscale select bits
           0000 = 1:1 Postscale
           0001 = 1:2 Postscale
           0010 = 1:3 Postscale
             .
             .
             .
           1111 = 1:16 Postscale

bit 2      TMR2ON: Timer2 ON
           1 = Timer2 is on
           0 = Timer2 is off

bit 1-0    T2KPS1:T2KPS0: Prescale select bits
           00 = Prescale is 1
           01 = Prescale is 4
           1x = Prescale is 16
```

Figure 9.6 *Timer2 control register (T2CON) bitmap.*

Timer Clock Source

The Timer2 module has a single source of input clock, which is the device clock (FOSC/4). However, the clock speed can be controlled by selecting one of the three prescale options (1:1, 1:4, or 1:16). This is accomplished by means of the control bits T2CKPS1:T2CKPS0 in the T2CON register (see Figure 9.6).

TMR2 and PR2 Registers

The TMR2 register is readable and writable, and is cleared on all device resets. Timer2 increments from 0x00 until it matches the period register (PR2) and then resets to 0x00 on the next increment cycle. PR2 is also a readable and writable register. The TMR2 register is cleared and the PR2 register is set when a WDT, POR, MCLR, or a BOR reset occurs.

Programming Timer2 is simplified by means of the interaction between the TMR2 and the PR2 register. On timer start-up the TMR2 register is initialized to 0x00 and the PR2 register to 0xff. In this state, TMR2 operates as a simple counter and resets as it reaches 0xff. Application code can define a timing period by setting these registers accordingly. For example, to obtain a time period of 50 cycles an application can set the PR2 register to 50 and then monitor when the TMR2 register overflows (TMR2IF flag is set) or when a Timer2 interrupt is generated.

If the PR2 register is cleared (set to 0x00), the TMR2 register will not increment and Timer2 will be disabled. Timer2 can also be shut off by clearing the TMR2ON control bit (T2CON register). When Timer2 is not used by an application it is recommended to turn it off because this minimizes the power consumption of the module.

Prescaler and Postscaler

Four bits serve to select the postscaler. This allows the postscaler rate from 1:1 to 1:16. After the postscaler overflows, the TMR2 interrupt flag bit (TMR2IF) is set to indicate the Timer2 overflow. This is useful in reducing the software overhead of the Timer2 interrupt service routine, because it will only execute when the postscaler is matched. The prescaler and postscaler counters are cleared when any of the following occurs:

1. Aa write to the TMR2 register

2. A write to the T2CON register

3. Any device reset (Power-on Reset, MCLR reset, Watchdog Timer Reset, Brown-out Reset)

During sleep, TMR2 will not increment. The prescaler will retain the last prescale count, ready for operation to resume after the device wakes from sleep.

Timer2 Initialization

The following code fragment shows the initialization of the Timer2 module, including the prescaler and postscaler:

```
    clrf      T2CON      ; stop timer2, prescaler = 1:1,
                         ; postscaler = 1:1
    clrf      TMR2       ; clear timer2 register
    clrf      INTCON     ; disable interrupts
    lrf       PIE1       ; disable peripheral interrupts
    clrf      PIR1       ; clear peripheral interrupts flags
    movlw     0x72       ; postscaler = 1:15
                         ; prescaler = 1:16
    movwf     T2CON      ; timer2 is off
    movlw     pr2value   ; value to load into the
    movwf     PR2        ; PR2 register.
    bsf       T2CON, TMR2CON ; timer2 starts to increment
; the timer2 interrupt is disabled, do polling on the
; overflow bit
t2_ovfl_wait
    btfss     PIR1, TMR2IF ; has tmr2 interrupt occurred?
    goto      t2_ovfl_wait ; no, continue loop
; timer has overflowed
    bcf       PIR1, TMR2IF ; yes, clear flag and continue.
```

9.5.3 Timer3 Module

The Timer3 module is a 16-bit timer/counter consisting of two 8-bit registers labeled TMR3H and TMR3L. Both registers are readable and writable. The register pair (TMR3H:TMR3L) increments from 0000h to FFFFh and rolls over to 0000h. The Timer3 Interrupt is generated on overflow and is latched in the TMR3IF interrupt flag bit. This interrupt can be enabled/disabled by setting/clearing the TMR3IE interrupt enable bit. Figure 9.7 is a block diagram of the Timer3 module.

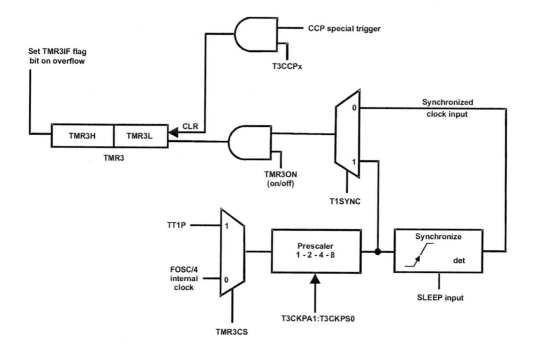

Figure 9.7 *Timer3 block diagram for 16-bit modes.*

Timer3 can operate in one of three modes:

1. As a synchronous timer

2. As a synchronous counter

3. As an asynchronous counter

The following features are characteristic of the Timer3 module:

- TMR3 also has an internal "reset input," that can be generated by a CCP module.

- TMR3 has the capability to operate off an external crystal/clock.

- TMR3 is the alternate time base for capture/compare.

Figure 9.8 is a bitmap of the Timer3 Control Register.

Timer3 increments every instruction cycle while in the timer mode. In counter mode, it increments on every rising edge of the external clock input. The Timer3 increment can be enabled or disabled by setting or clearing control bit TMR3ON (T3CON register in Figure 9.8). Timer3 also has an internal "reset input." This reset can be generated by a CCP special event trigger (Capture/Compare/PWM) module.

When the Timer1 oscillator is enabled (T1OSCEN, in T1CON, is set), the T1OSCI1 and T1OSO2 pins are configured as oscillator input and output, so the corresponding values in the TRIS register are ignored. The Timer3 module also has a software programmable prescaler. The operating mode is determined by clock select bit, TMR3CS (T3CON register), and the synchronization bit, T3SYNC (Figure 9.8).

bit 7							bit 0
RD16	T3CPP2	T3CKPS1	T3CKPS0	T3CCP1	T3SYNC	TMR3CS	TMR3ON

```
bit 7      Rd16: 16-bit Read/Write Mode Enable bit
             1 = Enables register Read/Write in one 16-bit operation
             0 = Enables register Read/Write in twso 8-bit operations

bit 6,3    T3CCP2:T3CCP1t: Timer3 and Timer1 CCPx Enable bits
             1x = Timer3 is clock source for capture/compare of CCP modules
             01 = Timer3 is clock source for capture/compare of CCP2
                  Timer1 is clock source for capture/compare of CCP1
             00 = Timer1 is clock source for capture/compare of CCP modules

bit 5:4    T3CKPS1:T3CKPS0: Timer3 Input Clock Prescale Select bits
             11 = 1:8 Prescale value
             10 = 1:4 Prescale value
             01 = 1:2 Prescale value
             00 = 1:1 Prescale value

bit 2      ~T3SYNC: Timer3 External Clock Input Synchronization Select
             When TMR3CS = 1:
             1 = Do not synchronize external clock input
             0 = Synchronize external clock input
             When TMR3CS = 0
             Timer1 uses the internal clock

bit 1      TMR3CS: Timer1 Clock Source Select bit
             1 = External clock form pin T1OSO/T13CKI (rising edge)
             0 = Internal clock at FOSC/4

bit 0      TMR3ON: Timer3 On bit
             1 = Timer3 enabled
             0 = Timer3 stopped
```

Figure 9.8 *Timer3 control register (T3CON) bitmap.*

Timer3 in Timer Mode

Timer mode is selected by clearing the TMR3CS bit (T3CON register in Figure 9.8). In this mode, the input clock to the timer is FOSC/4. The synchronize control bit, T3SYNC (T3CON register in Figure 9.8), has no effect because the internal clock is always synchronized.

Timer3 in Synchronized Counter Mode

The Timer3 counter mode is selected by setting bit TMR3CS (see Figure 9.8). In the counter mode, the timer increments on every rising edge of input on the T1OSI pin (when enable bit T1OSCEN is set) or the T13CKI pin (when bit T1OSCEN is cleared). If the T3SYNC bit is cleared, then the external clock input is synchronized with internal phase clocks. The synchronization is done after the prescaler stage, which operates asynchronously. Notice that Timer3 gets its external clock input from the same source as Timer1. The configuration of the Timer1 and Timer3 clock input will be controlled by the T1OSCEN bit in the Timer1 control register.

During SLEEP mode, Timer3 will not increment even if an external clock is present, as the synchronization circuit is shut off. The prescaler, however, will continue to increment.

External Clock Input Timing

The external clock input used for Timer3 in synchronized counter mode must meet certain requirements:

- When the prescaler is 1:1, the external clock input is the same as the prescaler output. In this case, there is synchronization of T1OSI/T13CKI with the internal phase clocks. Therefore, it is necessary for T1OSI/T13CKI to be high for at least 2TSCLK (and a small RC delay) and low for at least 2TSCLK.

- When a prescaler other than 1:1 is used, the external clock input is divided by the asynchronous prescaler. In this case, the prescaler output is symmetrical.

Timer3 in Asynchronous Counter Mode

When the ~T3SYNC bit is set, the external clock input is not synchronized. In this case, the timer continues to increment asynchronously to the internal phase clocks. The timer will continue to run during SLEEP and can generate an interrupt on overflow that will wake up the processor.

Because the counter can operate in sleep, Timer3 can be used to implement a true real-time clock. This also explains why in asynchronous counter mode, Timer3 cannot be used as a time base for capture or compare operations.

External Clock Input Timing with Unsynchronized Clock

If the T3SYNC control bit is set, the timer will increment completely asynchronously. Also note that the control bit T3SYNC is not usable when the system clock source comes from the same source as the Timer1/Timer3 clock input. This is because the T1CKI input will be sampled at one quarter the frequency of the incoming clock.

Timer3 Reading and Writing

Timer3 allows the 16-bit timer register to be read/written as two 8-bit registers or one 16-bit register. Which mode is selected is determined by the RD16 bit (see Figure 9.8). Timer3 is configured for 16-bit reads when the RD16 control bit (T3CON register) is set. In this case, the address for TMR3H is mapped to a buffer register. A read from TMR3L will load the contents of the high byte of Timer3 into the Timer3 high byte buffer. This scheme provides a mechanism to accurately read all 16 bits of Timer3 without having to determine whether a read of the high byte followed by a read of the low byte is valid due to a rollover between reads.

Writing in 16-Bit Mode

Writing the high byte of Timer3 must also take place through the TMR3H buffer register. In this case, the Timer3 high byte is updated with the contents of TMR3H when a write occurs to TMR3L. Here again, this allows writing all 16 bits to both the high and low bytes of Timer3 at once.

The high byte of Timer3 is not directly readable or writable in this mode. All reads and writes must take place through the Timer3 high byte buffer register. Writes to TMR3H do not clear the Timer3 prescaler. The prescaler is only cleared on writes to TMR3L.

16-Bit Read-Modify-Write Operation

Instructions that perform read-modify-write, such as BSF or BCF, first read the contents of a register, then make the appropriate changes, and finally place the result back into the register. When Timer3 is configured in 16-bit mode, the read portion of a read-modify-write instruction of TMR3L will not update the contents of the TMR3H buffer. In this case, the TMR3H buffer remains unchanged. However, when the write portion of the instruction takes place, the contents of TMR3H will be placed into the high byte of Timer3.

Reading in Asynchronous Counter Mode

The hardware ensures a valid read operation of TMR3H or TMR3L while the timer is running from an external asynchronous clock. However, reading the 16-bit timer in two 8-bit values poses problems because the timer may overflow between the reads.

Regarding write operations it is recommended that code stop the timer and write the desired values, although a write contention may occur by writing to the timer registers, while the register is incrementing. In this case an unpredictable value may result. Microchip provides no information on how to prevent or how to correct this possible error.

Reading the 16-bit value requires some care because two separate reads are required to read the entire 16-bits. The following code snippet shows reading a 16-bit timer value in cases when the timer cannot be stopped and while avoiding a timer rollover error.

```
; All interrupts are disabled
    movf     TMR3H, w  ; Read high byte
    movwf    TMPH
    movf     TMR3L, w  ; Read low byte
    movwf    TMPL
    movf     TMR3H, w  ; Read high byte
    subwf    TMPH, w   ; Sub 1st read with 2nd read
    btfsc    STATUS,Z  ; Is result = 0
    goto     CONTINUE  ; Good 16-bit read
; If the zero flag is not set, then TMR3L may have rolled
; over between the read of the high and low bytes. Reading
; the high and low bytes now will produce a valid value.
    movf     TMR3H, w  ; Read high byte
    movwf    TMPH
    movf     TMR3L, w  ; Read low byte
    movwf    TMPL
CONTINUE:
; Program continues at this label
```

To write a 16-bit value to the 16-bit TMR3 register is straightforward. First, the TMR3L register is cleared to ensure that there are many Timer3 clock/oscillator cycles before there is a rollover into the TMR3H register. The TMR3H register is then loaded, and finally, the TMR3L register is loaded. The following code snippet shows the sequence of operations.

```
; All interrupts are disabled
    clrf      TMR3L     ; Clear Low byte, Ensures no
; rollover into TMR3H
    movlw     HI_BYTE   ; Value to load into TMR3H
    movwf     TMR3H, F  ; Write High byte
    movlw     LO_BYTE   ; Value to load into TMR3L
    movwf     TMR3H, F  ; Write Low byte
CONTINUE;
; Program code continues at this label
```

Timer1 Oscillator in Timer3

The 18F452 PIC has an alternate crystal oscillator circuit that is built into the device and is labeled the Timer1 Oscillator. The output of this oscillator can be selected as the input into Timer3. The Timer1 Oscillator is primarily intended to operate as a time base for real-world timing operations, that is, the oscillator is primarily intended for a 32-kHz crystal, which is an ideal frequency for real-time keeping.

The fact that the SLEEP mode does not disable the Timer1 facilitates its use in keeping real-time. The Timer1 Oscillator is also designed to minimize power consumption, which can be a factor in real-time applications. The Timer1 Oscillator is enabled by setting the T1OSCEN control bit (T1CON register in Figure 9.8). After the Timer1 Oscillator is enabled, the user must provide a software time delay to ensure its proper start-up.

9.6 C-18 Timer Functions

The C18 Hardware Peripherals Library contains functions to enable, disable, configure, open, and close timers and to read and write to timer registers. The functions are furnished in four function groups:

- CloseTimerx, where x is any digit from 0 to 4, to disable a specific timer.
- OpenTimerx, where x is any digit from 0 to 4, to configure and enable a specific timer.
- ReadTimerx, where x is any digit from 0 to 4, to read the value currently in the timer registers.
- WriteTimerx, where x is any digit from 0 to 4, to write a value into a specified timer register.

The timer-related functions require including the timer.h header file. The functions are described in the following subsections.

9.6.1 CloseTimerx Function

This function disables the interrupt and the specified timer; for example,

```
CloseTimer0();
```

closes the Timer0 module.

9.6.2 OpenTimerx Function

This function opens and configures a specific Timer device available in the hardware. The arguments are bits that are logically anded to obtain the desired timer configuration. The following arguments are found in the timers.h file:

```
Enable Timer0 Interrupt:
    TIMER_INT_ON Interrupt enabled
    TIMER_INT_OFF Interrupt disabled
Timer Width:
    T0_8BIT 8-bit mode
    T0_16BIT 16-bit mode
Clock Source:
    T0_SOURCE_EXT External clock source (I/O pin)
    T0_SOURCE_INT Internal clock source (TOSC)
    External Clock Trigger (for T0_SOURCE_EXT):
    T0_EDGE_FALL External clock on falling edge
    T0_EDGE_RISE External clock on rising edge
Prescale Value:
    T0_PS_1_1 1:1 prescale
    T0_PS_1_2 1:2 prescale
    T0_PS_1_4 1:4 prescale
    T0_PS_1_8 1:8 prescale
    T0_PS_1_16 1:16 prescale
    T0_PS_1_32 1:32 prescale
    T0_PS_1_64 1:64 prescale
    T0_PS_1_128 1:128 prescale
    T0_PS_1_256 1:256 prescale
```

The following code snippet opens and configures Timer0 to disable interrupts, enable the 8-bit data mode, use the internal clock source, and select the 1:32 prescale.

```
// configure timer0
OpenTimer0( TIMER_INT_OFF &
        T0_8BIT &
        T0_SOURCE_INT &
        T0_PS_1_32 );
```

The C-18 functions for specific timers may contain support for other hardware devices. For example, the arguments in OpenTimer1, OpenTimer2, and OpenTimer3 functions include interaction with CCP modules.

9.6.3 ReadTimerx Function

The ReadTimerx functions allow reading the value of the specified timer register. The x parameter can take values representing any of the available timer modules, such as ReadTimer0 to ReadTimer4. The function's prototype is as follows:

```
unsigned int ReadTimerx (void);
```

The function takes data from the available timer registers as follows:

```
Timer0: TMR0L,TMR0H
Timer1: TMR1L,TMR1H
Timer2: TMR2
Timer3: TMR3L,TMR3H
```

```
Timer4: TMR4
```

When the ReadTimerx function is used in the 8-bit mode for a timer module that may be configured in 16-bit mode (for example, timer0, timer1, and Timer3)), the read operation does not ensure that the high-order byte will be zero. In this case, code may cast the result to a char for correct results. For example,

```
// Reading a 16-bit result from a 16-bit timer
// operating in 8-bit mode:
    unsigned int result;
    result = (unsigned char) ReadTimer0();
```

9.6.4 WriteTimerx Function

The WriteTimerx functions allow writing a value to the specified timer register. The x parameter can take values representing any of the available timer modules, such as WriteTimer0 to WriteTimer4. The function's prototype is as follows:

```
void WriteTimerx (unsigned int);
```

The function places data in the available timer registers as follows:

```
Timer0: TMR0L,TMR0H
Timer1: TMR1L,TMR1H
Timer2: TMR2
Timer3: TMR3L,TMR3H
Timer4: TMR4
```

For example:

```
WriteTimer0(32795);
```

9.7 Sample Programs

The following programs demonstrate the programming discussed in this chapter.

9.7.1 Timer0_as_Counter Program

```
; File name: Timer0_as_Counter.asm
; Date:       October 3, 2012
; No copyright
; Processor: PIC 18F452
;
; Port direction and wiring for this program:
; PORT        PINS       DIRECTION       DEVICE
;  C          0-6        Output          7-segment LED
;  A          3          Input           DIP Sw
;
; Description:
; A demonstration program to count actions on DIP switch
; # 3 (wired to RA4/T0CKI pin) and display count on the
; seven segment LED wired to Port C.
; Circuit is DemoBoard 18F452-A or equivalent.
;
```

```
;==========================================================
;              definition and include files
;==========================================================
     processor     18F452           ; Define processor
     #include   <p18F452.inc>


; ==========================================================
;                   configuration bits
;==========================================================
     config OSC = HS          ; Assumes high-speed resonator
     config WDT = OFF         ; No watchdog timer
     config LVP = OFF         ; No low voltage protection
     config DEBUG = OFF       ; No background debugger
;
; Turn off banking error messages
     errorlevel     -302
;==========================================================
;                        program
;==========================================================
     org      0      ; start at address
     goto main
; Space for interrupt handlers
;=============================
;      interrupt intercept
;=============================
     org      0x008      ; High-priority vector
     retfie
     org      0x018      ; Low-priority vector
     retfie
;================================
;   Table to returns 7-segment
;            codes
;================================
     org     $+2
; Note: Table is placed in low program memory at
;       an address where PCL = 0x1A. This provides space
;       for 115 retlw instructions (at 2 bytes per
;       instruction). 18 entries are actually used in
;       this example. By knowing the location of the
;       table (at 0x1A in this case) we make sure that
;       a code page boundary is not straddled while
;       accessing table entries because the instruction:
;                 addwf     PCL,F
;       does not update the PCH register.
;       Addresses (PCL value) increment by two for
;       each sequential instruction in the table.
codeTable:
     addwf     PCL,F           ; PCL is program counter latch
     retlw     0x3f ; 0 code
     retlw     0x06 ; 1
     retlw     0x5b ; 2
     retlw     0x4f ; 3
     retlw     0x66 ; 4
     retlw     0x6d ; 5
     retlw     0x7d ; 6
     retlw     0x07 ; 7
     retlw     0x7f ; 8
     retlw     0x6f ; 9
     retlw     0x00 ; Padding
```

```
;=============================================================
;                      main program entry point
;=============================================================
main:
; Set BSR for bank 0 operations
     movlb    0                ; Bank 0
; Init Port A for digital operation
     clrf     PORTA,0
     clrf     LATA,0
; ADCON1 is the configuration register for the A/D
; functions in Port A. A value of 0b011x sets all
; lines for digital operation
     movlw    B'00000110'  ; Digital mode
     movwf    ADCON1,0
; Port A. Set lines 2 to 5 for input
     movlw    B'00111100'   ; w = 00111100 binary
     movwf    TRISA,0       ; port A (lines 2 to 5) to input
; Initialize all lines in PORT C for output
     movlw    B'00000000' ; 0 = output
     movwf    TRISC,0       ; Port C tris register
;==============================
;  setup Timer0 as counter
;==============================
     clrf     TMR0L
     clrwdt                  ; Clear watchdog timer
; Setup the T0CON register
;                 |------------- On/Off control
;                 |              1 = Timer0 enabled
;                 ||----------- 8/16 bit mode select
;                 ||             1 = 8-bit mode
;                 |||---------- Clock source
;                 |||            1 = T0CKI pin
;                 ||||---------- Source edge select
;                 ||||           1 = high-to-low
;                 |||||--------- Prescaler assignment
;                 |||||          1 = prescaler NOT assigned
;                 ||||||||| ----- Prescaler select
;                 |||||||||
     movlw    b'11111000'
     movwf    T0CON
;=================================
; Check value in TMR0L and display
;=================================
; Every closing of DIP switch # 3 (connected to line
; RA4/TOCKI) adds one to the value in the TMR0L register.
; Loop checks this value, adjusts to the range 0 to 15
; and displays the result in the seven-segment LED on
; port B
checkTmr0:
     movf TMR0L,w        ; Timer register to w
; Elimate four high order bits
     andlw    b'00001111' ; Mask off high bits
; At this point the w register contains a 4-bit value
; in the range 0 to 0xf. Use this value (in w) to
; obtain seven-segment display code
     call codeTable
     movwf    PORTC               ; Display switch bits
     goto checkTmr0
     end
```

9.7.2   Timer0_Delay Program

```
; File name: Timer0_Delay.asm
; Date:       October 4, 2012
; No copyright
; Processor: PIC 18F452
;
; Port direction and wiring for this program:
; PORT       PINS       DIRECTION      DEVICE
;  C         0-7        Output         LEDs
;
; Description:
; Program to demonstrate programming of the 18F452 Timer0
; module. Program flashes eight LEDs in sequence counting
; from 0 to 0xff. Timer0 is used to delay the count.
; Circuit is DemoBoard 18F452-A or equivalent.
;
;=========================================================
;              definition and include files
;=========================================================
    processor     18F452         ; Define processor
    #include <p18F452.inc>
; =========================================================
;                  configuration bits
;=========================================================
    config OSC = HS          ; Assumes high-speed resonator
    config WDT = OFF         ; No watchdog timer
    config LVP = OFF         ; No low voltage protection
    config DEBUG = OFF       ; No background debugger
;
; Turn off banking error messages
    errorlevel    -302
;
;=========================================================
;                         program
;=========================================================
    org      0      ; start at address
    goto     main
; Space for interrupt handlers
;============================
;     interrupt intercept
;============================
    org      0x008     ; High-priority vector
    retfie
    org      0x018     ; Low-priority vector
    retfie
;=========================================================
;                 main program entry point
;=========================================================
main:
; Set BSR for bank 0 operations
    movlb     0              ; Bank 0
; Init Port A for digital operation
    clrf      PORTA,0
    clrf      LATA,0
; ADCON1 is the configuration register for the A/D
; functions in Port A. A value of 0b011x sets all
; lines for digital operation
    movlw  B'00000110'  ; Digital mode
```

```
      movwf    ADCON1,0
; Port A. Set lines 2 to 5 for input
      movlw    B'00111100'   ; w = 00111100 binary
      movwf    TRISA,0        ; port A (lines 2 to 5) to input
; Initialize all lines in PORT C for output
      movlw    B'00000000' ; 0 = output
      movwf    TRISC,0        ; Port C tris register
      clrf PORTC          ; Turn off all LEDs
;=============================
;  setup Timer0 as delay timer
;=============================
      clrf    TMR0H        ; Clear high latch
      clrf    TMR0L         ; Write both bytes
      clrwdt              ; Clear watchdog timer
; Setup the T0CON register
;                |------------ On/Off control
;                |               1 = Timer0 enabled
;                ||------------ 8/16 bit mode select
;                ||              0 = 16-bit mode
;                |||----------- Clock source
;                |||             0 = Internal clock
;                ||||---------- Source edge select
;                ||||            1 = high-to-low
;                |||||--------- Prescaler assignment
;                |||||           1 = prescaler not assigned
;                |||||||||| ----- No prescaler
;                ||||||||||
      movlw  b'10011000'
      movwf    T0CON
;==================================
;      endless loop calling
;        delay routiney
;==================================
; Display Port C count on LEDs
showLEDs:
      incf     PORTC,f        ; Add one to register
      call     tmr0_delay   ; Delay routine
      goto     showLEDs
;==================================
;      Timer0 delay routine
;==================================
tmr0_delay:
cycle:
      movf     TMR0L,w      ; Read low byte to latch
                             ; high byte
      movf     TMR0H,w      ; Now read high byte
      sublw    0xff          ; Subtract maximum count
      btfss    STATUS,Z     ; Test zero flag
      goto     cycle
; Reset counter
      clrf     TMR0H         ; Clear high byte buffer
      clrf     TMR0L         ; Write both low and high
      return
      end
```

9.7.3 Timer0_VarDelay Program

```
; File name: Timer0_VarDelay.asm
; Date:        October 5, 2012
; No copyright
```

```
; Processor: PIC 18F452
;
; Port direction and wiring for this program:
; PORT        PINS       DIRECTION      DEVICE
;  C          0-7        Output         LEDs
;
; Description:
; Using timer0 to produce a variable-lapse delay.
; The delay is calculated based on the number of machine
; cycles necessary for the desired wait period. For
; example, a machine running at a 4 MHz clock rate
; executes 1,000,000 instructions per second. In this
; case a 1/2 second delay requires 500,000 instructions.
; The wait period is passed to the delay routine in three
; editable constants which hold the high-, middle-, and
; low-order bytes of the counter.
; The routine uses Timer0 in 8-bit mode.
;
;===========================================================
;                  definition and include files
;===========================================================
      processor      18F452          ; Define processor
      #include <p18F452.inc>
; ===========================================================
;                  configuration bits
;===========================================================
      config OSC = HS        ; Assumes high-speed resonator
      config WDT = OFF       ; No watchdog timer
      config LVP = OFF       ; No low voltage protection
      config DEBUG = OFF     ; No background debugger
;
;=================================================
;        timer constant definitions
;=================================================
; Three timer constants are defined in order to implement
; a given delay. For example, a delay of one-half second
; in a 4MHz machine, requires a count of 500,000, while
; a delay of one-tenth second requires a count of 10,000.
; These numbers are converted to hexadecimal so they can
; be installed in three constants, for example:
;       1,000,000 = 0x0f4240 = one second at 4MHz
;         500,000 = 0x07a120 = one-half second at 4MHz
;          10,000 = 0x002710 = one-tenth second at 4MHz
; Values for one-half second installed in constants
; as follows:
; 500,000 = 0x07 0xa1 0x20
;           ---- ---- ----
;             |    |    |___ lowCnt
;             |    |_____ midCnt
;             |_____ highCnt
;
#define highCnt 0x07
#define midCnt 0xa1
#define lowCnt 0x20
; Constants can be edited for different delays
;=================================================
;                  variables in PIC RAM
;=================================================
; Local variables
      cblock    0x00 ; Start of block
```

```
        ; 3-byte auxiliary counter for delay.
        countH          ; High-order byte
        countM          ; Medium-order byte
        countL          ; Low-order byte
        endc

; Turn off banking error messages
        errorlevel    -302
;
;=============================================================
;                          program
;=============================================================
        org      0      ; start at address
        goto     main
; Space for interrupt handlers
;=============================
;      interrupt intercept
;=============================
        org      0x008      ; High-priority vector
        retfie
        org      0x018      ; Low-priority vector
        retfie
;=============================================================
;                     main program entry point
;=============================================================
main:
; Set BSR for bank 0 operations
        movlb         0              ; Bank 0
; Initialize all lines in PORT C for output
        movlw         B'00000000' ; 0 = output
        movwf         TRISC,0        ; Port C tris register
;=============================
;  setup Timer0 as counter
;       8-bit mode
;=============================
; Prescaler is assigned to Timer0 and initialzed
; to 2:1 rate

; Setup the T0CON register
;                |------------- On/Off control
;                |              1 = Timer0 enabled
;                ||------------ 8/16 bit mode select
;                ||             1 = 8-bit mode
;                |||----------- Clock source
;                |||            0 = internal clock
;                ||||---------- Source edge select
;                ||||           1 = high-to-low
;                |||||--------- Prescaler assignment
;                |||||          0 = prescaler assigned
;                ||||||||| ----- Prescaler select
;                |||||||||       1:2 rate
        movlw  b'11010000'
        movwf  T0CON
; Clear registers
        clrf     TMR0L
        clrwdt                  ; Clear watchdog timer
;=============================
;     display loop
;=============================
mloop:
```

```
; Turn on LED
      bsf        PORTC,0
; Initialize counters and delay
      call       setVars
      call       TM0delay
; Turn off LED
      bcf        PORTC,0
; Re-initialize counter and delay
      call            setVars
      call       TM0delay
      goto       mloop
;=================================
;  variable-lapse delay procedure
;         using Timer0
;=================================
; ON ENTRY:
;         Variables countL, countM, and countH hold
;         the low-, middle-, and high-order bytes
;         of the delay period, in timer units
; Routine logic:
; The prescaler is assigned to timer0 and setup so
; that the timer runs at 1:2 rate. This means that
; every time the counter reaches 128 (0x80) a total
; of 256 machine cycles have elapsed. The value 0x80
; is detected by testing bit 7 of the counter
; register.
TM0delay:
; Note:
;     The TMR0L register provides the low-order level
; of the count. Because the counter counts up from zero,
; code must pre-install a value in the counter register
; that represents the one-half the number of timer
; iterations (pre-scaler is in 1:2 mode) required to
; reach a count of 128. For example: if the value in
; the low counter variable is 140
; then 140/2 = 70. 128 - 70 = 58
; In other words, when the timer counter reaches 128,
; 70 * 2 (140) timer beats would have elapsed.
; Formula:
;         Value in TMR0L = 128 - (x/2)
; where x is the number of iterations in the low-level
; counter variable
; First calculate xx/2 by bit shifting
      rrncf           countL,f  ; Divide by 2
; now subtract 128 - (x/2)
      movlw      d'128'
; Clear the borrow bit (mapped to Carry bit)
      bcf        STATUS,C
      subfwb     countL,w
; Now w has adjusted result. Store in TMR0L
      movwf      TMR0L
; Routine tests timer overflow by testing bit 7 of
; the TMR0L register.
cycle:
      btfss           TMR0L,7        ; Is bit 7 set?
      goto       cycle          ; Wait if not set
; At this point TMR0 bit 7 is set
; Clear the bit
      bcf        TMR0L,7        ; All other bits are preserved
; Subtract 256 from beat counter by decrementing the
```

```
; mid-order byte
     decfsz     countM,f
     goto       cycle          ; Continue if mid-byte not zero
; At this point the mid-order byte has overflowed.
; High-order byte must be decremented.
     decfsz     countH,f
     goto       cycle
; At this point the time cycle has elapsed
     return
;==============================
;  set register variables
;==============================
; Procedure to initialize local variables for a
; delay period defined in local constants highCnt,
; midCnt, and lowCnt.
setVars:
     movlw           highCnt      ; From constants
     movwf           countH
     movlw           midCnt
     movwf           countM
     movlw           lowCnt
     movwf           countL
     return
     end
```

9.7.4 Timer0_VarInt Program

```
; File name: Timer0_VarInt.asm
; Date:        October 7, 2012
; No copyright
; Processor: PIC 18F452
;
; Port direction and wiring for this program:
; PORT        PINS       DIRECTION       DEVICE
;  C          0-7        Output          LEDs
;
; Description:
; Using timer0 to produce an interrupt-driven variable
; lapse delay. The delay is calculated based on the
; number of machine cycles necessary for the desired
; wait period as in the program Timer0_VariLapse.asm.
; The routine uses Timer0 in 8-bit mode.
;
;=========================================================
;              definition and include files
;=========================================================
     processor     18F452          ; Define processor
     #include <p18F452.inc>
; =========================================================
;                  configuration bits
;=========================================================
     config OSC = HS          ; Assumes high-speed resonator
     config WDT = OFF         ; No watchdog timer
     config LVP = OFF         ; No low voltage protection
     config DEBUG = OFF       ; No background debugger
;
;=========================================================
;        timer constant definitions
;=========================================================
```

```
; Three timer constants are defined in order to implement
; a given delay. For example, a delay of one-half second
; in a 4MHz machine, requires a count of 500,000, while
; a delay of one-tenth second requires a count of 10,000.
; These numbers are converted to hexadecimal so they can
; be installed in three constants, for example:
;        1,000,000 = 0x0f4240 = one second at 4MHz
;          500,000 = 0x07a120 = one-half second at 4MHz
;           10,000 = 0x002710 = one-tenth second at 4MHz
; Values for one-half second installed in constants
; as follows:
; 500,000 = 0x07 0xa1 0x20
;           ---- ---- ----
;            |    |    |___ lowCnt
;            |    |_____ midCnt
;            |_____ highCnt
;
#define highCnt 0x07
#define midCnt 0xa1
#define lowCnt 0x20
; Constants can be edited for different delays
;=====================================================
;                 variables in PIC RAM
;=====================================================
; Local variables
     cblock    0x00 ; Start of block
     ; 3-byte auxiliary counter for delay.
     countH        ; High-order byte
         countM            ; Medium-order byte
         countL            ; Low-order byte
     endc

; Turn off banking error messages
     errorlevel    -302
;
;=============================================================
;                            program
;=============================================================
     org       0       ; start at address
     goto      main
; Space for interrupt handlers
;============================
;      interrupt intercept
;============================
     org       0x008     ; High-priority vector
     goto      IntServ
;
     org       0x018     ; Low-priority vector
     retfie
;=============================================================
;                     main program entry point
;=============================================================
main:
; Set BSR for bank 0 operations
     movlb     0                ; Bank 0
; Initialize all lines in PORT C for output
     movlw     B'00000000' ; 0 = output
     movwf     TRISC             ; Port C tris register
     movwf     PORTC
;===============================
```

```
;   setup Timer0 as counter
;         8-bit mode
;==============================
     bcf      INTCON,TMR0IE
; Setup the T0CON register
;                |------------ On/Off control
;                |                 1 = Timer0 enabled
;                ||------------ 8/16 bit mode select
;                ||                1 = 8-bit mode
;                |||----------- Clock source
;                |||               0 = internal clock
;                ||||---------- Source edge select
;                ||||              1 = high-to-low
;                |||||--------- Prescaler assignment
;                |||||             1 = prescaler not assigned
;                ||||||||| ----- Prescaler select
;                |||||||||         1:2 rate
     movlw   b'11011000'
     movwf   T0CON
; Clear registers
     clrf    TMR0L
     clrwdt               ; Clear watchdog timer
;==============================
; Set up for Timer0 interupt
;==============================
; Disable interrupt priority levels in the RCON register
; setting up the mid range compatibility mode
     bsf     RCON,IPEN        ; Enable interrupt priorities
; INTCON register initialized as follows:
; (IPEN bit is clear)
;                |----------- high-priority interrupts
;                ||---------- low-priority peripheral
;                |||--------- timer0 overflow interrupt
;                ||||-------- external interrupt
;                |||||------- port change interrupt
;                ||||||------ overflow interrupt flag
;                |||||||----- external interrupt flag
;                ||||||||----- RB4:RB7 interrupt flag
     movlw   b'10100000'
     movwf   INTCON
; Set INTCON2 for falling edge operation
     bcf      INTCON2,INTEDG0
; Re-enable timer0 interrupt
         bsf     INTCON,TMR0IE   ; Activate Timer0 interrupt
         bcf     INTCON,TMR0IF  ; Clear interrupt flag
;==============================
;     display loop
;==============================
; Re-initialize counter and delay
;     call      setDelay
     movlw   b'00000001'
     movwf   PORTC
     call    setDelay
mloop:
     nop
     goto mloop


;==============================
;   set register variables
;==============================
```

```
; Procedure to initialize local variables for a
; delay period defined in local constants highCnt,
; midCnt, and lowCnt.
setDelay:
      movlw       highCnt      ; From constants
      movwf       countH
      movlw       midCnt
      movwf       countM
      movlw       lowCnt
      movwf       countL
; The timer0 register provides the low-order level
; of the count. Because the counter counts up from zero,
; in order to ensure that the initial low-level delay
; count is correct the value 256 - xx must be calculated
; where xx is the value in the original countL register.
      movf countL,w ; w holds low-order byte
      sublw      .255
; Now w has adjusted result. Store in TMR0
      movwf      TMR0L
      return


      return
;=========================================================
;             Interrupt Service Routine
;=========================================================
; This is a high-priority interrupt so critical registers
; are saved and restore automatically
; Service routine receives control when the timer
; register TMR0 overflows, that is, when 256 timer beats
; have elapsed
IntServ:
; First test if source is a timer0 interrupt
      btfss      INTCON,TMR0IF ; T0IF is timer0 interrupt
      goto notTOIF       ; Go if not RB0 origin
; If so clear the timer interrupt flag so that count continues
      bcf        INTCON,TMR0IF ; Clear interrupt flag
;==========================
;    interrupt action
;==========================
; Subtract 256 from beat counter by decrementing the
; mid-order byte
      decfsz     countM,f
      goto exitISR        ; Continue if mid-byte not zero
; At this point the mid-order byte has overflowed.
; High-order byte must be decremented.
      decfsz     countH,f
      goto exitISR
; At this point count has expired so the programmed time
; has elapsed. Service routine turns the LED on line 0,
; port B on and off at every conclusion of the count.
; This is done by XORing a mask with a one-bit at the
; port C line 0 position
      movlw      b'00000001'          ; Xoring with a 1-bit produces
                                      ; the complement

      xorwf      PORTC,f      ; Complement bit 0, port C
; Reset delay constants
      call       setDelay
;==========================
;          exit ISR
;==========================
```

```
exitISR:
notTOIF:
    retfie    0x01
    end       ; END OF PROGRAM
```

9.7.5 C_Timer_Show Program

```c
// Project name: C_Timer_Show
// Source files: C_Timer_Show.c
// Date: January 17, 2013

// Processor: PIC 18F452
// Environment:    MPLAB IDE Version 8.86
//                 MPLAB C-18 Compiler
//
// TEST CIRCUIT: Demo Board 18F452A or circuit wired as
// follows:
// PORT       PINS      DIRECTION      DEVICE
// C          0-3       Output         Green LEDs
// C          4-7       Output         Red LEDs
//
// Description:
// A demonstration program to display the low-order byte of
// the timer register on the LEDs wired to PORT C.
//
// INCLUDED CODE
#include <p18f452.h>
#include <timers.h>

#pragma config WDT = OFF
#pragma config OSC = HS
#pragma config LVP = OFF
#pragma config DEBUG = OFF

// Prototype
void TimerDelay (unsigned int);

/***********************************************************
                      main program
***********************************************************/
void main(void)
{
    unsigned char timerVar = 0;
    /* Initalize direction registers */
    TRISC = 0;
    PORTC = 0;
    /* Configure Timer0 to no interrupts, 16-bit data,
       internal clock source and 1:1 prescaler */
    OpenTimer0(
        TIMER_INT_OFF &
        T0_16BIT &
        T0_SOURCE_INT &
        T0_PS_1_1 );

    while(1) {
        TimerDelay(40000);
        PORTC = timerVar;
        timerVar++;
```

```
        }
}

void TimerDelay (unsigned int period) {
        unsigned int timerCnt = 0;
        // Reset the timer
        WriteTimer0(0x0);
        while (timerCnt < period) {
            timerCnt = ReadTimer0();
        }
        return;
}
```

Chapter 10

Data EEPROM

10.1 EEPROM on the PIC18 Microcontrollers

Electrically Erasable Programmable Read-Only Memory (EEPROM) is used in digital devices as nonvolatile storage for data. EEPROM (usually pronounced double-e prom or e-e prom) is found in flash drives, BIOS chips, in memory storage devices in many types of microcontrollers and other digital devices. EEPROM is a semi-permanent storage that can be erased and reprogrammed electrically without removing the chip from its socket. The technology used before the development of EEPROM, named EPROM, required that the chip be removed from the circuit and placed under ultraviolet light in order to erase it. In addition, EPROM required higher-than-TTL voltages for reprogramming while EEPROM does not. To the programmer, EEPROM memory in a microcontroller can be thought of as a very small hard disk drive or a nonremovable flash drive. We start this chapter with an overview of data EEPROM as implemented in the 18F452 PICs.

10.1.2 On-Board Data EEPROM

To the PIC programmer, EEPROM data memory can refer either to on-board EEPROM memory and to EEPROM memory ICs that are furnished as separate circuit components. EEPROM on-board data extends to either 256 or 1024 bytes in the F18 PIC family. Specifically in the 18F2XX and 18F4XX devices, there are 256 bytes of EEPROM memory. EEPROM memory is not mapped to the processor's code or data space but is addressed through special function registers.

EEPROM elements are classified according to their electrical interfaces into serial and parallel. In the present context we deal only with serial EEPROM because this is the one used in the 18F series microcontrollers. The storage capacity of Serial EEPROMs range from a few bytes to 128 kilobytes. In PIC technology, the typical use of serial EEPROM on-board memory and EEPROM ICs is in the storage of passwords, codes, configuration settings, and other information to be remembered after the system is turned off. For example, a PIC-based automated environment sensor can use EEPROM memory to store daily temperatures, humidity, air pressure, and other values. Later on, this information can be downloaded to a PC and the EEPROM storage erased and reused for new data. In personal computers, EEPROM memory is used to store BIOS code, passwords, and other system data.

Some early EEPROMs could be erased and rewritten about 100 times before failing, but more recent EEPROM tolerate thousands of erase-write cycles. EEPROM memory is different from Random Access Memory (RAM) in that RAM can be rewritten millions of times. Also, RAM is generally faster to write than EEPROM and considerably cheaper per unit of storage. On the other hand, RAM is volatile, which means that the contents are lost when power is removed.

PICs also use EEPROM-type memory internally as flash program memory and as data memory. In the present context we deal with EEPROM data memory. Serial EEPROM memory is also available as separate ICs that can be placed on the circuit board and accessed through PIC ports. For example, the Microchip 24LC04B EEPROM IC is a 4K electrically erasable PROM with a 2-wire serial interface that follows the I2C convention. Programming serial EEPROM ICs is not discussed in this book.

10.2 EEPROM Programming

As previously stated, the 18F242, 18F252, 18F442, and 18F452 all contain 256 bytes of on-board data EEPROM. This memory is both readable and writable during normal operation. EEPROM memory is not mapped in the register file space but is indirectly addressed through the Special Function Registers EECON1, EECON2, EEDATA, and EEADR. The address of EEPROM memory starts at location 0x00 and extends to the maximum contained in the PIC, in the case of the 18F452, 0xff. The following registers relate to EEPROM operations:

- EEDATA holds the data byte to be read or written.
- EEADR contains the EEPROM address to be accessed by the read or write operation.
- EECON1 contains the control bits for EEPROM operations.
- EECON2 protects EEPROM memory from accidental access. This is not a physical register.

The CPU may continue to access EEPROM memory even if the device is code protected, but in this case the device programmer cannot access EEPROM memory. Figure 10.1 is a bitmap of the EECON1 register in the 18F452.

10.2.1 Reading EEPROM Data

To read a data memory location in the 18F452 code must perform the following operations on the EECON1 register:

1. Wwrite the address to the EEADR register
2. Clear the EEPGD control bit
3. Clear the CFGS control bit
4. Set the control bit RD

bit 7

EEPGD: FLASH Program or Data EEPROM Memory Select bit
1 = Access FLASH Program memory
0 = Access Data EEPROM memory

bit 6

CFGS: FLASH Program/Data EE or Configuration Select bit
1 = Access Configuration or Calibration registers
0 = Access FLASH Program or Data EEPROM memory

bit 5

Unimplemented: Read as '0'

bit 4

FREE: FLASH Row Erase Enable bit
1 = Erase the program memory row addressed by TBLPTR
on the next WR command (cleared by completion of
erase operation)
0 = Perform write only

bit 3

WRERR: FLASH Program/Data EE Error Flag bit
1 = A write operation is prematurely terminated
(any MCLR or any WDT Reset during self-timed
programming in normal operation)
0 = The write operation completed
Note: When a WRERR occurs, the EEPGD or FREE bits are
not cleared. This allows tracing of the error
condition.

bit 2

WREN: FLASH Program/Data EE Write Enable bit
1 = Allows write cycles
0 = Inhibits write to the EEPROM

bit 1

WR: Write Control bit
1 = Initiates a data EEPROM erase/write cycle or a
program memory erase cycle or write cycle.
0 = Write cycle to the EEPROM is complete

bit 0

RD: Read Control bit
1 = Initiates an EEPROM read
0 = Does not initiate an EEPROM read

Figure 10.1 *18F452 EECON1 register bitmap.*

The EEDATA register can be read by the next instruction. EEDATA will hold this value until another read operation, or until it is written to by the user (during a write operation). The following procedure can be used to read EEPROM data.

```
;=======================
;  procedure to read
;     EEPROM data
;=======================
Read_EEPROM:
; On entry global variable ee_address contains address
; from which to read. On exit the W register contains
; value retrieved from EEPROM
      movff     ee_address, EEADR  ; Store in address
                                   ; register
      bcf       EECON1, EEPGD      ; point to data memory
```

```
    bcf        EECON1, CFGS   ; access program flash or data
                              ; EEPROM memory
    bsf        EECON1, RD     ; EEPROM read
    nop
    movff      EEDATA, WREG   ; W = EEDATA
    return
```

10.2.2 Writing EEPROM Data

Writing to an EEPROM data location requires the following operations:

1. The address is first written to the EEADR register

2. Data is then written to the EEDATA register

The actual writing sequence requires first writing the value 0x55 to EECON2, then writing 0xaa also to EECON2. In addition, the WREN bit in EECON1 must have been previously set to enable write operations. The write will not initiate if the above sequence is not exactly followed. It is recommended that interrupts be disabled during the write operation.

In this sequence, notice that the WREN bit in EECON1 must be set to enable writes. This prevents accidental writes to data EEPROM due to unexpected code execution (as would be the case with a runaway program). The WREN bit should be kept clear at all times, except when updating the EEPROM register. After a write sequence has been initiated, EECON1, EEADR, and EDATA cannot be modified. The WR bit cannot be set by code unless the WREN bit is set. The WREN bit must be set on a previous instruction because both WR and WREN cannot be set with the same instruction. The WREN bit is not cleared by hardware.

The WR (write control) bit of EECON1 plays an important role in the write operation. Setting the WR bit initiates a data EEPROM write cycle. The operation is self-timed and the bit is cleared by hardware once write is complete. Because the WR bit can only be set (not cleared) in software, its status indicates if a write operation can take place. In other words, code can make certain that the WR bit is cleared in order to proceed with the write operation. The following procedure can be used to write one byte of data to an EEPROM address.

```
;===========================
;  procedure to write one
;     value to EEPROM
;===========================
; On entry global variables contain the following data:
;     ee_digit = value to be stored
;     ee_address = EEPROM address in which to store
Write_EEPROM:
; Wait until WR bit in EECON1 register clears before
; beginning write operation
    nop
    btfsc      EECON1, WR        ; Test bit
    goto       Write_EEPROM      ; Loop if not cleared
; Write can now proceed
    movff      ee_address, EEADR   ; Get address
    movff      ee_digit, EEDATA    ; Data byte to write
    bcf        EECON1, EEPGD ; point to data memory
```

```
    bcf        EECON1, CFGS  ; access program flash or data
                             ;  EEPROM
    bsf        EECON1, WREN  ; enable writes
    bcf        INTCON, GIE   ; disable interrupts
; EEPROM required write sequence
    movlw      0x55
    movwf      EECON2        ; write 55h
    movlw      0xaa
    movwf      EECON2        ; write aah
    bsf        EECON1, WR    ; set WR bit to begin write
    nop
;
; Write completed
    bsf        INTCON, GIE   ; enable interrupts
    bcf        EECON1, WREN  ; disable writes on write complete
    return
```

At the completion of the write cycle, the WR bit is cleared in hardware and the EEPROM Write Complete Interrupt Flag bit (EEIF) is set. Code may either enable this interrupt, or poll this bit. EEIF must be cleared by software. Microchip recommends that critical applications should verify the write operation by reading EEPROM memory after the write operation has taken place in order to make sure that the correct value was stored. In this case, the read operation cannot take place until the WR bit is clear.

10.3 Data EEPROM Programming in C Language

The C18 compiler for PIC18 devices provides support for programming on-board EEPROM. The header file containing the necessary links is named eep.h and must be included in the code.

For the 18F family of microcntrollers, three different versions of the library functions for EEPROM support are provided with the compiler. These are labeled EEP_V1, EEP_V2, and EEP_V3, respectively. During program build, the compiler selects the correct version according to the hardware. Table 10.1 lists the processors compatible with each version.

Table 10.1

Versions of the EEPROM Library for 18F Devices

EEP_V1	18F1230, 18F1330
EEP_V2	18C242, 18C252, 18C442, 18C452, 18F242, 18F252, 18F442, 18F452, 18F248, 18F258, 18F448, 18F458, 18F2439, 18F2539, 18F4439, 18F4539, 18F1220, 18F1320, 18F2220, 18F2320, 18F4220, 18F4320, 8F2420, 18F2520, 18F4420, 18F4520, 18F2423, 18F2523, 18F4423, 18F4523, 18F2455, 18F2550, 18F4455, 18F4550, 18F2480, 18F2580, 18F4480, 18F4580, 18F2221, 18F2321, 18F4221, 18F4321, 18F2331, 18F2431, 18F4331, 18F4431, 18F23K20, 18F24K20, 18F25K20, 18F43K20, 18F44K20, 18F45K20, 18F13K50, 18LF13K50, 18F14K50, 18LF14K50
EEP_V3	18F1220, 18F1320, 18F6585, 18F6680, 18F8585, 18F8680, 18F2331, 18F2431, 18F4331, 18F4431

10.3.1 EEPROM Library Functions

Three functions are provided in the EEPROM C18 library:

Read_b_eep

```
Function:      Read single byte from Internal EEP
Include:       eep.h
Prototype:        unsigned char Read_b_eep( unsigned int
               badd );
Action:        Returns a value after reading the location
               passed as parameter.
Returns:       Returns the value read at the location.
Code Example:  Temp = Read_b_eep(0x0000);
```

Write_b_eep

```
Function:      Write single byte to Internal EEP
Include:       eep.h
Prototype:        void Write_b_eep(unsigned int badd,unsigned
               char bdata);
Action:        Writes data in to the specified location
               in the EEPROM
Code Example:  Write_b_eep (0x0000,0x20);
```

Busy_eep

```
Function:      Checks & waits the status of ER bit in EECON1
               register
Include:       eep.h
Prototype:        void Busy_eep ( void );
Action:        Waits until Write cycle to the EEPROM is
               complete.
Code Example:  Busy_eep ();
```

10.3.2 Sample Code

The following routine writes the binary digits 1, 3, 5, 7, and 9 to EEPROM memory address from 0x00 to 0x04.

```
eeaddress = 0x00;
eedata = 1;

while(eeaddress < 0x06) {
    Busy_eep();                        // Wait for ready
    Write_b_eep (eeaddress, eedata);
    eeaddress++;
    eedata += 2;
    }
```

Notice that the routine includes a call to the Busy_eep() function in order to make sure that the device is ready before writing each digit. Busy_eep() does not return until the previous write operation has completed. Omitting this call in a loop that writes several data items will probably result in a defective operation.

The following routine recovers the five digits stored in EEPROM memory and displays them consecutively in PORT C.

```
eeaddress = 0x00;          // Reset address pointer
while(eeaddress <= 0x04) {
    digit = Read_b_eep (eeaddress);
    count = 0x00;
    PORTC = digit;
        while (count <= 50000){
            count++;
        }
    eeaddress++;
}
```

The sample program C_EEPROM_Demo listed later in this chapter is a demonstration of EEPROM operation in C.

10.4 EEPROM Demonstration Programs

The program EEPROM_to_7Seg.asm listed later in this chapter and contained in the book's software package, is a demonstration of EEPROM memory access on the 18F452 PIC. The program stores five digits (1, 3, 5, 7, and 9) in the first 5 bytes of EEPROM memory and then retrieves these digits and displays them in the seven-segment LED. The program can be tested in Demo Board 18F452A or an equivalent circuit.

The program C_EEPROM_Demo.c also listed in this chapter and contained in the book's oftware package is the C18 version of the program EEPROM_to_7Seg.

10.4.1 EEPROM_to_7Seg Program

```
; File name: EEPROM_to_7Seg.asm
; Date:      January 23, 2013
; No copyright
; Processor: PIC 18F452
;
; Port direction and wiring for this program:
; PORT       PINS      DIRECTION     DEVICE
;  C         0-6       Output        7-segment LED
;
; Description:
; A demonstration program to store digits in EEPROM memory,
; read them, and display them in the 7-segment LED.
; Test circuit is DemoBoard 18F452-A or equivalent.
;
;=========================================================
;            definition and include files
;=========================================================
    processor     18F452        ; Define processor
    #include  <p18F452.inc>


; =========================================================
;                configuration bits
;=========================================================
    config OSC = HS          ; Assumes high-speed resonator
    config WDT = OFF         ; No watchdog timer
    config LVP = OFF         ; No low voltage protection
    config DEBUG = OFF       ; No background debugger
;
; Turn off banking error messages
```

```
      errorlevel      -302
;
;=============================================================
;                      variables in PIC RAM
;=============================================================
; Access RAM locations from 0x00 to 0x7F
TEMP            equ       0x000        ; Temporary register
ee_digit       equ       0x001
ee_address     equ       0x002
counter        equ       0x003
j              equ       0x004
k              equ       0x005
;=============================================================
;                            program
;=============================================================
      org       0     ; start at address
      goto main
; Space for interrupt handlers
;==============================
;      interrupt intercept
;==============================
      org       0x008       ; High-priority vector
      retfie
      org       0x018       ; Low-priority vector
      retfie
;=================================
;  Table to returns 7-segment
;            codes
;=================================
      org       $+2
; Note: Table is placed in low program memory at
;         an address where PCL = 0x1A. This provides space
;         for 115 retlw instructions (at 2 bytes per
;         instruction). 18 entries are actually used in
;         this example. By knowing the location of the
;         table (at 0x1A in this case) we make sure that
;         a code page boundary is not straddled while
;         accessing table entries because the instruction:
;                      addwf    PCL,F
;         does not update the PCH register.
;         Addresses (PCL value) increment by two for
;         each sequential instruction in the table.
codeTable:
      addwf     PCL,F           ; PCL is program counter latch
      retlw     0x3f ; 0 code
      retlw     0x06 ; 1
      retlw     0x5b ; 2
      retlw     0x4f ; 3
      retlw     0x66 ; 4
      retlw     0x6d ; 5
      retlw     0x7d ; 6
      retlw     0x07 ; 7
      retlw     0x7f ; 8
      retlw     0x6f ; 9
      retlw     0x77 ; A
      retlw     0x7c ; B
      retlw     0x39 ; C
      retlw     0x5b ; D
      retlw     0x79 ; E
      retlw     0x71 ; F
```

```
        retlw     0x00 ; Padding
;============================================================
;                   main program entry point
;============================================================
main:
; Set BSR for bank 0 operations
      movlb     0                   ; Bank 0
; Init Port A for digital operation
      clrf      PORTA,0
      clrf      LATA,0
; ADCON1 is the configuration register for the A/D
; functions in Port A. A value of 0b011x sets all
; lines for digital operation
      movlw     B'00000110'   ; Digital mode
      movwf     ADCON1,0
; Port A. Set lines 2 to 5 for input
      movlw     B'00111100'    ; w = 00111100 binary
      movwf     TRISA,0        ; port A (lines 2 to 5) to input
; Initialize all lines in PORT C for output
      movlw     B'00000000' ; 0 = output
      movwf     TRISC,0        ; Port C tris register
;================================
;     Write digits to EEPROM
;================================
; The binary values for the digits 1, 3, 5, 7, and 9 are
; stored in EEPROM memory addresses as follows:
;       ADDRESS         DIGIT
;        0x00             1
;        0x01             3
;        0x03             5
;        0x04             7
;        0x05             9
; Prime iterations counter
GET_AND_SHOW:
      movlw     .5             ; Five digits to write
      movwf     counter
      movlw     0x01           ; First value to store
      movwf     ee_digit       ; To local variable
      movlw     0x00           ; First address
      movwf     ee_address
NEXT_VALUE:
      call      Write_EEPROM     ; Local procedure
; Next digit and address
      incf      ee_digit,f
      incf      ee_digit,f
      incf      ee_address,f
      decfsz    counter,f
      goto      NEXT_VALUE
; At this point all 5 values are written to EEPROM
; Now retrieve and display the stored values
      call      Delay    ; Local procedure
; Prime iterations counter
      movlw     .5             ; Five digits to write
      movwf     counter
      movlw     0x00           ; First address
      movwf     ee_address
SHOW_NEXT:
      call      Read_EEPROM ; Local procedure
; WREG contains value read from EEPROM
; Store in TEMP variable
```

```
        movff    WREG,TEMP
; At this point the TEMP register contains one of the
; binary digits stored in EEPROM. In PIC18 devices this
; value must be doubled to obtain offset into table
; because the program counter increments by 2 to access
; sequential instructions
        movff    TEMP,WREG       ; Offset to W
        addwf    TEMP,f          ; Add to TEMP
; Use value in TEMP to obtain Seven-Segment display code
        movf     TEMP,W               ; TEMP to WREG
        call     codeTable
        movff    WREG,PORTC           ; Display switch bits
        call     Delay
        call     Delay
        call     Delay
        call     Delay
; Bump to next value stored
; Next digit and address
        incf     ee_address,f,0
        decfsz   counter,f,0
        goto     SHOW_NEXT
; At this point all 5 values have been read from EEPROM
; Loop to start of routine
        goto     GET_AND_SHOW


;===============================================================
;                       LOCAL PROCEDURES
;===============================================================
;==========================
;  procedure to write one
;     value to EEPROM
;==========================
; On entry global variables contain the following data:
;    ee_digit = value to be stored
;    ee_address = EEPROM address in which to store
Write_EEPROM:
; Wait until WR bit in EECON1 register clears before
; beginning write operation
        nop
        btfsc    EECON1, WR      ; Test bit
        goto     Write_EEPROM    ; Loop if not cleared
; Write can now proceed
        movff    ee_address, EEADR    ; Get address
        movff    ee_digit, EEDATA     ; Data byte to write
        bcf      EECON1, EEPGD ; point to data memory
        bcf      EECON1, CFGS  ; access program flash or data
                                     ; EEPROM
        bsf      EECON1, WREN  ; enable writes
        bcf      INTCON, GIE   ; disable interrupts
; EEPROM required write sequence
        movlw    0x55
        movwf    EECON2               ; write 55h
        movlw    0xaa
        movwf    EECON2               ; write aah
        bsf      EECON1, WR          ; set WR bit to begin write
        nop
; Write completed
        bsf      INTCON, GIE   ; enable interrupts
        bcf      EECON1, WREN  ; disable writes on write
                                     ; complete(eeif set)
```

```
        return
;=======================
;  procedure to read
;     EEPROM data
;=======================
Read_EEPROM:
; On entry global variable ee_address contains address
; from which to read. On exit the w register contains
; value retrieved from EEPROM
        movff    ee_address, EEADR  ; Store in address register
        bcf      EECON1, EEPGD      ; point to data memory
        bcf      EECON1, CFGS       ; access program flash or data
                                    ; EEPROM memory
        bsf      EECON1, RD         ; EEPROM read
        nop
        movff    EEDATA, WREG       ; W = EEDATA
        return


;===========================
;     delay sub-routine
;===========================
Delay:
    movlw .200 ; w = 200 decimal
    movwf    j,0       ; j = w
jloop:
    movwf    k,0       ; k = w
kloop:
    decfsz   k,f       ; k = k-1, skip next if zero
    goto     kloop
    decfsz   j,f       ; j = j-1, skip next if zero
    goto     jloop
    return
    end
```

10.4.2 C_EEPROM_Demo Program

```
// Project name: C_EEPROM_Demo
// Source files: C_EEPROM_Demo.c
// Date: January 26, 2013

// Processor: PIC 18F452
// Environment:    MPLAB IDE Version 8.86
//        MPLAB C-18 Compiler
//
// TEST CIRCUIT: Demo Board 18F452A or circuit wired as
// follows:
// PORT       PINS      DIRECTION      DEVICE
//   C        0-6       Output         7 SEGMENT LED
//
// Description:
// A demonstration program to store digits in EEPROM memory,
// read them, and display them in the 7-segment LED.
// Test circuit is DemoBoard 18F452-A or equivalent.
// This program is the C 18 version of EEPROM_to_7Seg.asm
//
// INCLUDED CODE
#include <p18f452.h>
#include <eep.h>

// DATA VARIABLES AND CONSTANTS
```

```
unsigned char eedata, digit;
unsigned int eeaddress, count;
#define MAX_COUNT 50000

#pragma config WDT = OFF
#pragma config OSC = HS
#pragma config LVP = OFF
#pragma config DEBUG = OFF

// Define table in program memory using the rom qualifier
rom unsigned char codeTable[]={0x3f, 0x06, 0x5b, 0x4f,
                               0x66, 0x6d, 0x7d, 0x07,
                               0x7f, 0x6f, 0x77, 0x7c,
                               0x39, 0x5b, 0x79, 0x71};

/**************************************************************
                        main program
**************************************************************/
void main(void)
{
    /* Initialize direction registers */
    TRISC = 0;
    PORTC = 0;
// Write digits to EEPROM memory addresses 0x00 to 0x05
//===============================
//     Write digits to EEPROM
//===============================
// The binary values for the digits 1, 3, 5, 7, and 9 are
// stored in EEPROM memory addresses as follows:
//         eeaddress      eedata
//          0x00            1
//          0x01            3
//          0x02            5
//          0x03            7
//          0x04            9
eeaddress = 0x00;
eedata = 1;

while(eeaddress < 0x06) {
    Busy_eep();
    Write_b_eep (eeaddress, eedata);
    eeaddress++;
    eedata += 2;
    }
// Data is stored in memory. Recover and display
while(1){
    eeaddress = 0x00;        // Reset address pointer
    while(eeaddress <= 0x04) {
        digit = Read_b_eep (eeaddress);
        count = 0x00;
        PORTC = codeTable[digit];
            while (count <= MAX_COUNT){
                count++;
            }
        eeaddress++;
    }
}
}
```

Chapter 11

Liquid Crystal Displays

11.1 LCD

A Liquid Crystal Display (LCD) is a hardware device frequently used for alphanumeric output in microcontroller-based embedded systems. LCDs are popular because of their reduced size, moderate cost, and because most LCDs can be mounted directly on the circuit board. According to their interface, LCDs are classified into serial and parallel. Serial LCDs require less I/O resources but execute slower than their parallel counterparts. Although serial LCDs require less control lines, they are considerably more expensive than the parallel type. In this chapter we discuss parallel-driven LCD devices based on the Hitachi HD44780 character-based controller, which is by far the most popular controller for PIC-driven LCDs.

11.1.1 LCD Features and Architecture

The HD44780 is a dot-matrix LCD controller and driver. The device displays ASCII alphanumeric characters, Japanese kana characters, and some symbols. A single HD44780 can display up to one 8-character line or two 8-character lines. An available extension driver makes possible addressing up to 80 characters.

The HD4478 contains a 9,920-bit character generator ROM that can produce a total of 240 characters;. 208 characters with a 5 times 8 dot resolution and 32 characters at a 5×10 dot resolution. The device is also capable of storing 64 times 8-bit characters data in its character generator RAM. This corresponds to 8 custom characters in a 5 times 8 dot resolution or 4 characters in a 5 times 10 dot resolution.

The controller is programmable in three different duty cycles: 1/8 for one line of 5 $\times$ 8 dots with cursor, 1/11 for one line of 5×10 dots with cursor, and 1/16 for two lines of 5×8 dots with cursor. The built-in commands include clearing the display, homing the cursor, turning the display on and off, turning the cursor on and off, setting display characters to blink, shifting the cursor and the display left-to-right or right-to-left, and reading and writing data to the character generator and to display data ROM.

11.1.2 LCD Functions and Components

The following hardware elements form part of the HD44780 controller:

- Two internal registers, labeled the data register and the instruction register
- Busy flag
- An address counter
- RAM area of display data (DDRAM)
- Character generator ROM
- Character generator RAM
- Timing generation circuit
- Liquid crystal display driver circuit
- Cursor and blink control circuit

The controller itself is often referred to as the MPU in the Hitachi literature.

Internal Registers

The HD44780 contains an instruction register (IR) and a data register (DR). The IR is used to store instruction codes, such as those to clear the display, define an address, or store a bitmap in character generator RAM. The IR can only be written from the controller. The data register (DR) is used to temporarily store data to be written into DDRAM or CGRAM as well as temporarily store data read from DDRAM or CGRAM. Data placed in the data register is automatically written into DDRAM or CGRAM by an internal operation.

Busy Flag

When the busy flag (BF) is 1, the HD44780U is in the internal operation mode, and the next instruction will not be accepted. The busy flag is mapped to data bit 7. Software must insure that the busy flag is reset (BF = 0) before the next instruction is entered.

Address Counter

The address counter (AC) stores the current address used in operations that access DDRAM or CGRAM. When an instruction contains address information, the address is stored in the address counter. Which RAM area is accessed, DDRAM or CGRAM, is also determined by the instruction that stores the address in the AC. The AC is automatically incremented or decremented after each instruction that writes or reads DDRAM or CGRAM data. The variations and options in operations that change the AC are described later in this chapter.

Display Data RAM (DDRAM)

The display data RAM area (DDRAM) is used to store the 8-bit bitmaps that represent the characters and graphics that are displayed. Display data is represented in 8-bit character codes. When equipped with the extension ,its capacity is 80 times 8 bits, or 80 characters. The memory not used for storing display characters can be used by software for storing any other 8-bit data. The mapping of DDRAM locations to the LCD display is discussed in Section 11.2.4.

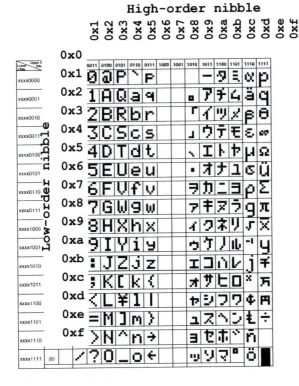

Figure 11.1 *HD44780 character set.*

Character Generator ROM (CGROM)

The character generator is a ROM that contains the bitmaps for 208 characters in 5 times 8 dot resolution or 32 characters in 5 times 10 dot resolution. Figure 11.1 shows the standard character set in the HD44780.

With a few exceptions, the characters in the range 0x20 to 0x7f correspond with those of the ASCII character set. The remaining characters are Japanese kana characters and special symbols. The characters in the range 0x0 to 0x1f which are the ASCII control characters do not respond as such in the HD44780. So sending a backspace (0x08), a bell (0x07), or a carriage return (0x0d) code to the controller has no effect.

Character Generator RAM (CGRAM)

The character generator RAM allows the creation of customized characters by defining the corresponding 5 times 8 bitmaps. Eight custom characters can be stored in the 5 times 8 dot resolution and four in the 5 times 10 resolution. The creation and use of custom characters is addressed later in this chapter.

Timing Generation Circuit

This circuit produces the timing signals for the operation of internal components circuits such as DDRAM, CGROM, and CGRAM. The timing generation circuit is not accessible to the program.

Liquid Crystal Display Driver Circuit

The liquid crystal display driver circuit consists of 16 common signal drivers and 40 segment signal drivers. The circuit responds to the number of lines and the character font selected. Once this is done, the circuit performs automatically and is not otherwise accessible to the program.

Cursor/Blink Control Circuit

The cursor and blink control circuit generates both the cursor and the character blinking. The cursor or the character blinking is applied to the character located in the data RAM address referenced in the address counter (AC).

11.1.3 Connectivity and Pinout

LCDs are powerful and complex devices. Fortunately, the programmer does not have to deal with all the complexities of LCD displays because the display hardware is usually furnished in a module that includes an LCD controller chip. Fortunately, most LCDs used in microcontroller circuits are equipped with the same controller: the Hitachi HD44780 and its upgraded versions up to the HD44780U. This controller provides a relatively simple interface between a microcontroller and the display hardware.

The fact that the HD44780 has become almost ubiquitous in LCD controller technology does not mean that these devices are without complications. The first difficulty confronted by the circuit designer is selecting the most appropriate LCD for the application, among dozens (perhaps hundreds) of available configurations, each one with its own resolution, interface technology, size, graphics options, pin patterns, and many other features. For this reason it is usually better to experiment with a simple LCD in a breadboard circuit before attempting a final circuit in hardware. The two common connectors used with the 44780-based LCDs have either 14 pins in a single row, each pin spaced 0.100 inches apart, or two rows of eight pins each, also spaced 0.100" apart. In both cases the pins are labeled in the LCD board. These connectors are shown in Figure 11.2.

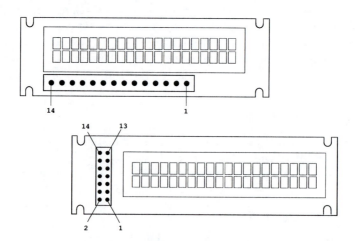

Figure 11.2 *Typical HD44780 connector pinouts.*

In LCDs with a backlight options sometimes the connectors to have two extra pins, usually numbered 15 and 16. Pin number 15 is connected to a 5V source for the backlight and pin number 16 to ground. Typical LCD wiring is shown in Table 11.1.

Table 11.1

Hitachi HD44780 LCD Controller Pin Out (80 characters or less)

PIN NO.	SYMBOL	DESCRIPTION
1	Vss	Ground
2	Vcc	Vcc (Power supply +5V)
3	Vee	Contrast control
4	RS	Set/reset 0 = instruction input 1 = data input
5	R/W	R/W (read/write select) 0 = write to LCD 1 = read LCD data
6	E	Enable. Clock signal to initiate data transfer
7	DB0	Data bus line 0
8	DB1	Data bus line 1
9	DB2	Data bus line 2
10	DB3	Data bus line 3
11	DB4	Data bus line 4
12	DB5	Data bus line 5
13	DB6	Data bus line 5
14	DB7	Data bus line 7

The pinout in Table 11.1 refers to controllers that address 80 characters or less. From the pinout in Table 11.1 it is evident that the interface to the LCD uses eight parallel lines (lines 7 to 14). However, it is also possible to drive the LCD using just four lines, which saves connections on limited circuits.

The reader should beware that LCDs are often furnished in custom boards that may or may not have other auxiliary components. These boards are often wired differently from the examples shown in Figure 11.2. In all cases the device' documentation and the corresponding data sheets should provide the appropriate wiring information.

11.2 Interfacing with the HD44780

The Hitachi 44780 controller allows parallel interfacing either using 4- or 8-bit data paths. In the 4-bit mode, each data byte is divided into a high-order and a low-order nibble and are transmitted sequentially, the high-nibble first. In the 8-bit parallel mode each data byte is transmitted from the PIC to the controller as a unit. The advantage of the 4-bit mode is greater economy of I/O lines on the PIC side. The disadvantage are slightly more complicated programming and minimally slower execution speed. Our first example and circuit uses the 8-bit data mode so as to avoid complications. Once the main processing routines are developed, we will make the necessary modifications so as to make possible the 4-bit data mode.

In addition to the data transmission mode, there are other circuit options to be considered. Two control lines between the microcontroller and the HD44780-driven

LCD are necessary in all cases: one to the RS line to select between data and instruction input modes, and another one to the E line to provide the pulse that initiates the data transfer. A third line, called the R/W control line, selects between the read and write modes of the LCD controller. This line can be either connected or grounded. If the R/W line is not connected to a microcontroller port, then the HD44780 operates only in the write data mode and all read operations will be unavailable.

11.2.1 Busy Flag and Timed Delay Options

Because many applications need not read text data from controller memory, the write-only mode is often an option, especially considering that microcontroller I/O ports are often in short supply and that this option saves one port for other duties. However, there is a less apparent drawback to no being able to read LCD data, which is that the application will not be able to monitor the busy flag. The HD44780 busy flag is mapped to data bit 7. If the busy flag is set then the controller is busy internally processing data. The busy flag clear indicates that the controller has concluded its operation and is ready to proceed with another data operation or command. Testing the BF flag requires reading bit 7, which means that not connecting the R/W line has the effect that applications cannot use the busy flag. In this case, programs can use time-delay routines to ensure that each operation completes before the next one begins.

To the circuit designer, to read or not to read controller data is a decision with several trade-offs. Using time delay routines to ensure that each controller operation has concluded is a viable option that, as already mentioned, saves one interface line. On the other hand, code that relies on timing routines is externally dependant on the clocks and timer hardware. This means that when code that relies on timing routines is ported to another circuit, with a different microcontroller, clocks, or timer hardware, the actual delays could change and the routines may fail. Furthermore, the use of delay routines is not very efficient because controller operations usually terminate before the timed delay has expired.

On the other hand, code that reads the busy flag to determine the termination of a controller operation is not without dangers. If, for any reason, the controller or the circuit fails, then the program can hangup in an endless loop waiting for the busy flag to clear. To be absolutely safe, the code would have to contain an external delay loop when testing the busy flag, so that if the external loop expires, then the processing can assume that there is a hardware problem and break out of the flag test loop. The programmer must decide whether this safety mechanism for reading the busy flag is necessary or not because its implementation requires an additional exception handler.

Another consideration is that circuits designed to read the HD44780 busy flag have good drive capabilities when sinking the line, but not when sourcing the line. This means that routines that read the state of the busy bit must introduce a 10-ms delay in order to ensure that the signal has reached its logic high. Otherwise code could read a false "not busy" state.

11.2.2 Contrast Control

In addition to the control lines that require processor interface, the HD44780 contains other control lines. One such line is used for the LCD contrast. The contrast control line (usually labeled Vee) is connected to pin number 3 (see Table 11.1). The actual implementation of the contrast control function varies with the manufacturer. In general, a for an LCD with a normal temperature range, the contrast control line can be wired as shown in Figure 11.3.

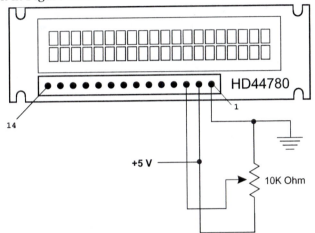

Figure 11.3 *Typical contrast adjustment circuit wiring.*

11.2.3 Display Backlight

Some LCDs are equipped with an LED backlight that serves to make the displayed characters more visible. In different LCDs the backlight is implemented differently. Some manufacturers wire the backlight directly to the LCD power supply (pin 1 and 2) while others provide additional pins that allow turning the backlight on or off independently of the LCD display. Backlit displays with fourteen pins usually belong to the first type, while those with sixteen pins have independent backlight control. If the backlight pins are adjacent to the other display pins then they are numbered 15 and 16. In this case pin number 15 is wired, through a current limiting resistor, to the +5V source and pin 16 to ground. It is also possible that the current-limiting resistor is built into the display. This information is available in the device's data sheet.

One special case to be aware of is that of some four-line displays that use pins 14 and 15 for other purposes. In these systems, backlight control, if available, is provided by other pins.

11.2.4 Display Memory Mapping

The Hitachi HD44780 is a memory mapped system in which characters are displayed by storing their ASCII codes in the memory cell associated with each digit's display area. The area of controller RAM mapped to character display memory has a capacity of 80 characters. This area is known as display data RAM or DDRAM.

In order to save circuitry, the common lines of the controller outputs to the liquid crystal display hardware are multiplexed. The duty ratio of a system is the number of multiplexed common lines, the most common being 1/16, although 1/8 and 1/11 duty ratios are also found in some systems. Because the duty ratio measures the number of multiplexed lines, it also determines the display mapping. For example, in a single line by sixteen character display with a 1/16 duty ratio, the first eight characters are mapped to one set of consecutive memory addresses and the second eight characters to another set of addresses. The reason is that in every display line, sixteen common access lines are multiplexed, instead of eight. By the same token, a two line by sixteen-character display with a 1/16 duty ratio requires sixteen common lines. In this case the address of the second lines is not a continuation of the address of the first line, but are in another address block not contiguous to the first one. Table 11.2 lists the memory address mapping of some common LCD configurations.

Table 11.2

Seven-Bit DDRAM Address Mapping for Common LCDs

CHARACTERS/ ROW	LINE NUMBER	CHARACTER NUMBER	FIRST IN GROUP	NEXT IN GROUP	LAST IN GROUP
8/1	1	1	0x00	0x01	0x07
8/2	1	1	0x00	0x01	0x07
	2	1	0x40	0x41	0x47
16/1	1	1	0x00	0x01	0x07
	1	9	0x40	0x41	0x47
16/2	1	1	0x00	0x01	0x0f
	2	1	0x40	0x41	0x4f
20/2	1	1	0x00	0x01	0x13
	2	1	0x40	0x41	0x53
24/2	1	1	0x00	0x01	0x17
	2	1	0x40	0x41	0x57
16/4	1	1	0x00	0X01	0x0f
	2	1	0x40	0x41	0x4f
	3	1	0x10	0x11	0x1f
	4	1	0x50	0x51	0x5f
20/4	1	1	0x00	0x01	0x13
	2	1	0x40	0x41	0x53
	3	1	0x14	0x15	0x27
	4	1	0x54	0x55	0x67

For example, in a typical two-line by sixteen character display, the addresses of the sixteen characters in the first line are from 0x00 to 0x0F, while the addresses of the characters in the second line are from 0x40 to 0x4F. Because there are 80 memory locations in the controller's DDRAM, each line contains storage for a total of forty characters. The range of the entire first line is from 0x00 to 0x27 (forty characters total) but of these only sixteen are actually displayed. The same applies to the second line of sixteen characters. In this case, the storage area is in the range 0x28 to 0x4f, but here again, only 16 characters are displayed. In the single-line by sixteen character display mentioned first, the addresses of the first eight characters would be a set from 0x00 to 0x07 and the addresses of the second eight characters in the line are from 0x40 to 0x47.

Notice that systems that exceed a total of 80 characters require two or more HD44780 controllers. Although the information provided in Table 11.2 corresponds with the mapping in most LCDs, it is a good idea to consult the data sheet of the specific hardware in order to corroborate the addresses mapping in a particular device before building the circuit.

Also important is to notice that Table 11.2 contains the seven low-order bits of the DDRAM addresses. HD44780 commands to set the DDRAM address for read or write operations require that the high-order bit (bit number 7) be set. Therefore, to write to DDRAM memory address 0x07, code will actually use the value 0x87; and to write to DDRAM address 0x43, code will use 0xc3 as the instruction operand.

11.3 HD44780 Instruction Set

The HD44780 instruction set includes operators to initialize the system and set operational modes, to clear the display; to manipulate the cursor; to set, reset, and control automatic display address shift; to set and reset the interface parameters; to poll the busy flag; and to read and write to CGRAM and DDRAM memory.

11.3.1 Instruction Set Overview

Pin number 4 in Table 11.1 selects two modes of operation on the HD44780 controller: instruction and data input. When the instruction mode is enabled (RS pin is set low), the controller receives commands that setup the hardware and determine its configuration and mode of operation. These commands are part of the HD44780 instruction set shown in Table 11.3.

Table 11.3

HD44780 Instruction Set in 8-Bit Data Mode

INSTRUCTION	RS	R/W	B7	B6	B5	B4	B3	B2	B1	B0	TIME (MS)
Clear Display	0	0	0	0	0	0	0	0	0	1	1..64
Return home	0	0	0	0	0	0	0	0	1	#	1.64
Entry mode set	0	0	0	0	0	0	0	1	I/D	S	37
Display/Cursor ON/OFF	0	0	0	0	0	0	1	D	C	B	37
Cursor/display shift	0	0	0	0	0	1	S/C	R/L	#	#	37
Function set	0	0	0	0	1	DL	N	F	#	#	37
Set CGRAM address	0	0	0	1	--- ----- address ----------						37
Set DDRAM address	0	0	1	-------------- address ----------							37
Read busy flag and address register	0	1	BF----- -------- address ----------								0
Write data	1	0	------------------ data -------------								37
Read data	0	1	------------------ data -------------								37

Note: Bits labeled # have no effect.

Clearing the Display

The instruction clears the display with blanks by writing the code 0x20 into all DDRAM addresses. It also returns the cursor to the home position (top-left display corner) and sets address 0 in the DDRAM address counter. After this command executes, the display disappears and the cursor goes to the left edge of the display.

Return Home

Returns the cursor to home position, which is the upper-leftmost position of the first character line. Sets DDRAM address 0 in address counter. Sets the display to its default status if it was shifted. DDRAM contents remain unchanged.

Entry Mode Set

Sets the direction of cursor movement and the display shift mode. If B1 bit is set (I/D bit in Table 11.3)t, cursor handing is set to the increment mode, that is, left-to-right. If this bit is clear then cursor movement is set to the decrement mode, that is, right-to-left. If B0 bit is set (S bit in Table 11.3), display shift is enabled. In the display shift more, it appears as if the display moves instead of the cursor, Otherwise display shift is disabled. Operations that read or write to CGRAM and operations that read DDRAM will not shift the display.

Display and Cursor ON/OFF

If B2 bit is set (D in Table 11.3), display is turned on. Otherwise it is turned off. When the display is turned off data in DDRAM is not changed. If B1 bit is set (C in Table 11.3), the cursor is turned on, otherwise it is turned off. Operations that change the current address in the DDRAM address register, like those to automatically increment or decrement the address, are not affected by turning off the cursor. The cursor is displayed as the eight line in the 5 x 8 character matrix. If B0 (B in Table 11.3), the character at the current cursor position blinks, otherwise the character does not blink. Note that character blinking and cursor are independent operations and that both can be set to work simultaneously.

Cursor/Display Shift

Moves the cursor or shifts the display according to the selected mode. The operation does not change the DDRAM contents. Because the cursor position always coincides with the value in the address register, the instruction provides software with a mechanism for making DDRAM corrections or to retrieve display data at specific DDRAM locations. Table 11.4 lists the four available options:

Function Set

Sets the parallel interface data length, the number of display lines, and the character font. If B4 bit is set (DL bit in Table 11.3), then the interface is set to 8 bits. Otherwise it is set to 4 bits. If B3 bit is zero (N bit in Table 11.3), the display is initialized for 1/8 or 1/11 duty cycle. When the N bit is set, the display is set to 1/16 duty cycle. Displays with multiple lines typically use the 1/16 duty cycle. The 1/16 duty cycle on a one-line display appears as if it were a two-line display, that is, the line consists of two separate address groups (see Table 11.2).

Table 11.4

Cursor/Display Shift Options

BITS S/C	R/L	OPERATION
0	0	Cursor position is shifted left. Address counter is decremented by one.
0	1	Cursor position is shifted right. Address counter is incremented by one.
1	0	Cursor and display are shifted left.
1	1	Cursor and display are shifted right.

If B2 bit is set (F in Table 11.3), then the display resolution is 5 times 10 pixels. Otherwise, the resolution is 5 times 8 pixels. This bit is not significant when the 1/16 duty cycle is selected, that is, when the N bit is set. The function set instruction should be issued during controller initialization. No other instruction can be executed before this one, except changing the interface data length.

Set CGRAM Address

Sets the CGRAM (character generator RAM) address to which data is sent or received after this operation. The CGRAM address is a six-bit field in the range 0 to 64 decimal. Once a value is entered in the CGRAM address register, data can be read or written from CGRAM.

Set DDRAM Address

Sets the DDRAM (display data RAM) address to which data is sent or received after this operation. The DDRAM address is a seven-bit field in the range 0 to 127 decimal. Once a value is entered in the DDRAM address register, data can be read or written from CGRAM. DDRAM address mapping was discussed previously in this chapter.

Read Busy Flag and Address Register

Reads the busy flag to determine if an internal operation is in progress and reads the address counter contents. The value in the address register is reported in bits 0 to 6. Bit 7 (BF) is the busy flag bit. This bit is read only. The address counter is incremented or decremented by 1 (according to the mode set) after the execution of a data write or read instruction.

Write Data

Writes eight data bits to CGRAM or DDRAM. Before data is written to either controller RAM area, software must first issue a set DDRAM address or set CGRAM address instruction (described previously). These two instructions not only set the next valid address in the address register, but also select either CGRAM or DDRAM for writing operations. What other actions take place as data is written to the controller depends on the settings selected by the entry mode set instruction. If the direction of cursor movement or data shift is in the increment mode, then the data write operation will add one to the value in the address register. If the cursor movement is enabled, then the cursor will be moved accordingly after the data write takes place. If the display shift mode is active, then the displayed characters will be shifted either right or left.

Read Data

Reads eight data bits to CGRAM or DDRAM. Before data is read from either controller RAM area, software must first issue a set DDRAM address or set CGRAM address instruction. These instructions not only set the next valid address in the address register, but also select either CGRAM or DDRAM for writing operations. Failing to set the corresponding RAM area results in reading invalid data.

What other actions take place as data is read from the controller RAM depends on the settings selected by the entry mode set instruction. If the direction of cursor movement or data shift is in the increment mode, then the data read operation will add one to the value in the address register. However, the display is not shifted by a read operation even if the display shift is active.

The cursor shift instruction has the effect of changing the contents of the address register. So if a cursor shift precedes a data read instruction, there is no need to reset the address by means of an address set command.

11.3.2 18F452 8-Bit Data Mode Circuit

The purpose of the circuit in Figure 11.4 is to provide a simple hardware that can be used to develop and exercise LCD display functions.

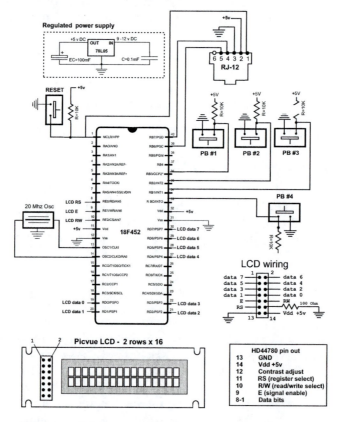

Figure 11.4 *18F452 to LCD 8-bit mode circuit.*

The circuit in Figure 11.4 uses 8-bit parallel data transmission interfacing with an 18F452 microcontroller. Three control lines interface the microcontroller and the LCD. As previously mentioned, the R/W line is not considered necessary because it is possible to build a system that does not read LCD data. Nevertheless, we have included the R/W line because it will allow us to read the busy flag in synchronizing operations. Table 11.5 shows the control and data connections for the circuit in Figure 11.4.

Table 11.5

Connections for 18F452/LCD 8-Bit Data Mode Circuit

PIN	18F452 PORT BIT	LCD PIN	NAME	LINE FUNCTION
1	RE0	4	RS	Select instruction/data register
2	RE2	5	R/W	Read/write select
18	RE1	6	E	Enable signal
13	RD7	14	BF	Busy flag
6–13	RD0-RD7	7–14	Data	Data lines

11.4 LCD Programming

Programming an LCD is a device-specific operation. Before attempting to write code, the programmer should become familiar with the circuit wiring diagram, the set-up parameters, and the specific hardware requirements. It is risky to make assumptions that a particular device conforms exactly to the HD44780 interface; fortunately, a style sheet will usually contain specifications that may not be in strict conformance with the standard. In addition to the PIC setup and initialization functions, code to display a simple text message on the LCD screen consists of the following display-related functions:

1. Define the required constants, variables, and buffers.

2. Set up and initialize ports used by the LCD.

3. Initialize the LCD to circuit and software specifications.

4. Select DDRAM start address on LCD.

5. Display text by transferring characters to LCD DDRAM.

If the LCD display consists of multiple lines, then Steps 4 and 5 are repeated for each line. LCD initialization and display operations vary according to whether the interface is 4- or 8-bits and whether the code uses delay loops of busy flag monitoring to synchronize operations. All of these variations are considered in the examples in this chapter.

11.4.1 Defining Constants and Variables

In any program, defining and documenting constants and fixed parameters should be done centrally, rather than hard-coded through the code. Centralizing these elements makes it possible to adapt code to circuit and hardware changes.

Constants

The 18F family PICs provide two directives for defining constants: the C-like #define directive and the equ (equate) directive. Which one to use is, in most cases, a matter of personal preference. Perhaps a general guideline is to use the #define statement to create literal constants, that is, constants that are not associated with program registers or variables. The equate directive is then used to define registers, flags, and local variables.

According to this scheme, an LCD display driver program could use #define statements to create literals that are related to the wiring diagram or the specific LCD values obtained form the data sheet, such as the DDRAM addresses for each display line, as in the following code fragment:

```
;=======================================================
;                   constant definitions
;   for PIC-to-LCD pin wiring and LCD line addresses
;=======================================================
;                      PORTE bit
#define E_line  1              1      ; |
#define RS_line 0              0      ; |  -- from wiring diagram
#define RW_line 2              2      ; |
; LCD line addresses (from LCD data sheet)
#define LCD_1 0x80 ; First LCD line constant
#define LCD_2 0xc0 ; Second LCD line constant
```

By the same token, the values associated with PIC register addresses and bit flags are defined using equates, as follows:

```
;=======================================================
;                   PIC register equates
;=======================================================
PORTA       equ        0x05
PORTB       equ        0x06
fsr         equ        0x04
status      equ        0x03
indf        equ        0x00
z           equ        0x02
```

One advantage of this scheme is that constants are easier to locate because they are grouped by device. Those for the LCD are in #define directives area and those for the PIC hardware in an area of equate directives.

There are also drawbacks to this approach, as symbols created in the #define directive are not available for viewing in the MPLAB debuggers. However, if the use of the #define directive is restricted to literal constants, then their viewing during a debugging session will not be missed because it will not change during program execution. MPLAB also supports the constant directive for creating a constant symbol. Its use is identical to the equ directive but the latter is more commonly found in code.

11.4.2 Using MPLAB Data Directives

The definition of data items changes when programming in absolute or relocatable modes. Relocatable mode is discussed in Appendix B, Section B.4.1.

Data Definition in Absolute Mode

Often a program executing in absolute mode needs to define a block of sequential symbols and assign to each one a corresponding name. In the PIC 18F452, the address space allocated to general-purpose registers allocated by the user is 4096 bytes, starting at address 0x00. One possible way of allocating user-defined registers is to use the equ directive to assign addresses in the PIC SRAM space; for example,

```
Var1    equ     0x00
Var2    equ     0x01
Var3    equ     0x02
Buf1    equ     0x03 ; 10-byte buffer space
Var4    equ     0x0d ; Next variable
```

Although this method is functional, it depends on the programmer calculating the location of each variable in the PIC's available SRAM space. Alternatively, MPLAP provides a cblock directive that allows defining a group of consecutive sequential symbols while referring only to the address of the first element in the group. If no address is entered in a cblock, then the assembler will assign an address of one higher than the final one in the previous cblock. Each cblock ends with the endc directive. The following code fragment showing the use of the cblock directive is from one of the sample programs for this chapter.

```
;=======================================================
;               variables in PIC RAM
;=======================================================
; Reserve 16 bytes for string buffer
    cblock    0x00
    strData
    endc
; Leave 16 bytes and continue with local variables
    cblock    0x10 ; Start of block
    count1         ; Counter # 1
    count2         ; Counter # 2
    count3         ; Counter # 3
    pic_ad         ; Storage for start of text area
    J          ; counter J
    K          ; counter K
    index          ; Index into text table
    endc
```

Notice in the preceding code fragment that the allocation for the 16-byte buffer space named strData is ensured by entering the corresponding start address in the second cblock. The PIC microcontrollers do not contain a directive for reserving memory areas inside a cblock, although the res directive can be used to reserve memory for individual variables.

Relocatable Code

Relocatable code builds correctly if data is defined using the equ directives previously mentioned. However, the equ directive is likely to generate linker errors. Additionally, variables defined with the equ directive are not visible to hardware debuggers (see Appendix B). MPLAB MASM supports several directives that are compatible with relocatable code and that make the variable names visible at debug time. The ones most often used are

> **udata** defines a section of uninitialized data. Items defined in a udata section are not initialized and can be accessed only through their names.

> **udata_acs** defines a section of uninitialized data that is placed in the access area. In PIC 18 devices access RAM is always used for data defined with the udada_acs directive. Applications use this area for the data items most often used.

> **udata_ovr** defines a section of ovr uninitialized, overlaid data. This data section is used for variables that can be declared at the same address as other variables in the same module or in other linked modules, such as temporary variables.

> **udata_shr** defines a section of uninitialized, shared data. This directive is used in defining data sections for PIC12/16 devices.

> **idata** defines a section of initialized data. This directive forces the linker to generate a lookup table that can be used to initialize the variables in this section to the specified values. When linked with MPLAB C18 code, these locations are initialized during execution of the start-up code. Issues regarding the idata and idata_acs directives aare discussed in the next subsection.

> **idata_acs** defines a section of initialized data that is placed in the access area. In PIC 18 devices, access RAM is always used for data defined with the idada_acs directive.

The following example shows the use of several RAM allocation directives:

```
     udata_acs 0x10 ; Allocated at address 0x10
j    res 1              ; Data in access bank
temp res 1
     idata
ThisV    db   0x29      ; Initialized data
Aword    dw   0xfe01
*    udata              ; Allocated by the Linker
varx res 1              ; One byte reserved
vary res 1              ; Another byte
```

The location of a section may be fixed in memory by supplying the optional address, as in the udata_acs example listed previously. If more than one of a section type is specified, each one must have a unique name. If a name is not provided, the default section names are .idata, .udata, .udata_acs, .udata_shr and .udata_ovr.

Issues with Initialized Data

The directives db, dw, data, and others can be used when defining initialized data in an idata or idata_acs section of an assembly language program. The db directive defines successive bytes of data while the dw directive defines successive words. The defined data can be initialized; for example,

```
     idata_acs  0x010
Var1 db    1,2,3
Var2 dw    0x1234,0x5678
String     data       "This is a test",0
```

The previous data definition seem to imply that an assembly language program can use the idata or idata_acs directives to define data items and strings that can later be accessed by their defined names. In reality this is not the case because the various versions of the idata directive do not initialize data in the PIC's data memory. Furthermore, MPLAB documentation states that the linker will generate a lookup table that can be used to initialize variables defined in the idata section but the location of this data and access to it is not documented.

The issue may be a moot one because the 18F PICs have available one Mbyte of program memory space, sufficient to holds 1,048,575 data bytes, while there are only 4,095 bytes of data memory. This means that applications may usually find that program memory is more abundant for storing program strings, and other items, than data memory.

11.4.3 LCD Initialization

LCD initialization depends on the specific hardware in use and on the circuit wiring. Information regarding the specific LCD can be obtained from the device's data sheet. Sometimes the data sheet includes examples of initialization values for different conditions and, in some cases, code listings. The information is usually sufficient to ensure correct initialization.

Popular LCD literature found online often contains initialization "myths" for requiring that a certain mystery code be used for no documented reason, or that a certain function be repeated a number of times. These code myths often result from trial-and-error programming and are not based in fact. The programmer should make sure that the code is rational and that every operation is actually required and documented.

Reset Function

The HD44780 is equipped with an internal reset circuit that performs an initial and automatic initialization on power up. Only if the minimal power supply conditions are not met will the automatic initialization fail. In the discussions that follow we assume that the internal reset circuit is operating properly and that the HD44780 has been initialized. The following operating conditions are enabled by the reset function:

- Display is cleared
- Function set to 8-bit interface, one line display, 5 times 8 dot character font
- Display is off, cursor is off, blinking is off
- Entry mode is set to increment by 1 and no shift

The busy flag is kept in the busy state (BF = 1) during device initialization. One way to determine if the automatic initialization has concluded is by testing the busy flag, which will be cleared in this case. If the internal reset did not take place, it is possible to perform a complete initialization by instruction, which includes a few more steps than if the internal reset had taken place. The Hitachi HD44780U data sheet provides a listing of the necessary operations.

Initialization Commands

If the internal reset took place it is necessary to set the communications line before the remaining initialization commands are presented to the LCD. This means that the E line, RS line and RW line must be all be set low. After the lines are set accordingly there should be a 125 millisecond delay. The following code fragment shows the processing:

```
bcf   PORTA,E_line   ; E line low
bcf   PORTA,RS_line  ; RS line low for command
bcf   PORTA,RW_line   ; Write mode
call delay_125        ;delay 125 microseconds
```

The procedure delay_125 in the previous code fragment is described later in this chapter.

Function Preset Command

Because the reset function has set the device in 8-bit data mode, the first command consists of writing 8 bits to the data register. If the low-order bits are not wired to the controller, as is the case in the 4-bit data mode, then these bits are ignored. In other words, the first initialization instruction completes with a single byte in either display mode. The Function Preset command can be coded as follows:

```
;*********************|
;    Preset command    |
;*********************|
    movlw    B'00100000'
    movwf    PORTD
    call     pulseE
```

The pulseE procedure contains code to bring the E line low and then high (pulse), which is required after each data read, write or command operation. The procedure is listed later in this section.

Function Set Command

Function set is the next initialization command sent to the LCD. The command determines whether the display font consists of 5 x 10 or 5 x 7 pixels. The latter is by far the more common. Also the duty cycle, which is typically 1/8 or 1/11 for single-line displays and 1/16 for multiple lines. The interface width is also determined in the Function Set command. It can be 4-bits or 8-bits. The following code fragment shows the commented code for the Function Set command.

```
;*********************|
;    Function Set      |
;*********************|
    movlw    0x38 ; 0 0 1 1 1 0 0 0 (FUNCTION SET)
             ;        | | | |__ font select:
             ;        | | |    1 = 5x10 in 1/8 or 1/11 dc
             ;        | | |    0 = 1/16 dc
             ;        | | |___ Duty cycle select
```

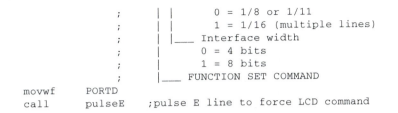

```
         ;       | |        0 = 1/8 or 1/11
         ;       | |        1 = 1/16 (multiple lines)
         ;       | |___ Interface width
         ;       |        0 = 4 bits
         ;       |        1 = 8 bits
         ;       |___ FUNCTION SET COMMAND
    movwf    PORTD
    call     pulseE      ;pulse E line to force LCD command
```

In the preceding code fragment, the LCD is initialized to multiple lines, 5 times 7 font, and 8-bit interface, as in the program LCD_18F_HelloWorld listed later in this chapter and found in the book's online software package. Note that bit number 4 will be cleared if initializing for a 4-bit data mode.

The procedure named pulseE sets the E line bit off and on to force command recognition by the LCD. The procedure is detailed later in this chapter.

Display Off

Some initialization routines in LCD documentation and data sheets require that the display be turned off following the Function Set command. If so, the Display Off command can be executed as follows:

```
;********************|
;    Display Off     |
;********************|
    movlw    0x08 ; 0 0 0 0 1 0 0 0 (DISPLAY ON/OFF)
             ;             | | | |___ Blink character at cursor
             ;             | | |      1 = on, 0 = off
             ;             | | |___ Curson on/off
             ;             | |      1 = on, 0 = off
             ;             | |____ Display on/off
             ;             |       1 = on, 0 = off
             ;             |____ COMMAND BIT
    movwf    PORTB
    call pulseE    ; pulse E line to force LCD command
```

Display and Cursor On

Whether or not the display is first turned off, it must be turned on. Also code must select if the cursor is on or off, or whether the character at the cursor position is to blink. The following command sets the cursor and the display on and the character blink off.

```
;********************|
; Display and Cursor On |
;********************|
    movlw    0x0e ; 0 0 0 0 1 1 1 0 (DISPLAY ON/OFF)
             ;             | | | |___ Blink character at cursor
             ;             | | |      1 = on, 0 = off
             ;             | | |___ Curson on/off
             ;             | |      1 = on, 0 = off
             ;             | |____ Display on/off
             ;             |       1 = on, 0 = off
             ;             |____ COMMAND BIT
    movwf    PORTD
    call     pulseE    ; pulse E line to force LCD command
```

Set Entry Mode

The Entry Mode command sets the direction of cursor movement or display shift mode. Normally the display is set to the increment mode when writing in Western European languages. The Entry Mode command also controls the display shift. If enabled, the displayed characters appear to scroll. This mode can be used to simulate an electronic billboard by storing more than one line of characters in DDRAM and the then scrolling the characters left-to-right. The following code sets the entry mode to the increment mode and no shift.

```
;*********************|
;    Set Entry Mode   |
;*********************|
     movlw     0x06 ; 0 0 0 0 0 1 1 0 (ENTRY MODE SET)
               ;              | | |___ display shift
               ;              | |      1 = shift
               ;              | |      0 = no shift
               ;              | |____ cursor increment mode
               ;              |       1 = left-to-right
               ;              |       0 = right-to-left
               ;              |___ COMMAND BIT

     movwf     PORTD
     call      pulseE
```

Cursor and Display Shift

This commands determines whether the cursor or the display shifts according to the selected mode. Shifting the cursor or the display provides a software mechanism for making DDRAM corrections or for retrieving display data at specific DDRAM locations. The four available options appear in Table 11.4 previously in this chapter. The following instructions set the cursor to shift right and disables display shift.

```
;*********************|
; Cursor/Display Shift |
;*********************|
     movlw     0x14 ; 0 0 0 1 0 1 0 0 (CURSOR/DISPLAY SHIFT)
               ;            | | | |_|___ don't care
               ;            | |_|__ cursor/display shift
               ;            |       00 = cursor shift left
               ;            |       01 = cursor shift right
               ;            |       10 = cursor and display
               ;            |            shifted left
               ;            |       11 = cursor and display
               ;            |            shifted right
               ;            |___ COMMAND BIT
     movwf     PORTD    ;0001 1111
     call      pulseE
```

Clear Display

The final initialization command is usually one to clear the display. It is entered as follows:

```
;*********************|
;    Clear Display    |
;*********************|
```

```
movlw      0x01 ; 0 0 0 0 0 0 0 1 (CLEAR DISPLAY)
           ;                  |___ COMMAND BIT
movwf      PORTD      ;0000 0001
call       pulseE
call       delay_5    ;delay 5 milliseconds after init
```

Notice that the last command is followed by a 5 millisecond delay. The delay procedure delay_5 is listed and described later in this chapter.

11.4.4 Auxiliary Operations

Several support routines are required for effective text display in LCD devices. These include time delay routines for timed access, a routine to pulse the E line in order to force the LCD to execute a command or to read or write text data, a routine to read the busy flag when this is the method used for processor/LCD synchronization, and routines to merge data with port bits so as to preserve the status of port lines not being addressed by code.

Time Delay Routine

In Chapter 9 we discussed several ways of producing time delays in the 18F PIC family of microcontrollers. The present task is to develop a software routine that ensures the time delay that must take place in LCD programming, as shown in Table 11.3.

One delay mechanism uses the TIMER0 module, which is a built-in 16-bit timer counter (discussed in Chapter 9). Once enabled, port A pin 4, labeled the TOCKI bit and associated with TMR0, can be used to time processor operations. In the particular case of LCD timing routines, using the TIMER0 module seems somewhat of an overkill, in addition to the fact that it requires the use of a port A line, which is often needed for other purposes.

Alternatively, timing routines that serve the purpose at hand can be developed using simple delay loops. In this case, no port line is sacrificed, and coding is considerably simplified. These routines are generically labeled software timers, in contrast to the hardware timers that depend on the PIC timer/counter devices. Software timers provide the necessary delay by means of program loops, that is, by wasting time. How long a delay is provided by the routine depends on the execution time of each instruction and on the number of repeated instructions.

Recall now that instructions on the PIC 18F452 consume four clock cycles. This means that if the processor clock is running at 4 MHz, then one-fourth of 4 MHz will be the execution time for each instruction, which is 1 µs. So if each instruction requires 1 µs, repeating 1,000 instructions will produce a delay of 1 ms. The following routines provide convenient delays for LCD interfacing.

```
;========================
;  Procedure to delay
;    125 microseconds
;========================
delay_125mics:
     movlw      D'42'           ; Repeat 42 machine cycles
     movwf      count1          ; Store value in counter
repeat:
```

```
    decfsz     count1,f    ; Decrement counter (1 cycle)
    goto       repeat      ; Continue if not 0 (2 cycles)
                           ; 42 * 3 = 126
    return                 ; End of delay
;=======================
;  Procedure to delay
;    5 milliseconds
;=======================
delay_5ms:
    movlw      D'41'       ; Counter = 41
    movwf      count2      ; Store in variable
delay:
    call       delay_125mics ; Delay 41 microseconds
    decfsz     count2,f    ; 41 times 125 = 5125 micro sec.
                           ; or approximately 5 ms
    goto delay
    return                 ; End of delay
```

Actually, the delay loop of the procedure named delay_5ms is not exactly the product of 41 iterations times 125 µs, as the instruction to decrement the counter and the goto to the label delay are also inside the loop. That means that three instruction cycles must be added to those consumed by the delay_125mics procedure. This results in a total of 41 * 3 or 123 instruction cycles that must be added to the 5,125 consumed by delay_125mics. In fact, there are several other minor delays by the instructions to initialize the counters that are not included in the calculation. In reality, the delay loops required for LCD interfacing need not be exact, as long as they are not shorter than the recommended minima.

In calculating software delays, it is important to recall that, in the case of the 18F452 the instruction execution time is determined by an external clock either in the form of an oscillator crystal, a resonator, an RC oscillator, or another compatible timing device in the circuit. The following are 18F452 oscillator modes:

The PIC 18F452 is available in various processor speeds, from 4 MHz to 20 MHz. These speeds describe the maximum capacity of the PIC hardware. The actual instruction speed is determined by the clocking device, so a 20-MHz 18F452 using a 4-MHz oscillator effectively runs at 4 MHz.

Pulsing the E Line

The LCD hardware does not recognize data as it is placed in the input lines. When the various control and data pins of the LCD are connected to ports in the PIC and data is placed in the port bits, no action takes place in the LCD controller. In order for the controller to respond to commands or to perform read or write operations, it must be activated by pulsating (sometimes called "strobing") the E line. The pulsing or strobing mechanism requires that the E line be kept low, then raised momentarily. The LCD checks the state of its lines on the raising edge of the E line. Once the command has completed, the E line is brought low again. The following code fragment pulses the E line in the manner described:

```
;=======================
;     pulse E line
;=======================
pulseE
```

```
        bsf       PORTA,E_line   ; pulse E line
        bcf       PORTA,E_line
        call      delay_125mics  ; delay 125 microseconds
        return
```

Notice that the pulsing of the E line is followed by a 125-microsecond delay. This delay is not part of the pulse function but is required by most LCD hardware. Also noticeable is that some pulse functions in the popular PIC literature include a no operation opcode (nop) between the commands to set and clear the E line. In most cases, this short delay does not hurt, but some LCDs require a minimum time lapse during the pulse and will not function correctly if the nop is inserted in the code.

Reading the Busy Flag

We have mentioned that synchronization between LCD commands and between data access operations can be based on time delay loops or on reading the LCD busy flag. The busy flag, which is in the same pin as the bit 7 data line, is cleared when the LCD is ready to receive the next command, read or write operation, and set if the device is not ready. Reading the state of the busy flag code can accomplish more effective synchronization than time delay loops. The sample program named LCD_18F_MsgFlag, in the book's online software package, performs LCD display using the busy flag method. The following procedure shows busy flag synchronization.

```
;========================
; busy flag test routine
;========================
; Procedure to test the HD44780 busy flag
; Execution returns when flag is clear
; Tech note:
; The HD44770 requires a 10 ms delay to set the busy flag
; when the line is being sourced. Reading the flag before
; this time has elapsed may result in a false "not busy"
; report. The code forces a 15 machine cycle delay to
; prevent this error.
busyTest:
        movlw     .15                 ; Repeat 15 machine cycles
        movwf     count1              ; Store value in counter
delay15:
        decfsz    count1,f            ; Decrement counter
        goto      delay15             ; Continue if not 0
; Delay concluded
        movlw     B'11111111'
        movwf     TRISD
        bcf       PORTE,RS_line ; RS line low for control
        bsf       PORTE,RW_line ; Read mode
is_busy:
        bsf       PORTE,E_line  ; E line high
        movff     PORTD,WREG    ; Read port D into W
                                ; Port D bit 7 is busy flag
        bcf       PORTE,E_line  ; E line low
        andlw     0x80          ; Test bit 7, high is busy
; Anding with a the mask 10000000 sets the Z flag if
; preserves the busy bit was clear:
;       Bxxx xxxx |
;                 | logical AND
;       1000 0000 |
;-------------------
```

```
;       0000 0000 if B bit was 0 (Z flag is set)
;       1000 0000 if B bit was 1 (Z flag is clear)
    btfss     STATUS,Z       ; Test zero bit in STATUS
    goto      is_busy        ; Repeat if Z flag clear
;                            ; indicating not busy
; At this point busy flag is clear
; Reset R/W line and port D to output
    bcf       PORTE,RW_line  ; Clear R/W line
    bsf       PORTE,RS_line  ; Setup for data
    clrf      TRISD
    return
```

Note that testing the busy flag requires setting the LCD in read mode, which in turn requires that there be a connection between a PIC port and the R/W line. The logic used by the procedure could appear a little convoluted because code is testing not the busy flag of the LCD controller, but the Z flag of the PIC after an AND instruction with a mask. Because the mask has the high bit set, the AND operation sets the Z flag in the STATUS register if the busy flag was clear. Otherwise, the Z flag is clear. In other words, after the AND operation, the Z flag is opposite to the state of the busy bit. Alternatively, the logic can be coded as follows:

```
is_busy:
    bsf  PORTE,E_line   ; E line high
    movff     PORTD,WREG      ; Read port D into W
                             ; Port D bit 7 is busy flag
    bcf       PORTE,E_line    ; E line low
    btfsc     PORTD,7         ; Test bit directly
    goto      is_busy         ; Repeat if Z flag clear
; Execution continues if not busy
```

Also note that the listed procedure contains no safety mechanism for detecting a hardware error condition in which the busy flag never clears. If such were the case, the program would hang in a forever loop. To detect and recover from this error, a clean code routine would have to include an external timing loop or some other means of recovering a possible hardware error.

Bit Merging Operations

PIC/LCD circuits often use some of the lines in an individual port while leaving others for other purposes. In this case, it is convenient that the routines that manipulate PIC/LCD port access do not change the settings of other port bits. Note that this situation is not exclusive of LCD interfacing and that the discussion that follows has general application in PIC programming.

A processing routine can be developed in order to change one or more port lines without affecting the remaining ones. For example, an application that uses a 4-bit interface between the PIC and the LCD typically leaves four unused lines in the access port, or uses some of these lines for interface connections. In this case, the programming problem can be described as merging bits of the data byte to be written to the port with some existing port bits. In this case, one operand is the access port value and the other one is the new value to write to this port. If the operation at hand uses the four high-order port bits, then its four low-order bits must be preserved. The logic required is simple: AND the corresponding operands with masks that clear the unneeded bits and preserve the significant ones, then OR the two operands. The following procedure shows the required processing:

```
;==================
;   merge bits
;==================
; Routine to merge the 4 high-order bits of the
; value to send with the contents of port B
; so as to preserve the 4 low-bits in port B
; Logic:
;      AND value with 1111 0000 mask
;      AND port B with 0000 1111 mask
;      At this point low nibble in the value and high
;      nibble in port B are all 0 bits:
;          value = vvvv 0000
;         port B = 0000 bbbb
;      OR value and port B resulting in:
;                  vvvv bbbb
; ON ENTRY:
;      w contains value bits
; ON EXIT:
;      w contains merged bits
merge4:
      andlw    b'11110000'    ; ANDing with 0 clears the
                              ; bit. ANDing with 1 preserves
                              ; the original value
      movwf    store2         ; Save result in variable
      movf     PORTB,w        ; port B to w register
      andlw    b'00001111'    ; Clear high nibble in port B
                              ; and preserve low nibble
      iorwf    store2,w       ; OR two operands in w
      return
```

Note that this particular example refers to merging two operand nibbles. The code can be adapted to merging other size bit fields by modifying the corresponding masks. For example, the following routine merges the high-order bit of one operand with the seven low-order bits of the second one:

```
; Routine to merge the high-order bit of the first operand with
; the seven low-order bits of the second operand
; ON ENTRY:
;          w contains value bits of first operand
;          port B is the second operand
merge1:
      andlw    b'10000000'    ; ANDing with 0 clears the
                              ; bit. ANDing with 1 preserves
                              ; the original value
      movwf    store2         ; Save result in variable
      movf     PORTB,w        ; port B to w register
      andlw    b'01111111'    ; Clear high-order  bit in
                              ; port B and preserve the seven
                              ; low order bits
      iorwf    store2,w       ; OR two operands in w
      return
```

The popular PIC literature contains routines to merge bit fields by assuming certain conditions in the destination operand, then testing the first operand bit to determine if the assumed condition should be preserved or changed. This type of operation has sometimes been called "bit flipping"; for example,

```
flipBit7:
; Code fragment to test the high-order bit in the variable named
; oprnd1 and preserve its status in the register variable PORTB
    bcf         PORTB,7        ; Assume oprnd1 bit is reset
    btfsc       oprnd1,7       ; Test operand bit and skip if
                               ; clear (assumption valid)
    bsf         PORTB,7        ; Set bit if necessary
    return
```

The logic in bit flipping routines has one critical flaw: if the assumed condition is false, then the second operand is changed improperly, alas for only a few microseconds. However, the incorrect value can produce errors in execution if it is used by another device during this period. Because there is no such objection to the merge routines based on masking, the programmer should always prefer them.

11.4.5 Text Data Storage and Display

Text display operations require some way of generating the ASCII characters that are to be stored in DDRAM memory. In Section 11.4.2 we discussed the use of several data directives and observed that, although the PIC Assembler contains several operators to generate ASCII data in program memory, there is no convenient way of storing a string in the general-purpose register area. Even if this were possible, SRAM is typically in short supply and text strings gobble up considerable data space.

Several possible approaches are available; the one most suitable usually depends on the total string length to be generated or stored, whether the strings are reused in the code, and to other program-related circumstances.

In this sense, short text strings can be produced character-by-character and sent sequentially to DDRAM memory by placing the characters in the corresponding port and pulsing the E line. The following code fragment consecutively displays the characters in the word "Hello" in this manner. Code assumes that the command to set the Address Register has previously been entered.

```
; Generate characters and send directly to DDRAM
    movlw       'H'            ; ASCII for H in w
    movwf       PORTB          ; Store code in port B
    call        pulseE         ; Pulse E line
    movlw       'e'            ; Continues
    movwf       PORTB
    call        pulseE
    movlw       'l'
    movwf       PORTB
    call        pulseE
    movlw       'l'
    movwf       PORTB
    call        pulseE
    movlw       'o'
    movwf       PORTB
    call        pulseE
    call        delay_5
```

Notice in the preceding fragment that the code assumes that the LCD has been initialized to automatically increment the Address Register left-to-right. This explains why the Address Register is bumped by the code to the next address with each port access.

The sample program LCD_18F_HelloWorld in this book's software package uses the direct display technique described in this subsection.

Generating and Storing a Text String

An alternative approach suitable for generating and displaying longer strings consists of storing the string data in a local variable (sometimes called a buffer) and then transferring the characters, one by one, from the buffer to DDRAM. This kind of processing has the advantage of allowing the reuse of the same string and the disadvantage of using up scarce data memory. The logic for one possible routine consists of first generating and storing the character string in program memory, then retrieving the characters one-by-one and displaying them. The character generation and storage logic is shown in Figure 11.5.

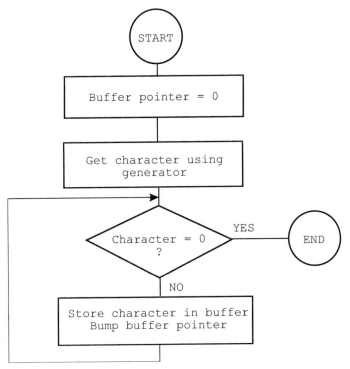

Figure 11.5 *Flowchart showing string generation logic.*

Applications that add characters or modify the text string during program execution often use this technique because the data memory string is easily accessed at run time.

Data in Program Memory

The mid-range PIC family of microcontrollers (such as the popular 16F series) provides a way of storing and retrieving data in program memory using a data table and indirect addressing. The retlw instruction furnishes a way of returning a table character in the W register. This mechanism works with the 18F PIC family, but not without complications, as is the case when the table data steps over a code segment boundary. The new instructions of the 18F PICs are more convenient and can easily access data located anywhere in the processor's address space.

Because the program memory space of the 18F devices extends to 21 bits and storage and control registers are 8-bits wide, three such registers are required for a complete address. The two low-order registers will have 8 bits capacity each, while the highest-order register (referred to as the upper register) holds the remaining 5 bits of the 21-bit address. The following code snippet from the sample program LCD_18F_4line_Reloc.asm shows storage and access to a text string in program memory:

```
;==========================
;    get table character
;==========================
; Local procedure to get a single character from a local
; table (msgTable) in program memory. Variable index holds
; offset into table
tableChar:
    movlw   UPPER msgTab
    movwf   TBLPTRU
    movlw   HIGH msgTab       ; Get address of Table
    movwf   TBLPTRH           ; Store in table pointer low register
    movlw   LOW msgTab        ; Get address of Table
    movwf   TBLPTRL
    movff   index,WREG        ; index to W
    addwf   TBLPTRL,f         ; Add index to table pointer low
    clrf    WREG              ; Clear register
    addwfc  TBLPTRH,F         ; Add possible carry
    addwfc  TBLPTRU,F         ; To both registers
    tblrd   *                 ; Read byte from table (into TABLAT)
    movff   TABLAT,WREG       ; Move TABLAT to W
    return

; Define table in code memory
msgTab      db    "4-line LCD Demo "
            db    "Relocatable mode"
```

The code uses the operators UPPER, HIGH, and LOWER to obtain the three corresponding elements of the 21-bit address. These are stored in the special function registers named TBLPTRU, TBLPTRH, and TBLPTRL. Because the table can straddle an 8-bit boundary, the possible carry must be added with the addwfc instructions. The table character is returned in WREG.

Displaying the Text String

A character retrieved from its storage (a local data memory buffer or a code memory string) can be displayed by moving each ASCII code from WREG into LCD DDRAM. The code to display the text character assumes that the LCD has previously been set in the auto increment mode during initialization and that the Address Register has been properly initialized with the corresponding DDRAM address. The following procedure demonstrates initialization of the DDRAM Address Register to the value defined in the constant named LCD_1.

```
;========================
; Set address register
;     to LCD line 1
;========================
; ON ENTRY:
;        Address of LCD line 1 in constant LCD_1
line1:
     bcf        PORTA,E_line  ; E line low
     bcf        PORTA,RS_line ; RS line low, set up for control
     call       delay_125     ; delay 125 microseconds
;
; Set to second display line
     movlw      LCD_1         ; Address and command bit
     movwf      PORTD
     call       pulseE        ; Pulse and delay
;
; Set RS line for data
     bsf        PORTA,RS_line ; Setup for data
     call       delay_125mics ; Delay
     return
```

Once the Address Register has been set up, the display operation consists of transferring each characters returned by the table access routine into LCD DDRAM. The following procedure can be used for this purpose:

```
;==============================
;    display 16 characters
;==============================
; Procedure to display one line of 16 characters in
; a 16 x 2 line LCD device
; On entry:
; Variable index holds offset into message text.
; LCD Address Counter register has been set to the
; line to be displayed
; Variable charCount is set to the number of characters
; to be displayed.
msgLine:
     call       tableChar     ; Get character
     movwf      PORTD         ; Store in port
     call       pulseE        ; Write to LCD
     incf       index         ; Bump index into table
     decfsz     charCount,f   ; Decrement counter
     goto       msgLine       ; Continue if not zero
     return
```

Notice that the procedure named msgLine, previously listed, assumes that the Address Counter register in the HD44780 has been previously set to the address of the corresponding test line. Note also that the variable charCount has been initialized to the number of characters to be displayed, and the variable index to the offset in the text string. This allows reusing the procedure to display text beginning at other locations in the LCD screen and of various lengths and offsets.

The previously listed procedures demonstrate just one of many possible variations of this technique. Another approach is to store the characters directly in DDRAM memory as they are produced by the message returning routine, thus avoiding the display procedure entirely. In this last case, the programming saves some data memory space at the expense of having to generate the message characters each time they are needed. Which approach is the most suitable one depends on the application.

Sample Program LCD_18F_MsgFlag

The sample program LCD_18f_MsgFlag listed later in this subsection demonstrates access to a string table, use of Timer0 to produce a variable delay, and testing the busy flag to determine if LCD is ready.

```
; File name: LCD_18F_MsgFlag.asm
; Date: February 22, 2013
; Author: Julio Sanchez
;
; STATE: Feb 22/13:
;         demo board: OK
;         breadboard: OK
;
; Program for PIC 18F452 and LCD display
; Executes in Demo Board B or compatible circuit
;
; Description:
; Display a message stored in program memory. Tests the busy
; flag at the conclusion of critical operations.
;
; A debugging technique:
; The program demonstrates a simple debugging technique that
; allows determining if program execution has proceeded normally
; up to a certain point in the code. This is done by calling a
; simple routine that sets one of the LEDs wired to PORT C
; in Demo Board B.
;=========================================================
;                         Circuit
;=========================================================
;                        18F452
;                +------------------+
;+5v-res0-ICD2 mc -| 1 !MCLR    PGD 40|
;                 | 2 RA0      PGC 39|
;                 | 3 RA1      RB5 38|
;                 | 4 RA2      5B4 37|
;                 | 5 RA3      RB3 36|
;                 | 6 RA4      RB2 35|
;                 | 7 RA5      RB1 34|
;     LCD RS  <==| 8 RE0      RB0 33|
;     LCD E   <==| 9 RE1          32|-------+5v
;     LCD RW  ==>|10 RE2          31|--------GR
;     +5v--------|11          RD7 30|==> LCD data 7 (44780 B flag)
;     GR---------|12          RD6 29|==> LCD data 6
;           osc ---|13 OSC1   RD5 28|==> LCD data 5
;           osc ---|14 OSC2   RD4 27|==> LCD data 4
;       LED0 <== |15 RC0      RC7 26|==> LED7
;       LED1 <== |16 RC1      RC6 25|==> LED6
;       LED2 <== |17 RC2      RC5 24|==> LED5
;       LED3 <== |18 RC3      RC4 23|==> LED4
;  LCD data 0 <==|19 RD0      RD3 22|==> LCD data 3
;  LCD data 1 <==|20 RD1      RD2 21|==> LCD data 2
;                 +------------------+
; Legend:
; E = LCD signal enable
; RW = LCD read/write
```

```
; RS = LCD register select
; GR = ground
;
;===========================
;        LCD wiring
;===========================
; LCD is wired in parallel to 16F877 as follows:
; DATA LINES:
;     |------- F87x -------|----- LCD -------|
;          port    PIN          line    PIN
;          RD0     19           DB0     8
;          RD1     20           DB1     7
;          RD2     21           DB2     10
;          RD3     22           DB3     9
;          RD4     27           DB4     12
;          RD5     28           DB5     11
;          RD6     29           DB6     14
;          RD7     30           DB7     13
; CONTROL LINES:
;     |------- F87x -------|----- LCD -------|
;          port    PIN          line    PIN
;          RE0     8            RS      3
;          RE1     9            E       5
;          RE2     10           RW      6
;


;============================================================
;               definition and include files
;============================================================
     processor      18F452          ; Define processor
     #include <p18F452.inc>
;
; ============================================================
;                    configuration bits
;============================================================
     config OSC = HS           ; Assumes high-speed resonator
     config WDT = OFF          ; No watchdog timer
     config LVP = OFF          ; No low voltage protection
     config DEBUG = OFF        ; No background debugger
;
; Turn off banking error messages
     errorlevel     -302


;============================================================
;                    constant definitions
;   for PIC-to-LCD pin wiring and LCD line addresses
;============================================================
; LCD used in the demo board is 2 lines by 16 characters
#define E_line 1    ;|
#define RS_line 0   ;| -- from wiring diagram
#define RW_line 2   ;|
; LCD line addresses (from LCD data sheet)
#define LCD_1 0x80 ; First LCD line constant
#define LCD_2 0xc0 ; Second LCD line constant
;======================
;    timer constants
;======================
; Three timer constants are defined in order to implement
; a given delay. For example, a delay of one-half second
; in a 4MHz machine requires a count of 500,000, while
```

```
; a delay of one-tenth second requires a count of 10,000.
; These numbers are converted to hexadecimal so they can
; be installed in three constants, for example:
;       1,000,000 = 0x0f4240 = one second at 4MHz
;         500,000 = 0x07a120 = one-half second
;         250,000 = 0x03d090 = one-quarter second
;         100,000 = 0x0186a0 = one-tenth second at 4MHz
; Note: The constants that define the LCD display line
;       addresses have the high-order bit set in
;       order to faciliate the controller command
; Values for one-half second installed in constants
; as follows:
; 100,000 = 0x01 0x86 0xa0
;           ---- ---- ----
;             |    |    |___ lowCnt
;             |    |_____ midCnt
;             |_____ highCnt
;
#define highCnt 0x01
#define midCnt 0x86
#define lowCnt 0xa0
;======================================================
;                variables in PIC RAM
;======================================================
; Continue with local variables
     cblock   0x00      ; Start of block
     count1             ; Counter # 1
     count2             ; Counter # 2
     com_code
     ; 3-byte auxiliary counter for delay.
     countH             ; High-order byte
     countM             ; Medium-order byte
     countL             ; Low-order byte
     index              ; Offset into message text
     charCount          ; Characters per line
     endc
;==========================================================
;                        program
;==========================================================
     org      0       ; start at address
     goto     main
; Space for interrupt handlers
;==============================
;      interrupt intercept
;==============================
     org      0x008     ; High-priority vector
     retfie
     org      0x018     ; Low-priority vector
     retfie
;==================================================================
;          M A I N    P R O G R A M    C O D E
;==================================================================
main:
     nop
     nop
; Set BSR for bank 0 operations
     movlb    0                 ; Bank 0
; Init Port A for digital operation
     clrf     PORTA,0
     clrf     LATA,0
```

```
; Port summary:
; PORTD   0-7                        OUTPUT
; PORTE 0 1 2              OUTPUT
; ADCON1 is the configuration register for the A/D
; functions in Port A. A value of 0b011x sets all
; lines for digital operation
     movlw     B'00000110'  ; Digital mode
     movwf     ADCON1,0
; Initialize all lines in PORT D and E for output
     clrf      TRISC
     clrf      TRISD          ; Port C tris register
     clrf      TRISE
; Clear all output lines
     clrf      PORTC
     clrf      PORTD
     clrf      PORTE
;==============================
;   setup Timer0 as counter
;        8-bit mode
;==============================
; Prescaler is assigned to Timer0 and initialzed
; to 2:1 rate
; Setup the T0CON register
;                |------------ On/Off control
;                |                    1 = Timer0 enabled
;                ||------------ 8/16 bit mode select
;                ||                   1 = 8-bit mode
;                |||---------- Clock source
;                |||               0 = internal clock
;                ||||--------- Source edge select
;                ||||              1 = high-to-low
;                |||||-------- Prescaler assignment
;                |||||             0 = prescaler assigned
;                ||||||||| ----- Prescaler select
;                ||||||||          1:2 rate
     movlw   b'11010000'
     movwf   T0CON
; Clear registers
     clrf      TMR0L
     clrwdt                   ; Clear watchdog timer
; Clear variables
     clrf      com_code
;=========================
;        init LCD
;=========================
; Wait and initialize HD44780
     call    Delay          ; 1/10 second at 4MHz
     call initLCD            ; Do forced initialization
     call    busyTest
     call    led0             ; Good so far
; Set controller to first display line
     movlw     LCD_1          ; Address + offset into line
     movwf   com_code
     call    LCD_command    ; Local procedure
     clrf    index          ; Offset into text string
     movlw   .16            ; Number of characters
     movff   WREG,charCount  ; To variable
     call    msgLine        ; Local LCD line procedure
     call    busyTest
     call    led1
```

```
; First line displayed
; Set controller to second display line
     movlw     LCD_2                    ; Address + offset into line
     movwf     com_code
     call      LCD_command      ; Local procedure
     movlw     .16              ; Offset into text message
     movff     WREG,index       ; To variable
     movff     WREG,charCount   ; and to counter
     call      msgLine          ; Local procedure
     call      busyTest
     call      led2
; Move off second line
     movlw     LCD_2+18          ; Address + offset into line
     movwf     com_code
     call      LCD_command       ; Local procedure
     call      busyTest
     call      led3
skip2:
     goto      skip2


;===========================================================
;===========================================================
;                      P r o c e d u r e s
;===========================================================
;===========================================================
;==================================
;          INITIALIZE LCD
;==================================
initLCD
; Initialization for Densitron LCD module as follows:
;    8-bit interface
;    2 display lines of 16 characters each
;    cursor on
;    left-to-right increment
;    cursor shift right
;    no display shift
;*********************|
;     COMMAND MODE    |
;*********************|
     bcf       PORTE,E_line    ; E line low
     bcf       PORTE,RS_line   ; RS line low for command
     bcf       PORTE,RW_line    ; Write mode
     call      delay_168        ;delay 125 microseconds
;*********************|
;     FUNCTION SET    |
;*********************|
     movlw     0x38        ; 0 0 1 1 1 0 0 0 (FUNCTION SET)
                           ;         | | | |__ font select:
                           ;         | | |    1 = 5x10 in 1/8 or 1/11 dc
                           ;         | | |    0 = 1/16 dc
                           ;         | | |___ Duty cycle select
                           ;         | |     0 = 1/8 or 1/11
                           ;         | |     1 = 1/16 (multiple lines)
                           ;         | |___ Interface width
                           ;         |     0 = 4 bits
                           ;         |     1 = 8 bits
                           ;         |___ FUNCTION SET COMMAND
     movwf     PORTD     ;0011 1000
     call pulseE   ;pulseE and delay
```

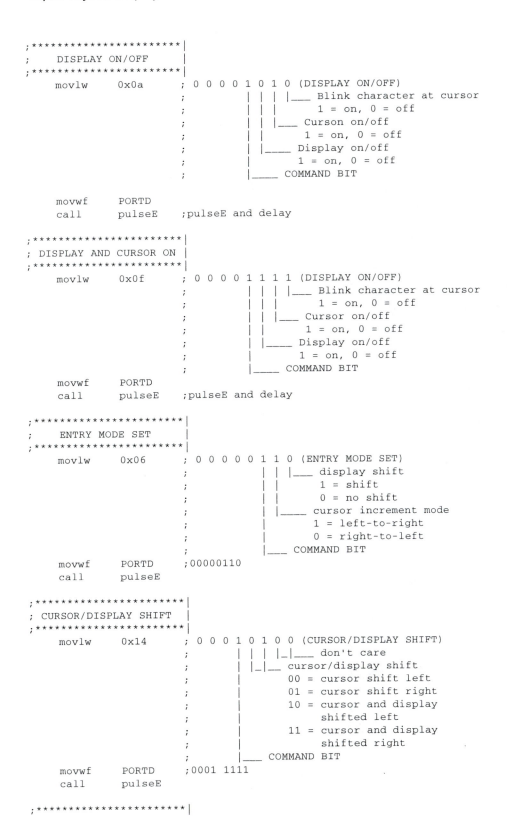

```
;**********************|
;    DISPLAY ON/OFF    |
;**********************|
     movlw    0x0a      ; 0 0 0 0 1 0 1 0 (DISPLAY ON/OFF)
                        ;             | | | |___ Blink character at cursor
                        ;             | | |        1 = on, 0 = off
                        ;             | | |___ Curson on/off
                        ;             | |        1 = on, 0 = off
                        ;             | |____ Display on/off
                        ;             |          1 = on, 0 = off
                        ;             |____ COMMAND BIT

     movwf    PORTD
     call     pulseE    ;pulseE and delay

;***********************|
; DISPLAY AND CURSOR ON |
;***********************|
     movlw    0x0f      ; 0 0 0 0 1 1 1 1 (DISPLAY ON/OFF)
                        ;             | | | |___ Blink character at cursor
                        ;             | | |        1 = on, 0 = off
                        ;             | | |___ Cursor on/off
                        ;             | |        1 = on, 0 = off
                        ;             | |____ Display on/off
                        ;             |          1 = on, 0 = off
                        ;             |____ COMMAND BIT

     movwf    PORTD
     call     pulseE    ;pulseE and delay

;***********************|
;    ENTRY MODE SET     |
;***********************|
     movlw    0x06      ; 0 0 0 0 0 1 1 0 (ENTRY MODE SET)
                        ;             | | |___ display shift
                        ;             | |        1 = shift
                        ;             | |        0 = no shift
                        ;             | |____ cursor increment mode
                        ;             |          1 = left-to-right
                        ;             |          0 = right-to-left
                        ;             |___ COMMAND BIT
     movwf    PORTD     ;00000110
     call     pulseE

;***********************|
; CURSOR/DISPLAY SHIFT  |
;***********************|
     movlw    0x14      ; 0 0 0 1 0 1 0 0 (CURSOR/DISPLAY SHIFT)
                        ;           | | | |_|___ don't care
                        ;           | |_|__ cursor/display shift
                        ;           |          00 = cursor shift left
                        ;           |          01 = cursor shift right
                        ;           |          10 = cursor and display
                        ;           |               shifted left
                        ;           |          11 = cursor and display
                        ;           |               shifted right
                        ;           |___ COMMAND BIT
     movwf    PORTD     ;0001 1111
     call     pulseE

;***********************|
```

```
;    CLEAR DISPLAY        |
;*********************** |
    movlw      0x01      ; 0 0 0 0 0 0 0 1 (CLEAR DISPLAY)
                         ;               |___ COMMAND BIT
    movwf      PORTD     ;0000 0001
;
    call pulseE
    call delay_28ms      ;delay 5 milliseconds after init
    return


;===============================================================
;              Time Delay and Pulse Procedures
;===============================================================
; Procedure to delay 42 x 4 = 168 machine cycles
; On a 4MHz clock the instruction rate is 1 microsecond
; 42 x 4 x 1 = 168 microseconds
delay_168
    movlw      D'42'                 ; Repeat 42 machine cycles
    movwf      count1                ; Store value in counter
repeat
    decfsz     count1,f              ; Decrement counter
    goto       repeat                ; Continue if not 0
    return                           ; End of delay
;
; Procedure to delay 168 x 168 microseconds
; = 28.224 milliseconds
delay_28ms
    movlw      D'42'                 ; Counter = 41
    movwf      count2                ; Store in variable
delay
    call       delay_168             ; Delay
    decfsz     count2,f              ; 40 times = 5 milliseconds
    goto       delay
    return                           ; End of delay
;=======================
;     pulse E line
;=======================
pulseE
    bsf        PORTE,E_line          ;pulse E line
    nop
    bcf        PORTE,E_line
    nop
    call       delay_168             ; Delay
    return
;
;=======================
; busy flag test routine
;=======================
; Procedure to test the HD44780 busy flag
; Execution returns when flag is clear
; Tech note:
; The HD44770 requires a 10 ms delay to set the busy flag
; when the line is being sourced. Reading the flag before
; this time has elapsed may result in a false "not busy"
; report. The code forces a 15 machine cycle delay to
; prevent this error.
busyTest:
    movlw      .15                   ; Repeat 15 machine cycles
    movwf      count1                ; Store value in counter
delay15:
```

```
        decfsz     count1,f         ; Decrement counter
        goto delay15                ; Continue if not 0
; Delay concluded
        movlw      B'11111111'
        movwf      TRISD
        bcf        PORTE,RS_line     ; RS line low for control
        bsf        PORTE,RW_line     ; Read mode
is_busy:
        bsf        PORTE,E_line      ; E line high
        movff      PORTD,WREG        ; Read port D into W
                                     ; Port D bit 7 is busy flag
        bcf        PORTE,E_line      ; E line low
        btfsc      PORTD,7           ; Test bit directly
        goto       is_busy           ; Repeat if busy flag set
; At this point busy flag is clear
; Reset R/W line and port D to output
        bcf        PORTE,RW_line ; Clear R/W line
        bsf        PORTE,RS_line ; Setup for data
        clrf       TRISD
        return
;=========================
;     LCD command
;=========================
LCD_command:
; On entry:
;          variable com_code cntains command code for LCD
; Set up for write operation
        bcf        PORTE,E_line      ; E line low
        bcf        PORTE,RS_line     ; RS line low, set up for control
        call       delay_168         ; delay 125 microseconds
; Write command to data port
        movf       com_code,0        ; Command code to W
        movwf      PORTD
        call       pulseE            ; Pulse and delay
; Set RS line for data
        bsf        PORTE,RS_line     ; Setup for data
        return


;===================================
;  variable-lapse delay procedure
;        using Timer0
;===================================
; ON ENTRY:
;        Variables countL, countM, and countH hold
;        the low-, middle-, and high-order bytes
;        of the delay period, in timer units
; Routine logic:
; The prescaler is assigned to timer0 and setup so
; that the timer runs at 1:2 rate. This means that
; every time the counter reaches 128 (0x80) a total
; of 256 machine cycles have elapsed. The value 0x80
; is detected by testing bit 7 of the counter
; register.
Delay:
        call    setVars
; Note:
;     The TMR0L register provides the low-order level
; of the count. Because the counter counts up from zero,
; code must pre-install a value in the counter register
; that represents one-half the number of timer
```

```
; iterations (pre-scaler is in 1:2 mode) required to
; reach a count of 128. For example: if the value in
; the low counter variable is 140
; then 140/2 = 70. 128 - 70 = 58
; In other words, when the timer counter reaches 128,
; 70 * 2 (140) timer beats would have elapsed.
; Formula:
;              Value in TMR0L = 128 - (x/2)
; where x is the number of iterations in the low-level
; counter variable
; First calculate xx/2 by bit shifting
      rrncf      countL,f   ; Divide by 2
; now subtract 128 - (x/2)
      movlw      d'128'
; Clear the borrow bit (mapped to Carry bit)
      bcf        STATUS,C
      subfwb     countL,w
; Now w has adjusted result. Store in TMR0L
      movwf      TMR0L
; Routine tests timer overflow by testing bit 7 of
; the TMR0L register.
cycle:
      btfss      TMR0L,7       ; Is bit 7 set?
      goto       cycle         ; Wait if not set
; At this point TMR0 bit 7 is set
; Clear the bit
      bcf        TMR0L,7       ; All other bits are preserved
; Subtract 256 from beat counter by decrementing the
; mid-order byte
      decfsz     countM,f
      goto       cycle         ; Continue if mid-byte not zero
; At this point the mid-order byte has overflowed.
; High-order byte must be decremented.
      decfsz     countH,f
      goto       cycle
; At this point the time cycle has elapsed
      return
;===============================
;   set register variables
;===============================
; Procedure to initialize local variables for a
; delay period defined in local constants highCnt,
; midCnt, and lowCnt.
setVars:
      movlw      highCnt      ; From constants
      movwf      countH
      movlw      midCnt
      movwf      countM
      movlw      lowCnt
      movwf      countL
      return


;===============================
;    display 16 characters
;===============================
; Procedure to display one line of 16 characters in
; a 16 x 2 line LCD device
; On entry:
; Variable index holds offset into message text.
; LCD Address Counter register has been set to the
```

```
;   line to be displayed
;   Variable charCount is set to the number of characters
;   to be displayed.
msgLine:
        call    tableChar       ; Get character
        movwf   PORTD           ; Store in port
        call    pulseE          ; Write to LCD
        incf    index           ; Bump index into table
        decfsz  charCount,f     ; Decrement counter
        goto    msgLine         ; Continue if not zero
        return
;==========================
;     get table character
;==========================
;   Local procedure to get a single character from a local
;   table (msgTable) in program memory. Variable index holds
;   offset into table
tableChar:
        movlw   UPPER msgTable
        movwf   TBLPTRU
        movlw   HIGH msgTable   ; Get address of Table
        movwf   TBLPTRH         ; Store in table pointer low register
        movlw   LOW msgTable    ; Get address of Table
        movwf   TBLPTRL
        movff   index,WREG      ; index to W
        addwf   TBLPTRL,f       ; Add index to table pointer low
        clrf    WREG            ; Clear register
        addwfc  TBLPTRH,F       ; Add possible carry
        addwfc  TBLPTRU,F       ; To both registers
        tblrd   *               ; Read byte from table (into TABLAT)
        movff   TABLAT,WREG     ; Move TABLAT to W
        return
;==========================
;     debugging routines
;==========================
led0:
; Light up LED 0 in PORC
        movlw   B'00000001'
        iorwf   PORTC
        return

led1:
; Light up LED 1 in PORC
        movlw   B'00000011'
        iorwf   PORTC
        return

led2:
; Light up LED 2 in PORC
        movlw   B'00000111'
        iorwf   PORTC
        return

led3:
; Light up LED 3 in PORC
        movlw   B'00001111'
        iorwf   PORTC
        return
;=================================================================
;                       code memory text
```

```
; ================================================================
    org     $+2
msgTable:
    db      "  LCD display   "        ; offset 0
    db      " reading B flag "        ; offset 16
    end
```

11.5 Data Compression Techniques

Circuits based on the parallel data transfer of 8 data bits require eight port lines devoted to this purpose. Assuming that three other lines are required for LCD commands and interfacing (RS, E, and R/W lines), that adds-up to eleven PIC-to-LCD lines. Several possible solutions allow compressing the data transfer function. The most obvious one is to use the 4-bit data transfer mode to free four port lines. Other solutions are based on dedicating other hardware components to the LCD function. Figure 11.6 shows a 4-bit data mode circuit.

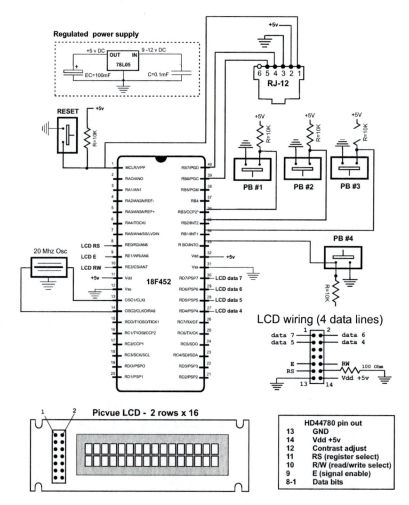

Figure 11.6 *PIC/LCD circuit for 4-bit data mode.*

Notice in Figure 11.6 that the four high-order lines of PORT D are connected to the LCD function while the four low-order lines are unused.

11.5.1 4-Bit Data Transfer Mode

Data compression can be achieved using the parallel interface capability of the Hitachi HD44780 controller. In this case, data transfers require just four data paths instead of eight. The one objection is that programming in 4-bit mode is slightly more complicated and that there will be a minor performance penalty. The slower execution speed results from the fact that in 4-bit mode data must be sent one nibble at a time. But because the delay is only required after the second nibble, the execution time penalty for 4-bit transfers is not very large.

Many of the routines developed for 8-bit data mode previously in this chapter can be reused without modification in the 4-bit mode. Others require minor changes, and there is one specific display procedure that must be specially developed. The first required change will be in the LCD initialization because bit 4 in the Function Set command must be clear for a 4-bit interface. The remaining initialization commands require no further change, although it is a good idea to consult the data sheet for the LCD hardware in use.

11.5.2 Preserving Port Data

Displaying data using a 4-bit interface consists of sending the high-order nibble followed by the low-order nibble, through the LCD 4-high order data lines, usually labeled DB5 to DB7. The pulsing of line E follows the last nibble sent. This means that software must provide a way of reading and writing to the appropriate port lines (the ones used in the data transfer) without altering the value stored in the port bits dedicated to other uses. In Section 11.3 we discussed bit merging routines, which are quite suitable for the purpose at hand.

However, bit merging techniques assume that all eight lines in the port register have been trissed for output. In this case, if one or more of the port lines not connected to the LCD are trissed for input then the bit-merging manipulations will fail because code cannot write data to an input line. In any case, bit-merging requires writing new data to a port, an operation that is unacceptable with some devices. The ideal solution for sharing a port between the LCD with another device or task would be that the LCD-related operations do not change the values in the remaining port lines, whether they are trissed for input or output. Unfortunately, there will always be cases in which a shared port could malfunction. The solutions proposed here are the best we could devise while still holding the complications to a reasonable state.

The following procedure, from the program LCD_18F_4line_Reloc.asm from the book's software package and listed later in this chapter, shows one possible accommodation.

```
;**********************|
;    write two nibbles |
;**********************|
; Routine to write the value in WREG to PORTD while
; preserving port data
```

```
write_nibs:
    movff   WREG,thischar    ; Save character
; Store values in PORTD and TRISD registers
    movff   TRISD,old_trisd
    movff   PORTD,old_portd
; Set low nibble of TRISD for input
    movlw   B'00001111'
    movff   WREG,TRISD
; Get character
    movff   thischar,WREG    ; Character to write
    movff   WREG,PORTD       ; Store high nibble
    call    pulseE           ; Pulse to send
    swapf   thischar         ; Swap character nibbles
    movff   thischar,PORTD   ; Store low nibble
    call    pulseE           ; Pulse to send
; Restore PORTD and TRISD
    movff   old_trisd,TRISD
    movff   old_portd,PORTD
    return
```

The write_nibs procedure requires two local variables in which to store the value in the TRIS and the PORT register at entry time. Then the code changes the port lines not used by the LCD function (typically the low-order nibble) to input by modifying the TRIS register. Because writing a a line trissed for input has no effect, the existing values in the non-LCD lines are preserved unchanged. On exit the original tris and port values are restored.

11.5.3 Master/Slave Systems

Up to this point we have assumed that driving the LCD is one of the functions performed by the PIC microcontroller, which also executes the other circuit functions. In practice, such a scheme is not always viable, either due to the high number of interface lines required or due to the amount of PIC code space used up by the LCD driver routines. An alternative approach is to dedicate a PIC exclusively to controlling the LCD hardware, while one or more other PICs perform the main circuit functions. In this scheme, the PIC devoted to the LCD function is referred to as a slave, while the one that sends the display commands is called the master.

When a sufficient number of interface lines are available, the connection between master and slave can be simplified using a parallel interface. For example, if four port lines can be used to interconnect the two PICs, then sixteen different command codes can be sent to the slave. The slave reads the communications lines much like it would read a multiple toggle switch. A simple protocol can be devised so that the slave uses these same interface lines to provide feedback to the master. For example, the slave sets one or all four lines low to indicate that it is ready for the next command, and sets them high to indicate that command execution is in progress and that no new commands can be received. The master, in turn, reads the communications lines to determine when it can send another command to the slave.

But using parallel communications between master and slave can be a self-defeating proposition, because it requires at least seven interface lines to be able to send ASCII characters. Because the scarcity of port lines is the original reason for using a master/slave setup, parallel communications may not be a good solution in many cases. On the other hand, communications between master and slave can take place serially, using a single interface line. The discussion of using serial interface between a master and an LCD slave driver PIC is left for the chapter on serial communications.

11.5.4 4-Bit LCD Interface Sample Programs

The sample program LCD_18F_4line_Reloc.asm program executes in relocatable code and demonstrates many of the topics discussed in this chapter. The program listing follows:

```
; File name: LCD_18F_4line_Reloc.asm
; Project: LCD_18F_4line_Reloc.mcp
; Date: February 26, 2013
; Author: Julio Sanchez
;
; STATE: Feb 26/13:
;         Demo board: OK
;         Breadboard: OK
;
; Program for PIC 18F452 and LCD display
; Executes in Demo Board B or compatible circuit
;
; Description:
; Display a message stored in a program memory table
;========================================================
;                      Circuit
;                wiring for 4-bit mode
;========================================================
;                        18F452
;                 +------------------+
;+5v-res0-ICD2 mc -| 1 !MCLR    PGD 40|
;                 | 2 RA0      PGC 39|
;                 | 3 RA1      RB5 38|
;                 | 4 RA2      5B4 37|
;                 | 5 RA3      RB3 36|
;                 | 6 RA4      RB2 35|
;                 | 7 RA5      RB1 34|
;     LCD RS  <==| 8 RE0      RB0 33|
;     LCD E   <==| 9 RE1          32|-------+5v
;     LCD RW  ==>|10 RE2          31|--------GR
;     +5v--------|11          RD7 30|==> LCD data 7
;     GR---------|12          RD6 29|==> LCD data 6
;          osc ---|13 OSC1    RD5 28|==> LCD data 5
;          osc ---|14 OSC2    RD4 27|==> LCD data 4
;                 |15 RC0     RC7 26|
;                 |16 RC1     RC6 25|
;                 |17 RC2     RC5 24|
;                 |18 RC3     RC4 23|
;                 |19 RD0     RD3 22|
;                 |20 RD1     RD2 21|
;                 +------------------+
;
; Legend:
; E = LCD signal enable
; RW = LCD read/write
; RS = LCD register select
; GR = ground
;
```

```
;==============================
;       LCD wiring
;==============================
; LCD is wired in for 4-bit data to 18F452 as follows:
; DATA LINES:
;    |------- F87x -------|----- LCD -------|
;         port      PIN            line     PIN
;         RD4            27          DB4      12
;         RD5            28          DB5      11
;         RD6            29          DB6      14
;         RD7            30          DB7      13
; CONTROL LINES:
;    |------- 16F452 -------|----- LCD -------|
;         port      PIN            line     PIN
;         RE0            8           RS       3
;         RE1            9           E        5
;         RE2            10          RW       6
;
    list p=18f452
    ; Include file, change directory if needed
    include "p18f452.inc"
; ============================================================
;                    configuration bits
; ============================================================
; Configuration bits set as required for MPLAB ICD 2
    config OSC = XT          ; Assumes high-speed resonator
    config WDT = OFF         ; No watchdog timer
    config LVP = OFF         ; No low voltage protection
    config DEBUG = OFF       ; No background debugger
    config PWRT = ON         ; Power on timer enabled
    config CP0 = OFF         ; Code protection block x = 0-3
    config CP1 = OFF
    config CP2 = OFF
    config CP3 = OFF
    config WRT0 = OFF        ; Write protection block x = 0-3
    config WRT1 = OFF
    config WRT2 = OFF
    config WRT3 = OFF
    config EBTR0 = OFF       ; Table read protection block x = 0-3
    config EBTR1 = OFF
    config EBTR2 = OFF
    config EBTR3 = OFF
;
; Turn off banking error messages
    errorlevel    -302


;============================================================
;                    constant definitions
;   for PIC-to-LCD pin wiring and LCD line addresses
;============================================================
; LCD used in the demo board is 2 lines by 16 characters
#define E_line 1     ; |
#define RS_line 0    ; | -- from wiring diagram
#define RW_line 2    ; |
; LCD line addresses (from LCD data sheet)
#define LCD_line1 0x80  ; First LCD line constant
#define LCD_line2 0xc0  ; Second LCD line constant
;=====================
;   timer constants
;=====================
```

```
; Three timer constants are defined in order to implement
; a given delay. For example, a delay of one-half second
; in a 4MHz machine requires a count of 500,000, while
; a delay of one-tenth second requires a count of 10,000.
; These numbers are converted to hexadecimal so they can
; be installed in three constants, for example:
;       1,000,000 = 0x0f4240 = one second at 4MHz
;         500,000 = 0x07a120 = one-half second
;         250,000 = 0x03d090 = one-quarter second
;         100,000 = 0x0186a0 = one-tenth second at 4MHz
; Note: The constant that define the LCD display line
;       addresses have the high-order bit set in
;       order to faciliate the controller command
; Values for one-half second installed in constants
; as follows:
; 500,000 = 0x07 0xa1 0x20
;            ---- ---- ----
;             |    |    |___ lowCnt
;             |    |_____ midCnt
;             |_____ highCnt
;
#define highCnt 0x07
#define midCnt 0xa1
#define lowCnt 0x20
;====================================================
;                 variables in PIC RAM
;====================================================
    udata_acs   0x000
count1      res 1           ; Counter # 1
com_code    res 1
; 3-byte auxiliary counter for delay.
countH      res 1           ; High-order byte
countM      res 1           ; Medium-order byte
countL      res 1           ; Low-order byte
index       res 1           ; Offset into message text
charCount   res 1           ; Characters per line
bflag       res 1           ; Busy flag nibble storage
thischar    res 1           ; Remember character
; Storage for TRIS D and PORT D
old_trisd   res 1           ; To preserve PORT D low
old_portd   res 1           ; order data
;=============================================================
;                          program
;=============================================================
    ; Start at the reset vector
Reset_Vector  code 0x000
    goto Start

; Start application beyond vector area

    code 0x030
Start:
; Set BSR for bank 0 operations
    movlb   0               ; Bank 0

; Init Port A for digital operation
    clrf    PORTA,0
    clrf    LATA,0
; Port summary:
; PORTD  0-7                     OUTPUT
```

```
; PORTE 0 1 2              OUTPUT
; ADCON1 is the configuration register for the A/D
; functions in Port A. A value of 0b011x sets all
; lines for digital operation
    movlw   B'00000110'  ; Digital mode
    movwf   ADCON1,0
; Initialize all lines in PORT D and E for output
    clrf    TRISD          ; Port D tris register
    clrf    TRISE
; Clear all output lines
    clrf    PORTD
    clrf    PORTE
;==============================
;  setup Timer0 as counter
;       8-bit mode
;==============================
; Prescaler is assigned to Timer0 and initialized
; to 2:1 rate
; Setup the T0CON register
;                |------------ On/Off control
;                |             1 = Timer0 enabled
;                ||----------- 8/16 bit mode select
;                ||            1 = 8-bit mode
;                |||---------- Clock source
;                |||           0 = internal clock
;                ||||--------- Source edge select
;                ||||          1 = high-to-low
;                |||||-------- Prescaler assignment
;                |||||         0 = prescaler assigned
;                ||||||||| ----- Prescaler select
;                |||||||||     1:2 rate
    movlw   b'11010000'
    movwf   T0CON
; Clear registers
    clrf    TMR0L
    clrwdt                   ; Clear watchdog timer
;=========================
;       init LCD
;=========================
    call    DelayOneHalfSec
; Wait and initialize HD44780
    call    initLCD_4bit   ; Do forced initialization
 ; Set controller to first display line
    call    busyTest         ; Test busy flag
;=========================
;   display text line # 1
;=========================
; Set DDRAM address
    call    CommandMode
    movlw   LCD_line1      ; Address + offset into line
    call    write_nibs     ; Local nibble write procedure
; Prepare to send text line
    call    DataMode       ; Set controller for data mode
    clrf    index          ; Point to first character
    movlw   .16            ; Characters per line
    movwf   charCount      ; Char count = 16
; Get character from table
char_line1:
    call    tableChar      ; Character to WREG
    call    write_nibs     ; Write two nibbles
```

```
    incf    index          ; Bump index into message
    decfsz  charCount,f    ; Decrement character count
    goto    char_line1
;============================
;   display text line # 2
;============================
; Set DDRAM address
    call    CommandMode
    movlw   LCD_line2      ; Address + offset into line
    call    write_nibs     ; Local nibble write procedure
; Prepare to send text line
    call    DataMode       ; Set controller for data mode
    movlw   .16            ; Characters per line
    movwf   charCount      ; Char count = 16
    movwf   index          ; 16th caracter in table
; Get character from table
char_line2:
    call    tableChar      ; Character to WREG
    call    write_nibs
    incf    index          ; Bump index into message
    decfsz  charCount,f    ; Decrement character count
    goto    char_line2
; Hang up to end execution
WaitHere:
    goto    WaitHere
;==========================================================
;==========================================================
;                    P r o c e d u r e s
;==========================================================
;==========================================================
;==============================
;          INITIALIZE LCD
;==============================
initLCD_4bit
; Initialization for Densitron LCD module as follows:
;   4-bit interface
;   2 display lines of 16 characters each
;   cursor on
;   left-to-right increment
;   cursor shift right
;   no display shift
; Store values in PORTD and TRISD registers
    movff   TRISD,old_trisd
    movff   PORTD,old_portd
; Set low nibble of TRISD for input
    movlw   B'00001111'
    movff   WREG,TRISD
;*********************|
;     COMMAND MODE    |
;*********************|
    bcf     PORTE,E_line    ; E line low
    bcf     PORTE,RS_line   ; RS line low for command
    bcf     PORTE,RW_line   ; Write mode
    call    Delay_28Ms      ; delay 28 milliseconds
;*********************|
;     8-bit command   |
;*********************|
;                        ; 0 0 1 0 0 0 0 0 (Select 4-bit)
                         ;     | |
                         ;     | |___ Interface width
```

```
                                    ;          |       0 = 4 bits
                                    ;          |       1 = 8 bits
                                    ;          |___ FUNCTION SET COMMAND
        movlw    0x20
        movff    WREG,PORTD
        call     pulseE

;**********************|
;     FUNCTION SET     |
;**********************|
;          0x28          ; 0 0 1 0 1 0 0 0 (FUNCTION SET 0x28)
                         ;          | | | |__ font select:
                         ;          | | |      1 = 5x10 in 1/8 or 1/11 dc
                         ;          | | |      0 = 1/16 dc
                         ;          | | |___ Duty cycle select
                         ;          | |       0 = 1/8 or 1/11
                         ;          | |       1 = 1/16 (multiple lines)
                         ;          | |___ Interface width
                         ;          |       0 = 4 bits
                         ;          |       1 = 8 bits
                         ;          |___ FUNCTION SET COMMAND
        movlw    0x20            ; High nibble first
        movff    WREG,PORTD      ; To port
        call     pulseE          ; Pulse and delay
        movlw    0x80            ; Low nibble
        movff    WREG,PORTD      ; To port
        call     pulseE          ; Pulse and delay
; Repeat command
        movlw    0x20            ; High nibble first
        movff    WREG,PORTD      ; To port
        call     pulseE          ; Pulse and delay
        movlw    0x80            ; Low nibble
        movff    WREG,PORTD      ; To port
        call     pulseE          ; Pulse and delay

;**********************|
;    DISPLAY ON/OFF    |
;**********************|
;          0x0a          ; 0 0 0 0 1 0 1 0 (DISPLAY ON/OFF)
                         ;          | | | |___ Blink character at cursor
                         ;          | | |       1 = on, 0 = off
                         ;          | | |___ Curson on/off
                         ;          | |       1 = on, 0 = off
                         ;          | |____ Display on/off
                         ;          |        1 = on, 0 = off
                         ;          |____ COMMAND BIT
        movlw    0x00            ; High nibble first
        movff    WREG,PORTD      ; To port
        call     pulseE          ; Pulse and delay
        movlw    0xa0            ; Low nibble
        movff    WREG,PORTD      ; To port
        call     pulseE          ; Pulse and delay

;**********************|
; DISPLAY AND CURSOR ON |
;**********************|
;          0x0f          ; 0 0 0 0 1 1 1 1 (DISPLAY ON/OFF)
                         ;          | | | |___ Blink character at cursor
                         ;          | | |       1 = on, 0 = off
                         ;          | | |___ Curson on/off
```

```
                                  ;              | |        1 = on,  0 = off
                                  ;              | |____ Display on/off
                                  ;              |        1 = on,  0 = off
                                  ;              |____ COMMAND BIT

        movlw    0x00             ; High nibble first
        movff    WREG,PORTD       ; To port
        call     pulseE           ; Pulse and delay
        movlw    0xf0             ; Low nibble
        movff    WREG,PORTD       ; To port
        call     pulseE         ; Pulse and delay

;*********************|
;    ENTRY MODE SET   |
;*********************|
;         0x06            ; 0 0 0 0 0 1 1 0 (ENTRY MODE SET)
                          ;             | | |___ display shift
                          ;             | |       1 = shift
                          ;             | |       0 = no shift
                          ;             | |____ cursor increment mode
                          ;             |       1 = left-to-right
                          ;             |       0 = right-to-left
                          ;             |___ COMMAND BIT

        movlw    0x00             ; High nibble first
        movff    WREG,PORTD       ; To port
        call     pulseE           ; Pulse and delay
        movlw    0x60             ; Low nibble
        movff    WREG,PORTD       ; To port
        call     pulseE           ; Pulse and delay

;*********************|
; CURSOR/DISPLAY SHIFT |
;*********************|
;         0x14            ; 0 0 0 1 0 1 0 0 (CURSOR/DISPLAY SHIFT)
                          ;         | | | |_|___ don't care
                          ;         | |_|__ cursor/display shift
                          ;         |       00 = cursor shift left
                          ;         |       01 = cursor shift right
                          ;         |       10 = cursor and display
                          ;         |            shifted left
                          ;         |       11 = cursor and display
                          ;         |            shifted right
                          ;         |___ COMMAND BIT

        movlw    0x10             ; High nibble first
        movff    WREG,PORTD       ; To port
        call     pulseE           ; Pulse and delay
        movlw    0x40             ; Low nibble
        movff    WREG,PORTD       ; To port
        call     pulseE         ; Pulse and delay

;*********************|
;    CLEAR DISPLAY    |
;*********************|
;         0x01       ; 0 0 0 0 0 0 0 1 (CLEAR DISPLAY)
                     ;               |___ COMMAND BIT
        movlw    0x00             ; High nibble first
        movff    WREG,PORTD       ; To port
        call     pulseE           ; Pulse and delay
```

```
        movlw    0x10              ; Low nibble
        movff    WREG,PORTD        ; To port
        call     pulseE            ; PulseE and delay
        call     Delay_28Ms        ; Delay 28 milliseconds after init
; Restore PORTD and TRISD
        movff    old_trisd,TRISD
        movff    old_portd,PORTD
        return


;========================
;     pulse E line
;========================
pulseE
        bsf      PORTE,E_line          ;pulse E line
        bcf      PORTE,E_line
        call     Delay_28Ms       ; Delay
        return
;
;========================
; busy flag test routine
;========================
; Procedure to test the HD44780 busy flag
; Execution returns when flag is clear
; Tech note:
; The HD44770 requires a 10 ms delay to set the busy flag
; when the line is being sourced. Reading the flag before
; this time has elapsed may result in a false "not busy"
; report. The code forces a 15 machine cycle delay to
; prevent this error.
busyTest:
        movlw    .15               ; Repeat 15 machine cycles
        movwf    count1            ; Store value in counter
delay15:
        decfsz   count1,f          ; Decrement counter
        goto delay15               ; Continue if not 0
; Delay concluded
        movlw    B'11111111'
        movwf    TRISD
        bcf      PORTE,RS_line ; RS line low for control
        bsf      PORTE,RW_line ; Read mode
is_busy:
        bsf      PORTE,E_line
        nop
        movff    PORTD,bflag       ; Read nibble from LCD
        nop
        bcf      PORTE,E_line      ; bring low
        nop
        bsf      PORTE,E_line
        nop
        movff    PORTD,WREG        ; Discard low nibble into W
        nop
        bcf      PORTE,E_line
; Test high bit of bflag variable
        btfsc    bflag,7
        goto     is_busy
; At this point busy flag is clear
; Reset R/W line and port D to output
        bcf      PORTE,RW_line ; Clear R/W line
        bsf      PORTE,RS_line ; Setup for data
        clrf     TRISD
```

```
      return

;===================================
;   variable-lapse delay procedure
;        using Timer0
;===================================
Delay:
; ON ENTRY:
;           Variables countL, countM, and countH hold
;           the low-, middle-, and high-order bytes
;           of the delay period, in timer units
; Routine logic:
; The prescaler is assigned to timer0 and setup so
; that the timer runs at 1:2 rate. This means that
; every time the counter reaches 128 (0x80) a total
; of 256 machine cycles have elapsed. The value 0x80
; is detected by testing bit 7 of the counter
; register.
;
; Note:
;     The TMR0L register provides the low-order level
; of the count. Because the counter counts up from zero,
; code must pre-install a value in the counter register
; that represents one-half the number of timer
; iterations (pre-scaler is in 1:2 mode) required to
; reach a count of 128. For example: if the value in
; the low counter variable is 140
; then 140/2 = 70. 128 - 70 = 58
; In other words, when the timer counter reaches 128,
; 70 * 2 (140) timer beats would have elapsed.
; Formula:
;           Value in TMR0L = 128 - (x/2)
; where x is the number of iterations in the low-level
; counter variable
; First calculate xx/2 by bit shifting
    rrncf      countL,f  ; Divide by 2
; now subtract 128 - (x/2)
    movlw      d'128'
; Clear the borrow bit (mapped to Carry bit)
    bcf        STATUS,C
    subfwb     countL,w
; Now w has adjusted result. Store in TMR0L
    movwf      TMR0L
; Routine tests timer overflow by testing bit 7 of
; the TMR0L register.
cycle:
    btfss      TMR0L,7       ; Is bit 7 set?
    goto       cycle         ; Wait if not set
; At this point TMR0 bit 7 is set
; Clear the bit
    bcf        TMR0L,7       ; All other bits are preserved
; Subtract 256 from beat counter by decrementing the
; mid-order byte
    decfsz     countM,f
    goto       cycle         ; Continue if mid-byte not zero
; At this point the mid-order byte has overflowed.
; High-order byte must be decremented.
    decfsz     countH,f
    goto       cycle
; At this point the time cycle has elapsed
```

```
        return

;==========================
; one-half second timer
;==========================
DelayOneHalfSec:
; 500,000 = 0x07 0xa1 0x20
;              ---- ---- ----
;               |    |    |___ lowCnt
;               |    |_____ midCnt
;               |_____ highCnt
        movlw     0x07          ; From values above
        movwf     countH
        movlw     0xa1
        movwf     countM
        movlw     0x20
        movwf     countL
        call      Delay
        return


;==========================
;    28 Ms timer
;==========================
Delay_28Ms:
; 28/1000
; 000,028 = 0x00 0x00 0x28
;              ---- ---- ----
;               |    |    |___ lowCnt
;               |    |_____ midCnt
;               |_____ highCnt
; Note: because counters are decremented AFTER loop
;       a zero value would wrap-around
        movlw     0x01          ; From values above
        movwf     countH
        movlw     0x01
        movwf     countM
        movlw     0x20
        movwf     countL
        call      Delay
        return


;==========================
;    get table character
;==========================
; Local procedure to get a single character from a local
; table (msgTable) in program memory. Variable index holds
; offset into table
tableChar:
        movlw     UPPER msgTab
        movwf     TBLPTRU
        movlw     HIGH msgTab      ; Get address of Table
        movwf     TBLPTRH          ; Store in table pointer low register
        movlw     LOW msgTab       ; Get address of Table
        movwf     TBLPTRL
        movff     index,WREG       ; index to W
        addwf     TBLPTRL,f        ; Add index to table pointer low
        clrf      WREG             ; Clear register
        addwfc    TBLPTRH,F        ; Add possible carry
        addwfc    TBLPTRU,F        ; To both registers
        tblrd     *                ; Read byte from table (into TABLAT)
```

```
        movff    TABLAT,WREG      ; Move TABLAT to W
        return

;***********************|
;     COMMAND MODE      |
;***********************|
; Procedure to set the controller in command mode
CommandMode:
        bcf      PORTE,E_line    ; E line low
        bcf      PORTE,RS_line   ; RS line low for command
        bcf      PORTE,RW_line   ; Write mode
        call     Delay_28Ms           ; delay 28 milliseconds
        return
;***********************|
;        DATA MODE      |
;***********************|
; Procedure to set the controller in data mode
DataMode:
        bcf      PORTE,E_line    ; E line low
        bsf      PORTE,RS_line   ; RS line high for read/write
        bcf      PORTE,RW_line   ; Write mode
        call     Delay_28Ms           ; delay 28 milliseconds
        return
;***********************|
;     write two nibbles |
;***********************|
; Routine to write the value in WREG to PORTD while
; preserving port data
write_nibs:
        movff    WREG,thischar   ; Save character
; Store values in PORTD and TRISD registers
        movff    TRISD,old_trisd
        movff    PORTD,old_portd
; Set low nibble of TRISD for input
        movlw    B'00001111'
        movff    WREG,TRISD
; Get character
        movff    thischar,WREG    ; Character to write
        movff    WREG,PORTD       ; Store high nibble
        call     pulseE           ; Pulse to send
        swapf    thischar         ; Swap character nibbles
        movff    thischar,PORTD   ; Store low nibble
        call     pulseE           ; Pulse to send
; Restore PORTD and TRISD
        movff    old_trisd,TRISD
        movff    old_portd,PORTD
        return

; Define table in code memory
msgTab       db    "4-line LCD Demo "
             db    "Relocatable mode"
        end
```

11.6 LCD Programming in C18

The C18 language includes support for programming LCD devices. This support comes in a set of external LCD functions that is part of the Software Peripherals Library.
The library includes eleven primitive functions as well as a header file named xlcd.h

found in the MCC18 compiler installation folder. To determine the path to the xlcd.h file,you can navigate to the Project>Set Language Tool Locations command in MPLAB. Click the [+] button in the Microchip C18 Toolsuite entry and then the [+] button on Default Search Paths & Directories. Selecting the Include Search Path option will display in the location window the current path to the C18 header files. Figure 11.7 is a screen snapshot of the final screen.

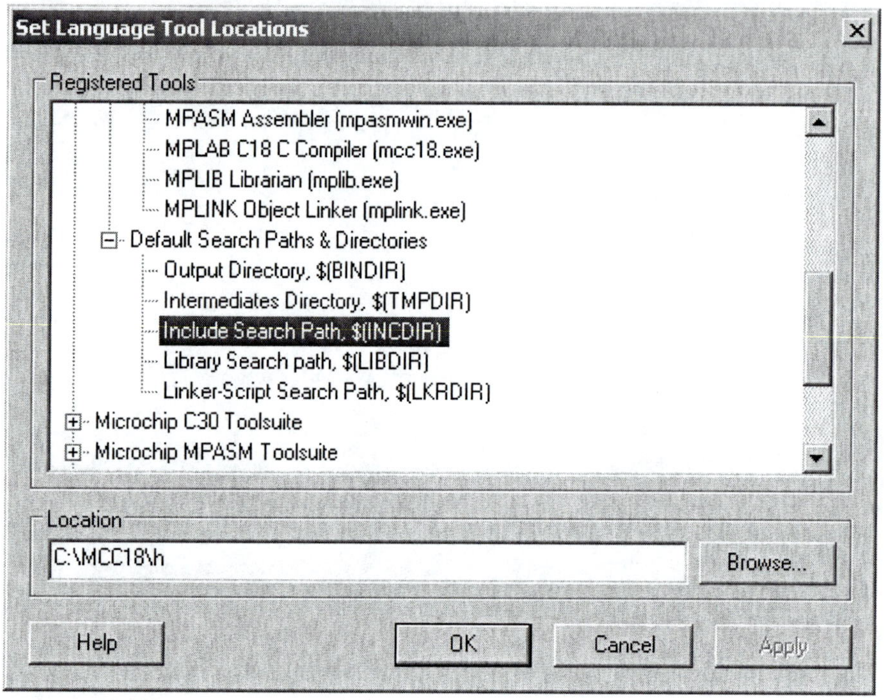

Figure 11.7 *Finding the path to C18 header files.*

It is quite possible that there would be several copies of the xlcd.h file in a particular system. It is important that the file edited is the one in the C18 compiler search path. Before making changes in the file, it is a good idea to make a copy of the original one for future reference.

11.6.1 Editing xlcd.h

The xlcd.h file is furnished as a template and defaults to register and hardware assignments that would coincide with the user's system only by pure chance. At the start of the file, there is an include statement that references the generic processor header file named p18cxxx.h. This file contains a cascade of tests for the include statement that defines the specific F18 device in your system. The cldc.h file provides instructions on the required editing.

Defining the Interface

The LCD support includes both the 4-bit and the 8-bit data modes described earlier in this chapter. The default support is for 4-bit operation. The file contains the following lines:

```
/* Interface type 8-bit or 4-bit
 * For 8-bit operation uncomment the #define BIT8
 */
/* #define BIT8 */

/* When in 4-bit interface define if the data is in the upper
 * or lower nibble.  For lower nibble, comment the #define UPPER
 */
/* #define UPPER */
```

For 8-bit operation, the comments are removed from the first #define statement. If in 4-bit mode, then the second #define statement is uncommented if the data is to be presented in the upper port nibble. Otherwise, the data is expected in the lower nibble.

Defining the Data Port and Tris Register

The xlcd.h file contains two #define statements that select the port and tris register associated with the LCD data path, as follows:

```
/* DATA_PORT defines the port to which the LCD data lines are connected */
#define DATA_PORT        PORTB
#define TRIS_DATA_PORT TRISB

The operands must be edited if the data port is any other than PORT B.
Defining the Control Lines
The control lines determine the wiring between the microcontroller and the
device. These lines are conventionally labled RS, RW, and E. Their function
was described in Section
11.4.3. The port lines and corresponding tris registers are defined in the
header file as follows:

/* CTRL_PORT defines the port where the control lines are
 * connected.
   These are just samples, change to match your application.
 */
#define RW_PIN    LATBbits.LATB6        /* PORT for RW */
#define TRIS_RW   TRISBbits.TRISB6      /* TRIS for RW */
#define RS_PIN    LATBbits.LATB5        /* PORT for RS */
#define TRIS_RS   TRISBbits.TRISB5      /* TRIS for RS */
#define E_PIN     LATBbits.LATB4        /* PORT for D  */
#define TRIS_E    TRISBbits.TRISB4      /* TRIS for E  */
```

For example, to match the wiring of the circuit in Figure 11.4, the #define statements would be edited as follows:

```
#define RW_PIN    LATEbits.LATE2        /* PORTE line 2 for RW */
#define TRIS_RW   TRISEbits.TRISE2      /* TRIS for RW */
#define RS_PIN    LATEbits.LATE0        /* PORTE line 0 for RS */
#define TRIS_RS   TRISEbits.TRISE0      /* TRIS for RS */
#define E_PIN     LATEbits.LATE1        /* PORTE line 1 for E  */
#define TRIS_E    TRISEbits.TRISE1      /* TRIS for E  */
```

The remaining elements in the file are constants that do not normally require changing. The file assumes that the memory model of the program is far and the parameter class is auto.

11.6.2 Timing Routines

The XLCD libraries in C18 require that the user furnish three delay functions. This is necessary in order to accommodate the processor speed in the system. The delay routines must use specific names, return types, and parameters as follows:

- extern void DelayFor18TCY(void) for 18 cycles delay
- extern void DelayPORXLCD(void) for 15 ms delay
- extern void DelayXLCD(void) for 5 ms delay

Because the routines must be visible to the C18 library modules, they must be prototyped as external in the file that contains them, as in the previous declarations. The following code sample is suitable for a system running at 4 MHz.

```
//****************************************************************
//              timer funtions required by C18
//****************************************************************
void DelayFor18TCY( void )
{
     Nop();
     Nop();
     Nop();
     Nop();
     Nop();
     Nop();
     Nop();
     Nop();
     Nop();
     Nop();
     Nop();
     Nop();
// 12 operations plus call and return times
}
void DelayPORXLCD (void)
{
// Calculations for a 4 MHz oscillator
     Delay1KTCYx(15);    // Delay of 15ms
                         // Cycles = (TimeDelay * Fosc) / 4
                         // Cycles = (15ms * 4MHz) / 4 = 15
                         // Cycles = 15,000
     return;
}

void DelayXLCD (void)
{
     Delay1KTCYx(5); // Delay of 5ms
         // Cycles = (TimeDelay * Fosc) / 4)
         // Cycles = (5ms * 4MHz) / 4 = 5
         // Cycles = 5,000
     return;
}
```

Microchip furnishes a code sample for LCD programming named EXTERNAL LCD, which includes a C language file for implementing the three required delays. The file name is delay_xlcd.c. Code listing is as follows:

```
// File with delay functions from Microchip's
// EXTERNAL XLCD code sample
// Author: Harsha J.M. (04/04/09)

#include "xlcd.h"

void DelayXLCD(void)
{
    unsigned char i=0;
    for(i=0;i<25;i++);
}

void DelayFor18TCY(void)
{
    unsigned char i=0;
    for(i=0;i<10;i++);
}

void DelayPORXLCD(void)
{
    unsigned char i=0;
    for(i=0;i<10;i++);
}
```

11.6.3 XLCD Library Functions

The following C language primitives are included in the XLCD C18 library are listed in Table 11.6.

Table 11.6

LCD Primitives in XLCD

BusyXLCD	Test for LCD controller busy?
OpenXLCD	Configure the I/O lines used for controlling the LCD and initialize the LCD.
putcXLCD	Write a byte to the LCD controller.
putsXLCD	Write a string from data memory to the LCD.
putrsXLCD	Write a string from program memory to the LCD.
ReadAddrXLCD	Read address byte from the LCD controller.
ReadDataXLCD	Read a byte from the LCD controller.
SetCGRamAddr	Set the character generator address.
SetDDRamAddr	Set the display data address.
WriteCmdXLCD	Write a command to the LCD controller.
WriteDataXLCD	Write a byte to the LCD controller.

The functions are summarized in the following subsections.

BusyXLCD

Determine if the LCD controller is busy. Returns 1 if the controllwe is busy and 0 otherwise.

```
Include: p18cxxx.h
         xlcd.h
Prototype: unsigned char BusyXLCD( void );
File Name: busyxlcd.c
Code Example: while( BusyXLCD() ); // Waits for LCD not busy.
```

OpenXLCD

```
Function: Initialize the Hitachi HD4478 according to I/O pins definitions in
xldc.h
Include: xlcd.h
Prototype: void OpenXLCD( unsigned char lcdtype );
```

Arguments: unsigned char lcdtype is a bitmask created by performing a bitwise AND operation ('&') with a value from each of the following operands defined in the file xlcd.h.

```
Data Interface:
    FOUR_BIT 4-bit Data Interface mode
    EIGHT_BIT 8-bit Data Interface mode
    LCD Configuration:
    LINE_5X7 5x7 characters, single line display
    LINE_5X10 5x10 characters display
    LINES_5X7 5x7 characters, multiple line display
File Name: openxlcd.c
Code Example:
    OpenXLCD( EIGHT_BIT & LINES_5X7 );
```

putrXLCD

Calls WriteDataXLCD (* buffer). See below.

putsXLCD

This function writes a string of characters located in buffer to the Hitachi HD44780 LCD controller. It stops transmission when a null character is encountered. The null character is not transmitted. The caller must check that the LCD controller is not busy before calling BusyXLCD. The data is written to the character generator RAM or the display data RAM depending on what the previous call to setcgram() or setddram().

```
Include: xlcd.h
Prototype: void putsXLCD( char *buffer );
void putrsXLCD( const rom char *buffer );
Arguments: const rom char *buffer.
```

Strings located in data memory should be used with the "puts" versions of these functions. Strings located in program memory, including string literals, should be used with the "putrs" versions of these functions.

```
File Name: putsxlcd.c
putrxlcd.c
Code Example:
    char mybuff [20];
    putrsXLCD( "Hello World" );
    putsXLCD( mybuff );
```

ReadAddr

Reads the address byte from the Hitachi HD44780 LCD controller. A previous call to BusyXLCD determines if the controller is busy.

```
Include: xlcd.h
Prototype: unsigned char ReadAddrXLCD( void );
Remarks: This function reads the address byte from the Hitachi HD44780 LCD
         controller. The address is from the character generator RAM or the
```

```
                display data RAM depending on the previous call to setcgram() or
                setddram().
Return Value:
                This function returns an 8-bit quantity. The address is contained
                in the lower order 7 bits and the BUSY status flag in the Most
                Significant bit.
Filename: readaddr.c
Code Example:
        char addr;
        .
        .
        .
        while ( BusyXLCD() );
        addr = ReadAddrXLCD();
```

ReadDataXLCD

Reads a data byte from the Hitachi HD44780 LCD controller. The caller must check that the LCD controller is not busy by calling BusyXLCD. The data is read from the character generator RAM or the display data RAM depending on the previous call to setcgram() or set ddram().

```
Include: xlcd.h
Prototype: char ReadDataXLCD( void );
Filename: readdata.c
Remarks: This function reads a data byte from the Hitachi HD44780 LCD
         controller.
Code Example:
        char data achar;
        .
        .
        .
        while ( BusyXLCD() );
        achar = ReadAddrXLCD();
```

SetDDRamAddr

Sets the display data address of the Hitachi HD44780 LCD controller. The caller must first check to see if the LCD controller is busy by calling BusyXLCD.

```
Include: xlcd.h
Prototype: void SetCGRamAddr( unsigned char addr );
Arguments: addr
File Name: setddram.c
Code Example:
        char cgaddr = 0xc0;
        .
        .
        .
        while( BusyXLCD() );
        SetCGRamAddr( cgaddr );
```

SetCGRamAddr

This routine sets the character generator address of the Hitachi HD44780 LCD controller. The caller must first check the state of the LCD controller by calling BusyXLCD.

```
Include: xlcd.h
Prototype: void SetDDRamAddr( unsigned char addr );
Arguments: unsigned char addr
File Name: setcgram.c
Code Example:
    char ddaddr = 0x10;
    .
    .
    .
    while( BusyXLCD() );
    SetDDRamAddr( ddaddr );
```

WriteCmdXLCD

Writes a command to the Hitachi HD44780 LCD controller. The caller must first check the LCD controller by calling BusyXLCD.

```
Include: xlcd.h
Prototype: void WriteCmdXLCD( unsigned char cmd );
Arguments: unsigned char cmd
The following values are defined in xlcd.h:
    DOFF Turn display off
    DON Turn display on
    CURSOR_OFF Enable display with no cursor
    BLINK_ON Enable display with blinking cursor
    BLINK_OFF Enable display with unblinking cursor
    SHIFT_CUR_LEFT Cursor shifts to the left
    SHIFT_CUR_RIGHT Cursor shifts to the right
    SHIFT_DISP_LEFT Display shifts to the left
    SHIFT_DISP_RIGHT Display shifts to the right
```

Alternatively, the command may be a bitmask that is created by performing a bitwise AND operation ('&') with a value from each of the categories defined in the file xlcd.h.

```
Data Transfer Mode:
    FOUR_BIT 4-bit Data Interface mode
    EIGHT_BIT 8-bit Data Interface mode
    Display Type:
    LINE_5X7 5x7 characters, single line
    LINE_5X10 5x10 characters display
    LINES_5X7 5x7 characters, multiple lines
File Name: wcmdxlcd.c
Code Example:
    while( BusyXLCD() );
    WriteCmdXLCD( EIGHT_BIT & LINES_5X7 );
    WriteCmdXLCD( BLINK_ON );
    WriteCmdXLCD( SHIFT_DISP_LEFT );
    putcXLCD
```

WriteDataXLCD

Writes a data byte to the Hitachi HD44780 LCD controller. The caller must first check the state of the LCD controller by calling BusyXLCD. The data is written to the character generator RAM or the display data RAM, depending on the previous call to setddram or setcgram.

```
Include: xlcd.h
```

```
Prototype: void WriteDataXLCD( char data );
Arguments: data
           The value of data can be any 8-bit value, but should correspond
           to a valid character in the HD44780 LCD controller RAM table.
File Name: writdata.c
Code Example:
    unsigned char achar = 'A';
    .
    .
    .
    WriteDataXLCD( achar );
```

11.7 LCD Application Development in C18

The development of a program that contains LCD functions follows the same steps as any other C18 application; however, certain precautions and customizations make the process easier. To this effect, in the present section we present the development of a C18 application that contains LCD operations. In this presentation we do not revisit the process of creating a C18 program described in Chapter 5. The process described does not follow the use of code-generating utilities such as Application Maestro. We believe the programmer benefits from creating his own code, even at the price of a little more effort. The walkthrough was done using MPLAB version 8.86 under Windows XP.

11.7.1 Using the Project Wizard

Although the Project Wizard is not necessary in creating a C18 application in MPLAB, it is useful because it provides a sequence of steps while not hiding details from the developer. In the following exercise we will create a project with a single workspace. The project name is C_LCD_Example.mcw. The project files can be found in the book's online software package.

The Project Wizard command is located in the MPLAB Project menu. The initial screen, labeled Step 1, is shown in Figure 5.4 and consists of selecting the PIC device used for the project, in our case the 18F452.

The second step allows us to select the active toolsuite, which is actually the programming language and development tools used in the project. We navigate through the options offered in the top window to Microchip C18 Toolsuite, as shown in Figure 5.5.

The third step consists of defining and naming the project file for the application, as seen in Figure 5.6. At this point we can used the browse button to navigate to an existing folder, or create a new folder anywhere in the system's memory space. In this walkthrough we create a new folder named C_LCD_Example and use this same name for the project.

The fourth step consists of adding the necessary files to the project. If we have a principal program file (often named main.c), we can include this file in the project at this time, or the main program file can be created and included in the project at a later date. Although any file required by the application can be included later, it is usually convenient to include those that we are certain will be needed at this time.

One such case comprises the LCD support files that is furnished with the C18 package as the Software Peripherals Library.

In a normal compiler installation, these files are located in MCC18/src/pmc_common/XLCD. If at this time we know which support files will be used by our program, we can select those and add them. Otherwise we can select all the support files in the directory in the knowledge that those not used can be later eliminated from the project. We usually prefer this last option.

Once the files are added, they are preceded by a letter that determines the file-addition mode. The options are A (automatic), U (user project), S (system path), and C (copy to project). This last option incorporates the file into the project without requiring an absolute path. It is usually the safest one. The addition option is changed by clicking the identifier letter.

While in the fourth step we can also select the linker script required by the application for the particular processor. In a default installation, the linker scripts will be found in the MCC18/lkr directory. We must locate the file named l18452.lkr; click the Add>> button and optionally change the addition path as previously discussed.

A header file that must be included in the project is named xlcd.h and is found in the MCC18/h directory. As with the other support files it, can be included at this time or later.

Main Program File

At this point we have created a project and its workspace but we have not yet created or added the main program file in C language. We will name this file with the same name as the project: C_LCD_Example.c.

The most convenient way of starting application development is by means of a template file. In the case of LCD programming, the template file is actually a minimal LCD application. The file is named C_LCD_Template.c and is found in this book's online software package. At this time we can copy the file to our Project directory and rename it accordingly. The template file code follows:

```
// Project name:
// Source files: XLCD support files
// Header file:  xlcd.h
// Date:
//
// Processor: PIC 18F452
// Environment:    MPLAB IDE Version 8.xx
//       MPLAB C-18 Compiler
//
// TEST CIRCUIT:
//
// PORT       PINS      DIRECTION      DEVICE
// Description:
// Template file for LCD applications.
// Assumes the following:
//       Hitachi HD44780 LCD controller
//       Two lines by 16 characters each
//       Wired for 8-bit data
```

```
//          Uses C18 LCD Software Peripheral Library
//
//
#include <p18cxxx.h>
#include "xlcd.h"
//
// Configuration bits set as required for MPLAB ICD 2
#pragma config OSC = XT          // Assumes high-speed resonator
#pragma   config WDT = OFF       // No watchdog timer
#pragma   config LVP = OFF       // No low voltage protection
#pragma   config DEBUG = OFF     // No background debugger
#pragma config PWRT = ON         // Power on timer enabled
#pragma config CP0 = OFF         // Code protection block x = 0-3
#pragma config CP1 = OFF
#pragma config CP2 = OFF
#pragma config CP3 = OFF
#pragma config WRT0 = OFF        // Write protection block x = 0-3
#pragma config WRT1 = OFF
#pragma config WRT2 = OFF
#pragma config WRT3 = OFF
#pragma config EBTR0 = OFF       // Table read protection block x = 0-3
#pragma config EBTR1 = OFF
#pragma config EBTR2 = OFF
#pragma config EBTR3 = OFF

// Global data
char XLCD_Disp1[] = "  Testing LCD    ";
char XLCD_Disp2[] = "    display      ";

void main(void)
{
    unsigned char config=0xff;

    ADCON1 = 0xFF;

    // Define configuration for 8 bits and two 5 X 7 lines
    config = EIGHT_BIT  & LINES_5X7;
    // Initialize LCD
    OpenXLCD(config);
    // Test for busy
    while( BusyXLCD() );           // Wait until LCD not busy
    // Set starting address in the LCD RAM for display.
    // Can be edited to match system
    SetDDRamAddr(0x80);
    while( BusyXLCD() );           // Wait until LCD not busy
    putsXLCD(XLCD_Disp1);          // Display first string
    while( BusyXLCD() );           // Wait until LCD not busy
    // Set address for display second line
    SetDDRamAddr(0xC0);
    while( BusyXLCD() );           // Wait until LCD not busy
    putsXLCD(XLCD_Disp2);          // Display second string
    while( BusyXLCD() );           // Wait until LCD not busy

    while(1);                      //end of template

}
```

Once we have edited, renamed, and included the template file in the project's source files, we can attempt to compile the application using the Build All command in the MPLAB Project menu. Make sure that the desired build configuration (debug or release) has been selected in the corresponding window. The build tab in the MPLAB Output window will record the build process and list any compile- or link-time errors. The Build All command recompiles all the source files in the project and thus serves as a check of the Software Peripherals Library modules.

Chapter 12

Real-Time Clocks

12.1 Measuring Time

This chapter deals with the measurement of time in discrete, digital units. In this context we speak of "real-time" as years, days, hours, minutes, and so on. A real-time clock (RTC) is one that measures time in hours, minutes, and seconds, and a real-time calendar is in years, months, weeks, and days. Because time is a continuum that escapes our comprehension, we must divide it into measurable chunks that can be manipulated and calculated. However, not all time units are in proportional relations to one another. There are 60 seconds in a minute and 60 minutes in an hour, but 24 hours in a day and 28, 29, 30, or 31 days in a month. Furthermore, the months and the days of the week have traditional names. Finally, the Gregorian calendar requires adding a 29th day to February on any year that is evenly divisible by 4. The device or software to perform all of these time calculations is referred to as a real-time clock. In this chapter we discuss the use of real-time clocks in PIC 18F circuits.

Note that the notion of real-time programming does not coincide with the programming of real-time clocks. Real-time computing (also called reactive computing) refers to systems that are subject to a time constraint. For example, a program that must ensure that a deadline is met between a world event and a system response. Non-real-time systems cannot guarantee a response even if it is the usual result. In this sense, the topic of this chapter is limited to clocking devices and routines that measure time in conventional units, not to real-time computing.

12.1.1 Clock Signal Source

All digital systems require a timing device that produces the clock cycles by which the hardware operates. The timer or clock is the beating heart of a digital device. Pulse-generating technologies include oscillators, resonators, resistor-capacitor circuits, piezoelectric crystals, and others. It is in timing operations that the accuracy of the pulse-generating device becomes a critical issue. In this context, a digital clock or watch will be as accurate as the frequency of the signal source. If a commercial oscillator has a 10 % variation in accuracy, a timing routine that uses this oscillator can ensure no more than 10 % accuracy.

Crystal oscillators use the mechanical resonance of a vibrating piezoelectric material to create a very precise frequency. One common type is quartz crystals. They provide a stable and accurate signal that is used in clocks, quartz watches, integrated circuits, radio transmitters and receivers, and various test and measuring instruments. Quartz crystals are manufactured in frequencies from one tenth of a kilohertz to several megahertz. In embedded applications the crystal often used for real-time calculations oscillates at a rate of 32.768 kHz and is usually called a 32-kHz crystal. Because 32,768 is a power of 2, it provides a convenient source for binary numerical time calculations. For example, 65,535 oscillations (0xffff in hex) take place in 2 seconds. A counter set to count down from 0x8000 oscillations of a 32-kHz crystal will rollover every one second. By the same token, 327 oscillations of this crystal will measure one hundredth of a second.

32-kHz Crystal Circuit

The reason that an external crystal can be used directly with the 18F452 is that the microcontroller provides support for an external clock source. In our discussion of the Timer1 module (see Section 9.5.1), we saw that Timer1 can be set to use an external clock in its asynchronous counter mode. In this case, the 18F452 RC0 and RC1 pins are multiplexed as input from the clock source, typically a 32-kHz quartz crystal. In this function, the pins are labeled T1OSI and T1OSO. T1OSI is the amplifier input line and T1OSO the amplifier output.

The version of the crystal most often used in embedded applications takes the form of a small, silvery cylinder with two wire connectors. The circuit also requires two capacitors, one on each terminal of the 32-kHz crystal. Figure 12.1 shows the typical wiring of this circuit.

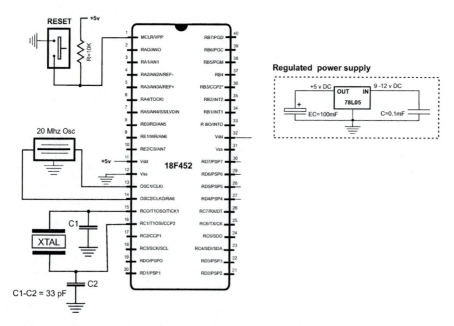

Figure 12.1 *18F452 circuit with a 32 kHz crystal in Port C.*

The value of these capacitors has been a topic of some debate on the Internet because Microchip recommends 33 pF but suggests that the user should consult the manufacturer of the crystal. We have used 22- and 33-pF capacitors in our experimental circuits without any observable difference.

12.1.2 Programming the Timer1 Clock

Programming the real-time clock based on Timer 1 is straightforward and simple. The timer can be initialized to use an 8-bit or 16-bit counter, but in real-time clock applications the 16-bit option is easier to implement. Once the timer is initialized and assuming a 32-kHz crystal, the 16-bit counter overflows every 2 seconds. Applications can either poll the timer registers (TMR1H and TMR1L) or set up an interrupt that is triggered by the overflow. Interrupt-based handlers work well using the 18F452 low- or high-priority interrupt, which means that most applications can be designed to use one of the interrupts.

Reading and writing the 16-bit timer registers was discussed in Section 9.5.1. In this application, code selects the 8- or the 16-bit mode by means of the RD16 bit in the T1CON register. The address for TMR1H is mapped to a buffer register for the high byte of Timer1. This determines that a read from TMR1L will load the contents of the high byte of Timer1 into the Timer1 high-byte buffer. This scheme makes it possible to accurately read all 16 bits of Timer1 without having to determine if a rollover took place while the read was in progress. This is also the case with a write operation. Writing the high byte of Timer1 must also take place through the TMR1H buffer register. In other words, Timer1 high byte is updated with the contents of TMR1H when a write occurs to TMR1L. This allows writing all 16 bits to both the high and low bytes of Timer1 in a single operation.

Setting Up Timer1 Hardware

The sequence of operations required to set up a 16-bit real-time clock based on Timer 1 interrupt on overflow is the following:

- Turn off Timer1 interrupt during setup operations.
- Set up the T1CON register to enable 16-bit operation, enable the Timer1 oscillator, select not-synchronized operation, select external clock source, and enable the Timer1 interrupt on overflow.
- Set up priority levels on interrupts.
- Assign a priority to the Timer1 interrupt.
- Set up the Timer1 interrupt in the INTCON register.
- Activate falling edge mode in INTCON2 register.
- Clear the interrupt flag and reenable the Timer1 interrupt.

The code for this initialization is found in the sample program RTC_18F_Timer1 Code listed later in this section.

Coding the Interrupt Handler

In 16-bit mode, the interrupt handler receives control every time the TMR1H register overflows. Applications can set an initial value in the counter registers in order to control the time period of the interrupt. For example, if the 16-bit counter is preset to 0x8000 then the register will overflow in one-half the default time. The 1-second time lapse is very common in real-time clock code. The interrupt latency and the timer required for resetting the counter registers will affect the accuracy of the clock. In many cases, the error introduced is tolerable for the particular application. Otherwise it is possible to calculate the time error and compensate for it by presetting the low-order counter register. Once the timer registers are preset, code can clear the interrupt flag to restart the count.

Sample Program RTC_18F_Timer1.asm

The sample program RTC_18F_Timer1.asm is found in the project of the same name in this book's online software package. The program uses the high-priority interrupt to operate a real-time clock by the Timer1 module. Code assumes that an LED is wired to port C, line 2. The LED is toggled on and off by the interrupt handler at 1-second intervals. The program runs in the circuit in Figure 12.1. Code is as follows:

```
; File name: RTC_18F_Timer1.asm
; Project: RTC_18F_Timer1.mcp
; Date: March 4, 2013
; Author: Julio Sanchez
;
; STATE: Tested OK
;
; Program for PIC 18F452 using Timer1 and 32kHz crystal
; to provide real-time operations.
;
; Executes in compatible circuit
;
; Description:
; Demonstration of the Timer1 external clock using a 32
; kHz quart crystal wired to port C lines 0 and 1.
; Application toggles on and off an LED wired to port
; C, line 2, at one second intervals. Code uses the high-
; priority interrupt as a time keeper.
;
;============================================================
;                         Circuit
;============================================================
;                        18F452
;              +------------------+
;+5v-res0-ICD2 mc -| 1 !MCLR   PGD 40|
;                  | 2 RA0     PGC 39|
;                  | 3 RA1     RB5 38|
;                  | 4 RA2     5B4 37|
;                  | 5 RA3     RB3 36|
;                  | 6 RA4     RB2 35|
;                  | 7 RA5     RB1 34|
;                  | 8 RE0     RB0 33|
;                  | 9 RE1         32|-------+5v
;                  |10 RE2         31|--------GR
;        +5v--------|11        RD7 30|
;        GR---------|12        RD6 29|
```

```
;             osc ---|13 OSC1      RD5 28|
;             osc ---|14 OSC2      RD4 27|
; 32khz crystal ---|15 RC0        RC7 26|
; 32khz crystal ---|16 RC1        RC6 25|
;         LED <==|17 RC2          RC5 24|
;               |18 RC3           RC4 23|
;               |19 RD0           RD3 22|
;               |20 RD1           RD2 21|
;               +-----------------+
;
;

    list p=18f452
    ; Include file, change directory if needed
    include "p18f452.inc"
; ==========================================================
;                 configuration bits
;==========================================================
; Configuration bits set as required for MPLAB ICD 2
    config OSC = XT          ; Assumes high-speed resonator
    config WDT = OFF         ; No watchdog timer
    config LVP = OFF         ; No low voltage protection
    config DEBUG = OFF       ; No background debugger
    config PWRT = ON         ; Power on timer enabled
    config CP0 = OFF         ; Code protection block x = 0-3
    config CP1 = OFF
    config CP2 = OFF
    config CP3 = OFF
    config WRT0 = OFF        ; Write protection block x = 0-3
    config WRT1 = OFF
    config WRT2 = OFF
    config WRT3 = OFF
    config EBTR0 = OFF       ; Table read protection block x = 0-3
    config EBTR1 = OFF
    config EBTR2 = OFF
    config EBTR3 = OFF
;
; Turn off banking error messages
    errorlevel   -302
;==========================================================
;                      program
;==========================================================
    ; Start at the reset vector
Reset_Vector  code 0x000
    goto Start
;==========================================================
;                interrupt intercepts
;==========================================================
; HIGH PRIORITY INTERRUPT VECTOR
    org       0x008
    goto      ServiceRtn

; LOW-PRIORITY HANDLER NOT IMPLEMENTED IN THIS EXAMPLE
    org       0x018          ; LOW PRIORITY VECTOR
    retfie    0x00      ; No context save return


;==========================================================
;               m a i n   p r o g r a m
;==========================================================
    code      0x030
Start:
```

```
; Set BSR for bank 0 operations
    movlb    0                ; Bank 0

; Init Port A for digital operation
    clrf    PORTA,0
    clrf    LATA,0
; ADCON1 is the configuration register for the A/D
; functions in Port A. A value of 0b011x sets all
; lines for digital operation
    movlw   B'00000110'  ; Digital mode
    movwf   ADCON1,0
; Initialize PORT C for output.
; Setting of the tris register does not affect lines
; wired to crystal
    clrf TRISC           ; Output to LED. TRIS bits
                              ; are don't care for crystal
; Clear output lines
    clrf PORTC           ; Turn off LED
;
;=========================
; T1CON in oscillator mode
;=========================
    bcf  PIE1,TMR1IE       ; Turn off Timer1 interrupt
                              ; during setup
; Setup the T1CON register
;                |------------ READ/WRITE
;                |             16-bit operation
;                ||----------- UNIMPLEMENTED
;                |||---------- Clock source
;                ||||--------- PRESCALE BITS
;                ||||               00 = no prescale
;                |||||-------- T1OSCEN - Enable oscillator
;                ||||||------- !T1SYNC - Not synchronized
;                |||||||------ TMR1CS - External clock
;                |||||||| ----- TMR1ON - Timer 1 interrupt
;                ||||||||
    movlw    b'10001111'
    movff    WREG,T1CON
; Setup Timer1 interrupt
    bsf      RCON,IPEN     ; Priority levels on interrupts
    bsf      IPR1,TMR1IP      ; High priority Tmr1 overflow
; Clear Timer1 registers
    clrf     TMR1H
    clrf     TMR1L
    clrwdt
;===============================
; Set up for Timer1 interrupt
;===============================
; INTCON register initialized as follows:
; (IPEN bit is clear)
;                |------------ GIE/GIEH - high-priority interrupts
;                |                  enabled
;                ||----------- PEIE/GEIL - Peripheral enabled
;                |||---------- timer0 overflow interrupt
;                ||||--------- external interrupt
;                |||||-------- port change interrupt
;                ||||||------- overflow interrupt flag
;                |||||||------ external interrupt flag
;                ||||||||----- RB4:RB7 interrupt flag
    movlw    b'11000000'
```

```
    movwf    INTCON
; Set INTCON2 for falling edge operation
    bcf      INTCON2,INTEDG0
; Re-enable timer1 interrupt
    bcf      INTCON,TMR0IF    ; Clear interrupt flag
    bsf      PIE1,TMR1IE      ; Turn on Timer1 interrupt
                              ; after setup


;==========================
;   loop doing nothing
;==========================
endless:
    nop
    nop
    goto     endless


;===========================================================
;                  interrupt service routine
;===========================================================
ServiceRtn:
; HIGH PRIORITY INTERRUPT HANDLER
; Sets Timer1 counter to roll over every 32,768 beats
; Resets Timer1 interrupt flag
; Toggles state of LED on line 2, PORT C
; Returns saving context
    movlw    0x80       ; High bit set
    movff    WREG,TMR1H     ; To high order register
    clrf     TMR1L          ; Clear low and write both
    bcf      PIR1,TMR1IF    ; Clear Timer1 interrupt flag
; Test status of PORTC line 2 LED
    btfss    PORTC,2        ; Is LED on ?
    goto turnON         ; Turn it ON
; LED is on. Turn it off
    bcf      PORTC,2        ; Clear bit to turn off
    retfie   0x01       ; Return saving context
turnON:
    bsf      PORTC,2        ; Turn on LED
    retfie   0x01       ; Return saving context


        end
```

12.2 Real-Time Clock ICs

An alternative to using the PIC's hardware support for real-time clocks, as shown in the preceding section, is to include a real-time clock IC in the circuit. These dedicated ICs, usually called RTCs, are integrated circuits designed to keep track of time in conventional units, that is, in years, days, hours, minutes, and seconds. Many real-time clock ICs are available with different characteristics, data formats, modes of operation, and interfaces. Most of the ones used in PIC circuits have a serial interface in order to save access ports. Most RTC chips provide a battery connection so that the time can be kept when the system is turned off.

In the sections that follow, we discuss one popular RTC chip: the NJU6355, but this is by no means the only option for embedded systems. Demo Board B contains a real-time clock IC (6355) that can be used in testing the programming in the sections that follow..

12.2.1 NJU6355

The NJU6355 series is a serial I/O real-time clock suitable for microcontroller-based embedded systems. The IC includes a counter, shift register, voltage regulator, and the interface controller. It is usually wired to an external 32-kHz quartz crystal The interface to a PIC requires four lines. The operating voltage is the TTL level so it can be wired directly on the typical PIC circuit. The output data includes year, month, day-of-week, hour, minutes, and seconds. Figure 12.2 is the pin diagram for the chip.

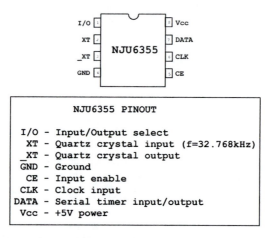

Figure 12.2 *NJU6355 pin diagram.*

12.2.2 6355 Data Formatting

The NJU6355 output is in packed BCD format, that is, each decimal digit is represented by a 4-bit binary number. The chip's logic correctly calculates the number of days in each month as well as the leap years. All unused bits are reported as binary 0. Figure 12.3 is a bitmap of the formatted timer data.

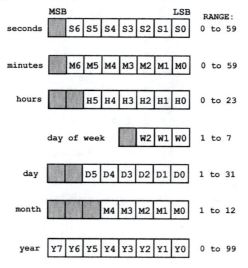

Figure 12.3 *NJU6355 timer data format*

Timer data can be read when the I/O line is low and the CE line is high. Output from the 6355 is LSB first. A total of 52 significant bits are read in bottom-up order as the data is shown in Figure 12.2. That is, the first bit received is the least-significant bit of the year, then the month, after that the day, and so forth. All date items are 8 bits, except the day of week, which is 4 bits. Non-significant bits in each field are reported as zero. This means that the value for the 10th month (October) are the binary digits 00001012. Reporting unused digits as zero simplifies the conversion into ASCII.

The NJU6355 does not report valid time data until after it has been initialized, even if there is power and clock signals into the chip. Initialization requires writing data into the 6355 registers. In order to write to the IC, code must set the I/O and the CE lines high. At this moment, all clock updates stop and the RTC goes into the write mode. Here again, the input data is latched in LSB first, starting with the year and concluding with the minutes. There is no provision for writing seconds into the RTC, so the total number of bits written is 44.

The 6355 contains a mechanism for detecting conditions that could compromise the clock's operation, such as low power. In this case, the special value 0xee is written into each digit of the internal registers so that processing routines can become aware that the timer is in error.

The NJU6355 requires the installation of an external crystal oscillator. The crystal must have a frequency of 32.768 kHz. The time-keeping accuracy of the RTC is determined by this oscillator. The capacity of the oscillator must match that of the RTC and of the circuit. A standard crystal with a capacitance of 12.5pF works well for applications that do not demand high clock accuracy. For more exacting applications the 6355 can programmed to check the clock frequency and determine its error. The chip's frequency checking mode is described in an NJU6355 Application Note available from NJR.

12.2.3 Initialization and Clock Primitives

Several core procedures (we could call them clock primitives) are necessary for developing RTC applications. The first one initializes the clock hardware. Two other primitives read the current time and write to the clock registers. Because clock data can be in 8- or 4-bit formats, it is useful to design the procedures so that they provide separate entry points to handle the 4- and the 8-bit options.

Reading and Writing Clock Data

Once the 6355 has been initialized, the application can read or write to the chip's registers. The core procedure for the read operation has two entry points: read4RTC returns 4 bits from an RTC register and readRTC returns the full 8 bits. Code is as follows:

```
;=============================
;  read 4/8 bits from RTC
;=============================
; Procedure to read 4/8 bits stored in 6355 registers
; Value returned in w register
```

```
read4RTC
    movlw    .4                      ; 4 bit read
    goto     anyBits
readRTC
    movlw    .8                      ; 8 bits read
anyBits:
    movwf    counter
; Read 6355 read operation requires the IO line be set low
; and the CE line high. Data is read in the following order:
; year, month, day, day-of-week, hour, minutes, seconds
readBits:
    bsf      PORTB,CLK               ; Set CLK high to validate data
    bsf      STATUS,C                ; Set the carry flag (bit = 1)
; Operation:
;    If data line is high, then bit read is a 1-bit
;    otherwise bit read is a 0-bit
    btfss    PORTB,DAT               ; Is data line high?
                                     ; Leave carry set (1 bit) if high
    bcf      STATUS,C                ; Clear the carry bit (make bit 0)
; At this point the carry bit matches the data line
    bcf      PORTB,CLK               ; Set CLK low to end read
; The carry bit is now rotated into the temp1 register
    rrcf     temp1,1
    decfsz   counter,1               ; Decrement the bit counter
    goto     readBits                ; Continue if not last bit
; At this point all bits have been read (8 or 4)
    movf     temp1,0                     ; Result to w
    return
```

Notice that the procedure's two entry points load the w register with the number of bits to be read, the value of which is stored in a local register named counter. When counter goes to zero, the read operation terminates. The actual reading of clock data consists of testing the 6355 data line (labeled DAT) for a 0 or a 1 bit. The result is stored in the carry flag, which is then rotated into the local variable named temp1. The rrcf instruction (rotate right file register through carry) accomplishes this. Notice that the 6355 requires that the clock line (labeled CLK) be held high during a read.

Equipped with the 4- and 8-bit read primitives, reading 6355 data consists of calling the two routines and storing each result in a local variable. The variables are named year, month, day, dayOfWeek, hour, minutes, and seconds. The read operation requires that the clock's IO line be held low and the CE line high. The CLK line is initially set low and then high by the actual read operation. Code is as follows:

```
;=============================
;       read RTC data
;=============================
; Procedure to read the current time from the RTC and store
; data (in packed BCD format) in local time registers.
; According to wiring diagram
; NJU6355 Interface for read operations:
; DAT        PORTB,0          Input
; CLK        PORTB,1          Output
; CE         PORTB,2          Output
; IO         PORTB,3          Output
Get_Time
```

```
; Clear port B
    movlw    b'00000000'
    movwf    PORTB
; Make data line input
    Bank1
    movlw    b'00000001'
    movwf    TRISB
    Bank0
; Reading RTC data requires that the IO line be low and the
; CE line be high. CLK line is held low
    bcf      PORTB,CLK      ; CLK low
    call     LCD_delay_1
    bcf      PORTB,IO       ; IO line low
    call     LCD_delay_1
    bsf      PORTB,CE       ; and CE line high
; Data is read from RTC as follows:
;  year        8 bits (0 to 99)
;  month       8 bits (1 to 12)
;  day         8 bits (1 to 31)
;  dayOfWeek 4 bits (1 to 7)
;  hour        8 bits (0 to 23)
;  minutes     8 bits (0 to 59)
;  seconds     8 bits (0 to 59)
;              ======
;   Total      52 bits
    call     readRTC
    movwf    year
    call     LCD_delay_1
;
    call     readRTC
    movwf    month
    call     LCD_delay_1
;
    call     readRTC
    movwf    day
    call     LCD_delay_1
;
; dayOfWeek of week is a 4-bit value
    call     read4RTC
    movwf    dayOfWeek
    call     LCD_delay_1
;
    call     readRTC
    movwf    hour
    call     LCD_delay_1
;
    call     readRTC
    movwf    minutes
    call     LCD_delay_1
;
    call     readRTC
    movwf    seconds
; Done
    bcf      PORTB,CE            ; CE line low to end output
    return
```

Notice that when the routine ends, the CE line is set low again to terminate output from the clock IC. Also notice that the read operation returns data sequentially, starting with the year and ending with seconds. If an application requires one of the elements in the list but not others, it must still read the preceding ones and ignore these values. Data that follows the desired entry need not be read because setting the clock's CS line low ends the read operation and resets the clock's internal data pointer.

Writing to the 6355 follows the same sequential pattern as the read operation: data is first written to the year register, then to the month, and lastly to the minutes. The seconds value in the 6355 cannot be written. Here again, two core routines are required: one to write 8-bit registers and another one to write to the 4-bit register. The entry points are named write4RTC and writeRTC, respectively. Code is as follows:

```
;============================
;    write 4/8 bits to RTC
;============================
; Procedure to write 4 or 8 bits to the RTC registers
; ON ENTRY:
;     temp1 register holds value to be written
; ON EXIT:
;     nothing
write4RTC
        movlw    .4             ; Init for 4 bits
        goto     allBits
writeRTC
        movlw    .8             ; Init for 8 bits
allBits:
        movwf    counter        ; Store in bit counter
        clrf     temp1          ; Clear local register
writeBits:
        bcf      PORTB,CLK      ; Clear the CLK line
        call     LCD_delay_2     ; Wait
        bsf      PORTB,DAT      ; Set the data line to RTC
        btfss    temp1,0        ; Send LSB
        bcf      PORTB,DAT      ; Clear data line
        call     LCD_delay_2    ; Wait for operation to complete
        bsf      PORTB,CLK      ; Bring CLK line high to validate
        rrcf     temp1,f        ; Rotate bits in storage
        decfsz   counter,f      ; Decrement bit counter
        goto     writeBits      ; Continue if not last bit
        return
```

The actual write operation is a loop in which the register named counter determines the number of iterations. For each bit, code clears the CLK line, then sets the DAT line high. If the low order bit of temp1 is set then the DAT line is turned from high to low. Otherwise it is kept high. The bit write takes place when the CLK line is again turned high. Then the bits in temp1 are rotated right and the loop continues until the counter register goes to 0.

Initialize RTC

The procedure to initialize the 6355 consists of CLK line low, and the IO and CE lines high. Then data must be written to all 6355 registers. Code is as follows:

```
;============================
;        init RTC
;============================
```

```
; Procedure to initialize the real-time clock chip. If chip
; is not initialized it will not operate and the values
; read will be invalid.
; Because the 6355 operates in BCD format the stored values must
; be converted to packed BCD.
; According to wiring diagram
; NJU6355 Interface for setting time:
; DAT             PORTB,0              Output
; CLK             PORTB,1              Output
; CE         PORTB,2          Output
; IO         PORTB,3          Output
setRTC:
      clrf    TRISB
; Writing to the 6355 requires that the CLK bit be held
; low while the IO and CE lines are high
      bcf     PORTB,CLK    ; CLK low
      call    delay_5
      bsf     PORTB,IO     ; IO high
      call    delay_5
      bsf     PORTB,CE     ; CE high
; Data is stored in RTC as follows:
;  year       8 bits (0 to 99)
;  month      8 bits (1 to 12)
;  day             8 bits (1 to 31)
;  dayOfWeek   4 bits (1 to 7)
;  hour            8 bits (0 to 23)
;  minutes    8 bits (0 to 59)
;                  ======
;   Total          44 bits
; Seconds cannot be written to RTC. RTC seconds register
; is automatically initialized to zero
      movf    year,w                ; Get item from storage
      call    bin2bcd               ; Convert to BCD
      movwf   temp1
      call    writeRTC

      movf    month,w
      call    bin2bcd
      movwf   temp1
      call    writeRTC

      movf    day,w
      call    bin2bcd
      movwf   .temp1
      call    writeRTC

      movf    dayOfWeek,w           ; dayOfWeek of week is 4-bits
      call    bin2bcd
      movwf   temp1
      call    write4RTC

      movf    hour,w
      call    bin2bcd
      movwf   temp1
      call    writeRTC

      movf    minutes,w
      call    bin2bcd
      movwf   temp1
      call    writeRTC
```

```
; Done
    bcf        PORTB,CLK            ; Hold CLK line low
    call       delay_5
    bcf        PORTB,CE             ; and the CE line
                                    ; to the RTC
    call       delay_5
    bcf        PORTB,IO             ; RTC in output mode
    return
```

12.2.4 BCD Conversions

In addition to the RTC procedures to initialize the clock registers and to read clock
data, the application requires auxiliary procedures to manipulate and display data in
BCD format. The 6355 uses the packed BCD format and in order to display clock data
on the LCD,a BCD-to-ASCII conversion is required. Additionally, because program
data is stored in binary form, it is also necessary to have a routine to convert binary
data into BCD form. A simple algorithm for converting binary to BCD is as follows:

1. The value 10 is subtracted from the source operand until the reminder is less than
   0 (carry cleared). The number of subtractions is the high-order BCD digit.

2. The value 10 is then added back to the subtrahend to compensate for the last subtraction.

3. The final reminder is the low-order BCD digit.

The binary to BCD conversion procedure is coded as follows:

```
;=============================
;  binary to BCD conversion
;=============================
; Convert a binary number into two packed BCD digits
; ON ENTRY:
;          w register has binary value in range 0 to 99
; ON EXIT:
;          output variables bcdLow and bcdHigh contain two
;          unpacked BCD digits
;          w contains two packed BCD digits
; Routine logic:
;    The value 10 is subtracted from the source operand
;    until the reminder is < 0 (carry cleared). The number
;    of subtractions is the high-order BCD digit. 10 is
;    then added back to the subtrahend to compensate
;    for the last subtraction. The final reminder is the
;    low-order BCD digit
; Variables:
;     inNum       storage for source operand
;     bcdHigh     storage for high-order nibble
;     bcdLow      storage for low-order nibble
;     thisDig        Digit counter
bin2bcd:
    movwf      inNum            ; Save copy of source value
    clrf       bcdHigh              ; Clear storage
    clrf       bcdLow
    clrf       thisDig
min10:
    movlw      .10
    subwf      inNum,f          ; Subtract 10
```

```
        btfsc     STATUS,C      ; Did subtract overflow?
        goto      sum10         ; No. Count subtraction
        goto      fin10
sum10:
        incf      thisDig,f     ; Increment digit counter
        goto      min10
; Store 10th digit
fin10:
        movlw     .10
        addwf     inNum,f       ; Adjust
        movf      thisDig,w     ; Get digit counter contents
        movwf     bcdHigh       ; Store it
; Calculate and store low-order BCD digit
        movf      inNum,w       ; Store units value
        movwf     bcdLow        ; Store digit
; Combine both digits
        swapf     bcdHigh,w     ; High nibble to HOBs
        iorwf     bcdLow,w      ; ORin low nibble
        return
```

Because the program requires displaying values stored in BCD format by the 6355 hardware, a routine is necessary to convert two packed BCD digits into two ASCII decimal digits. The conversion logic is quite simple: the BCD digit is converted to ASCII by adding 0x30 to its value. All that is necessary is to shift and mask-out bits in the packed BCD operand so as to isolate each digit and then add 0x30 to each one. The code is as follows:

```
;===============================
;     BCD to ASCII decimal
;            conversion
;===============================
; ON ENTRY:
;        w register has two packed BCD digits
; ON EXIT:
;        output variables asc10, and asc1 have
;        two ASCII decimal digits
;
; Routine logic:
;   The low order nibble is isolated and the value 0x30
;   added to convert to ASCII. The result is stored in
;   the variable asc1. Then the same is done to the
;   high-order nibble and the result is stored in the
;   variable asc10
;
Bcd2asc:
        movwf     store1        ; Save input
        andlw     b'00001111'   ; Clear high nibble
        addlw     0x30              ; Convert to ASCII
        movwf     asc1              ; Store result
        swapf     store1,w          ; Recover input and swap digits
        andlw     b'00001111'   ; Clear high nibble
        addlw     0x30              ; Convert to ASCII
        movwf     asc10         ; Store result
        return
```

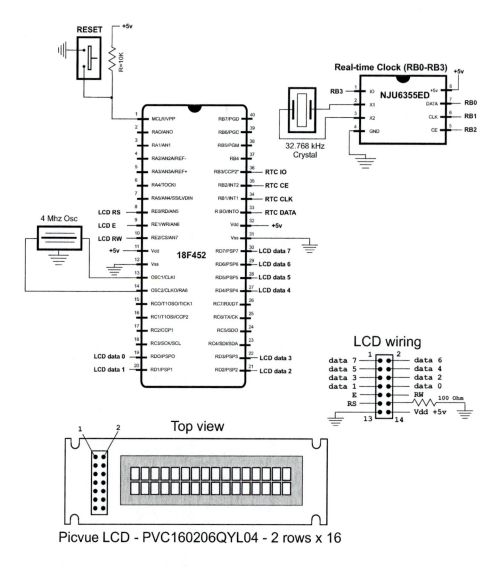

Figure 12.4 *Real-time clock demonstration circuit.*

12.3 RTC Demonstration Circuit and Program

Notice that the circuit in Figure 12.4 is a subset of Demo Board 18F452, which can be used to test the sample code listed in the following section.

12.3.1 RTC_F18_6355.asm Program

The demonstration program named RTC_F18_6355.asm listed in this section and found in the book's software package demonstrates programming of the RTC 6355 IC. Code reads the clock data in an endless loop. The hours, minutes, and seconds are displayed on the second line of the LCD as follows:

```
                          H:xx M:xx S:xx
```

Where xx represents the two BCD digits read from the clock and converted to ASCII decimal for display.

Code Details

The most important operations performed by the RTC_F18_6355 program were discussed earlier or were covered in previous chapters. Nevertheless, the code contains a few details that merit additional comment.

The program uses several #define statements to create constants that are later referenced in the code. For example,

```
; Defines from real-time clock wiring diagram
; all lines in port B
#define   DAT      0    ;|
#define   CLK      1    ;| -- from circuit wiring diagram
#define   CE       2    ;|
#define   IO       3    ;|
```

These statements associate numeric values from the wiring diagram with the standard names used by the real-time clock documentation. Henceforth the numeric values can be substituted with the symbolic names, thereby improving the readability and reliability of the code, as in the following fragment:

```
; Writing to the 6355 requires that the CLK bit be held
; low while the IO and CE lines are high
    bcf  PORTB,CLK ; CLK low
    call LCD_delay_2
    bsf  PORTB,IO      ; IO high
    call LCD_delay_2
    bsf  PORTB,CE      ; CE high
```

The application also uses #define statements to create other constants, such as the LCD lines and addresses and the values used in the initialization of the RTC. Notice that the use of #define statements (or of equivalent equates) makes the code easier to test and to modify because all instances of the constant or label are found in a single location in the code.

Code Listing

```
; File name: RTC_18F_6355.asm
; Last Update: March 8, 2013
; Author: Julio Sanchez
; Processor: 18F452
; State:
;      March 8/13 - Demo Board 18F452 OK
;
; Description:
; Program to demonstrate the NJU6355 Real Time Clock IC.
; Program uses LCD to display results of hours, minutes,
; and seconds, as follows:
;
;                      |=====================|
;                      |   Real time clock   |
```

```
;                      |    H:03 M:02 S:01    |
;                      |=====================|
;
; Initialization values are in #define statements that start
; with i, such as iYear, iMonth, etc.
;
; WARNING:
; Code assumes 4MHz clock. Delay routines must be
; edited for faster clock
;==========================================================
;                            Circuit
;==========================================================
;                         18F452
;                    +------------------+
;+5v-res0-ICD2 mc -|  1 !MCLR      PGD 40|
;                   |  2 RA0       PGC 39|
;                   |  3 RA1       RB5 38|
;                   |  4 RA2       5B4 37|
;                   |  5 RA3       RB3 36|==> RTC IO
;                   |  6 RA4       RB2 35|==> RTC CE
;                   |  7 RA5       RB1 34|<== RTC CLK
;       LCD RS   <==|  8 RE0       RB0 33|<== RTC DATA
;       LCD E    <==|  9 RE1           32|-------+5v
;       LCD RW   ==>|10 RE2           31|--------GR
;       +5v--------|11           RD7 30|==> LCD data 7
;       GR---------|12           RD6 29|==> LCD data 6
;            osc ---|13 OSC1      RD5 28|==> LCD data 5
;            osc ---|14 OSC2      RD4 27|==> LCD data 4
;                   |15 RC0       RC7 26|
;                   |16 RC1       RC6 25|
;                   |17 RC2       RC5 24|==>
;                   |18 RC3       RC4 23|==>
;    LCD data 0 <==|19 RD0       RD3 22|==> LCD data 3
;    LCD data 1 <==|20 RD1       RD2 21|==> LCD data 2
;                    +------------------+
; Legend:
; LCD E = LCD signal enable
; LCD RW = LCD read/write
; LCD RS = LCD register select
; GR = ground
; RTC DATA = serial timer I/O
; RTC CLK = Clock input
; RTC CE = Input enable
; RTC IO = Input/output select
;============================================================
    list p=18f452
    ; Include file, change directory if needed
    include "p18f452.inc"
; ==========================================================
;                  configuration bits
;==========================================================
; Configuration bits set as required for MPLAB ICD 2
    config OSC = XT              ; Assumes high-speed resonator
    config WDT = OFF             ; No watchdog timer
    config LVP = OFF             ; No low voltage protection
    config DEBUG = OFF           ; No background debugger
    config PWRT = ON          ; Power on timer enabled
    config CP0 = OFF          ; Code protection block x = 0-3
    config CP1 = OFF
    config CP2 = OFF
```

```
      config CP3 = OFF
      config WRT0 = OFF          ; Write protection block x = 0-3
      config WRT1 = OFF
      config WRT2 = OFF
      config WRT3 = OFF
      config EBTR0 = OFF         ; Table read protection block x = 0-3
      config EBTR1 = OFF
      config EBTR2 = OFF
      config EBTR3 = OFF
;
; Turn off banking error messages
      errorlevel      -302
;============================================================
;                   constant definitions
;   for PIC-to-LCD pin wiring and LCD line addresses
;============================================================
; LCD used in the demo board is 2 lines by 16 characters
#define E_line 1    ;|
#define RS_line 0   ;| -- from wiring diagram
#define RW_line 2   ;|
; LCD line addresses (from LCD data sheet)
#define LCD_1 0x80 ; First LCD line constant
#define LCD_2 0xc0 ; Second LCD line constant
; Defines from real-time clock wiring diagram
; all lines in port B
#define       DAT              0   ;|
#define       CLK              1   ;| -- from circuit wiring diagram
#define       CE               2   ;|
#define       IO               3   ;|
;
; Defines for RTC initialization (values are arbitrary)
#define iYear     .7
#define iMonth    .6
#define iDay      .5
#define iDoW      .4
#define iHour     .3
#define iMin      .2
#define iSec      .1
;=======================
;   timer constants
;=======================
; Three timer constants are defined in order to implement
; a given delay. For example, a delay of one-half second
; in a 4MHz machine requires a count of 500,000, while
; a delay of one-tenth second requires a count of 10,000.
; These numbers are converted to hexadecimal so they can
; be installed in three constants, for example:
;       1,000,000 = 0x0f4240 = one second at 4MHz
;         500,000 = 0x07a120 = one-half second
;         250,000 = 0x03d090 = one-quarter second
;         100,000 = 0x0186a0 = one-tenth second at 4MHz
; Note: The constant that define the LCD display line
;       addresses have the high-order bit set in
;       order to faciliate the controller command
; Values for one-tenth second installed in constants
; as follows:
; 500,000 = 0x01 0x86 0xa0
;               ---- ---- ----
;               |    |    |___ lowCnt
;               |    |_____ midCnt
```

```
;                    |_____ highCnt
;
#define highCnt 0x01
#define midCnt 0x86
#define lowCnt 0xa0


;========================================================
;                  variables in PIC RAM
;========================================================
; Reserve 32 bytes for string buffer
     cblock   0x000
     strData
     endc
; Reserve three bytes for ASCII digits
     cblock   0x22
     asc100
     asc10
     asc1
; Continue with local variables
     countH
     countM
     countL
     com_code
     char_count
     char_save
     count1          ; Counter # 1
     count2          ; Counter # 2
     count3          ; Counter # 3
     index           ; Index into text table (also used
                     ;     for auxiliary storage)
     store1          ; Local temporary storage
     store2          ; Storage # 2
; Storage for BCD digits
     bcdLow          ; low-order nibble of packed BCD
     bcdHigh         ; High-order nibble
; Variables for Real-Time Clock
     year
     month
     day
     dayOfWeek       ; Sunday to Saturday (1 to 7)
     hour
     minutes
     seconds
     temp1
     counter
; Storage for BCD conversion routine
     inNum           ; Source operand
     thisDig         ; Digit counter
     endc


;========================================================
;                         program
;========================================================
; Start at the reset vector
     org      0x000
     goto     main
; No interrupts in this application
     org      0x008
     retfie
     org      0x018
```

```
    retfie
;=============================
;   table in program memory
;=============================
    org             0x100
msgTable:
    db              "Real time clock "      ; offset 0
    db              "H:   M:   S:      "    ; offset 16
;offsets                 |    |     |
;                        |    |     |_____ msgTable + 29
;                        |    |_____ msgTable + 24
;                        |_____ msgTable + 19
; Start application beyond vector area
    org             0x200
main:
    nop
    nop
; Set BSR for bank 0 operations
    movlb   0                   ; Bank 0
; Init Port A for digital operation
    clrf    PORTA,0
    clrf    LATA,0
; Port summary:
; PORTD       0-7                       OUTPUT
; PORTE 0 1 2                   OUTPUT
; ADCON1 is the configuration register for the A/D
; functions in Port A. A value of 0b011x sets all
; lines for digital operation
    movlw   B'00000110'  ; Digital mode
    movwf   ADCON1,0
; Initialize all lines in PORT D and E for output
    clrf    TRISD           ; Port C tris register
    clrf    TRISE
; Clear all output lines
    clrf    PORTD
    clrf    PORTE
;=============================
;   setup Timer0 as counter
;       8-bit mode
;=============================
; Prescaler is assigned to Timer0 and initialzed
; to 2:1 rate
; Setup the T0CON register
;                   |------------- On/Off control
;                   |              1 = Timer0 enabled
;                   ||------------ 8/16 bit mode select
;                   ||             1 = 8-bit mode
;                   |||----------- Clock source
;                   |||            0 = internal clock
;                   ||||---------- Source edge select
;                   ||||           1 = high-to-low
;                   |||||--------- Prescaler assignment
;                   |||||          0 = prescaler assigned
;                   |||||||| ----- Prescaler select
;                   ||||||||        1:2 rate
    movlw   b'11010000'
    movwf   T0CON
; Clear registers
    clrf    TMR0L
    clrwdt                      ; Clear watchdog timer
```

```
; Clear variables
    clrf    com_code
;===========================
;        init LCD
;===========================
; Wait and initialize HD44780
    call    Delay
    call    InitLCD              ; Do forced initialization
    call    Delay
; Move program memory table msgTable to RAM buffer
; named strData located at RAM 0x000
    call    Msg2Data
; Set RS line for data
    bsf              PORTE,RS_line         ; Setup for data
    call    Delay           ; Delay
    call    RAM2LCDLine1 ; Display top LCD line
    call    RAM2LCDLine2    ; Second LCD line
;===============================
;    real-time clock processing
;===============================
; Initialize real-time clock
    call    initRTC              ; Initialize variables
    call    setRTC          ; Start clock
    call    delay_5              ; Wait for operation to conclude
newTime:
; Get variables from RTC
    call    Get_Time
    call    delay_5              ; Wait
;==========================
;         hours
;==========================
    movf    hour,w       ; Get hours
    call    Bcd2asc                  ; Conversion routine
; At this point three ASCII digits are stored in local
; variables. Move digits to display area
    movf    asc1,w       ; Unit digit
    movwf   .19                  ; Store in buffer
    movf    asc10,w              ; Same with other digit
    movwf   .18
    call    delay_5
;==========================
;     minutes
;==========================
    movf    minutes,w
    call    Bcd2asc                  ; Conversion routine
; At this point three ASCII digits are stored in local
; variables. Move two digits to display area
    movf    asc1,w       ; Unit digit
    movwf   .24                  ; Store in buffer
    movf    asc10,w              ; same with other digit
    movwf   .23
    call    delay_5
;==========================
;     seconds
;==========================
    movf    seconds,w
    call    Bcd2asc                  ; Conversion routine
; Move digits to display area
    movf    asc1,w       ; Unit digit
    movwf   .29                  ; Store in buffer
```

```
        movf    asc10,w              ; same with other digit
        movwf   .28
        call    delay_5
; Display seconds LCD line
        call    RAM2LCDLine2    ; Second LCD line
        goto    newTime


;===========================================================
;===========================================================
;                    P r o c e d u r e s
;===========================================================
;===========================================================
;===============================
;           INITIALIZE LCD
;===============================
InitLCD
; Initialization for Densitron LCD module as follows:
;    8-bit interface
;    2 display lines of 16 characters each
;    cursor on
;    left-to-right increment
;.   cursor shift right
;    no display shift
;*********************|
;     COMMAND MODE    |
;*********************|
        bcf             PORTE,E_line ; E line low
        bcf             PORTE,RS_line      ; RS line low for command
        bcf             PORTE,RW_line   ; Write mode
        call    delay_168           ;delay 125 microseconds
;*********************|
;     FUNCTION SET    |
;*********************|
        movlw   0x38                    ; 0 0 1 1 1 0 0 0 (FUNCTION SET)
                                        ;       | | | |__ font select:
                                        ;       | | |    1 = 5x10 in 1/8 or 1/11 dc
                                        ;       | | |    0 = 1/16 dc
                                        ;       | | |___ Duty cycle select
                                        ;       | |      0 = 1/8 or 1/11
                                        ;       | |      1 = 1/16 (multiple lines)
                                        ;       | |___ Interface width
                                        ;       |      0 = 4 bits
                                        ;       |      1 = 8 bits
                                        ;       |___ FUNCTION SET COMMAND
        movwf   PORTD ;0011 1000
        call    pulseE ;pulseE and delay

;*********************|
;     DISPLAY ON/OFF  |
;*********************|
        movlw   0x0a                    ; 0 0 0 0 1 0 1 0 (DISPLAY ON/OFF)
                                        ;         | | | |___ Blink character
                                        ;         | | |     1 = on, 0 = off
                                        ;         | | |___ Curson on/off
                                        ;         | |      1 = on, 0 = off
                                        ;         | |____ Display on/off
                                        ;         |       1 = on, 0 = off
                                        ;         |____ COMMAND BIT


        movwf   PORTD
```

```
    call      pulseE ;pulseE and delay

;***********************|
; DISPLAY AND CURSOR ON |
;***********************|
    movlw     0x0c              ; 0 0 0 0 1 1 0 0 (DISPLAY ON/OFF)
                                ;             | | | |___ Blink character
                                ;             | | |      1 = on, 0 = off
                                ;             | | |___ Cursor on/off
                                ;             | |      1 = on, 0 = off
                                ;             | |____ Display on/off
                                ;             |        1 = on, 0 = off
                                ;             |____ COMMAND BIT
    movwf     PORTD
    call      pulseE ;pulseE and delay

;***********************|
;     ENTRY MODE SET    |
;***********************|
    movlw     0x06              ; 0 0 0 0 0 1 1 0 (ENTRY MODE SET)
                                ;             | | |___ display shift
                                ;             | |      1 = shift
                                ;             | |      0 = no shift
                                ;             | |____ cursor increment mode
                                ;             |        1 = left-to-right
                                ;             |        0 = right-to-left
                                ;             |___ COMMAND BIT
    movwf     PORTD ;00000110
    call      pulseE

;***********************|
; CURSOR/DISPLAY SHIFT  |
;***********************|
    movlw     0x14              ; 0 0 0 1 0 1 0 0 (CURSOR/DISPLAY SHIFT)
                                ;         | | | |_|___ don't care
                                ;         | |_|__ cursor/display shift
                                ;         |        00 = cursor shift left
                                ;         |        01 = cursor shift right
                                ;         |        10 = cursor and display
                                ;         |              shifted left
                                ;         |        11 = cursor and display
                                ;         |              shifted right
                                ;         |___ COMMAND BIT
    movwf     PORTD ;0001 1111
    call      pulseE

;***********************|
;    CLEAR DISPLAY      |
;***********************|
    movlw     0x01              ; 0 0 0 0 0 0 0 1 (CLEAR DISPLAY)
                                ;                 |___ COMMAND BIT
    movwf     PORTD ;0000 0001
;
    call      pulseE
    call      delay_28ms    ;delay 5 milliseconds after init
    return


;================================================================
;            Time Delay and Pulse Procedures
```

```
;=================================================================
; Procedure to delay 42 x 4 = 168 machine cycles
; On a 4MHz clock the instruction rate is 1 microsecond
; 42 x 4 x 1 = 168 microseconds
delay_168
     movlw    D'42'                  ; Repeat 42 machine cycles
     movwf    count1                 ; Store value in counter
repeat
     decfsz   count1,f               ; Decrement counter
     goto     repeat                 ; Continue if not 0
     return                               ; End of delay
;
; Procedure to delay 168 x 168 microseconds
; = 28.224 milliseconds
delay_28ms
     movlw    D'42'                  ; Counter = 41
     movwf    count2                 ; Store in variable
delay
     call     delay_168              ; Delay
     decfsz   count2,f       ; 40 times = 5 milliseconds
     goto     delay
     return                          ; End of delay
;========================
;     pulse E line
;========================
pulseE
     bsf      PORTE,E_line ;pulse E line
     bcf      PORTE,E_line
     call     delay_168              ;delay 168 microseconds
     return


;==========================
;     LCD command
;==========================
LCD_command:
; On entry:
;          variable com_code cntains command code for LCD
; Set up for write operation
     bcf      PORTE,E_line           ; E line low
     bcf      PORTE,RS_line          ; RS line low, set up for control
     call     delay_168              ; delay 125 microseconds
; Write command to data port
     movf     com_code,0     ; Command code to W
     movwf    PORTD
     call     pulseE                 ; Pulse and delay
; Set RS line for data
     bsf      PORTE,RS_line          ; Setup for data
     return


;===================================
;  variable-lapse delay procedure
;          using Timer0
;===================================
; ON ENTRY:
;          Variables countL, countM, and countH hold
;          the low-, middle-, and high-order bytes
;          of the delay period, in timer units
; Routine logic:
; The prescaler is assigned to timer0 and setup so
```

```
; that the timer runs at 1:2 rate. This means that
; every time the counter reaches 128 (0x80) a total
; of 256 machine cycles have elapsed. The value 0x80
; is detected by testing bit 7 of the counter
; register.
Delay:
    call    setVars
; Note:
;     The TMR0L register provides the low-order level
; of the count. Because the counter counts up from zero,
; code must pre-install a value in the counter register
; that represents one-half the number of timer
; iterations (pre-scaler is in 1:2 mode) required to
; reach a count of 128. For example: if the value in
; the low counter variable is 140
; then 140/2 = 70. 128 - 70 = 58
; In other words, when the timer counter reaches 128,
; 70 * 2 (140) timer beats would have elapsed.
; Formula:
;              Value in TMR0L = 128 - (x/2)
; where x is the number of iterations in the low-level
; counter variable
; First calculate xx/2 by bit shifting
    rrcf    countL,f      ; Divide by 2
; now subtract 128 - (x/2)
    movlw   d'128'
; Clear the borrow bit (mapped to Carry bit)
    bcf     STATUS,C
    subfwb  countL,w
; Now w has adjusted result. Store in TMR0L
    movwf   TMR0L
; Routine tests timer overflow by testing bit 7 of
; the TMR0L register.
cycle:
    btfss   TMR0L,7              ; Is bit 7 set?
    goto    cycle         ; Wait if not set
; At this point TMR0 bit 7 is set
; Clear the bit
    bcf     TMR0L,7              ; All other bits are preserved
; Subtract 256 from beat counter by decrementing the
; mid-order byte
    decfsz  countM,f
    goto    cycle         ; Continue if mid-byte not zero
; At this point the mid-order byte has overflowed.
; High-order byte must be decremented.
    decfsz  countH,f
    goto    cycle
; At this point the time cycle has elapsed
    return
;==============================
;  set register variables
;==============================
; Procedure to initialize local variables for a
; delay period defined in local constants highCnt,
; midCnt, and lowCnt.
setVars:
    movlw   highCnt       ; From constants
    movwf   countH
    movlw   midCnt
    movwf   countM
```

```
        movlw    lowCnt
        movwf    countL
        return

;=======================
;   Procedure to delay
;    42 microseconds
;=======================
delay_125
        movlw    D'42'                  ; Repeat 42 machine cycles
        movwf    count1                 ; Store value in counter
repeat2
        decfsz   count1,f               ; Decrement counter
        goto     repeat2                     ; Continue if not 0
        return                               ; End of delay


;=======================
;   Procedure to delay
;    5 milliseconds
;=======================
delay_5
        movlw    D'41'                  ; Counter = 41
        movwf    count2                 ; Store in variable
delay2
        call     delay_125              ; Delay
        decfsz   count2,f       ; 40 times = 5 milliseconds
        goto     delay2
        return                          ; End of delay
;=======================


;==========================
;    get table character
;==========================
; Local procedure to get a single character from a local
; table (msgTable) in program memory. Variable index holds
; offset into table
tableReadChar:
        movlw    UPPER msgTable
        movwf    TBLPTRU
        movlw    HIGH msgTable  ; Get address of Table
        movwf    TBLPTRH        ; Store in table pointer low register
        movlw    LOW msgTable   ; Get address of Table
        movwf    TBLPTRL
        movff    index,WREG     ; index to W
        addwf    TBLPTRL,f      ; Add index to table pointer low
        clrf     WREG           ; Clear register
        addwfc   TBLPTRH,F      ; Add possible carry
        addwfc   TBLPTRU,F      ; To both registers
        tblrd    *              ; Read byte from table (into TABLAT)
        movff    TABLAT,WREG    ; Move TABLAT to W
        return

;=================================================================
;                    conversion procedures
;=================================================================
;=============================
;    BCD to ASCII decimal
;=============================
; ON ENTRY:
;        WREG has two packed BCD digits
```

```
; ON EXIT:
;           output variables asc10, and asc1 have
;           two ASCII decimal digits
; Routine logic:
;   The low order nibble is isolated and the value 30H
;   added to convert to ASCII. The result is stored in
;   the variable asc1. Then the same is done to the
;   high-order nibble and the result is stored in the
;   variable asc10

Bcd2asc:
    movwf    store1       ; Save input
    andlw    b'00001111'  ; Clear high nibble
    addlw    0x30         ; Convert to ASCII
    movwf    asc1         ; Store result
    swapf    store1,w     ; Recover input and swap digits
    andlw    b'00001111'  ; Clear high nibble
    addlw    0x30         ; Convert to ASCII
    movwf    asc10        ; Store result
    return
;
;=============================
;  binary to BCD conversion
;=============================
; Convert a binary number into two packed BCD digits
; ON ENTRY:
;           w register has binary value in range 0 to 99
; ON EXIT:
;           output variables bcdLow and bcdHigh contain two
;           packed unpacked BCD digits
;           w contains two packed BCD digits
; Routine logic:
;   The value 10 is subtracted from the source operand
;   until the remainder is < 0 (carry cleared). The number
;   of subtractions is the high-order BCD digit. 10 is
;   then added back to the subtrahend to compensate
;   for the last subtraction. The final remainder is the
;   low-order BCD digit
; Variables:
;     inNum      storage for source operand
;     bcdHigh    storage for high-order nibble
;     bcdLow     storage for low-order nibble
;     thisDig       Digit counter
bin2bcd:
    movwf    inNum   ; Save copy of source value
    clrf     bcdHigh ; Clear storage
    clrf     bcdLow
    clrf     thisDig
min10:
    movlw    .10
    subwf    inNum,f      ; Subtract 10
    btfsc    STATUS,C     ; Did subtract overflow?
    goto     sum10        ; No. Count subtraction
    goto     fin10
sum10:
    incf     thisDig,f    ;increment digit counter
    goto     min10
; Store 10th digit
fin10:
    movlw    .10
```

```
        addwf    inNum,f              ; Adjust for last subtract
        movf     thisDig,w     ; get digit counter contents
        movwf    bcdHigh              ; Store it
; Calculate and store low-order BCD digit
        movf     inNum,w       ;       Store units value
        movwf    bcdLow        ; Store digit
; Combine both digits
        swapf    bcdHigh,w     ; High nibble to HOBs
        iorwf    bcdLow,w      ; ORin low nibble
        return
;
;=================================================================
;                       6355 RTC procedures
;=================================================================
;===========================
;          init RTC
;===========================
; Procedure to initialize the real-time clock chip. If chip
; is not initialized it will not operate and the values
; read will be invalid.
; Because the 6355 operates in BCD format the stored values must
; be converted to packed BCD.
; According to wiring diagram
; NJU6355 Interface for setting time:
; DAT              PORTB,0                  Output
; CLK              PORTB,1                  Output
; CE       PORTB,2                  Output
; IO       PORTB,3               Output
setRTC:
        clrf     TRISB
; Writing to the 6355 requires that the CLK bit be held
; low while the IO and CE lines are high
        bcf            PORTB,CLK     ; CLK low
        call     delay_5
        bsf            PORTB,IO      ; IO high
        call     delay_5
        bsf            PORTB,CE      ; CE high
; Data is stored in RTC as follows:
;  year             8 bits (0 to 99)
;  month            8 bits (1 to 12)
;  day              8 bits (1 to 31)
;  dayOfWeek     4 bits (1 to 7)
;  hour             8 bits (0 to 23)
;  minutes          8 bits (0 to 59)
;                   ======
;   Total           44 bits
; Seconds cannot be written to RTC. RTC seconds register
; is automatically initialized to zero
        movf     year,w        ; Get item from storage
        call     bin2bcd                 ; Convert to BCD
        movwf    temp1
        call     writeRTC

        movf     month,w
        call     bin2bcd
        movwf    temp1
        call     writeRTC

        movf     day,w
        call     bin2bcd
```

```
        movwf     temp1
        call      writeRTC

        movf      dayOfWeek,w          ; dayOfWeek of week is 4-bits
        call      bin2bcd
        movwf     temp1
        call      write4RTC

        movf      hour,w
        call      bin2bcd
        movwf     temp1
        call      writeRTC

        movf      minutes,w
        call      bin2bcd
        movwf     temp1
        call      writeRTC
; Done
        bcf       PORTB,CLK            ; Hold CLK line low
        call      delay_5
        bcf       PORTB,CE             ; and the CE line
                                       ; to the RTC
        call      delay_5
        bcf       PORTB,IO             ; RTC in output mode
        return
;==============================
;        read RTC data
;==============================
; Procedure to read the current time from the RTC and store
; data (in packed BCD format) in local time registers.
; According to wiring diagram
; NJU6355 Interface for read operations:
; DAT         PORTB,0              Input
; CLK         PORTB,1              Output
; CE          PORTB,2              Output
; IO          PORTB,3              Output
Get_Time
; Clear port B
        movlw     b'00000000'
        movwf     PORTB
; Make data line input
        movlw     b'00000001'
        movwf     TRISB
; Reading RTC data requires that the IO line be low and the
; CE line be high. CLK line is held low
        bcf       PORTB,CLK    ; CLK low
        call      delay_125
        bcf       PORTB,IO     ; IO line low
        call      delay_125
        bsf       PORTB,CE     ; and CE line high
; Data is read from RTC as follows:
;   year              8 bits (0 to 99)
;   month             8 bits (1 to 12)
;   day               8 bits (1 to 31)
;   dayOfWeek     4 bits (1 to 7)
;   hour              8 bits (0 to 23)
;   minutes           8 bits (0 to 59)
;   seconds           8 bits (0 to 59)
;                     ======
;   Total             52 bits
```

```
        call    readRTC
        movwf   year
        call    delay_125

        call    readRTC
        movwf   month
        call    delay_125

        call    readRTC
        movwf   day
        call    delay_125

; dayOfWeek of week is a 4-bit value
        call    read4RTC
        movwf   dayOfWeek
        call    delay_125

        call    readRTC
        movwf   hour
        call    delay_125

        call    readRTC
        movwf   minutes
        call    delay_125

        call    readRTC
        movwf   seconds

        bcf             PORTB,CE            ; CE line low to end output
        return

;=============================
;   read 4/8 bits from RTC
;=============================
; Procedure to read 4/8 bits stored in 6355 registers
; Value returned in w register
read4RTC
        movlw   .4              ; 4 bit read
        goto    anyBits
readRTC
        movlw   .8              ; 8 bits read
anyBits:
        movwf   counter
; Read 6355 read operation requires the IO line be set low
; and the CE line high. Data is read in the following order:
; year, month, day, day-of-week, hour, minutes, seconds
readBits:
        bsf     PORTB,CLK       ; Set CLK high to validate data
        bsf     STATUS,C        ; Set the carry flag (bit = 1)
; Operation:
;   If data line is high, then bit read is a 1-bit
;   otherwise bit read is a 0-bit
        btfss   PORTB,DAT       ; Is data line high?
                                        ; Leave carry set (1 bit) if
high
        bcf     STATUS,C        ; Clear the carry bit (make bit 0)
; At this point the carry bit matches the data line
        bcf             PORTB,CLK       ; Set CLK low to end read
; The carry bit is now rotated into the temp1 register
        rrcf    temp1,1
```

```
        decfsz    counter,1          ; Decrement the bit counter
        goto      readBits           ; Continue if not last bit
; At this point all bits have been read (8 or 4)
        movf      temp1,0                    ; Result to w
        return

;=============================
;    write 4/8 bits to RTC
;=============================
; Procedure to write 4 or 8 bits to the RTC registers
; ON ENTRY:
;     temp1 register holds value to be written
; ON EXIT:
;     nothing
write4RTC
        movlw     .4                 ; Init for 4 bits
        goto      allBits
writeRTC
        movlw     .8                 ; Init for 8 bits
allBits:
        movwf     counter            ; Store in bit counter
writeBits:
        bcf       PORTB,CLK    ; Clear the CLK line
        call      delay_5      ; Wait
        bsf       PORTB,DAT    ; Set the data line to RTC
        btfss     temp1,0            ; Send LSB
        bcf       PORTB,DAT    ; Clear data line
        call      delay_5            ; Wait for operation to complete
        bsf       PORTB,CLK    ; Bring CLK line high to validate
        rrcf      temp1,f            ; Rotate bits in storage
        decfsz    counter,1    ; Decrement bit counter
        goto      writeBits    ; Continue if not last bit
        return

;=============================
;    init time variables
;=============================
; Procedure to initialize time variables for testing
; Constants used in ininitialization are located in
; #define statements.
initRTC:
        movlw     iYear
        movwf     year
        movlw     iMonth
        movwf     month
        movlw     iDay
        movwf     day
        movlw     iDoW
        movwf     dayOfWeek
        movlw     iHour
        movwf     hour
        movlw     iMin
        movwf     minutes
        movlw     iSec
        movwf     seconds
        return

;===============================
;    Test string from program
;       to data memory
```

```
;================================
Msg2Data:
; Procedure to store in PIC RAM buffer at address 0x000 the
; 32-byte message contained in the code area labeled
; msgTable
; ON ENTRY:
;           index is local variable that hold offset into
;           text table. This variable is also used for
;           temporary storage of offset into buffer
;           char_count is a counter for the 32 characters
;           to be moved
;           tableReadChar is a procedure that returns the
;           string at the offset stored in the index
;           variable
; ON EXIT:
;           Text message stored in buffer
;
; Store 12-bit address in FSR0
    lfsr    0,0x000              ; FSR0 = 0x000
; Initialize index for text string access
    clrf    index
    movlw   .32                           ; Characters to move'
    movff   WREG,char_count     ; To counter register
readThenWrite:
    call    tableReadChar       ; Local procedure
; WREG now holds character from table
    movff   WREG,POSTINC0       ; Indirect write and bump
                                              ; pointer
    incf    index               ; Next character
    decfsz  char_count          ; Decrement counter
    goto    readThenWrite
    return

;================================
; Display RAM table at LCD 1
;================================
; Routine to display 16 characters on LCD line 1
; from a RAM table starting at address 0x000
; ON ENTRY:
RAM2LCDLine1:
; Set variables
    movlw   .16                          ; Count
    movff   WREG,char_count     ; To counter
; Store 12-bit RAM table address in FSR0
    lfsr    0,0x000
; Set controller to first display line
    movlw   LCD_1               ; Address + offset into line
    movwf   com_code
    call    LCD_command         ; Local procedure
; Set RS line for data
    bsf     PORTE,RS_line       ; Setup for data
    call    Delay          ; Delay
; Retrieve message from program memory and store in LCD
ReadAndDisplay0:
    movff   POSTINC0,WREG       ; Read byte and bump pointer
; Character byte in WREG
    movwf   PORTD
    call    pulseE
    decfsz  char_count          ; Decrement counter
    goto    ReadAndDisplay0
```

```
        return

;===============================
; Display RAM table at LCD 2
;===============================
; Routine to display 16 characters on LCD line 2
; from a RAM table starting at address 0x010
; ON ENTRY:
; index holds the offset into the text string in RAM

RAM2LCDLine2:          ; Display second line
; Set variables
        movlw    .16                        ; Count
        movff    WREG,char_count    ; To counter
; Store 12-bit RAM table address in FSR0
        lfsr     0,0x010                     ; Test offset
; Set controller to second display line
        movlw    LCD_2                ; Address + offset into line
        movwf    com_code
        call     LCD_command      ; Local procedure
; Set RS line for data
        bsf      PORTE,RS_line       ; Setup for data
        call     Delay
;
ReadAndDisplay2:
        movff    POSTINC0,WREG        ; Read byte and bump pointer
; Character byte in WREG
        movwf    PORTD
        call     pulseE
        decfsz   char_count          ; Decrement counter
        goto     ReadAndDisplay2
        return
        end
```

12.4 Real-Time Clocks in C18

C18 provides support for Timer 1 operations that can be used to implement a real-time clock using a 32-kHz crystal as an external source, as described in Section 12.1. The timer primitives are available in the C18 Hardware Peripherals Library. These functions were described in Chapter 9, Section 9.6, and following. Setting up an interrupt-driven system in C18 is the topic of Sections 8.5 and following and are demonstrated in the program discussed in Section 12.3.1 later in this chapter. On the other hand, C18 does not provide support for the NJU6355 or any other RTC integrated circuit. The programmer needing to implement a real-time clock in C18 can either base the software on the Timer1 module with an external crystal or develop primitives in C18 and assembly language to operate a real-time clock IC such as the NJU6355.

12.4.1 Timer1-Based RTC in C18

If an application can devote Port C lines 0 and 1 to interfacing with a 32-kHz crystal, as well as one of the interrupt sources, then the real-time clock can be coded without great aggravation. The assembly language code in the program RTC_18F_Timer1.asm developed in Section 12.1.2 can serve as a model for the equivalent C18 version. A C language program using the low-priority vector of the 18F452 was developed in Section 8.5.2. With very few changes, the code can be modified to support the high-priority interrupt vector. The Timer 1 primitives in the C18 Hardware Peripherals Library can be used to set up Timer1 for this application. Code for setting up the system is as follows:

```
/************************************************************
                    main program
************************************************************/
void main(void)
{

    // Init Port A for digital operation
    PORTA = 0;              // Clear port
    LATA = 0;               // and latch register
    // ADCON1 is the configuration register for the A/D
    // functions in Port A. A value of 0b011x sets all
    // lines for digital operation
    ADCON1 = 0b00000110;// Code for digital mode
    // Initalize direction registers
    TRISC = 0x00;           // Port C lines for output
    PORTC = 0xff;           // Clear port register
    // Setup RB0 interrupt
    RCONbits.IPEN = 1;          // Set interrupt priority bit
    IPR1bits.TMR1IP;            // on Timer1 overflow
    TMR1H = 0x00;               // Clear counters
    TMR1L = 0x00;
    // Setup the INTCON register:
    //         |------------ GIE/GIEH - high-priority interrupts
    //         |                     enabled
    //         ||----------- PEIE/GEIL - Peripheral enabled
    //         |||---------- timer0 overflow interrupt
    //         ||||--------- external interrupt
    //         |||||-------- port change interrupt
    //         ||||||------- overflow interrupt flag
    //         |||||||------ external interrupt flag
    //         ||||||||----- RB4:RB7 interrupt flag
    //         11000000 = 0xc0
    INTCON = 0xc0;
    // Set INTCON2 for falling edge
    INTCON2bits.INTEDG0 = 0;
    // Configure Timer1 for interrupt on overflow, 16-bit
    // data, external clock source and 1:1 prescaler,
    // Timer1 oscillator on, and no synchronization
    OpenTimer1(
            TIMER_INT_ON &
            T1_16BIT_RW &
            T1_SOURCE_EXT &
            T1_PS_1_1 &
            T1_OSC1EN_ON &
            T1_SYNC_EXT_OFF );
    INTCONbits.TMR0IF = 0;              // Clear flag
    PIE1bits.TMR1IE = 1;                // Enable interrupt

    PORTC = 0xff;

    while(1) {
            Nop();
    }
}
```

Notice that the main() function in the C_Timer1_RTC.c program hangs up in an endless loop because all the work is done by the interrupt handler, which is coded as follows:

```
// Prototype for the high-priority ISR
void high_ISR(void);
// Locate the interrupt vector
#pragma code high_vector = 0x08
// Implement a jump to a handler named high_ISR
// Using inline assembly language
void high_interrupt(void)
{
    _asm
    goto      high_ISR
    _endasm
}
// Restore compiler addressing
#pragma code
// Define and code the handler
#pragma interrupt high_ISR
void high_ISR(void)
{
    // Set timer1 to roll over every seconds
    WriteTimer1(0x8000);
    PIR1bits.TMR1IF = 0; // Reset interrupt
    // Test Port C, line 2 to toggle LED
    if(PORTC & 0x4)
            PORTC = 0x00;                // Turn LEDs off
    else
            PORTC = 0xff;                // Turn LEDs on
}
C_Timer1_RTC.c Code Listing
// Project name: C_Timer1_RTC.mpc
// Source files: C_Timer1_RTC.c
// Date: March 9, 2013
//
// Processor: PIC 18F452
// Environment:    MPLAB IDE Version 8.86
//                 MPLAB C-18 Compiler
//
// Description:
// Demonstration of the Timer1 external clock using a
// 32 kHz quartz crystal wired to port C lines 0 and 1.
// Application toggles on and off the LEDs wired to port
// c lines 2 and 3, at one second intervals. Code uses
// the high-priority interrupt as a time keeper.
//
//============================================================
//                        Circuit
//============================================================
//                     18F452
//                  +------------------+
//+5v-res0-ICD2 mc -| 1 !MCLR    PGD 40|
//                  | 2 RA0      PGC 39|
//                  | 3 RA1      RB5 38|
//                  | 4 RA2      5B4 37|
//                  | 5 RA3      RB3 36|
//                  | 6 RA4      RB2 35|
//                  | 7 RA5      RB1 34|
```

```
//                           |  8 RE0      RB0 33|
//                           |  9 RE1          32|-------+5v
//                           |10 RE2           31|--------GR
//         +5v--------|11                 RD7 30|
//         GR---------|12                 RD6 29|
//               osc ---|13 OSC1          RD5 28|
//               osc ---|14 OSC2          RD4 27|
// 32khz crystal ---|15 RC0              RC7 26|
// 32khz crystal ---|16 RC1              RC6 25|
//            LED <==|17 RC2             RC5 24|
//            LED <==|18 RC3             RC4 23|
//                           |19 RD0             RD3 22|
//                           |20 RD1             RD2 21|
//                           +------------------+
// Legend:
// E = LCD signal enable
// RW = LCD read/write
// RS = LCD register select
// GR = ground
//
// CRYSTAL FOR TIMER1 OSCILLATOR:
//       RC0
//       RC1
//
// INCLUDED CODE
#include <p18f452.h>
#include <timers.h>
#include <eep.h>


#pragma config OSC = XT          // Assumes high-speed resonator
#pragma config WDT = OFF         // No watchdog timer
#pragma config LVP = OFF         // No low voltage protection
#pragma config DEBUG = OFF              // No background debugger
#pragma config PWRT = ON         // Power on timer enabled
#pragma config CP0 = OFF         // Code protection block x = 0-3
#pragma config CP1 = OFF
#pragma config CP2 = OFF
#pragma config CP3 = OFF
#pragma config WRT0 = OFF        // Write protection block x = 0-3
#pragma config WRT1 = OFF
#pragma config WRT2 = OFF
#pragma config WRT3 = OFF
#pragma config EBTR0 = OFF       // Table read protection block x = 0-3
#pragma config EBTR1 = OFF
#pragma config EBTR2 = OFF
#pragma config EBTR3 = OFF


// Prototype for the high-priority ISR
void high_ISR(void);
// Locate the interrupt vector
#pragma code high_vector = 0x08
// Implement a jump to a handler named high_ISR
// Using inline assembly language
void high_interrupt(void)
{
    _asm
    goto     high_ISR
    _endasm
}
```

```
// Restore compiler addressing
#pragma code
// Define and code the handler
#pragma interrupt high_ISR
void high_ISR(void)
{
    // Set timer1 to roll over every seconds
    WriteTimer1(0x8000);
    PIR1bits.TMR1IF = 0; // Reset interrupt
    // Test Port C, line 2 to toggle LED
    if(PORTC & 0x4)
            PORTC = 0x00;                    // Turn LEDs off
    else
            PORTC = 0xff;                    // Turn LEDs on
}

/***********************************************************
                      main program
***********************************************************/
void main(void)
{

    // Init Port A for digital operation
    PORTA = 0;              // Clear port
    LATA = 0;               // and latch register
    // ADCON1 is the configuration register for the A/D
    // functions in Port A. A value of 0b011x sets all
    // lines for digital operation
    ADCON1 = 0b00000110;// Code for digital mode
    // Initalize direction registers
    TRISC = 0x00;           // Port C lines for output
    PORTC = 0xff;           // Clear port register
    // Setup RB0 interrupt
    RCONbits.IPEN = 1;             // Set interrupt priority bit
    IPR1bits.TMR1IP;               // on Timer1 overflow
    TMR1H = 0x00;                  // Clear counters
    TMR1L = 0x00;
    // Setup the INTCON register:
    //           |------------ GIE/GIEH - high-priority interrupts
    //           |                     enabled
    //           ||----------- PEIE/GEIL - Peripheral enabled
    //           |||---------- timer0 overflow interrupt
    //           ||||--------- external interrupt
    //           |||||-------- port change interrupt
    //           ||||||------- overflow interrupt flag
    //           |||||||------ external interrupt flag
    //           ||||||||----- RB4:RB7 interrupt flag
    //       b'11000000' = 0xc0
    INTCON = 0xc0;
    // Set INTCON2 for falling edge
    INTCON2bits.INTEDG0 = 0;
    // Configure Timer1 for interrupt on overflow, 16-bit
    // data, external clock source and 1:1 prescaler,
    // Timer1 oscillator on, and no synchronization
    OpenTimer1(
            TIMER_INT_ON &
            T1_16BIT_RW &
            T1_SOURCE_EXT &
            T1_PS_1_1 &
            T1_OSC1EN_ON &
```

```
            T1_SYNC_EXT_OFF );
    INTCONbits.TMR0IF = 0;                    // Clear flag
    PIE1bits.TMR1IE = 1;           // Enable interrupt

    PORTC = 0xff;

    while(1) {
            Nop();
    }
}
```

Chapter 13

Analog Data and Devices

13.1 Operations on Computer Data

We measure natural forces and phenomena using digital representations, but natural events are actually continuous. Time, pressure, voltage, current, temperature, humidity, gravitational attraction — all exist as continuous entities that we represent in volts, pounds, hours, amperes, or degrees, in order to be able to perform numerical calculations. In other words, natural phenomena occur in analog quantities that we digitize in order to facilitate computerized measurements and data processing.

A potentiometer in an electrical circuit allows reducing the voltage level from the circuit maximum to ground, or zero level. If we were to measure and control the action of the potentiometer, we need to quantify its action into a digital value within the physical range of the circuit. In other words, we need to convert an analog quantity that varies continuously between 0 and 5 volts, to a discrete digital value range that we can store and possibly display. If the voltage range of the potentiometer is from 5 to 0 volts, we can digitize its action into a numeric range of 0 to 500 units, or we can measure the angle or rotation of the potentiometer disk in degrees from 0 to 180. The device that performs either conversion is called an A/D or analog-to-digital converter. The reverse process, digital-to-analog, is also sometimes necessary, although not as often as A/D. In this chapter we explore A/D conversions in PIC software and hardware.

13.2 18F452 A/D Hardware

Many PIC microcontrollers, including the 18F series, come with onboard A/D hardware. One of the advantages of using onboard A/D converters is saving interface lines. While a circuit that depends on an A/D conversion device typically requires three lines to interface with the microcontroller, a similar circuit can be implemented in a PIC with internal A/C conversion by simply connecting the analog device to the corresponding PIC port. In the PIC world, where I/O lines are often in short supply, this advantage is not insignificant.

The A/D module of 18F452 provides 8-bit conversion resolution and can receive analog input in up to sixteen different channels.The 10-bit Analog-to-Digital (A/D) Converter module can have up to sixteen analog inputs. The analog input charges a sample and hold capacitor that serves as input into the converter. The hardware then generates a digital result of this analog level via successive approximation. This A/D conversion of the analog input signal results in a corresponding 10-bit digital number. The analog reference voltages are software selectable to either the device's supply voltages (Vdd and Vss) or the voltage level on the AN3/VREF+ and AN2/VREF pins. The A/D converter continues to convert while the device is in SLEEP mode. The A/D module has four registers. These registers are:

13.2.1 A/D Module on the 18F452

The number of lines depends on the specific version of the F18 device. The descriptions that follow refer specifically to the 18F452 40-pin device. This implementation of the A/D module is compatible with the one in the mid-range PICs such as the 16F877. The converter uses a sample and hold capacitor to store the analog charge and performs a successive approximation algorithm to produce the digital result. The converter resolution is 10 bits, which are stored in two 8-bit registers. One of the registers has only 4 significant bits.

The A/D module has high- and low-voltage reference inputs that are selected by software. The module can operate while the processor is in SLEEP mode, but only if the A/D clock pulse is derived from its internal RC oscillator. The module contains four registers accessible to the application:

- ADRESH - Result High Register
- ADRESL - Result Low Register
- ADCON0 - Control Register 0
- ADCON1 - Control Register 1

Of these, it is the ADCON0 register that controls most of the operations of the module. Port A pins RA0 to RA5 and PORT E pins RE0 to RE2 are multiplexed as analog input pins into the A/C module. Figure 13.1 shows the registers associated with A/D module operations.

REGISTER NAME	7	6	5	4	3	2	1	0 bits
INTCON	GIE	PEIE						
PIR1		ADIF						
PIE1		ADIE						
ADRESH	A/D Result Register High Byte							
ADRESL	A/D Result Register Low Byte							
ADCON0	ADSC1	ADSC0	CHS2	CHS1	CHS0	GO/DONE		ADON
ADCON1	ADFM				PCFG3	PCFG2	PCFG1	PCFG0

Figure 13.1 *Registers related to A/C module operations.*

ADCON0 Register

The ADCON0 register is located at address 0xfc2h. Seven of the eight bits are meaningful in A/D control and status operations. Figure 13.2 is a bitmap of the ADCON0 register.

```
bits:    7        6        5       4        3        2       1        0

      | ADSC1  | ADSC0  | CHS2  | CHS1   | CHS0   | GO/DONE |░░░░░| ADON |
```

```
bit 7-6  ADCS1:ADCS0:  A/D Conversion Clock Select bits
              00 = FOSC/2
              01 = FOSC/8
              10 = FOSC/32
              11 = FRC (internal A/D module RC oscillator)
bit 5-3  CHS2:CHS0:  Analog Channel Select bits
              000 = channel 0, (RA0=AN0)
              001 = channel 1, (RA1=AN1)
              010 = channel 2, (RA2=AN2)
              011 = channel 3, (RA3=AN3)
              100 = channel 4, (RA5=AN4)
              101 = channel 5, (RE0=AN5) | not active
              110 = channel 6, (RE1=AN6) | in 28-pin
              111 = channel 7, (RE2=AN7) | 16F87x PICS
bit 2    GO/DONE:  A/D Conversion Status bit
              If ADON = 1:
              1 = A/D conversion in progress (setting this
                  bit starts the A/D conversion)
              0 = A/D conversion not in progress (this bit
                  is automatically cleared by hardware when
                  the A/D conversion is complete)
bit 1    Unimplemented: Read as '0'
bit 0    ADON:  A/D On bit
              1 = A/D converter module is operating
              0 = A/D converter module is shut-off and
                  consumes no power
```

Figure 13.2 *ADCON0 register bitmap.*

In Figure 13.2, bits 7 and 6, labeled ASCC1 and ADSC0, are the selection bits for the A/D conversion clock. The conversion time per bit is defined as TAD in PIC documentation. A/D conversion requires a minimum of 12 TAD in a 10-bit ADC. The source of the A/D conversion clock is software selected. The four possible options for TAD are

- Fosc/2

- Fosc/8

- Fosc/32

- Internal A/D module RC oscillator (varies between 2 and 6 μs)

The conversion time is the analog-to-digital clock period multiplied by the number of bits of resolution in the converter, plus the two to three additional clock periods for settling time, as specified in the data sheet of the specific device. The various sources for the analog-to-digital converter clock represent the main oscillator frequency divided by 2, 8, or 32. The third choice is the use of a dedicated internal RC clock that has a typical period of 2 to 6 µs. Because the conversion time is determined by the system clock, a faster clock results in a faster conversion time.

The A/D conversion clock must be selected to ensure a minimum Tad time of 1.6 µs. The formula for converting processor speed (in MHz) into Tad microseconds is as follows:

$$Tad = \frac{1}{\dfrac{Tosc}{Tdiv}}$$

where Tad is A/D conversion time, Tosc is the oscillator clock frequency in MHz, and Tdiv is the divisor determined by bits ADSC1 and ADSC0 of the ADCON0 register. For example, in a PIC running at 10 MHz, if we select the Tosc/8 option (divisor equal 8), the A/D conversion time per bit is calculated as follows:

$$Tad = \frac{1}{\dfrac{5\ MHz}{8}} = 1.6$$

In this case, the minimum recommended conversion speed of 1.6 µs is achieved. However, in a PIC with an oscillator speed of 10 MHz, this option produces a conversion speed of 0.8 µs, less than the recommended minimum. In this case, we would have to select the divisor 32 option, giving a conversion speed of 3.2 µs.

Table 13.1

A/C Converter Tad at Various Oscillator Speeds

OPERATION	ADCS1:ADCS0	TAD IN MICROSECONDS			
		20MHz	10MHz	5MHz	1.25MHz
Fosc/2	00	0.1	0.2	**0.4**	**1.6**
Fosc/8	01	0.4	**0.8**	**1.6**	**6.4**
Fosc/32	10	**1.6**	**3.2**	**6.4**	25.6
RC	**11**	**2-6**	**2-6**	2-6	2-6
Note: Values in bold are within the recommended limits					

In Table 13.1, converter speeds of less than 1.6 µs or higher than 10 µs are not recommended. Recall that the Tad speed of the converter is calculated per bit, so the total conversion time in a 10-bit device is approximately the Tad speed multiplied by 10 bits, plus three additional cycles; therefore, a device operating at a Tad speed of 1.6 µs requires 1.6 µs * 13, or 20.8 µs, for the entire conversion.

Bits CHS2 to CHS0 in the ADCON0 register (see Figure 13.2) determine which of the analog channels is selected. There are several channels for analog input but only one A/2 converter circuitry. So the setting of this bit field determines which of six or eight possible channels is currently read by the A/C converter. An application can change the setting of these bits in order to read several analog inputs in succession.

Bit 2 of the ADCON0 register, labeled GO/DONE, is both a control and a status bit. Setting the GO/DONE bit starts A/D conversion. Once conversion has started, the bit indicates if it is still in progress. Code can test the status of the GO/DONE bit in order to determine if conversion has concluded.

Bit 0 of the ADCON0 register turns the A/D module on and off. The initialization routine of an A/D-enabled application turns on this bit. Programs that do not use the A/D conversion module leave the bit off to conserve power.

ADCON1 Register

The ADCON1 register also plays an important role in programming the A/D module. Bit 7 of the ADCON1 register is used to determine the bit justification of the digital result. Because the 10-bit result is returned in two 8-bit registers, the six unused bits can be placed either on the left- or the right-hand side of the 16-bit result. If ADCON1 bit 7 is set, then the result is right-justified; otherwise it is left-justified. Figure 13.3 shows the location of the significant bits.

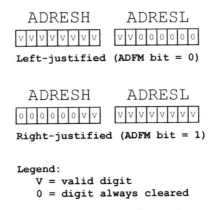

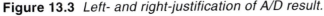

Figure 13.3 *Left- and right-justification of A/D result.*

One common use of right justification is to reduce the number of significant bits in the conversion result. For example, an application on the 18F452 uses the A/D conversion module but requires only 8-bit accuracy in the result. In this case, code can left-justify the conversion result, read the ADRESH register, and ignore the low-order bits in the ADRESL register. By ignoring the two low-order bits, the 10-bit accuracy of the A/D hardware is reduced to 8 bits and the converter performs as an 8-bit accuracy unit.

The bit field labeled PCFG3 to PCFG0 in the ADCON1 register determines port configuration as analog or digital and the mapping of the positive and negative voltage reference pins. The number of possible combinations is limited by the 4 bits allocated to this field, so the programmer and circuit designer must select the option that is most suited to the application when the ideal one is not available. Table 13.1 shows the port configuration options.

Table 13.1

A/D Converter Port Configuration Options

PCFG3: PCFG0	An7 Re2	An6 Re1	An5 Re0	An4 Ra5	An3 Ra3	An2 Ra2	An1 Ra1	An0 Ra0	Vref+	Vref-	CHAN/ Refs
0000	A	A	A	A	A	A	A	A	VDD	VSS	8/0
0001	A	A	A	A	Vre+	A	A	A	RA3	VSS	7/1
0010	D	D	D	A	A	A	A	A	VDD	VSS	5/0
0011	D	D	D	A	Vre+	A	A	A	RA3	VSS	4/1
0100	D	D	D	D	A	D	A	A	VDD	VSS	3/0
0101	D	D	D	D	Vre+	D	A	A	RA3	VSS	2/1
011x	D	D	D	D	D	D	D	D	VDD	VSS	0/0
1000	A	A	A	A	Vre+	Vre-	A	A	RA3	RA2	6/2
1001	D	D	A	A	A	A	A	A	VDD	VSS	6/0
1010	D	D	A	A	Vre+	A	A	A	RA3	VSS	5/1
1011	D	D	A	A	Vre+	Vre-	A	A	RA3	RA2	4/2
1100	D	D	D	A	Vre+	Vre-	A	A	RA3	RA2	3/2
1101	D	D	D	D	Vre+	Vre-	A	A	RA3	RA2	2/2
1110	D	D	D	D	D	D	D	A	VDD	VSS	1/0
1111	D	D	D	D	Vre+	Vre-	D	A	RA3	RA2	1/2

```
Legend:
    D = digital input
    A = analog input
    CHAN/Refs = analog channels/voltage reference inputs
```

Suppose a circuit that calls for two analog inputs, wired to ports RA0 and RA1, with no reference voltages. In Table 13.1 we can find two options that select ports RA0 and RA1 and are analog inputs: these are the ones selected with PCFG bits 0100 and 0101. The first option also selects port RA3 as analog input, even though not required in this case. The second one also selects port RA3 as a positive voltage reference, also not required.

Either option works in this example; however, any pin configured for analog input produces incorrect results if used as a digital source. Therefore, a channel configured for analog input cannot be used for non-analog purposes. By the same token, a channel configured for digital input should not be used for analog data because extra current is consumed by the hardware. Finally, channels to be used for analog-to-digital conversion must be configured for input in the corresponding TRIS register.

SLEEP Mode Operation

The A/D module can be made to operate in SLEEP mode. As mentioned previously, SLEEP mode operation requires that the A/D clock source be set to RC by setting both ADCS bits in the ADCON0 register. When the RC clock source is selected, the A/D module waits one instruction cycle before starting the conversion. During this period, the SLEEP instruction is executed, thus eliminating all digital switching noise from the conversion. The completion of the conversion is detected by testing the GO/DONE bit. If a different clock source is selected, then a SLEEP instruction causes the conversion-in-progress to be aborted and the A/D module to be turned off.

13.2.2 A/D Module Sample Circuit and Program

The circuit in Figure 13.4 can be used to demonstrate the A/D converter module in 18F452 PIC. The circuit is also used for demonstrating the LM335 temperature sensor later in this chapter.

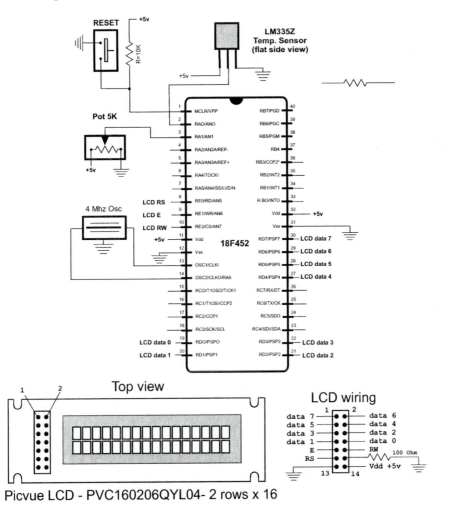

Figure 13.4 *Demonstration circuit for A/D conversion module.*

The circuit in Figure 13.4 contains a 5K potentiometer wired to analog port RA1 of an 18F452. The LCD display is used to show three digits, in the range 0 to 255, that represent the relative position of the potentiometer's disk. The program named A2D_Pot2LCD.asm, in this book's online software and listed later in this chapter, uses the built-in A/D module. Programming the A/D module consists of the following steps:

1. Configure the PIC I/O lines to be used in the conversion. All analog lines are initialized as input in the corresponding TRIS registers.

2. Select the ports to be used in the conversion by setting the PCFGx bits in the ADCON1 register. Selects right- or left-justification.

3. Select the analog channels, select the A/D conversion clock, and enable the A/D module.

4. Wait the acquisition time.

5. Initiate the conversion by setting the GO/DONE bit in the ADCON0 register.

6. Wait for the conversion to complete.

7. Read and store the digital result.

Initialize A/D Module

The following procedure from the A2D_Pot2LCD.asm program initializes the A/D module for the required processing:

```
;==============================
;      init A/D module
;==============================
; 1. Procedure to initialize the A/D module, as follows:
;     Configure the PIC I/O lines. Init analog lines as input.
; 2. Select ports to be used by setting the PCFGx bits in the
;     ADCON1 register. Select right- or left-justification.
; 3. Select the analog channels, select the A/D conversion
;     clock, and enable the A/D module.
; 4. Wait the acquisition time.
; 5. Initiate the conversion by setting the GO/DONE bit in the
;     ADCON0 register.
; 6. Wait for the conversion to complete.
; 7. Read and store the digital result.
InitA2D:
    movlw    b'00000010'
    movwf    TRISA ; Set PORT A, line 1, as input
; Select the format and A/D port configuration bits in
; the ADCON1 register
; Format is left-justified so that ADRESH bits are the
; most significant
;  0  x  x  x  1  1  1  0  <== value installed in ADCON1
;  7  6  5  4  3  2  1  0  <== ADCON1 bits
;  |              |__|__|__|____ RA0 is analog.
;  |                     Vref+ = Vdd
;  |                     Vref- = Vss
;  |_____ 0 = left-justified
;
    movlw    b'00001110'
    movwf    ADCON1          ; RA0 is analog. All others
                            ; digital
                            ; Vref+ = Vdd
; Select D/A options in ADCON0 register
; For a 10 MHz clock the Fosc32 option produces a conversion
; speed of 1/(10/32) = 3.2 microseconds, which is within the
; recommended range of 1.6 to 10 microseconds.
;  1  0  0  0  1  0  0  1  <== value installed in ADCON0
;  7  6  5  4  3  2  1  0  <== ADCON0 bits
```

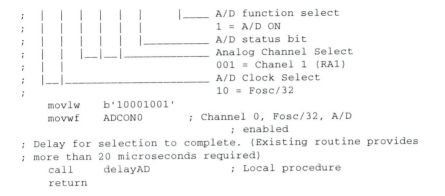

```
;   |   |   |   |   |   |       |____ A/D function select
;   |   |   |   |   |   |             1 = A/D ON
;   |   |   |   |   |   |_____ A/D status bit
;   |   |   |__|__|_____ Analog Channel Select
;   |   |                            001 = Chanel 1 (RA1)
;   |__|_____ A/D Clock Select
;                                     10 = Fosc/32
      movlw   b'10001001'
      movwf   ADCON0        ; Channel 0, Fosc/32, A/D
                            ;   enabled
; Delay for selection to complete. (Existing routine provides
; more than 20 microseconds required)
      call    delayAD               ; Local procedure
      return
```

A/D Conversion

Once the module is initialized, the A/D line or lines can be read by software. The conversion is initiated by setting the GO/DONE bit in the ADCON0 register. Software then tests the GO/DONE bit to determine when the conversion has ended. The following procedure performs the necessary operations.

```
;===============================
;         read A/D line
;===============================
; Procedure to read the value in the A/D line and convert
; to digital
ReadA2D:
; Initiate conversion
      bsf     ADCON0,GO     ; Set the GO/DONE bit
; GO/DONE bit is cleared automatically when conversion ends
convWait:
      btfsc   ADCON0,GO     ; Test bit
      goto    convWait              ; Wait if not clear
; At this point conversion has concluded
; ADRESH register (bank 0) holds 8 MSBs of result
; ADRESL register (bank 1) holds 4 LSBs.
; In this application value is left-justified. Only the
; MSBs are read
      movf    ADRESH,W            ; Digital value to w register
      return


;=======================
;    delay procedure
;=======================
; For a 10 MHz clock the Fosc32 option produces a conversion
; speed of 1/(10/32) = 3.2 microseconds. At 3.2 ms per bit
; 13 bits require approximately 41 ms. The instruction time
; at 10 MHz is 10 ms. 4/10 = 0.4 ms per insctruction. To delay
; 41 ms a 10 MHz PIC must execute 11 instructions. Add one
; more for safety.
delayAD:
      movlw   .12           ; Repeat 12 machine cycles
      movwf   count1        ; Store value in counter
repeat11:
      decfsz  count1,f      ; Decrement counter
      goto    repeat11              ; Continue if not 0
      return
```

13.2.3 A2D_Pot2LCD Program

```
; File name: A2D_Pot2LCD.asm
; Date: March 17, 2013
; Author: Julio Sanchez
;
; STATE:
;      Tested March 18/13: Demo board 18F452
;
; Description:
; Program to demonstrate use of the Analog to Digital
; Converter (A/D) module on the 18F452. Program reads the
; value of a potentionmeter connected to PORT A, line 1
; and displays resistance in the range 0 to 255 on the
; attached LCD.
;
; Executes in Demo Board 18F452 or compatible circuit
;
;==========================================================
;                         Circuit
;                   wiring for 8-bit mode
;==========================================================
;                         18F452
;                    +------------------+
;+5v-res0-ICD2 mc -| 1 !MCLR    PGD 40|
;                  | 2 RA0      PGC 39|
;        Pot 5K ==>| 3 RA1      RB5 38|
;                  | 4 RA2      5B4 37|
;                  | 5 RA3      RB3 36|
;                  | 6 RA4      RB2 35|
;                  | 7 RA5      RB1 34|
;     LCD RS   <==| 8 RE0      RB0 33|
;     LCD E    <==| 9 RE1          32|-------+5v
;     LCD RW   ==>|10 RE2          31|--------GR
;     +5v--------|11          RD7 30|==> LCD data 7
;     GR---------|12          RD6 29|==> LCD data 6
;           osc ---|13 OSC1    RD5 28|==> LCD data 5
;           osc ---|14 OSC2    RD4 27|==> LCD data 4
;                  |15 RC0      RC7 26|
;                  |16 RC1      RC6 25|
;                  |17 RC2      RC5 24|==>
;                  |18 RC3      RC4 23|==>
;   LCD data 0 <==|19 RD0      RD3 22|==> LCD data 3
;   LCD data 1 <==|20 RD1      RD2 21|==> LCD data 2
;                    +------------------+
;
; Legend:
; E = LCD signal enable
; RW = LCD read/write
; RS = LCD register select
; GR = ground
;
;==============================================================
    list p=18f452
    ; Include file, change directory if needed
    include "p18f452.inc"
;==========================================================
;                  configuration bits
;==========================================================
; Configuration bits set as required for MPLAB ICD 2
```

```
          config OSC = XT              ; Assumes high-speed resonator
          config WDT = OFF             ; No watchdog timer
          config LVP = OFF             ; No low voltage protection
          config DEBUG = OFF           ; No background debugger
          config PWRT = ON           ; Power on timer enabled
          config CP0 = OFF           ; Code protection block x = 0-3
          config CP1 = OFF
          config CP2 = OFF
          config CP3 = OFF
          config WRT0 = OFF          ; Write protection block x = 0-3
          config WRT1 = OFF
          config WRT2 = OFF
          config WRT3 = OFF
          config EBTR0 = OFF         ; Table read protection block x = 0-3
          config EBTR1 = OFF
          config EBTR2 = OFF
          config EBTR3 = OFF
;
; Turn off banking error messages
      errorlevel    -302
;=============================================================
;                  constant definitions
;   for PIC-to-LCD pin wiring and LCD line addresses
;=============================================================
; LCD used in the demo board is 2 lines by 16 characters
#define E_line 1    ;|
#define RS_line 0   ;| -- from wiring diagram
#define RW_line 2   ;|
; LCD line addresses (from LCD data sheet)
#define LCD_1 0x80 ; First LCD line constant
#define LCD_2 0xc0 ; Second LCD line constant
;
;======================
;   timer constants
;======================
; Three timer constants are defined in order to implement
; a given delay. For example, a delay of one-half second
; in a 4MHz machine, requires a count of 500,000, while
; a delay of one-tenth second requires a count of 10,000.
; These numbers are converted to hexadecimal so they can
; be installed in three constants, for example:
;      1,000,000 = 0x0f4240 = one second at 4MHz
;        500,000 = 0x07a120 = one-half second
;        250,000 = 0x03d090 = one-quarter second
;        100,000 = 0x0186a0 = one-tenth second at 4MHz
; Note: The constant that defines the LCD display line
;       addresses have the high-order bit set in
;       order to faciliate the controller command
; Values for one-tenth second installed in constants
; as follows:
; 500,000 = 0x01 0x86 0xa0
;           ---- ---- ----
;             |    |    |___ lowCnt
;             |    |_____ midCnt
;             |_____ highCnt
;
#define highCnt 0x01
#define midCnt 0x86
#define lowCnt 0xa0
```

```
;========================================================
;                  variables in PIC RAM
;========================================================
; Reserve 32 bytes for string buffer
    cblock   0x000
    strData                       ; Label for debugging
    endc
; Reserve three bytes for ASCII digits
    cblock   0x22
    asc100
    asc10
    asc1
; Continue with local variables
    countH
    countM
    countL
    com_code
    char_count
    char_save
    count1           ; Counter # 1
    count2           ; Counter # 2
    count3           ; Counter # 3
    index            ; Index into text table (also used
                           ; for auxiliary storage)
; Storage for ASCII decimal conversion and digits
    inNum            ; Source operand
    thisDig          ; Digit counter

    endc


;========================================================
;                            program
;========================================================
; Start at the reset vector
    org      0x000
    goto     main
; No interrupts used by this application
    org      0x008
    retfie
    org      0x018
    retfie
;===========================
;   table in program memory
;===========================
    org             0x100
msgTable:
    db              "Pot resistance  "      ; offset 0
    db              " (0-255): ***   "      ; offset 16
;offset                   |
;                         |____ msgTable + 26
;
    ; Start application beyond vector area
    org             0x200
main:
    nop
; Set BSR for bank 0 operations
    movlb    0                     ; Bank 0
; Init Port A for digital operation
    clrf     PORTA,0
    clrf     LATA,0
```

```
; Port summary:
; PORTD      0-7                          OUTPUT
; PORTE 0 1 2                    OUTPUT
; ADCON1 is the configuration register for the A/D
; functions in Port A. A value of 0b011x sets all
; lines for digital operation
    movlw   B'00000110'  ; Digital mode
    movwf   ADCON1,0
; Initialize all lines in PORT D and E for output
    clrf    TRISD        ; Port C tris register
    clrf    TRISE
; Clear all output lines
    clrf    PORTD
    clrf    PORTE
;==============================
;  setup Timer0 as counter
;        8-bit mode
;==============================
; Prescaler is assigned to Timer0 and initialzed
; to 2:1 rate
; Setup the T0CON register
;                 |------------- On/Off control
;                 |               1 = Timer0 enabled
;                 ||----------- 8/16 bit mode select
;                 ||             1 = 8-bit mode
;                 |||---------- Clock source
;                 |||            0 = internal clock
;                 ||||--------- Source edge select
;                 ||||          1 = high-to-low
;                 |||||-------- Prescaler assignment
;                 |||||          0 = prescaler assigned
;                 ||||||||| ----- Prescaler select
;                 |||||||||      1:2 rate
    movlw   b'11010000'
    movwf   T0CON
; Clear registers
    clrf    TMR0L
    clrwdt                 ; Clear watchdog timer
; Clear variables
    clrf    com_code
;=========================
;        init LCD
;=========================
; Wait and initialize HD44780
    call    Delay
    call    InitLCD            ; Do forced initialization
    call    Delay
; Move program memory table msgTable to RAM buffer
; named strData located at RAM 0x000
    call    Msg2Data
; Set RS line for data
    bsf     PORTE,RS_line      ; Setup for data
    call    Delay          ; Delay
    call    RAM2LCDLine1 ; Display top LCD line
    call    RAM2LCDLine2   ; Second LCD line
;=========================
;        A2D operations
;=========================
; Initialize A/D conversion lines
    call    InitA2D                ; Local procedure
```

```
;============================
;    read POT digital value
;============================
readPOT:
    call     ReadA2D              ; Local procedure
; w has digital value read from analog line RA1
; Display result
    call     bin2asc              ; Conversion routine
; At this point three ASCII digits are stored in local
; variables. Move digits to display area
    movff    asc1,.28     ; Unit digit
    movff    asc10,.27    ; same with other digits
    movff    asc100,.26
    call     delay_5
; Display line
    call     RAM2LCDLine2    ; Second LCD line
    goto     readPOT
;=========================================================
;=========================================================
;                        P r o c e d u r e s
;=========================================================
;=========================================================
;================================
;          INITIALIZE LCD
;================================
InitLCD
; Initialization for Densitron LCD module as follows:
;    8-bit interface
;    2 display lines of 16 characters each
;    cursor on
;    left-to-right increment
;    cursor shift right
;    no display shift
;**********************|
;    COMMAND MODE      |
;**********************|
    bcf      PORTE,E_line ; E line low
    bcf      PORTE,RS_line      ; RS line low for command
    bcf      PORTE,RW_line   ; Write mode
    call     delay_168           ;delay 125 microseconds
;**********************|
;    FUNCTION SET      |
;**********************|
    movlw    0x38   ; 0 0 1 1 1 0 0 0 (FUNCTION SET)
                    ;       | | | |__ font select:
                    ;       | | |    1 = 5x10 in 1/8 or 1/11 dc
                    ;       | | |    0 = 1/16 dc
                    ;       | | |___ Duty cycle select
                    ;       | |      0 = 1/8 or 1/11
                    ;       | |      1 = 1/16 (multiple lines)
                    ;       | |___ Interface width
                    ;       |      0 = 4 bits
                    ;       |      1 = 8 bits
                    ;       |___ FUNCTION SET COMMAND
    movwf    PORTD ;0011 1000
    call     pulseE ;pulseE and delay

;**********************|
;    DISPLAY ON/OFF    |
;**********************|
```

```
    movlw    0x0a   ; 0 0 0 0 1 0 1 0 (DISPLAY ON/OFF)
             ;                | | | |___ Blink character at cursor
             ;                | | |      1 = on, 0 = off
             ;                | | |___ Cursor on/off
             ;                | |      1 = on, 0 = off
             ;                | |____ Display on/off
             ;                |        1 = on, 0 = off
             ;                |____ COMMAND BIT

    movwf    PORTD
    call     pulseE ;pulseE and delay

;*********************|
; DISPLAY AND CURSOR ON |
;*********************|
    movlw    0x0c   ; 0 0 0 0 1 1 0 0 (DISPLAY ON/OFF)
             ;                | | | |___ Blink character at cursor
             ;                | | |      1 = on, 0 = off
             ;                | | |___ Cursor on/off
             ;                | |      1 = on, 0 = off
             ;                | |____ Display on/off
             ;                |        1 = on, 0 = off
             ;                |____ COMMAND BIT
    movwf    PORTD
    call     pulseE ;pulseE and delay

;*********************|
;    ENTRY MODE SET    |
;*********************|
    movlw    0x06   ; 0 0 0 0 0 1 1 0 (ENTRY MODE SET)
             ;                | | |___ display shift
             ;                | |      1 = shift
             ;                | |      0 = no shift
             ;                | |____ cursor increment mode
             ;                |        1 = left-to-right
             ;                |        0 = right-to-left
             ;                |___ COMMAND BIT
    movwf    PORTD  ;00000110
    call     pulseE

;*********************|
; CURSOR/DISPLAY SHIFT |
;*********************|
    movlw    0x14   ; 0 0 0 1 0 1 0 0 (CURSOR/DISPLAY SHIFT)
             ;                | | | |_|___ don't care
             ;                | |_|__ cursor/display shift
             ;                |        00 = cursor shift left
             ;                |        01 = cursor shift right
             ;                |        10 = cursor and display
             ;                |             shifted left
             ;                |        11 = cursor and display
             ;                |             shifted right
             ;                |___ COMMAND BIT
    movwf    PORTD  ;0001 1111
    call     pulseE

;*********************|
;    CLEAR DISPLAY     |
;*********************|
    movlw    0x01   ; 0 0 0 0 0 0 0 1 (CLEAR DISPLAY)
```

```
                              ;                    |___ COMMAND BIT
       movwf     PORTD  ;0000 0001
   ;
       call      pulseE
       call      delay_28ms   ;delay 5 milliseconds after init
       return


;================================================================
;              Time Delay and Pulse Procedures
;================================================================
; Procedure to delay 42 x 4 = 168 machine cycles
; On a 4MHz clock the instruction rate is 1 microsecond
; 42 x 4 x 1 = 168 microseconds
delay_168
       movlw     D'42'                ; Repeat 42 machine cycles
       movwf     count1               ; Store value in counter
repeat
       decfsz    count1,f             ; Decrement counter
       goto      repeat               ; Continue if not 0
       return                              ; End of delay
;
; Procedure to delay 168 x 168 microseconds
; = 28.224 milliseconds
delay_28ms
       movlw     D'42'                ; Counter = 41
       movwf     count2               ; Store in variable
delay
       call      delay_168            ; Delay
       decfsz    count2,f    ; 40 times = 5 milliseconds
       goto      delay
       return                         ; End of delay
;========================
;     pulse E line
;========================
pulseE
       bsf       PORTE,E_line ;pulse E line
       bcf       PORTE,E_line
       call      delay_168            ;delay 168 microseconds
       return


;========================
;     LCD command
;========================
LCD_command:
; On entry:
;          variable com_code cntains command code for LCD
; Set up for write operation
       bcf       PORTE,E_line         ; E line low
       bcf       PORTE,RS_line        ; RS line low, set up for control
       call      delay_168            ; delay 125 microseconds
; Write command to data port
       movf      com_code,0      ; Command code to W
       movwf     PORTD
       call      pulseE               ; Pulse and delay
; Set RS line for data
       bsf       PORTE,RS_line        ; Setup for data
       return


;===================================
```

```
;   variable-lapse delay procedure
;        using Timer0
;====================================
; ON ENTRY:
;          Variables countL, countM, and countH hold
;          the low-, middle-, and high-order bytes
;          of the delay period, in timer units
; Routine logic:
; The prescaler is assigned to timer0 and setup so
; that the timer runs at 1:2 rate. This means that
; every time the counter reaches 128 (0x80) a total
; of 256 machine cycles have elapsed. The value 0x80
; is detected by testing bit 7 of the counter
; register.
Delay:
    call     setVars
; Note:
;     The TMR0L register provides the low-order level
; of the count. Because the counter counts up from zero,
; code must pre-install a value in the counter register
; that represents the one-half the number of timer
; iterations (pre-scaler is in 1:2 mode) required to
; reach a count of 128. For example: if the value in
; the low counter variable is 140
; then 140/2 = 70. 128 - 70 = 58
; In other words, when the timer counter reaches 128,
; 70 * 2 (140) timer beats would have elapsed.
; Formula:
;          Value in TMR0L = 128 - (x/2)
; where x is the number of iterations in the low-level
; counter variable
; First calculate x/2 by bit shifting
    rrcf     countL,f      ; Divide by 2
; now subtract 128 - (x/2)
    movlw    d'128'
; Clear the borrow bit (mapped to Carry bit)
    bcf      STATUS,C
    subfwb   countL,w
; Now w has adjusted result. Store in TMR0L
    movwf    TMR0L
; Routine tests timer overflow by testing bit 7 of
; the TMR0L register.
cycle:
    btfss    TMR0L,7              ; Is bit 7 set?
    goto     cycle       ; Wait if not set
; At this point TMR0 bit 7 is set
; Clear the bit
    bcf      TMR0L,7              ; All other bits are preserved
; Subtract 256 from beat counter by decrementing the
; mid-order byte
    decfsz   countM,f
    goto     cycle        ; Continue if mid-byte not zero
; At this point the mid-order byte has overflowed.
; High-order byte must be decremented.
    decfsz   countH,f
    goto     cycle
; At this point the time cycle has elapsed
    return
;===============================
;   set register variables
```

```
;===============================
; Procedure to initialize local variables for a
; delay period defined in local constants highCnt,
; midCnt, and lowCnt.
setVars:
    movlw    highCnt      ; From constants
    movwf    countH
    movlw    midCnt
    movwf    countM
    movlw    lowCnt
    movwf    countL
    return

;======================
;  Procedure to delay
;    42 microseconds
;======================
delay_125
    movlw    D'42'              ; Repeat 42 machine cycles
    movwf    count1            ; Store value in counter
repeat2
    decfsz   count1,f          ; Decrement counter
    goto     repeat2               ; Continue if not 0
    return                        ; End of delay

;======================
;  Procedure to delay
;    5 milliseconds
;======================
delay_5
    movlw    D'41'             ; Counter = 41
    movwf    count2           ; Store in variable
delay2
    call     delay_125        ; Delay
    decfsz   count2,f      ; 40 times = 5 milliseconds
    goto     delay2
    return                        ; End of delay
;========================

;==========================
;    get table character
;==========================
; Local procedure to get a single character from a local
; table (msgTable) in program memory. Variable index holds
; offset into table
tableReadChar:
    movlw    UPPER msgTable
    movwf    TBLPTRU
    movlw    HIGH msgTable    ; Get address of Table
    movwf    TBLPTRH          ; Store in table pointer low register
    movlw    LOW msgTable     ; Get address of Table
    movwf    TBLPTRL
    movff    index,WREG       ; index to W
    addwf    TBLPTRL,f        ; Add index to table pointer low
    clrf     WREG             ; Clear register
    addwfc   TBLPTRH,F        ; Add possible carry
    addwfc   TBLPTRU,F        ; To both registers
    tblrd    *                ; Read byte from table (into TABLAT)
    movff    TABLAT,WREG      ; Move TABLAT to W
    return
```

```
;================================
;  Test string from program
;       to data memory
;================================
Msg2Data:
; Procedure to store in PIC RAM buffer at address 0x000 the
; 32-byte message contained in the code area labeled
; msgTable
; ON ENTRY:
;          index is local variable that hold offset into
;          text table. This variable is also used for
;          temporary storage of offset into buffer
;          char_count is a counter for the 32 characters
;          to be moved
;          tableReadChar is a procedure that returns the
;          string at the offset stored in the index
;          variable
; ON EXIT:
;          Text message stored in buffer
;
; Store 12-bit address in FSR0
      lfsr      0,0x000              ; FSR0 = 0x000
; Initialize index for text string access
      clrf      index
      movlw     .32                           ; Characters to move'
      movff     WREG,char_count     ; To counter register
readThenWrite:
      call      tableReadChar       ; Local procedure
; WREG now holds character from table
      movff     WREG,POSTINC0       ; Indirect write and bump
                                    ; pointer
      incf      index               ; Next character
      decfsz    char_count          ; Decrement counter
      goto      readThenWrite
      return

;================================
; Display RAM table at LCD 1
;================================
; Routine to display 16 characters on LCD line 1
; from a RAM table starting at address 0x000
; ON ENTRY:
RAM2LCDLine1:
; Set variables
      movlw     .16                           ; Count
      movff     WREG,char_count     ; To counter
; Store 12-bit RAM table address in FSR0
      lfsr      0,0x000
; Set controller to first display line
      movlw     LCD_1               ; Address + offset into line
      movwf     com_code
      call      LCD_command         ; Local procedure
; Set RS line for data
      bsf       PORTE,RS_line       ; Setup for data
      call      Delay               ; Delay
; Retrieve message from program memory and store in LCD
ReadAndDisplay0:
      movff     POSTINC0,WREG       ; Read byte and bump pointer
; Character byte in WREG
      movwf     PORTD
```

```
        call     pulseE
        decfsz   char_count          ; Decrement counter
        goto     ReadAndDisplay0
        return


;================================
; Display RAM table at LCD 2
;================================
; Routine to display 16 characters on LCD line 2
; from a RAM table starting at address 0x010
; ON ENTRY:
; index holds the offset into the text string in RAM
RAM2LCDLine2:; Display second line
; Set variables
        movlw    .16                        ; Count
        movff    WREG,char_count    ; To counter
; Store 12-bit RAM table address in FSR0
        lfsr     0,0x010                    ; Test offset
; Set controller to second display line
        movlw    LCD_2                ; Address + offset into line
        movwf    com_code
        call     LCD_command    ; Local procedure
; Set RS line for data
        bsf      PORTE,RS_line      ; Setup for data
        call     Delay
ReadAndDisplay2:
        movff    POSTINC0,WREG          ; Read byte and bump pointer
; Character byte in WREG
        movwf    PORTD
        call     pulseE
        decfsz   char_count          ; Decrement counter
        goto     ReadAndDisplay2
        return


;============================================================
;                 Analog to Digital Procedures
;============================================================
;==========================
;      init A/D module
;==========================
; 1. Procedure to initialize the A/D module, as follows:
;    Configure the PIC I/O lines. Init analog lines as input.
; 2. Select ports to be used by setting the PCFGx bits in the
;    ADCON1 register. Select right- or left-justification.
; 3. Select the analog channels, select the A/D conversion
;    clock, and enable the A/D module.
; 4. Wait the acquisition time.
; 5. Initiate the conversion by setting the GO/DONE bit in the
;    ADCON0 register.
; 6. Wait for the conversion to complete.
; 7. Read and store the digital result.
InitA2D:
        movlw    b'00000010'
        movwf    TRISA ; Set PORT A, line 1, as input
; Select the format and A/D port configuration bits in
; the ADCON1 register
; Format is left-justified so that ADRESH bits are the
; most significant
; 0  x  x  x  1  1  1  0  <== value installed in ADCON1
; 7  6  5  4  3  2  1  0  <== ADCON1 bits
```

```
;    |                |__|__|__|____ RA0 is analog.
;    |                              Vref+ = Vdd
;    |                              Vref- = Vss
;    |_____ 0 = left-justified
;
     movlw    b'00001110'
     movwf    ADCON1        ; RA0 is analog. All others
                            ; digital
                            ; Vref+ = Vdd
; Select D/A options in ADCON0 register
; For a 10 MHz clock the Fosc32 option produces a conversion
; speed of 1/(10/32) = 3.2 microseconds, which is within the
; recommended range of 1.6 to 10 microseconds.
;   1  0  0  0  1  0  0  1  <== value installed in ADCON0
;   7  6  5  4  3  2  1  0  <== ADCON0 bits
;   |  |  |  |  |  |     |____ A/D function select
;   |  |  |  |  |  |          1 = A/D ON
;   |  |  |  |  |  |_____ A/D status bit
;   |  |  |__|__|_____ Analog Channel Select
;   |  |                      001 = Channel 1 (RA1)
;   |__|_____ A/D Clock Select
;                             10 = Fosc/32
     movlw    b'10001001'
     movwf    ADCON0        ; Channel 0, Fosc/32, A/D
                                 ; enabled
; Delay for selection to complete. (Existing routine provides
; more than 20 microseconds required)
     call     delayAD             ; Local procedure
     return
;=============================
;         read A/D line
;=============================
; Procedure to read the value in the A/D line and convert
; to digital
ReadA2D:
; Initiate conversion
     bsf      ADCON0,GO      ; Set the GO/DONE bit
; GO/DONE bit is cleared automatically when conversion ends
convWait:
     btfsc    ADCON0,GO      ; Test bit
     goto     convWait           ; Wait if not clear
; At this point conversion has concluded
; ADRESH register (bank 0) holds 8 MSBs of result
; ADRESL register (bank 1) holds 4 LSBs.
; In this application value is left-justified. Only the
; MSBs are read
     movf     ADRESH,W            ; Digital value to w register
     return
;========================
;     delay procedure
;========================
; For a 10 MHz clock the Fosc32 option produces a conversion
; speed of 1/(10/32) = 3.2 microseconds. At 3.2 ms per bit
; 13 bits require approximately 41 ms. The instruction time
; at 10 MHz is 10 ms. 4/10 = 0.4 ms per insctruction. To delay
; 41 ms a 10 MHz PIC must execute 11 instructions. Add one
; more for safety.
delayAD:
     movlw    .12            ; Repeat 12 machine cycles
     movwf    count1         ; Store value in counter
```

```
repeat11:
    decfsz    count1,f      ; Decrement counter
    goto      repeat11              ; Continue if not 0
    return


;===============================
;   binary to ASCII decimal
;         conversion
;===============================
; ON ENTRY:
;         WREG has binary value in range 0 to 255
; ON EXIT:
;         output variables asc100, asc10, and asc1 have
;         three ASCII decimal digits
; Routine logic:
;   The value 100 is subtracted from the source operand
;   until the remainder is < 0 (carry set). The number
;   of subtractions is the decimal hundreds result. 100 is
;   then added back to the subtrahend to compensate
;   for the last subtraction. Now 10 is subtracted in the
;   same manner to determine the decimal tens result.
;   The final remainder is the decimal units result.
; Variables:
;     inNum      storage for source operand
;     asc100     storage for hundreds position result
;     asc10      storage for tens position result
;     asc1       storage for unit position reslt
;     thisDig       Digit counter
bin2asc:
    movwf     inNum         ; Save copy of source value
    clrf      asc100        ; Clear hundreds storage
    clrf      asc10         ; Tens
    clrf      asc1          ; Units
    clrf      thisDig
sub100:
    movlw     .100
    subwf     inNum,f       ; Subtract 100
    btfsc     STATUS,C      ; Did subtract overflow?
    goto      bump100              ; No. Count subtraction
    goto      end100
bump100:
    incf      thisDig,f     ; Increment digit counter
    goto      sub100
; Store 100th digit
end100:
    movf      thisDig,w     ; Adjusted digit counter
    addlw     0x30          ; Convert to ASCII
    movwf     asc100        ; Store it
; Calculate tens position value
    clrf      thisDig
; Adjust minuend
    movlw     .100          ; Minuend
    addwf     inNum,f       ; Add value to minuend to
                            ; Compensate for last operation
sub10:
    movlw     .10
    subwf     inNum,f       ; Subtract 10
    btfsc     STATUS,C      ; Did subtract overflow?
    goto      bump10        ; No. Count subtraction
    goto      end10
```

```
bump10:
    incf      thisDig,f      ;increment digit counter
    goto      sub10
;
; Store 10th digit
end10:
    movlw     .10
    addwf     inNum,f                  ; Adjust for last subtraction
    movf      thisDig,w      ; get digit counter contents
    addlw     0x30           ; Convert to ASCII
    movwf     asc10          ; Store it
; Calculate and store units digit
    movf      inNum,w        ; Store units value
    addlw     0x30           ; Convert to ASCII
    movwf     asc1           ; Store digit
    return

    end
```

13.3 A/D Conversion in C18

The C18 Hardware Peripherals Library includes support for analog-to-digital conversions by means of the following functions:

- BusyADC tests if the A/D converter is currently performing a conversion.
- CloseADC disables the A/D converter.
- ConvertADCstarts an A/D conversion.
- OpenADC configures and initializes the A/D converter.
- ReadADC reads the results of an A/D conversion.
- SetChanADC selects A/D channel to be used.

13.3.1 Conversion Primitives

The conversion functions are implemented differently for the various devices of the PIC 18F family. The descriptions that follow refer specifically to the 18F452

Busy ADC

This function tests if the ADC module is currently performing a conversion. It returns 1 in a char type if a conversion is in progress and 0 otherwise. Function parameters are

```
Include: adc.h
Prototype:    char BusyADC( void );
File Name: adcbusy.c
```

CloseADC

This function disables the A/D module an its associated interrupt mechanism. Function parameters are

```
Include: adc.h
Prototype: void CloseADC( void );
File Name: adcclose.c
```

ConvertADC

This function starts an A/D conversion. The BusyADC() function can be used to determine if the conversion has concluded. Function parameters are

```
Include: adc.h
Prototype: void ConvertADC( void );
File Name: adcconv.c
```

OpenADC

This function configures the A/D converter. The arguments passed to the function can change for different PIC 18F devices. The ones listed are for the 18F452. The OpenADC() function resets the A/D peripheral to the Power On state and configures the A/D-related Special Function Registers according to the options specified. Function parameters are

```
Include: adc.h
Prototype: void OpenADC( unsigned char config,
                         unsigned char config2 );
Arguments:
config
Defines a bitmask created by performing a bitwise AND operation ('&') with a
value from each of the categories listed below. These values are defined in
the file adc.h.
    A/D clock source:
        ADC_FOSC_2 FOSC / 2
        ADC_FOSC_4 FOSC / 4
        ADC_FOSC_8 FOSC / 8
        ADC_FOSC_16 FOSC / 16
        ADC_FOSC_32 FOSC / 32
        ADC_FOSC_64 FOSC / 64
        ADC_FOSC_RC Internal RC Oscillator
    A/D result justification:
        ADC_RIGHT_JUST Result in Least Significant bits
        ADC_LEFT_JUST Result in Most Significant bits
    A/D voltage reference source:
        ADC_8ANA_0REF VREF+=VDD, VREF-=VSS,
                      All analog channels
        ADC_7ANA_1REF AN3=VREF+, All analog
                      channels except AN3
        ADC_6ANA_2REF AN3=VREF+, AN2=VREF
        ADC_6ANA_0REF VREF+=VDD, VREF-=VSS
        ADC_5ANA_1REF AN3=VREF+, VREF-=VSS
        ADC_5ANA_0REF VREF+=VDD, VREF-=VSS
        ADC_4ANA_2REF AN3=VREF+, AN2=VREFADC_
        4ANA_1REF             AN3=VREF+
        ADC_3ANA_2REF AN3=VREF+, AN2=VREFADC_
        3ANA_0REF             VREF+=VDD, VREF-=VSS
        ADC_2ANA_2REF AN3=VREF+, AN2=VREFADC_
        2ANA_1REF             AN3=VREF+
        ADC_1ANA_2REF AN3=VREF+, AN2=VREF-,
                      AN0=A
        ADC_1ANA_0REF AN0 is analog input
        ADC_0ANA_0REF All digital I/O
config2
Defines a bitmask that is created by performing a bitwise AND operation
('&') with a value from each of the categories listed below. These values
are defined in the adc.h file.
    Channel:
```

```
        ADC_CH0 Channel 0
        ADC_CH1 Channel 1
        ADC_CH2 Channel 2
        ADC_CH3 Channel 3
        ADC_CH4 Channel 4
        ADC_CH5 Channel 5
        ADC_CH6 Channel 6
        ADC_CH7 Channel 7
    A/D Interrupts:
        ADC_INT_ON Interrupts enabled
        ADC_INT_OFF Interrupts disabled
File Name: adcopen.c
Code Example:
    OpenADC( ADC_FOSC_32 &
        ADC_RIGHT_JUST &
        ADC_1ANA_0REF,
        ADC_CH0 &
        ADC_INT_OFF );
```

ReadADC

This function reads the 10-bit signed result of an A/D conversion. Based on the configuration of the A/D converter by the last call to openADC(), the result will be contained in the Least Significant or Most Significant bits of the 10-bit result. The function parameters are

```
Include:      adc.h
Prototype:    int ReadADC( void );
File Name:    adcread.c
```

SetChanADC

This function selects the channel used as input to the A/D converter. The function parameters are

```
Include:      adc.h
Prototype:    void SetChanADC( unsigned char channel );
Arguments:    channel
              One of the following values (defined in adc.h):
                  ADC_CH0 Channel 0
                  ADC_CH1 Channel 1
                  ADC_CH2 Channel 2
                  ADC_CH3 Channel 3
                  ADC_CH4 Channel 4
                  ADC_CH5 Channel 5
                  ADC_CH6 Channel 6
                  ADC_CH7 Channel 7
File Name:    adcsetch.c
Code Example: SetChanADC( ADC_CH0 );
```

13.3.2 C_ADConvert.c Program

The sample program C_ADConvert.c in the book's software package and listed later in this section uses C language code to read the resistance of a 5K potentiometer wired to port RA1 as in the circuit in Figure 13.4 and in Demo Board 18F452. The C language program is the equivalent of the program A2D_Pot2LCD.asm presented earlier in this chapter.

The C language code uses the following C18 hardware libraries:

- xlcd.h
- adc.h
- atdlib.h
- delays.h

The program performs the following steps:

1. Opens the XLCD device and initializes it for 8-bit mode and 5 times 7 pixels. Turns off blink and flashing cursor.

2. Sets the DDRAM address and displays the first LCD text line.

3. Sets up an endless loop. Initializes the A/D device with the following processing: Fosc 32, left justification, eight analog channels, channel 1 is active for analog input, and turns off ADC interrupts.

4. After an acquisition delay, code starts the conversion process. Once conversion has concluded, the value in ADRESH register is stored in a local variable of type unsigned char. Alternatively, the resistance value could have been obtained by calling ReadLCD().

5. The ADC is closed.

6. The ASCII buffer for the resistance digits is cleared. The converted ASCII string is moved to a display buffer.

7. The DDRAM address for the second display line is set and the second LCD line is displayed.

8. The DDRAM address for the string buffer is set and the resistance digits are displayed in the second LCD line. A long delay ensures stability before the next reading.

9. Execution loops to Step 3.

C_ADConvert.c Code Listing

```
// Project name: C_ADConvert.mcp
// Source files: C_ADConvert.c
//               XLCD support files
// Header files:  xlcd.h
//                adc.h
//                stdlib.h
//                delays.h
//
// Date: March 20/2013
//
//
// Processor: PIC 18F452
// Environment:    MPLAB IDE Version 8.xx
//                 MPLAB C-18 Compiler
//
// TEST CIRCUIT:
// Executes in Demo Board 18F452 or compatible circuit
//
```

```
//===========================================================
//                         Circuit
//                  wiring for 8-bit mode
//===========================================================
//                      18F452
//                 +------------------+
//5v-res0-ICD2 mc -| 1 !MCLR    PGD 40|
//                 | 2 RA0      PGC 39|
//        Pot 5K ==>| 3 RA1      RB5 38|
//                 | 4 RA2      5B4 37|
//                 | 5 RA3      RB3 36|
//                 | 6 RA4      RB2 35|
//                 | 7 RA5      RB1 34|
//     LCD RS  <==| 8 RE0      RB0 33|
//     LCD E   <==| 9 RE1          32|-------+5v
//     LCD RW  ==>|10 RE2          31|--------GR
//     +5v--------|11          RD7 30|==> LCD data 7
//     GR---------|12          RD6 29|==> LCD data 6
//          osc ---|13 OSC1     RD5 28|==> LCD data 5
//          osc ---|14 OSC2     RD4 27|==> LCD data 4
//               |15 RC0      RC7 26|
//               |16 RC1      RC6 25|
//               |17 RC2      RC5 24|
//               |18 RC3      RC4 23|
//   LCD data 0 <==|19 RD0      RD3 22|==> LCD data 3
//   LCD data 1 <==|20 RD1      RD2 21|==> LCD data 2
//                 +------------------+
//
//
// Legend:
// E = LCD signal enable
// RW = LCD read/write
// RS = LCD register select
// GR = ground
//
// Application code assumes the following:
//          Hitachi HD44780 LCD controller
//          Two lines by 16 characters each
//          Wired for 8-bit data
//          Uses C18 LCD Software Peripheral Library
//          A/D Conversion functions
//
#include <p18cxxx.h>
#include "xlcd.h"
#include <adc.h>
#include <delays.h>
#include <stdlib.h>
//
// Configuration bits set as required for MPLAB ICD 2
#pragma config OSC = XT          // Assumes high-speed resonator
#pragma       config WDT = OFF   // No watchdog timer
#pragma       config LVP = OFF   // No low voltage protection
#pragma       config DEBUG = OFF // No background debugger
#pragma config PWRT = ON         // Power on timer enabled
#pragma config CP0 = OFF          // Code protection block x = 0-3
#pragma config CP1 = OFF
#pragma config CP2 = OFF
#pragma config CP3 = OFF
#pragma config WRT0 = OFF         // Write protection block x = 0-3
#pragma config WRT1 = OFF
```

```
#pragma config WRT2 = OFF
#pragma config WRT3 = OFF
#pragma config EBTR0 = OFF        // Table read protection block x = 0-3
#pragma config EBTR1 = OFF
#pragma config EBTR2 = OFF
#pragma config EBTR3 = OFF

// Global data
char XLCD_Disp1[] = "Pot resistance  ";
char XLCD_Disp2[] = " (0-255):        ";
char A2D_String[] = "        ";

void main(void)
{
    unsigned char config = 0x00;
    unsigned char blinkoff = 0x00;
    unsigned char charvalue = 0;
    int x;                                    // A counter
    unsigned int result;              // ADC converted value

    // Define configuration for 8 bits and two 5 X 7 lines
    config = EIGHT_BIT  & LINES_5X7;
    blinkoff = CURSOR_OFF & BLINK_OFF;

    //************************
    //     Initialize LCD
    //************************
    while( BusyXLCD() );         // Wait until LCD not busy
    OpenXLCD(config);                  // Intialize LCD
    // Test for busy
    while( BusyXLCD() );         // Wait until LCD not busy
    // Turn off cursor and blinking
    WriteCmdXLCD( blinkoff );
    // Set starting address in the LCD RAM for display.
    // Can be edited to match system
    while( BusyXLCD() );
    SetDDRamAddr(0x80);
    while( BusyXLCD() );         // Wait until LCD not busy
    putsXLCD(XLCD_Disp1);        // Display first string
    while( BusyXLCD() );         // Wait until LCD not busy

    //********************************************************
    //                  main program loop
    //********************************************************
    while(1) {
    //************************
    //    initialize A/D
    //************************
    // Configure converter
    OpenADC( ADC_FOSC_32 &
                    ADC_LEFT_JUST &
                    ADC_8ANA_0REF,
                    ADC_CH1 &
                    ADC_INT_OFF );

    Delay10TCYx(5);                      // Delay for acquisition
    ConvertADC();                // Start conversion
    while( BusyADC() );          // Wait for completion
    // Instead of calling ReadADC() the value can be
    // read directly off the ADRESH register
```

```
      charvalue = ADRESH;
//  Alternatively values can be obtained with ReadADC()
//  and typecase, as follows:
//    result = ReadADC();                    // Get value as int
//    charvalue = (unsigned char) result; // Typecast

      CloseADC();
      // Clear buffer
      for(x = 0; x < 7; x++)
            A2D_String[x] = 0x20;

      // Convert integer to ASCII string
      itoa( charvalue, A2D_String);

      // Set address for display second line
      SetDDRamAddr(0xC0);
      while( BusyXLCD() );        // Wait until LCD not busy
      putsXLCD(XLCD_Disp2);       // Display second LCD line
      // Set address for ASCII string line
      SetDDRamAddr(0xCA);
      while( BusyXLCD() );        // Wait until LCD not busy
      putsXLCD(A2D_String);       // Display second string
      while( BusyXLCD() );        // Wait until LCD not busy
      // Long delay before next read
      Delay10KTCYx(25);
      }

}
```

13.4 Interfacing with Analog Devices

Many devices available as integrated circuits provide an analog report of some physical data. These devices, usually referred to as transducers, are used to measure many physical quantities and phenomena, including voltage, current, resistance, light intensity, humidity, temperature, wind intensity, and atmospheric pressure among many others. A programmable thermostat for a home-based system would typically include several transducers including temperature and humidity levels. In the previous examples in this chapter we used the A/D module of the 18F452 to measure the resistance of a standard potentiometer. In the present we briefly discuss interfacing with a temperature sensor.

13.4.1 LM 34 Temperature Sensor

The LM35 is a popular family of temperature sensors that includes devices in various packaging formats and different functionality. The LM135 sub-family, presently discussed, is an integrated circuit with the following features:

- Easily calibrated
- Wide operating temperature range
- Directly calibrated in degrees Kelvin, Fahrenheit, or Centigrade.
- 1°C initial accuracy available
- 200°C range
- Operates from 400 mA to 5 mA

- Low cost

- Less than 1mA dynamic impedance

The LM135, LM235, and LM335 are available in hermetic TO transistor packages while the LM335 is also available in 8-pin SO-8 and plastic TO-92 packages.

The principal characteristic of the LM35 series sensors is their linear output. This means that the voltage generated by the sensor varies directly and linearly with the sensed quantity.

13.4.2 LM135 Circuits

Figure 13.5 shows the connection diagrams for the SO-8 and TO-92 packages of the LMx35 temperature sensors.

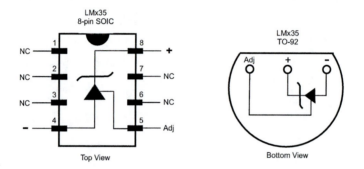

Figure 13.5 *Connection diagrams for LMx35 temperature sensors.*

The LMx35 family includes the LM135, LM235, LM335, LM135A, LM235A, and LM335A. The LM135 sensor operates over a range of -55°C to 150°C, the LM235 over a range of -40°C to 125°C, and the LM335 over a range of -40°C to 100°C.

Calibrating the Sensor

The ICs of the LM135 family can be calibrated in order to reduce the possible error to 1°C over a range of 100°C. The calibration is simplified by the fact that all LM135 sensors have linear output. Calibration is accomplished by means of a 10kOhm potentiometer wired to the Adj terminal of the device. The calibration circuit is shown in Figure 13.6.

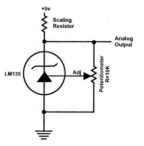

Figure 13.6 *LM135 calibration circuit.*

The resistor labeled Scaling Resistor in Figure 13.6 allows using the +5v supply line as a voltage reference into the LM135. This allows expanding the voltage output range to make better use of the A/D converter hardware.

13.4.3 C_ADC_LM35.c Program

The program C_ADC_LM35 in the book's software package and listed later in this section reads and displays in an LCD device the temperature value of an LM35 sensor. The program uses the A/D module of the 18F452 PIC. The value displayed is the raw reading of the sensor device. Applications will process this value according to the circuit's hardware and the data desired. A particular implementation of this code will take into account the specific LM35 device in the circuit, the value of the scaling resistor, and the desired format for the displayed results. Code listing is as follows:

```
// Project name: C_ADC_LM35.mcp
// Source files: C_ADC_LM35.c
//              XLCD support files
// Header files: xlcd.h
//               adc.h
//               stdlib.h
//               delays.h
//
// Date: March 20/2013
//
// Processor: PIC 18F452
// Environment:    MPLAB IDE Version 8.xx
//                 MPLAB C-18 Compiler
//
// Description:
// Read and display raw temperature value off an LM35 sensor.
//
// TEST CIRCUIT:
// Executes in Demo Board 18F452 or compatible circuit
//
//=========================================================
//                       Circuit
//                  wiring for 8-bit mode
//=========================================================
//                       18F452
//                 +-----------------+
//5v-res0-ICD2 mc -| 1 !MCLR   PGD 40|
//         LM35 ==>| 2 RA0     PGC 39|
//                 | 3 RA1     RB5 38|
//                 | 4 RA2     5B4 37|
//                 | 5 RA3     RB3 36|
//                 | 6 RA4     RB2 35|
//                 | 7 RA5     RB1 34|
//     LCD RS  <==| 8 RE0     RB0 33|
//     LCD E   <==| 9 RE1         32|-------+5v
//     LCD RW  ==>|10 RE2         31|--------GR
//     +5v--------|11         RD7 30|==> LCD data 7
//     GR---------|12         RD6 29|==> LCD data 6
//           osc ---|13 OSC1   RD5 28|==> LCD data 5
//           osc ---|14 OSC2   RD4 27|==> LCD data 4
//                 |15 RC0     RC7 26|
//                 |16 RC1     RC6 25|
//                 |17 RC2     RC5 24|
```

```
//                    |18 RC3        RC4 23|
//    LCD data 0 <==|19 RD0        RD3 22|==> LCD data 3
//    LCD data 1 <==|20 RD1        RD2 21|==> LCD data 2
//                    +------------------+
//
// Legend:
// E = LCD signal enable
// RW = LCD read/write
// RS = LCD register select
// GR = ground
//
// Application code assumes the following:
//           Hitachi HD44780 LCD controller
//       Two lines by 16 characters each
//           Wired for 8-bit data
//           Uses C18 LCD Software Peripheral Library
//       A/D Conversion functions
//
#include <p18cxxx.h>
#include "xlcd.h"
#include <adc.h>
#include <delays.h>
#include <stdlib.h>
//
// Configuration bits set as required for MPLAB ICD 2
#pragma config OSC = XT          // Assumes high-speed resonator
#pragma       config WDT = OFF        // No watchdog timer
#pragma       config LVP = OFF        // No low voltage protection
#pragma       config DEBUG = OFF      // No background debugger
#pragma config PWRT = ON      // Power on timer enabled
#pragma config CP0 = OFF      // Code protection block x = 0-3
#pragma config CP1 = OFF
#pragma config CP2 = OFF
#pragma config CP3 = OFF
#pragma config WRT0 = OFF      // Write protection block x = 0-3
#pragma config WRT1 = OFF
#pragma config WRT2 = OFF
#pragma config WRT3 = OFF
#pragma config EBTR0 = OFF     // Table read protection block x = 0-3
#pragma config EBTR1 = OFF
#pragma config EBTR2 = OFF
#pragma config EBTR3 = OFF

// Global data
char XLCD_Disp1[] = "LM35 temperature";
char XLCD_Disp2[] = "(raw val):      ";
char A2D_String[] = "        ";

void main(void)
{
    unsigned char config = 0x00;
    unsigned char blinkoff = 0x00;
    unsigned char charvalue = 0;
    int x;                                        // A counter
    unsigned int result;              // ADC converted value

    // Define configuration for 8 bits and two 5 X 7 lines
    config = EIGHT_BIT  & LINES_5X7;
    blinkoff = CURSOR_OFF & BLINK_OFF;
    //***********************
```

```
//      Initialize LCD
//************************
while( BusyXLCD() );              // Wait until LCD not busy
OpenXLCD(config);                    // Intialize LCD
// Test for busy
while( BusyXLCD() );              // Wait until LCD not busy
// Turn off cursor and blinking
WriteCmdXLCD( blinkoff );
// Set starting address in the LCD RAM for display.
// Can be edited to match system
while( BusyXLCD() );
SetDDRamAddr(0x80);
while( BusyXLCD() );              // Wait until LCD not busy
putsXLCD(XLCD_Disp1);            // Display first string
while( BusyXLCD() );              // Wait until LCD not busy
//*********************************************************
//                  main program loop
//*********************************************************
while(1) {
//************************
//    initialize A/D
//************************
// Configure converter
OpenADC( ADC_FOSC_32 &
                ADC_RIGHT_JUST &
                ADC_8ANA_0REF,
                ADC_CH0 &
                ADC_INT_OFF );
Delay10TCYx(5);                       // Delay for acquisition
ConvertADC();                    // Start conversion
while( BusyADC() );              // Wait for completion
// Instead of calling ReadADC() the value can be
// read directly off the ADRESH register
result = ReadADC();                           // Get value as int
CloseADC();
// Clear buffer
for(x = 0;  x < 7;  x++)
        A2D_String[x] = 0x20;
// Convert integer to ASCII string
itoa( result, A2D_String);
// Set address for display second line
SetDDRamAddr(0xC0);
while( BusyXLCD() );              // Wait until LCD not busy
putsXLCD(XLCD_Disp2);            // Display second LCD line
// Set address for ASCII string line
SetDDRamAddr(0xCA);
while( BusyXLCD() );              // Wait until LCD not busy
// Value displayed is the raw reading of the sensor device.
// Applications will process this value according to the
// circuit's hardware and the desired results.
putsXLCD(A2D_String);           // Display second string
while( BusyXLCD() );              // Wait until LCD not busy
// Long delay before next read
Delay10KTCYx(25);
}

}
```

Chapter 14

Operating Systems

14.1 Time-Critical Systems

Most major embedded system applications perform more than one activity. For example, a control system for a microwave oven must set the oven temperature, operate the timer, detect if the oven door is open, and monitor the keypad for input and commands. The more complex a system the more difficult it becomes to manage the simultaneous tasks it must perform. In this sense the software must be capable of dividing system resources adequately between the different tasks, setting priorities, and detecting error conditions that could be critical. In the microwave oven example previously mentioned, the software should be able to detect if the door is open so that dangerous radiation is not allowed to escape.

Embedded systems are often time-critical applications. Not only must the software ensure that events takes place, but also that they take place within a give time frame. As the number of simultaneous activities becomes larger, ensuring the timeliness of each activity becomes more difficult. In a sense, each activity competes for the CPU's attention, and the software determines which one gets it at any given instant.

The system designer must investigate the underlying requirements of each task in order to devise a strategy that satisfies the most important needs. It is the Real Time Operating System (RTOS) and not the applications themselves that determines the priorities. We can say that the operating system becomes the god of the machine. The following elements are important in this context.

- What is an operating system in the context of embedded programming.
- What is meant by "real-time".
- What is necessary in order to achieve genuine and simulated multi-tasking.
- What principles govern the design and operations of a RTOS and of real-time programming.

14.1.2   Multitasking in Real-Time

Multitasking is the technology by which multiple processes are able to share system resources and CPU. Barrring specialized hardware, it can be said that if a single CPU is present, then only one task can be executing at any point in time. The notion of multitasking solves the problem of executing several tasks by allocating CPU time to each task. The task that is executing is said to be running, and the other ones are said to be waiting or suspended. The processes by which CPU time is allocated to each task is called scheduling. The operation by which CPU time is reassigned from one task to another one is referred to as a context switch. Context switching produces the illusion that several tasks are executing simultaneously.

Scheduling strategies of operating systems can be classified as follows:

- Multiprogramming systems. In this model, each task keeps running until it reaches a point at which it must wait for an external event or until the scheduler forcibly suspends its execution. Multiprogramming systems tend to maximize CPU usage.

- Time-sharing systems. In time-sharing, the running task is allocated a time slot at the end of which it must relinquish CPU usage voluntarily or it will be forced to by the scheduler. Time-sharing systems allow several programs to execute apparently simultaneously.

- Priority-based systems. In this model, the scheduler is driven by the priorities assigned to each task. When an event of a higher priority requires service, the operating system preempts the one currently executing. Priority-based systems are usually event driven.

Several terms emerge from this classification: cooperative multitasking refers to a scheduling algorithm in which each task voluntarily cedes execution time to other tasks. This approach has been largely replaced by one referred to as preemptive multitasking. With preemptive approach, the operating system assigns a slice of CPU time to each task according to a set of priorities. If an external event takes place that requires immediate attention, then the running task can be preempted in order to respond to the event. Real-time operating systems are based on both cooperative and preemptive multitasking.

14.2 RTOS Scope

In the context of real-time programming and of embedded systems, we must define the scope of a software system that can be classified as an operating system. In a conventional computer system, it is clear that the operating system is a stand-alone application that controls devices, allocates resources, prioritizes operations requested by application, and performs other management and control duties. In the sense of Windows, Unix, and Mac OS, the operating system is a monitor program that executes in the foreground.

Real-time operating systems do not always conform to this model. Microcontroller-based circuits and devices often do not provide the support necessary for a stand-along control programs the likes of Windows or Unix. Although the definition of what constitutes and operating system is a matter of semantics, in the present context we adopt the widest possible description and include all control facilities used in scheduling and resource allocation operations. Thus, RTOS can be implemented in the following forms:

- As a conventional, stand-alone program that is furnished and loaded independently of application code

- As a collection or library of control routines that are appended to the application at compile time but that constitute a separate software entity

- As routines that perform control, resource allocation, and other operating system functions but that are part of the application itself

14.2.1 Tasks, Priorities, and Deadlines

Imagine a microcontroller-based circuit designed to operate the temperature and irrigation controls in a greenhouse. The circuit contains sensors that read the air temperature in the greenhouse as well as the soil humidity. If the air temperature exceeds 80°F, the air cooling system is turned on until the temperature reaches 70°F. If the soil humidity is lower than 40%, then the irrigation system is turned on until the humidity reaches 90%. The flowchart for the greenhouse controller is shown in Figure 14.1.

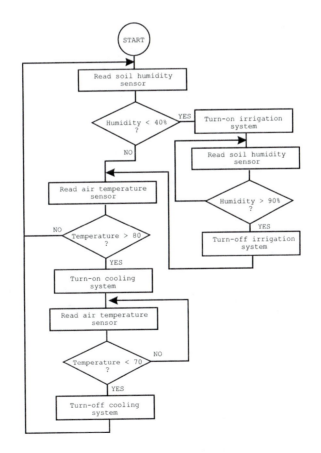

Figure 14.1 *Flowchart of a greenhouse control system.*

The flowchart of Figure 14.1 shows that the greenhouse control program must perform two distinct operations: one is to monitor the soil humidity and turn the irrigation system on and off, and another one is to monitor the air temperature and turn the cooling system on and off. These could be called the irrigation control task and the air temperature control task. However, in the flowchart of Figure 14.1, it is notable that the air temperature control task is suspended while the irrigation task is in progress. This could cause the air temperature to become exceedingly high while the system is busy monitoring the soil humidity during irrigation.

Some of the problems and challenges of multitasking in a real-time operating system become apparent even in this simple example. One solution to the problem mentioned in the preceding paragraph can be based on multitasking. For example, the software could be designed so that monitoring the two sensors takes place in a closed loop. If one of the sensors indicates that an irrigation or cooling operation is required, then the corresponding system is turned either on or off. The resulting flowchart is shown in Figure 14.2.

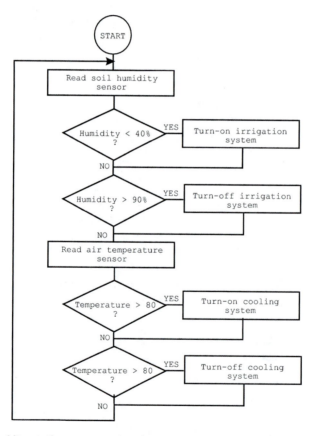

Figure 14.2 *Alternative processing for greenhouse control.*

The processing in the flowchart of Figure 14.2 can be visualized as consisting of several tasks. One of them would be monitoring the state of the sensors; a second task consists of controlling the irrigation system; and a third one of controlling the cooling system. In order to avoid redundant operations, such as turning on a system that is already on or turning off one that is turned off, the software can have global variables that are accessible to all tasks. In this manner, the sensor monitoring task would set a variable (probably a bit flag) to indicate the state of each system. Thus, the system control tasks could determine the state of a system and avoid trying to turn on an operation that is already active.

For the sake of experimentation, let us add an additional complication to the greenhouse control system: assume that the electric power available to the greenhouse only allows performing a single operation, that is, that the cooling and the irrigation systems cannot be on at the same time. This limitation leads to the establishment of priorities, for example, that irrigation should have preference over cooling or vice versa. Consequently, if the cooling system is operating at the time that the soil moisture reaches the minimum level, then cooling operation will be stopped so that irrigation can take place. The system can accommodate this requirement by establishing priorities that can be based or simple or complex rules.

Finally, one or more activities or tasks can be subject to specific deadlines. For example, in the greenhouse example, it may be the case that irrigation operations must end by 6 PM every day and not re-start until 8 AM the following morning independently of any other factor. In this case, the system must contain a real-time clock and checking this clock will be part of the scheduling. These constraints and limitations in the greenhouse control example are often present in most operating systems and give rise to the concepts of tasks, priorities, and deadlines. In the following sections we look at them in greater detail.

14.2.2 Executing in Real-Time

The real-time element in a Real-Time Operating System relates to the fact that the system must respond quickly and predictably to events. In the greenhouse example mentioned previously the system must control irrigation and cooling operations and must do so as rapidly as possible. A system that operates in real-time cannot usually postpone sensory or control activities for a later day or hour. In fact, we assume that the latency of a real-time system is minimal. This can be summarized by stating that a system operating in real-time must be able to provide adequate action within the required priorities and deadlines. This definition does not imply that real-time systems must operate at any given speed but that it must execute at sufficient speed so that the time-based requirements are met.

14.3 RTOS Programming

The most elementary programming model for any system or application, sometimes called sequential programming, implies a set of instructions that are executed one at a time. This sequential order of execution is only violated when the program branches to another location or when a sub-program executes. In the second version of the example greenhouse control program, we have attempted to implement multitasking by means of sequential programming. However, this approach contains several weak points:

- The execution time of program loops is not constant. That is, the routine that monitors the sensors may take different time during several iterations. For example, one iteration of the loop may require performing several control operations while another iteration may not. This variation may be tolerable in the greenhouse example but could be unacceptable in the control system of a nuclear power plant.

- Tasks may interfere with each other. In the greenhouse control system example, the task of controlling the irrigation system forces a delay in monitoring the temperature sensors.

- Complex priorities can be algorithmically difficult to enforce.

In the greenhouse control system example, some tasks are time-triggered while others are event-triggered. In this sense stopping irrigation activities at 8 PM is a time-triggered activity, while starting the cooling system when the ambient temperature reaches 70^OF is event-triggered.

14.3.1 Foreground and Background Tasks

One solution to the task prioritization problem is to allow some tasks to execute in the background. For example, the greenhouse control program needs to know the time-of-day in order to suspend irrigation during the night hours. If the time-of-day operations are handled by an interrupt service routine, then we can state that this task takes place in the background while a foreground loop takes care of reading the sensors and controlling irrigation and cooling processes. The foreground and background tasks can easily communicate through variables that are accessible to both routines. In this case, the interrupt handler could set a "do not irrigate" flag at 6 PM every evening and reset the flag at 8 AM every morning. The irrigation control routine would consult the state of this flag in order decide whether to start irrigation.

Interrupts in Tasking

In embedded systems, interrupts provide a viable mechanism for prioritizing tasks, especially those that take place in a time cycle or that are constrained by a deadline. The interrupt mechanism is in fact sequential because other CPU activity is suspended while the interrupt is in progress. However, the fact that the interrupt can be triggered by programmable external events makes it a power tool in simulating the parallel execution of multiple tasks. However, we should be careful not to assume that tasks with higher priorities should always be handled by the interrupt routine.

What tasks can be easily handled by interrupt usually depends on the system itself. For example, in an 18F family PIC-based system, it would be relatively easy to have a time-of-day clock executing in the background loop (see Section 12.1.2). On the other hand, setting up an interrupt to take place when a sensor is within a given range may be much more difficult. The requirements of each system and the hardware architecture determines which tasks should be handled in the foreground and which in the background.

In general, the 18F family PIC system makes it easy to assign time-triggered tasks to interrupts because several overflow timers are available in the hardware. Event-triggered tasks can be interrupt driven if the events are of a binary nature. For example, the action on a pushbutton switch can be vectored to an interrupt handler. On the other hand, it is more difficult to have range-based events trigger an interrupt, as is the case with the sensors of the greenhouse control system.

14.3.2 Task Loops

Real-time systems often contain conventional loops as a simple way of carrying out several tasks. The flowcharts in Figure 14.1 and Figure 14.2 both contain task loops. Tasks loops are useful in cases where scheduling can be accomplished linearly, that is, by running tasks successively while polling ports for events that signal the need for a specific task to be executed or terminated. Although polling routines usually waste considerable processor time in flag-testing operations, in many cases the tasks can be handled successfully in spite of this waste. The problem arises when a simple task loop cannot insure a timely response to events. In these cases the possible solutions are the offloading of one or more tasks to interrupt handlers or the implementation of preemptive multitasking.

14.3.3 Clock-Tick Interrupt

One of the simplest and most useful applications of interrupts in RTOS is to produce a timed beat. 18F family devices furnish hardware support for timers. The typical implementation is in the form of a count-up mechanism that overflows when it reaches the maximum capacity of the timer register, at which time an interrupt is generated. Software can be designed to use the interrupt to execute time-triggered tasks. In the previous example of a greenhouse irrigation system, we proposed a time-of-day clock interrupt that would handle turning off the irrigation system at dusk and back on in the morning.

In multitasking systems, the clock-tick interrupt can be used to assign slices of execution time to different tasks. In the cooperative model, each task would relinquish control back to the scheduler on completion of a cycle or phase of its execution. In the preemptive model, the interrupt itself will signal to the scheduler that a task needs to be suspended. In one case, the timer tick is used by the task itself to time its own activity. In the other one, the clock-tick is used by the scheduler routine to control the execution of several tasks.

Notice that although the clock signal is often derived from the CPU's clock or oscillator hardware, it does not necessarily take place at the clock's rate. Most microcontroller systems provide ways of manipulating the rate of the clock-tick interrupt by means of prescalers, preloading the counter register (as in the examples of Chapter 12), or using external clock sources.

14.3.4 Interrupts in Preemptive Multitasking

In a preemptive system, the scheduler must be capable of suspending a task in order reestablish priorities or to direct execution to another task. In microcontroller-based systems, the only available mechanism for suspending code execution is the interrupt. The classical approach for multitasking in real-time systems is by a timer interrupt that executes every so many cycles of the system clock. If the frequency of the interrupt can be set to conform to a unit of allowed task execution time, then the interrupt handler receives control at the conclusion of every time period and the task is effectively preempted. This amount of time is called a time slice. In this manner, the task scheduler controls the execution of running processes and is able to enforce the priority queue and ensure that processes meet their deadlines

The actual implementation of preemptive multitasking requires more than an interrupt system that can be programmed at a certain frequency of the system clock. In order to direct execution to various tasks, the scheduler must be able to save the context and the program counter of each interrupted task so that the task can be restarted at a future time. The context itself consists of the values of the critical registers and memories that are accessible to the task. The program counter (which we do not consider part of the context) is usually stored in the stack by the interrupt intercept. Both the program counter and the context are sometimes referred to as the machine state. A typical multitasking scheduler retrieves the program counter from the stack and stores it together with the context. A task is restarted by restoring the program counter and the context from this storage.

14.4 Constructing the Scheduler

The following concepts, discussed previously, are essential to the notion of a real-time, multitasking operating system:

- A task is a unit of program execution defined by a routine or subprogram that performs a specific chore. The notion of a task is related to real-time systems while the more term "process" is often used in relation to conventional computing.

- A time slice is the minimal unit of execution time allowed to each task.

- A context switch is the moment in time when a task is suspended so that priorities and task states can be examined. A context switch can take place at the conclusion of a task (or of a task phase), as is the case in cooperative scheduling. A context switch can also be forced at the conclusion of a time slice, as is the case in preemptive scheduling.

14.4.1 Cyclic Scheduling

The simplest model for a multitasking system is one that allows each task to run to completion. In this case, the scheduler is limited to performing the context switching at the conclusion of each task. The task loop, mentioned previously, can often be the base of a cyclic scheduler. Graphically, cyclic scheduling is represented in Figure 14.3.

Figure 14.3 *Cyclic scheduling.*

14.4.2 Round-Robin Scheduling

Another simple algorithm is round-robin scheduling. The typical implementation of round-robin scheduling is based on an interrupt-driven clock-tick. The round-robin model assigns equal priority to all tasks and gives each task the same time slice. Tasks are selected in a fixed order that does not change during execution. At the conclusion of each time slice the currently executing task is suspended and the next one in order is started. This implies that tasks are preempted at the end of each time slice, therefore round-robin scheduling assumes preemptive multitasking. Figure 14.4 depicts round-robin scheduling.

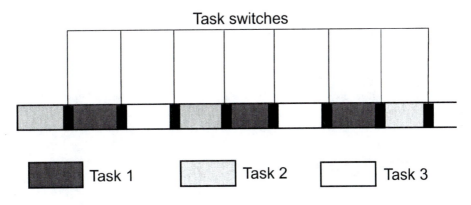

Figure 14.4 *Round-robin scheduling.*

Although the model in Figure 14.4 appears quite straightforward, round-robin is not without drawbacks. One of them is shown by the solid black lines separating each task in Figure 14.4. These lines indicate the task-switching operations, which can have a significant overhead. Another possible scenario is that a task completes before its time slice expires, in which case the system wastes execution time. Thirdl, round-robin does not take into account task priorities and priority changes.

14.4.3 Task States and Prioritized Scheduling

The flaws of the round-robin model make evident the need to establish conditions that reflect the state of each running task, as well as the possibility of tasks changing state during execution. The following task states are usually recognized:

- Task ready. In this state, a task is ready to start executing and change to the active state.

- Task active. Task is running and has been allocated CPU time. A task leaves the active state because it has completed or because it has been preempted by the operating system.

- Task blocked. A task that is ready is prevented from running. One reason could be that the task is waiting for an external event to take place or for data to be received. Also, a task can be blocked if a needed resource is being used by another task.

- Task stopped. A task is excluded from those that are granted CPU time.

- Task destroyed. The task no longer exists.

In order to avoid the limitations of conventional round-robin scheduling, it is possible to come up with a scheme in which each task is assigned a priority ranking. In this manner, tasks with higher priority are allowed to complete, or are assigned a larger time slice than tasks with lower priority. There are many algorithms for implementing priority scheduling. In one model, an executing task that is assigned a higher priority is allowed to hold the CPU until it has completed. In another variation of the algorithm, the priority rankings are used to proportionally assign time slices to each task, but tasks are preempted at the end of their allotted time. A more sophisticated approach is to determine each task's time slice not only according to the priority rankings, but also according to the task's complexity and requirements. The task's state (ready, active, blocked, stopped, or destroyed) also determines if and when the task is allowed CPU time.

Prioritized-preemptive scheduling also contains the inherent risk of a low-priority task not being allowed to execute at all. In this case, the task is said to be deadlocked. The scheduling algorithm determines how such cases are handled, that is, if deadlock is allowed or if all tasks are insured a minimum execution time slice.

14.5 A Small System Example

In small systems, purely preemptive schedulers are difficult to implement because it requires that the operating system have sufficient information to make all decisions that determine access to CPU time. Furthermore, preemptive methods must usually pay a price in lower performance due in part to the data requirements and the time consumed in context switching.

Often it is possible to combine preemptive scheduling with cooperative techniques in which tasks relinquish CPU time of their own accord. For example, a task that has run to completion may return control to the scheduler even if its time slice has not yet expired. Or a task may return control to the scheduler if it is in a noncritical stage of its execution cycle.

14.5.1 Task Structure

At the beginning of its design, the programmer chooses which activities of the system will be considered tasks. At this point it is important to avoid creating too many tasks because each context switch implies time and storage overheads. Sometimes the system deadlines can serve as a starting point for identifying tasks. In this sense related activities that are must meet similar deadlines can be grouped into a single task. By the same token, activities that are related in their function or their data requirements can often be bundled into a single task.

Although tasks may be preempted by the scheduler, they must be coded to run continuously. In other words, each task is a self-contained and semi-autonomous program. Tasks cannot access each other's code because this would create problems with reentrancy at task switch time. However, several tasks can access the same subprogram.

In certain preemptive models, each task must be associated with a priority ranking. This ranking can be a static value that remains unchanged during program execution, or a dynamic one which can change at runtime. Priorities should be based on the task's importance to the system, the user, and the program environment. The following ranks can be established:

- High-priority tasks are essential for system survival.

- Middle-priority tasks are necessary for correct system operation.

- Low-priority tasks are those that might be occasionally expendable or that can delay in their completion.

Priorities can also be influenced by the task's deadlines, because high-priority should be given to tasks with very tight time deadlines. However, a task can be given a lower priority even if it has a demanding deadline, if the task itself is of low importance.

14.5.2 Semaphore

It is often the case that several tasks may need access to the same resource. A resource could be a hardware component such as a peripheral or a disk storage area, a printer, a software module, or a subprogram. In any case, granting access to a resource requires special controls.

The data structure used is called a semaphore. Invented by Edsger Dijkstra, the semaphore is a control device that allows managing access to a shared resource in a multiprogramming environment. The semaphore records the number of units of a shared resource that are available. In the case of a single unit, control is by a binary semaphore. Otherwise, the semaphore can be a numeric value that is incremented and decremented as the resource is used or restored.

In a binary semaphore (the most common one), the task will be able to use the resource if its semaphore is in the GO state. In this case, the task will change the semaphore to the WAIT state while it is using the resource, and back to the GO state when it is finished. While a semaphore is in the WAIT state, any task needing the resource will go into a blocked state. This implementation corresponds with the concept of mutual exclusion, that is, all others are excluded when one task is accessing the resource.

The so-called counting semaphore is used if there is more than one resource of the same type. In this case, the semaphore initially holds the total number of available resources, and tasks increment and decrement the count as previously described. At any time, the counting semaphore holds the number of units of a resource currently available.

An effect of setting a semaphore to the WAIT state is that another task becomes blocked. In this manner, a semaphore can be used to provide time synchronization and signaling between different tasks. One task can block another by setting a semaphore, and can release it by clearing the semaphore.

14.6 Sample OS Application

```
; File name: 18F452_OS_Demo.asm
; Date: June 25, 2013
; Author: Julio Sanchez
; PIC 18F452
;
;==========================================================
;                         CPU pinout
;==========================================================
;                           18F452
;                 +------------------+
;      MCLR/Vpp ===>|  1          40 |===> RB7/PGD
;       RA0/AN0 <==>|  2          39 |<==> RB6/PGC
;       RA1/AM1 <==>|  3          38 |<==> RB5/PGM
;   RA2/AN2/REF- <==>|  4          37 |<==> RB4
;   RA3/AN3/REF+ <==>|  5          36 |<==> RB3/CCP2
;      RA4/TOCKI <==>|  6          35 |<==> RB2/INT2
; RA5/AN4/SS/LVDIN <==>|  7        34 |<==> RB1/INT1
;     RE0/RD/AN5 <==>|  8          33 |<==> RB0/INT0
;     RE1/WR/AN6 <==>|  9          32 |<=== Vdd
;     RE2/CS/AN7 <==>|10           31 |===> Vss
;            Vdd ===>|11           30 |<==> RD7/PSP7
;            Vss <== |11           29 |<==> RD6/PSP6
;      OSCI/CLKI ===>|13           28 |<==> RD5/PSP5
;  OSC2/CLK0/RA6 <==>|14           27 |<==> RD4/PSP4
; RC0/T1OSO/TICK1 <==>|15          26 |<==> RC7/RX/DT
; RC1/T1OS1/CCP2 <==>|16           25 |<==> RC6/TX/CK
;       RC2/CCP1 <==>|17           24 |<==> RC5/SD0
;    RC3/SCK/SCL <==>|18           23 |<==> RC4/SDI/SDA
;      RD0/PSP0 <==>|19            22 |<==> RD3/PSP3
;      RD1/PSP1 <==>|20            21 |<==> RD2/PSP2
;                 +------------------+
; Legend:
; Crys = 32.768 KHz crystal    DBx = LCD data byte 1-7
; cap = 22 Pf capacitor        E = LCD signal enable
; res0 = 10K resistor          RW = LCD read/write
; res1 = 470 Ohm resistor      RS = LCD register select
; res2 = 412 Ohm resistor      GR = ground


;
; Description:
; A RTOS demonstration program. Circuit contains three
; LEDs (yellow, red, and green) which can be enabled or
; disabled and flash at different rates. Each LCD is
; considered a task in a multitasking environment. An
; interrupt handler operates a timer tick in the
; background. Three pushbutton switches are used to activate
; the various modes and to set the parameters for each
; task. Three DIP switches allow deactivating any task
; without changing the task's delay value. The DIP switches
; simulate the assignment of some critical resource without
; which a task cannot run.
;
; WARNING:
; Code assumes 4MHz clock. Delay routines must be
; edited for faster clock.
; PROGRAM OPERATION:
; State diagram:
```

```
;                        +--------+
;                        |  INIT  |                +------------+
;                        +---|----+                |   TASK     |
;                            |                      |  STATUS    |
;                            |                      |  SWITCHES  |
;                            |                      +-----|------+
;       +----------+    +----------+          +----|------+
;       |  SETUP   |<=====>| COMMAND  |<=======>|   TASK     |
;       |  MODE    |    |  MODE    |          | SCHEDULER  |
;       +----|-----+    +----------+          +-----|----+
;            |          +----------------+          |
;            |_____| TIMER VARIABLES |_____|
;                       +----------------+
;
; The LCD displays the current delay values for each task.
; as follows:
;
;         0                          15
;         | | | | | | | | | | | | | | | |
;         T a s k :      1      2      3        <== line 1
;         D e l a y :    1      5      9        <== line 2
;                        |      |      |
;                        7     10     13 <== digit offset
;
; In the default start-up all tasks are running and the
; default delays are 1, 5, and 9 respectively.
;
;===========================
;       command mode
;===========================
; Tasks are running
; Button # 1 enters the setup mode
; Setup mode allows entering new values for the task
; delays (range 0 to 9). A task with a delay value of 0
; is blocked. All tasks are suspended.
;===========================
;       setup mode
;===========================
; SETUP MODE KEYS:
; Pushbutton # 1 toggles to next digit.
;           # 2 tabs to next digit field
;           # 3 ends setup mode and return to
;               command mode
;
;===========================
;       LCD wiring
;===========================
; LCD is wired in parallel to 16F877 as follows:
; DATA LINES:
;    |------- F87x -------|----- LCD -------|
;           port      PIN                 line    PIN
;           RD0        19                 DB0      8
;           RD1        20                 DB1      7
;           RD2        21                 DB2      10
;           RD3        22                 DB3      9
;           RD4        27                 DB4      12
;           RD5        28                 DB5      11
;           RD6        29                 DB6      14
;           RD7        30                 DB7      13
; CONTROL LINES:
```

```
;      |------- F87x -------|----- LCD -------|
;               port           PIN              line        PIN
;               RE0            8                RS          3
;               RE1            9                E           5
;               RE2            10               RW          6
;
;============================
;     Pushbutton switches
;     action and wiring
;============================
; All switches are wired active low, this means that
; port reads binary 0 when switch is closed
; x87 PORT    PIN    SWITCH FUNCTION
; STATUS MODE ACTION
;    RB0       33            SW1            Go to SETUP mode
;    RB2       34            SW2            NO ACTION
;    RB3       35            SW3            NO ACTION
; SETUP MODE ACTION:
;    RB0       33            SW1            Toggle digit (0 to 9)
;    RB2       34            SW2            Tab to next digit
;                                           Exit to STATUS mode if
;                                           at last digit
;    RB3       35            SW3     Return to STATUS mode
;============================
;         DIP switches
;     (resource assignment)
;============================
; x87 PORT    PIN    Function
;    RC4       23            Activate/de-activate Task 1
;    RC5       24               "             Task 2
;    RC6       25               "             Task 3
;
;============================
;         LED wiring
;============================
; x87 PORT    PIN    LED    Color      Function
;    RA0       2      1      green      Task 1
;    RA1       3      2      red        Task 2
;    RA2       4      3      yellow     Task 3
;
;============================================================
;               16F877 configuration options
;============================================================
; Switches used in __config directive:
;    _CP_ON          Code protection ON/OFF
; * _CP_OFF
; * _PWRTE_ON       Power-up timer ON/OFF
;    _PWRTE_OFF
;    _BODEN_ON       Brown-out reset enable ON/OFF
; * _BODEN_OFF
; * _PWRTE_ON       Power-up timer enable ON/OFF
;    _PWRTE_OFF
;    _WDT_ON         Watchdog timer ON/OFF
; * _WDT_OFF
;    _LPV_ON         Low voltage IC programming enable ON/OFF
; * _LPV_OFF
;    _CPD_ON         Data EE memory code protection ON/OFF
; * _CPD_OFF
; OSCILLATOR CONFIGURATIONS:
;    _LP_OSC         Low power crystal oscillator
```

```
;    _XT_OSC          External parallel resonator/crystal oscillator
; *  _HS_OSC          High speed crystal resonator
;    _RC_OSC          Resistor/capacitor oscillator
; |                   (simplest, 20% error)
; |
; |_____ * indicates setup values presently selected

     processor        18f452        ; Define processor
     #include <p18f452.inc>
     __CONFIG _CP_OFF & _WDT_OFF & _BODEN_OFF & _PWRTE_ON & _HS_OSC &
_WDT_OFF & _LVP_OFF & _CPD_OFF

; __CONFIG directive is used to embed configuration data
; within the source file. The labels following the directive
; are located in the corresponding .inc file.
;
; Turn off banking error messages
     errorlevel       -302
;
;============================================================
;                     constant definitions
;   for PIC-to-LCD pin wiring and LCD line addresses
;============================================================
; LCD used in the demo board is 2 lines by 16 characters
#define E_line 1     ; |
#define RS_line 0    ; | -- from wiring diagram
#define RW_line 2    ; |
; LCD line addresses (from LCD data sheet)
#define LCD_1 0x80 ; First LCD line constant
#define LCD_2 0xc0 ; Second LCD line constant
; Address of individual characters on second LCD line
#define     CHAR_1 0xc7 ; First character in second line
#define     CHAR_2 0xca ; Second character
#define CHAR_3 0xcd ; Third character
; Note: The constant that define the LCD display line
;       addresses have the high-order bit set in
;       order to facilitate the controller command
;
;===================================================
;                  variables in PIC RAM
;===================================================
; Reserve 16 bytes for string buffer
     cblock   0x20
     strData
     endc
; Continue with local variables
     cblock   0x30          ; Start of block
     count1                 ; Counter # 1
     count2                 ; Counter # 2
     count3                 ; Counter # 3
     pic_ad                 ; Storage for start of text area
                            ; (labeled strData) in PIC RAM
     J                      ; counter J
     K                      ; counter K
     index                  ; For LCD display
     temp                   ; Auxiliary storage for conversion
; Digit-generic data for general procedures
     thisField              ; Current digit (0 to 3)
     digAsc                 ; ASCII for currently selected field
                            ; comes from cycle1 to cycle3 variables
```

```
        charAdd          ; Address of current digit field
                         ;    CHAR_1 to CHAR_3
; Variables for preserving context during interrupt
        old_w            ; Context saving
        old_STATUS   ; Idem
;=========================
;   task-related variables
;=========================
        taskState        ; Bitmapped task state control
        delay1           ; Task delays
        delay2
        delay3
        status1          ; Task delay status
        status2
        status3
; Field-specific registers for ASCII digits
        cycle1           ; Task 1 delay (in ASCII)
        cycle2           ; Task 2 delay
        cycle3           ; Task 3 delay
        endc


;============================================================
;                          M A C R O S
;============================================================
; Macros to select the register banks
; Data memory bank selection bits:
; RP1:RP0         Bank
;   0:0            0      Ports A,B,C,D, and E
;   0:1            1      Tris A,B,C,D, and E
;   1:0            2
;   1:1            3
Bank0       MACRO                    ; Select RAM bank 0
            bcf    STATUS,RP0
            bcf    STATUS,RP1
            ENDM
Bank1       MACRO                    ; Select RAM bank 1
            bsf    STATUS,RP0
            bcf    STATUS,RP1
            ENDM
Bank2       MACRO                    ; Select RAM bank 2
            bcf    STATUS,RP0
            bsf    STATUS,RP1
            ENDM
Bank3       MACRO                    ; Select RAM bank 3
            bsf    STATUS,RP0
            bsf    STATUS,RP1
            ENDM
;============================================================
;                          program
;============================================================
            org        0       ; start at address
        NOP
```

```
            goto   main
; Space for interrupt handlers
;============================
;     interrupt intercept
;============================
    org            0x04
```

```
        goto      IntServ
main:
; Port setup:
; PORT A lines:
;   PORT              LINE   DEVICE       MODE
;   RA0               2                   LED 1       OUTPUT
;   RA1               3                   LED 2       OUTPUT
;   RA2               4                   LED 3       OUTPUT
; PORT B lines:
;   RB0               33                  SWITCH 1    INPUT
;   RB1               34                  SWITCH 2    INPUT
;   RB2               35                  SWITCH 3    INPUT
;   RB3               36                  SWITCH 3    INPUT
; PORT C lines:
;   RC4               23                  DIP 1       INPUT
;   RC5               24                  DIP 2       INPUT
;   RC6               25                  DIP 3       INPUT
; PORT D lines:
; RD0-RD7 19-22/27-30    LCD Data         OUTPUT
; Port E lines:
; RE0-RE2    8-10        LCD Ctrl         OUTPUT

      nop
      nop

; Select bank 1 to tris output registers
      Bank1
; Tris port D for output
      movlw    B'00000000'
      movwf    TRISD        ; and port D
      movwf    TRISE
; By default port A lines are analog. To configure them
; as digital code must set bits 1 and 2 of the ADCON1
; register (in bank 1)
      movlw    0x06         ; binary 0000 0110  is code to
                            ; make all port A lines digial
      movwf    ADCON1
; Tris port A for output
      movlw    B'00000000'
      movwf    TRISA        ; Tris port A
; Tris port B bits 0 to 3 for input
      movlw    B'00001111'
      movwf    TRISB
; Tris port C lines 4, 5, and 6 for input
      movlw    B'01110000'
      movwf    TRISC
; Back to Bank 0
      Bank0
; Clear all output lines
      movlw    B'00000000'
      movwf    PORTA
      movwf    PORTD
      movwf    PORTE
;=============================
;  setup timer1 and interrupt
;=============================
      call     timer1Off
;=========================
;  init task variables
;=========================
```

```
; delayx variables hold the delay values (range 0 to 9)
; statusx variables hold the current delay count
     clrf      status1
     clrf      status2
     clrf      status3
; Initialize  ASCII digits to default values
; task 1 = 1, Task 2 = 5, Task 3 = 9
     movlw     '1'
     movwf     cycle1
     movlw     '5'
     movwf     cycle2
     movlw     '9'
     movwf     cycle3
; Convert cycles to binary and store in variable
; delay1 to delay3
     call      cyc2bin
; Update taskState variable according to DIP switches
     call      updateState
; Select current field (range 0 to 2)
     clrf      thisField   ; Field 0 is first digit
; Set address of first digit in general variable
     movlw     CHAR_1      ; Address of first digit
     movwf     charAdd            ; To general variable
; Store base address of text buffer in PIC RAM
     movlw     0x20        ; Start address of text buffer
     movwf     pic_ad      ; to local variable
;=========================
;      init LCD
;=========================
; Wait and initialize HD44780
     call      delay_28ms       ; Allow LCD time to initialize itself
     call      initLCD          ; Then do forced initialization
     call      delay_28ms       ; (Wait probably not necessary)


;============================================================
;============================================================
;          c o m m a n d    p r o c e s i n g
;============================================================
;============================================================
; State diagram:
;                         +--------+
;                         |  INIT  |                +-------------+
;                         +---|----+                |    TASK     |
;                             |                     |   STATUS    |
;                             |                     |   SWITCHES  |
;                             |                     +-----|-------+
;                             |                     +----|-------+
;    +----------+       +----------+                +----|-------+
;    |  SETUP   |<=====>|  COMMAND |<========>|    TASK     |
;    |  MODE    |       |  MODE    |          |  SCHEDULER  |
;    +----|-----+       +----------+          +-----|----+
;         |             +-----------------+         |
;         |_____| TIMER VARIABLES |_____|
;                       |-----------------|
;
commandMode:
;===========================
;      display LCD text
;===========================
; Store 16 blanks in PIC RAM, starting at address stored
; in variable pic_ad
```

```
        call      blank16
; Call procedure to store ASCII characters for message
; in text buffer
        call      storeStatus
; Set DDRAM address to start of first line
        call      line1
; Call procedure to display 16 characters in LCD
        call      display16
;=======================
;   second LCD line
;=======================
        call      delay_168    ; Wait for termination
        call      blank16              ; Blank buffer
; Call procedure to store ASCII characters for message
; in text buffer
        clrw
        call      storeParams
        call      line2        ; DDRAM address of LCD line 2
        call      display16
; Now test switches
; PORT B lines:
;    RB0                 33              SWITCH 1     INPUT
;    RB1                 34              SWITCH 2     INPUT
;    RB2                 35              SWITCH 3     INPUT
; Switches are ACTIVE LOW
; In Status mode the following switches are active
;    RB0 = go to setup mode


;=========================================================
;=========================================================
;                 timer control and command mode
;=========================================================
;=========================================================
timerCtrl:
; Turn on interrupt
        call      timer1Off
        call      long_delay
        call      timer1On
        call      long_delay
; Reset status
        clrf      status1
        clrf      status2
        clrf      status3
;=========================================================
;                  command wait loop
;=========================================================
timerWait:
; Update state according to DIP swtiches settings
        call      updateState
; Test RB0 line
        btfss     PORTB,0                ; Test RB0 line
        goto      setupMode    ; Command handler
; Not RB0
        goto      timerWait
;=========================================================
;                   setup mode subcommand
;=========================================================
setupMode:
        call      long_delay
; Turn off clock
```

```
         call    timer1Off
; Turn off all LEDs
         call    ledsOff
; Call procedure to store ASCII characters for message
; in text buffer
         call    storeParams
         call    line2          ; DDRAM address of LCD line 2
         call    display16
; Reset field
         clrf    thisField
; Debounce last keystroke
         call    long_delay
         call    long_delay
; Update digit and display
         call    field2GR       ; Field digit to general register
         call    showDigit      ; Display digit
         call    GR2Field       ; General register to field
         goto    setupKey
         nop                                   ; Spacer
;=================================
; SETUP mode command processing
;=================================
; Key commands:
;   PB#/LINE  NAME   ACTION
;     1 RB0   ENTER  Toggle digit
;     2 RB1   TAB            Move to next digit
;     3 RB2   DONE   Go back to STATUS mode
; Switches are ACTIVE LOW
setupKey:
         btfss   PORTB,0             ; Test RB0 line
         goto    cmdENTER       ; Handler
; Not RB0
         nop
         btfss   PORTB,1             ; Test RB1 line
         goto    cmdTAB
; Not RB1
         nop
         btfss   PORTB,2             ; Test RB2 line
         goto    cmdDONE
         goto    setupKey
;=======================
;   DONE subcommand
;=======================
; Execution returns to main command processing
cmdDONE:
         goto    lastField      ; Reset to last field and
                                          ; return to timer control
                                          ; mode

;=======================
;   ENTER subcommand
;=======================
cmdENTER:
; In the SETUP mode the enter key toggles the digit
; in the currently selected display field. Field
; number is stored in thisField.
         call    field2GR       ; Field digit to general register
         call    nextDigit      ; Bump digit
         call    showDigit      ; Display it
         call    GR2Field       ; General register to field
         call    long_delay
```

```
        goto      setupKey      ; Monitor next keystroke

;========================
;     tab to next digit
;========================
; The variable thisField (range 0 to 2) contains the
; current digit field number corresponding to the
; digits cycle1, cycle2, and cycle3. Action on the
; tab command key toggles to the next field.
cmdTAB:
; First test if value is last field (thisField = 2)
        movlw     0x02          ; Value of third field
        subwf     thisField,w   ; Compare operation
        btfsc     STATUS,Z      ; Test Z flag
        goto      lastField     ; Go if last field
; At this point thisField is not = 3
        incf      thisField,f   ; Bump field
        call      field2GR      ; Field digit to general register
        call      showDigit     ; Display it
        call      GR2Field      ; General register to field
        call      long_delay
        goto      setupKey      ; Monitor next keystroke
; Wrap arround to first field
;============================
;       end of Setup command
;============================
lastField:
        call      long_delay
; The values are presently in the field digit
; variables cycle1, cycle2, dur1, and dur2.
        call      cyc2bin       ; ASCII cycle (in cyclex) to
                                ; binary (in delayx)
; Reset statusx variables
        clrf      status1
        clrf      status2
        clrf      status3
; Done. Go back to command mode
        goto      commandMode


;===========================================================
;===========================================================
;                     P r o c e d u r e s
;===========================================================
;===========================================================
;=====================
;    INITIALIZE LCD
;=====================
initLCD
; Initialization for Densitron LCD module as follows:
;    8-bit interface
;    2 display lines of 16 characters each
;    cursor on
;    left-to-right increment
;    cursor shift right
;    no display shift
;**********************|
;     COMMAND MODE     |
;**********************|
        bcf       PORTE,E_line  ; E line low
        bcf       PORTE,RS_line         ; RS line low for command
```

```
    bcf        PORTE,RW_line    ; Write mode
    call       delay_168             ;delay 125 microseconds
;***********************|
;      FUNCTION SET     |
;***********************|
    movlw      0x38                  ; 0 0 1 1 1 0 0 0 (FUNCTION SET)
                                     ;         | | | |__ font select:
                                     ;         | | |    1 = 5x10 in 1/8 or 1/11 dc
                                     ;         | | |    0 = 1/16 dc
                                     ;         | | |___ Duty cycle select
                                     ;         | |      0 = 1/8 or 1/11
                                     ;         | |      1 = 1/16 (multiple lines)
                                     ;         | |___ Interface width
                                     ;         |      0 = 4 bits
                                     ;         |      1 = 8 bits
                                     ;         |___ FUNCTION SET COMMAND
    movwf      PORTD ;0011 1000
    call       pulseE ;pulseE and delay


;***********************|
;    DISPLAY ON/OFF     |
;***********************|
    movlw      0x0a                  ; 0 0 0 0 1 0 1 0 (DISPLAY ON/OFF)
                                     ;         | | | |___ Blink character
                                     ;         | | |     1 = on, 0 = off
                                     ;         | | |___ Cursor on/off
                                     ;         | |       1 = on, 0 = off
                                     ;         | |____ Display on/off
                                     ;         |        1 = on, 0 = off
                                     ;         |____ COMMAND BIT

    movwf      PORTD
    call       pulseE ;pulseE and delay


;***********************|
; DISPLAY AND CURSOR ON |
;***********************|
    movlw      0x0f                  ; 0 0 0 0 1 1 1 1 (DISPLAY ON/OFF)
                                     ;         | | | |___ Blink character
                                     ;         | | |     1 = on, 0 = off
                                     ;         | | |___ Cursor on/off
                                     ;         | |       1 = on, 0 = off
                                     ;         | |____ Display on/off
                                     ;         |        1 = on, 0 = off
                                     ;         |____ COMMAND BIT
    movwf      PORTD
    call       pulseE ;pulseE and delay


;***********************|
;    ENTRY MODE SET     |
;***********************|
    movlw      0x06                  ; 0 0 0 0 0 1 1 0 (ENTRY MODE SET)
                                     ;         | | |___ display shift
                                     ;         | |      1 = shift
                                     ;         | |      0 = no shift
                                     ;         | |____ cursor increment mode
                                     ;         |       1 = left-to-right
                                     ;         |       0 = right-to-left
                                     ;         |___ COMMAND BIT
    movwf      PORTD ;00000110
```

```
    call    pulseE

;*********************|
; CURSOR/DISPLAY SHIFT |
;*********************|
    movlw   0x14                    ; 0 0 0 1 0 1 0 0 (CURSOR/DISPLAY SHIFT)
                                    ;       | | | |_|___ don't care
                                    ;       |  |_|__ cursor/display shift
                                    ;       |         00 = cursor shift left
                                    ;       |         01 = cursor shift right
                                    ;       |         10 = cursor and display
                                    ;       |              shifted left
                                    ;       |         11 = cursor and display
                                    ;       |              shifted right
                                    ;       |___ COMMAND BIT
    movwf   PORTD  ;0001 1111
    call    pulseE

;*********************|
;   CLEAR DISPLAY      |
;*********************|
    movlw   0x01  ; 0 0 0 0 0 0 0 1 (CLEAR DISPLAY)
                  ;               |___ COMMAND BIT
    movwf   PORTD ;0000 0001
;
    call    pulseE
    call    delay_28ms   ;delay 5 milliseconds after init
    return
;====================================
;     Time Delay and Pulse Procedures
;====================================
; Procedure to delay 42 x 4 = 168 machine cycles
; On a 4MHz clock the instruction rate is 1 microsecond
; 42 x 4 x 1 = 168 microseconds
delay_168
    movlw   D'42'           ; Repeat 42 machine cycles
    movwf   count1          ; Store value in counter
repeat
    decfsz  count1,f        ; Decrement counter
    goto    repeat          ; Continue if not 0
    return                  ; End of delay
;
; Procedure to delay 168 x 168 microseconds
; = 28.224 milliseconds
delay_28ms
    movlw   D'42'           ; Counter = 41
    movwf   count2          ; Store in variable
delay
    call    delay_168       ; Delay
    decfsz  count2,f    ; 40 times = 5 milliseconds
    goto    delay
    return                  ; End of delay
;=====================
;     pulse E line
;=====================
pulseE
    bsf     PORTE,E_line ;pulse E line
    bcf     PORTE,E_line
    call    delay_168           ;delay 168 microseconds
    return
```

```
;==============================
;       delay procedures
;==============================
; Approximately 160,000 machine cycles
; decfsz and goto are 2-cycle instructions
; so both loops take up 4 machine cycles per iteration.
; 40,000 x 4 = 160,000 cycles
; = 160 milliseconds at 4MHz.
long_delay
                movlw  D'200' ; w = 200 decimal
                movwf  J            ; J = w
jloop:          movwf  K            ; K = w

kloop:          decfsz K,f          ; K = K-1, skip next if zero
                goto   kloop ; 4 machine cycles loop
                decfsz J,f          ; J = J-1, skip next if zero
                goto   jloop ; 4 machine cycles loop
                return


;==========================================================
;                 data manipulation procedures
;==========================================================
;==============================
;   field data to general register
;==============================
; Data in the field variables (cycle1, cycle2, and cycle3
; is moved to the general register digAsc
; Variable thisField holds 0, 1, and 2 field number
field2GR:
    movlw    .0                   ; 0 to w register
    subwf    thisField,w  ; Subtract current field
    btfss    STATUS,Z     ; Test zero flag
    goto     testFV1              ; Go if not first field
; At this point first digit field is selected
; move data to general register (digAsc)
    movf     cycle1,w     ; First digit to w
    movwf    digAsc       ; To general register
; Set address of first digit in general variable
    movlw    CHAR_1       ; Address of first digit
    movwf    charAdd                ; To general variable
    goto     exitField    ; General processing
; Testing second digit field (field value = 1)
testFV1:
    movlw    .1                   ; 1 to w register
    subwf    thisField,w  ; Subtract current field
    btfss    STATUS,Z     ; Test zero flag
    goto     testFV2              ; Go if not field
; At this point second digit field is selected
    movf     cycle2,w     ; First digit to w
    movwf    digAsc       ; To general register
; Set address of second digit in general variable
    movlw    CHAR_2       ; Address of digit
    movwf    charAdd                ; To general variable
    goto     exitField    ; General processing
; Testing third digit field (field value = 2)
testFV2:
; At this point third digit field is selected
    movf     cycle3,w            ; Digit to w
    movwf    digAsc       ; To general register
; Set address of third digit in general variable
```

```
        movlw    CHAR_3        ; Address of digit
        movwf    charAdd              ; To general variable
exitField:
        return


;=================================
;    digit update procedure
;=================================
; ASCII digit is stored in variable digAsc.
; Digit is tested for '9' and rewound. Otherwise
; it is bumped top the next digit
nextDigit:
        movf     digAsc,w      ; Digit to w register
        sublw    0x39          ; ASCII for '9'
        btfsc    STATUS,Z      ; Test zero flag
        goto     is9                   ; Go if digit is '9'
; Digit is not '9'. Bump to next digit
        incf     digAsc,f      ; Bump to next digit
        return
; At this point digit storage is reset back to '0'
is9:
        movlw    '0'                   ; ASCII '0'
        movwf    digAsc        ; Digit to variable
        return
;=================================
;           display digit
;=================================
; Procedure to display digit
showDigit:
        call     char2LCD      ; Flash cycle1 digit
; Delay
        call     long_delay
        return


;=================================
; general data to field registers
;=================================
; Procedure to move data from the general register
; (digAsc) to the corresponding field register
; Field values are 0, 1, and 2
GR2Field:
        movlw    0x00          ; 0 to w register
        subwf    thisField,w   ; Subtract current field
        btfss    STATUS,2      ; Test zero flag be (2)
        goto     notFV1        ; Go if not first field
; At this point first digit field is selected
; move data to general register (digAsc)
        movf     digAsc,w      ; Digit to w
        movwf    cycle1        ; To field register
        goto     allFields     ; General processing
; Testing second digit field (field value = 1)
notFV1:
        movlw    0x01          ; 2 to w register
        subwf    thisField,w   ; Subtract current field
        btfss    STATUS,2      ; Test zero flag be (2)
        goto     notFV2        ; Go if not field
; At this point second digit field is selected
        movf     digAsc,w      ; Digit to w
        movwf    cycle2        ; To digit field
        goto     allFields     ; General processing
```

```
; Testing third digit field (field value = 2)
notFV2:
; At this point third digit field is selected
    movf      digAsc,w           ; Digit to w
    movwf     cycle3       ; To general register
allFields:
    call      long_delay
    return
;
;==============================
;      cycle to binary
;==============================
; ASCII values of delay time are stored in the variables
; cycle1 to cycle3. These values are converted to binary
; and stored in variables delay1 to delay3
cyc2bin:
    movf      cycle1,w     ; First cycle to w
    movwf     temp         ; temporary register
    movlw     0x30         ; 30 hex
    subwf     temp,w       ; Subtract and store in w
    movwf     delay1
; cycle2 to delay2
    movf      cycle2,w     ; First cycle to w
    movwf     temp         ; temporary register
    movlw     0x30         ; 30 hex
    subwf     temp,w       ; Subtract and store in w
    movwf     delay2
; cycle3 to delay3
    movf      cycle3,w     ; First cycle to w
    movwf     temp         ; temporary register
    movlw     0x30         ; 30 hex
    subwf     temp,w       ; Subtract and store in w
    movwf     delay3
    return
;==========================
;     update taskState
; according to DIP switches
;==========================
; Procedure to read task state switches wired to port C
; lines 4, 5, and 6 and update the corresponding bits
; in the taskState variable
; Task states are mapped to the three low-order bits
; taskState  7  6  5  4  3  2  1  0
;                           |  |  |___ task 3 (green)
;                           |  |_____ task 2 (red)
;                           |_____ task 1 (yellow)
updateState:
; Clear task switch variable
    clrf      taskState
; Test RC0 line
    btfsc     PORTC,4              ; Test RC4 line
    goto      stateBit1    ; Next task state bit
; At this point task state must be set
    bsf                taskState,0
stateBit1:
; Test RC1 line
    btfsc     PORTC,5              ; Test RC5 line
    goto      stateBit2    ; Next task state bit
; At this point task state must be set
    bsf       taskState,1
```

```
stateBit2:
; Test RC2 line
    btfsc   PORTC,6                 ; Test RC6 line
    goto    stateDone   ; Exit
; At this point task state must be set
    bsf     taskState,2
stateDone:
    return

;===========================================================
;                    LCD display procedures
;===========================================================
;===================================
;   display currnet field character
;            on LCD line 2
;===================================
char2LCD:
; ON ENTRY:
;  Address of character in LCD line 2 in variable charAdd
;  Character to be displayed in variable digAsc
; ON EXIT:
; Character is displayed and cursor is reset to character
; position
    bcf     PORTE,E_line ; E line low
    bcf     PORTE,RS_line       ; RS line low, setup for control
    call    delay_168           ; delay
; Set to character position in second display line
    movf    charAdd,w           ; Address with high-bit set
    movwf   PORTD
    call    pulseE              ; Pulse and delay
; Set RS line for data
    bsf     PORTE,RS_line       ; RS = 1 for data
    call    delay_168    ; delay
; Display character
    movf    digAsc,w            ; cycle1 digit to w
    movwf   PORTD               ; Write digit to LCD port
    call    pulseE
; Done. Reset cursor after write operation
    bcf     PORTE,E_line ; E line low
    bcf     PORTE,RS_line       ; RS line low, setup for control
    call    delay_168           ; delay
; Set to character position in second display line
    movf    charAdd,w           ; Address with high-bit set
    movwf   PORTD
    call    pulseE              ; Pulse and delay
    return

;===========================
;    display 16 characters
;===========================
display16:
; Sends 16 characters from PIC buffer with address stored
; in variable pic_ad to LCD line previously selected
; Set up for data
    bcf     PORTE,E_line ; E line low
    bsf     PORTE,RS_line       ; RS line low for control
    call    delay_168           ; Delay
; Set up counter for 16 characters
    movlw   D'16'               ; Counter = 16
    movwf   count3
```

```
; Get display address from local variable pic_ad
      movf      pic_ad,w              ; First display RAM address to W
      movwf     FSR                   ; W to FSR
getchar:
      movf      INDF,w                ; get character from display RAM
                                      ; location pointed to by file select
                                      ; register
      movwf     PORTD
      call      pulseE ;send data to display
; Test for 16 characters displayed
      decfsz    count3,f             ; Decrement counter
      goto      nextchar             ; Skipped if done
      return
nextchar:
      incf      FSR,f                ; Bump pointer
      goto      getchar


;========================
;      blank buffer
;========================
; Procedure to store 16 blank characters in PIC RAM
; buffer starting at address stored in the variable
; pic_ad
blank16:
      movlw     D'16'          ; Setup counter
      movwf     count1
      movf      pic_ad,w       ; First PIC RAM address
      movwf     FSR                  ; Indexed addressing
      movlw     0x20           ; ASCII space character
storeit:
      movwf     INDF           ; Store blank character in PIC RAM
                                         ; buffer using FSR register
      decfsz    count1,f       ; Done?
      goto      incFSR         ; no
      return                   ; yes
incFSR:
      incf      FSR,f          ; Bump FSR to next buffer space
      goto      storeit


;========================
; Set address register
;     to LCD line 1
;========================
; ON ENTRY:
;         Address of LCD line 1 in constant LCD_1
line1:
      bcf       PORTE,E_line         ; E line low
      bcf       PORTE,RS_line        ; RS line low, set up for control
      call      delay_168            ; delay 125 microseconds
; Set to second display line
      movlw     LCD_1                ; Address and command bit
      movwf     PORTD
      call      pulseE               ; Pulse and delay
; Set RS line for data
      bsf       PORTE,RS_line        ; Setup for data
      call      delay_168            ; Delay
      return
;========================
; Set address register
;     to LCD line 2
```

```
;========================
; ON ENTRY:
;           Address of LCD line 2 in constant LCD_2
line2:
       bcf      PORTE,E_line ; E line low
       bcf      PORTE,RS_line      ; RS line low, setup for control
       call     delay_168          ; delay
; Set to second display line
       movlw    LCD_2              ; Address with high-bit set
       movwf    PORTD
       call     pulseE             ; Pulse and delay
; Set RS line for data
       bsf      PORTE,RS_line      ; RS = 1 for data
       call     delay_168          ; delay
       return


;========================
;   Turn ON LED 1
;========================
led1_on:
; Procedure to turn on LED No. 1 wired to port RA0
       movf     PORTA,w            ; Port A to w
       iorlw    B'00000001'        ; Set LOB
;                        | | |____ LED 1
;                        | |_____ LED 2
;                        |_____ LED 3
       movwf    PORTA
       return
;========================
;   Turn OFF LED 1
;========================
led1_off:
; Procedure to turn off LED No. 1 wired to port RA0
       movf     PORTA,w            ; Port A to w
       andlw    B'11111110'        ; Clear LOB
;                        | | |____ LED 1
;                        | |_____ LED 2
;                        |_____ LED 3
       movwf    PORTA
       return
;========================
;   Turn ON LED 2
;========================
led2_on:
; Procedure to turn on LED No. 2 wired to port RA1
       movf     PORTA,w            ; Port A to w
       iorlw    B'00000010'        ; Set bit 1
;                        | | |____ LED 1
;                        | |_____ LED 2
;                        |_____ LED 3
       movwf    PORTA
       return
;========================
;   Turn OFF LED 2
;========================
led2_off:
; Procedure to turn off LED No. 2 wired to port RA1
       movf     PORTA,w            ; Port A to w
       andlw    B'11111101'        ; Clear bit 1
```

```
;                            |||____ LED 1
;                            ||_____ LED 2
;                            |_____ LED 3
     movwf     PORTA
     return
;========================
;    Turn ON LED 3
;========================
led3_on:
; Procedure to turn on LED No. 3 wired to port RA2
     movf      PORTA,w              ; Port A to w
     iorlw     B'00000100'          ; Set bit 2
;                       |||____ LED 1
;                       ||_____ LED 2
;                       |_____ LED 3
     movwf     PORTA
     return
;========================
;    Turn OFF LED 3
;========================
led3_off:
; Procedure to turn off LED No. 3 wired to port RA3
     movf      PORTA,w              ; Port A to w
     andlw     B'11111011'          ; Clear bit 2
;                       |||____ LED 1
;                       ||_____ LED 2
;                       |_____ LED 3
     movwf     PORTA
     return
;========================
;    Turn ON LED 4
;========================
led4_on:
; Procedure to turn on LED No. 4 wired to port RA3
     movf      PORTA,w              ; Port A to w
     iorlw     B'00001000'          ; Set LOB
;                  |_____ LED 4
     movwf     PORTA
     return
;========================
;    Turn OFF LED 1
;========================
led4_off:
; Procedure to turn off LED No. 1 wired to port RA0
     movf      PORTA,w              ; Port A to w
     andlw     B'11110111'          ; Clear LOB
;                  |_____ LED 4
     movwf     PORTA
     return
;========================
;    flash all 3 LEDs
;========================
flash3leds:
     call      led1_on
     call      long_delay
     call      led1_off
     call      long_delay
     call      led2_on
     call      long_delay
     call      led2_off
```

```
        call     long_delay
        call     led3_on
        call     long_delay
        call     led3_off
        call     long_delay
        return
;============================
; turn OFF all 3 LEDs
;============================
ledsOff:
        movf     PORTA,w                  ; Port A to w
        andlw    B'11110000'  ; Clear LOB
        movwf    PORTA
        return


;=========================================================
;=========================================================
;                    interrupt service routine
;=========================================================
;=========================================================
; Service routine receives control when the timer1
; registers TMR1H/TMR1L overflow, that is when the word
; value rolls over from 0xffff to 0x0000.
; The timer is setup to work off the system clock.
; At 4MHz the delay per timer1 cycles is of:
;    65,535 * (0.25 microsec) = 16,383 microsecs
;                             = 16.383 milliseconds
IntServ:
; Clear the timer interrupt flag so that count continues
        bcf              PIR1,0       ; Clear TIMER1 interrupt flag
; Save context
        movwf    old_w        ; Save w register
        swapf    STATUS,w     ; STATUS to w
        movwf    old_STATUS   ; Save STATUS
; At this point code must determine which (if any) of
; the three possible tasks is in a READY or RUNNING
; State. Task states are mapped to the three low-order
; bits of the taskState variable:
; taskState  7  6  5  4  3  2  1  0
;                             |  |  |___ task 3 (green)
;                             |  |_____ task 2 (red)
;                             |_____ task 1 (yellow)
        btfsc    taskState,2        ; Test task 3
        goto     task1              ; Task 1 routine
testTask2:
        btfsc    taskState,1        ; Task 2
        goto     task2
testTask3:
        btfsc    taskState,0        ; Task 3
        goto     task3
; At this point all tasks are in a NOT READY or
; NOT RUNNING state
        goto     exitISR                       ; Exit now
;========================
;        task handlers
;========================
;    line       color       task          timer delay       status
;    RA0        green       Task 1        delay1            status1
;    RA1        red         Task 2        delay2            status2
;    RA2        yellow      Task 3        delay3            status3
```

```
; The values in the delayx registers are in the range
; 0 to 9. A delay of 0 means that the LED flashes at
; the clock rate. A delay of 1 means that the LED
; flashes every other clock beat (skip 1). A delay of
; 5 that the LED skips 5 clock beats before turning on
; or off.
; The statusx registers control the number of beats
; currently skipped. Thus if delay2 = 5 then status2
; will hold values 0 to 5. When delayx = statusx an
; on/off operation is performed and statusx is reset.
;==========================
;     task1 = yellow LED
;==========================
; Related registers: delay 1 and status 1
task1:
     incf      status1,f    ; Next status
     movf      delay1,w     ; Delay to w
     subwf     status1,w    ; Compare
     btfss     STATUS,Z     ; Test Z flag
     goto      exitT1       ; Exit handler if not zero
; At this point the statusx register has reached the
; maximum count for the given delay.
; Clear statusx register
     clrf      status1
; Set/reset portA line 0 by xoring a mask with a one-bit
     movlw     b'00000001'  ; Xoring with a 1-bit produces
                            ; the complement
     xorwf     PORTA,f      ; Complement bit 2, port A
exitT1:
     goto      testTask2
;==========================
;     task 2 = red LED
;==========================
; Related registers: delay 2 and status 2
task2:
     incf      status2,f    ; Next status
     movf      delay2,w     ; Delay to w
     subwf     status2,w    ; Compare
     btfss     STATUS,Z     ; Test Z flag
     goto      exitT2       ; Exit handler if not zero
; At this point the statusx register has reached the
; maximum count for the given delay.
; Clear statusx register
     clrf      status2
; Set/reset portA line 0 by xoring a mask with a one-bit
     movlw     b'00000010'  ; Xoring with a 1-bit produces
                                    ; the complement
     xorwf     PORTA,f                ; Complement bit 2, port A
exitT2:
     goto      testTask3
;==========================
;     task 3 = green LED
;==========================
; Related registers: delay 3 and status 3
task3:
     incf      status3,f    ; Next status
     movf      delay3,w     ; Delay to w
     subwf     status3,w    ; Compare
     btfss     STATUS,Z     ; Test Z flag
     goto      exitT3       ; Exit handler if not zero
```

```
; At this point the statusx register has reached the
; maximum count for the given delay.
; Clear statusx register
    clrf     status3
; Set/reset portA line 0 by xoring a mask with a one-bit
    movlw    b'00000100'  ; Xoring with a 1-bit produces
                          ; the complement
    xorwf    PORTA,f      ; Complement bit 2, port A
exitT3:
    goto     exitISR              ; Done!
;=========================
;        exit ISR
;=========================
exitISR:
; Restore context
    swapf    old_STATUS,w ; Saved STATUS to w
    movfw    STATUS       ; To STATUS register
    swapf    old_w,f      ; Swap file register in itself
    swapf    old_w,w      ; re-swap back to w
; Reset,interrupt
    retfie


;========================================================
;          timer1 interrupt related procedures
;========================================================
timer1On:
; Setup timer1 high register for 32,768 iterations
; by setting the register's high-order bit.
; Note: timer1 was initialized during a previous call
;       to the timer1Off procedure.
    clrf     TMR1H ; Clear register
    bsf      TMR1H,7      ; Set bit 7
    clrf     TMR1L ; Counter low byte
; Enable interrupts
    bsf      INTCON,7     ; Enable global interrupts
    bsf      INTCON,6     ; And peripherals
    Bank1
    bsf      PIE1,0       ; Timer1 overflow interrupt on
; Start timer 1
    Bank0
    bsf      T1CON,0              ; TMR1ON bit
    return
;============================
;    turn off timer1
;============================
timer1Off:
    bcf      INTCON,7     ; Global interrupts off
    bcf      INTCON,6     ; Disable peripheral interrupts
    Bank1
    bcf      PIE1,0       ; Disable TIMER1 interrupt
    Bank0
    bcf      PIR1,0       ; Clear TIMER1 overflow flag
    movlw    B'00001000'  ; Setup TIMER1 control register
;   movlw    B'00001010'  ; Setup TIMER1 control register
    movwf    T1CON    ; lableled T1CON
; Bit map:
;   0  0  0  0  1  0  1  0
;   7  6  5  4  3  2  1  0  <== T1CON register
;   |  |  |  |  |  |  |  |
;   |  |  |  |  |  |  |  |___ TMR1ON
```

```
;    |   |   |   |   |   |   |        1 = enable timer1
;    |   |   |   |   |   |   |       *0 = stop timer1
;    |   |   |   |   |   |   |___ TMR1CS
;    |   |   |   |   |   |          *1 = external clock from
;    |   |   |   |   |   |               T1OSO/T1OS1 on rising edge
;    |   |   |   |   |   |           0 = internal clock
;    |   |   |   |   |   |___ T1SYNC
;    |   |   |   |   |           When TMR1CS = 1
;    |   |   |   |   |               1 = do not synchronize clock
;    |   |   |   |   |              *0 = synchronize clock
;    |   |   |   |   |           Bit ignored when TMR1CS = 0
;    |   |   |   |   |___ T1OSCEN
;    |   |   |   |          *1 = oscillator enabled
;    |   |   |   |           0 = oscillator off (no power drain)
;
;    |   |   |__|_____ T1CKPS1:T1CKPS0 (prescale values)
;    |   |            11 = 1:8        10 = 1:4
;    |   |            01 = 1:2       *00 = 1:1
;    |__|_____ UNIMPLEMENTED (read as 0)
;
     return


;=========================================================
;                    code resident tables
;=========================================================
; The tables are attached following the routines that use
; them. Also, to avoid crossing page boundaries each
; routine is orged at increments of 0x40 (64 bytes).
     org      0x400
;===============================
;  first text string procedure
;===============================
storeStatus:
; Procedure to store in PIC RAM buffer the message
; contained in the code area labeled msg1
; ON ENTRY:
;        variable pic_ad holds address of text buffer
;        in PIC RAM
;        w register hold offset into storage area
;        msg1 is routine that returns the string characters
;        an a zero terminator
; ON EXIT:
;        Text message stored in buffer
;
; Store base address of text buffer in FSR
     movf     pic_ad,w     ; first display RAM address to W
     movwf    FSR               ; W to FSR
; Initialize index for text string access
     clrf     index
get_msg_char:
     movlw    HIGH msg1     ; High 8-bit address
     movwf    PCLATH        ; To program counter high
     movf     index,w             ; Load character offset into W
     call     msg1          ; Get character from table
; Test for zero terminator
     andlw    0x0ff
     btfsc    STATUS,Z      ; Test zero flag
     goto     endstr1             ; End of string
; ASSERT: valid string character in w
;        store character in text buffer (by FSR)
```

```
    movwf   INDF         ; store in buffer by FSR
    incf    FSR,f        ; increment buffer pointer
; Restore table character counter from variable
    incf    index,f             ; Bump offset counter
    goto    get_msg_char ; Continue
endstr1:
    return
;         0                         15
;         | | | | | | | | | | | | | | | |
;         T a s k :    1    2    3          <== line 1
msg1:
    addwf   PCL,f        ; Access table
    retlw   'T'
    retlw   'a'
    retlw   's'
    retlw   'k'
    retlw   ':'
    retlw   ' '
    retlw   ' '
    retlw   '1'
    retlw   ' '
    retlw   ' '
    retlw   '2'
    retlw   ' '
    retlw   ' '
    retlw   '3'
    retlw   ' '
    retlw   ' '
    retlw   0


;=================================
;   second text string procedure
;=================================
; Procedure to to store paramenters string and ASCII digits
; in second LCD line
; ON ENTRY:
; ASCII digits are stored in variables cycle1, cycle2,
; dur1, and dur2. Digits are copied to display area
; after message
    org     0x440
storeParams:
; Store base address of text buffer in FSR
    movf    pic_ad,w     ; first display RAM address to W
    movwf   FSR                  ; W to FSR
; Initialize index for text string access
    clrf    index
get_msg_char1:
    movlw   HIGH msg2    ; High 8-bit address
    movwf   PCLATH       ; To program counter high
    movf    index,w              ; Load character offset into W
    call    msg2         ; Get character from table
; Test for zero terminator
    andlw   0x0ff
    btfsc   STATUS,Z     ; Test zero flag
    goto    endstr2              ; End of string
; ASSERT: valid string character in w
;         store character in text buffer (by FSR)
    movwf   INDF         ; store in buffer by FSR
    incf    FSR,f        ; increment buffer pointer
; Restore table character counter from variable
```

```
        incf      index,f              ; Bump offset counter
        goto      get_msg_char1        ; Continue
; Digits are now moved into text buffer located at
; offset 0x20, as follws:
;     cycle1 digit at 0x27
;     cycle2       at 0x2a
;     cycle3       at 0x2d
;         0                           15
;         | | | | | | | | | | | | | | | |
;         T a s k :    1    2    3          <== line 1
;         D e l a y :  1    5    9          <== line 2
;                      |    |    |
;                      7    10   13 <== digit offset
endstr2:
        movf      cycle1,w    ; First cycle digit
        movwf     0x27        ; Store in variable
        movf      cycle2,w
        movwf     0x2a
        movf      cycle3,w
        movwf     0x2d
        return
;         0                           15
;         | | | | | | | | | | | | | | | |
;         D e l a y :  1    5    9          <== line 2
msg2:
        addwf     PCL,f       ; Access table
        retlw     'D'
        retlw     'e'
        retlw     'l'
        retlw     'a'
        retlw     'y'
        retlw     ':'
        retlw     ' '
        retlw     'x'
        retlw     ' '
        retlw     ' '
        retlw     'x'
        retlw     ' '
        retlw     ' '
        retlw     'x'
        retlw     ' '
        retlw     ' '
        retlw     0


; End of program

        end
```

Appendix A

MPLAB C18 Language Tutorial

A.1 In This Appendix

This appendix is an overview of the general C language implemented in the MPLAB C18 Compiler. It is intended for readers who are unfamiliar with conventional C programming and provides a basic tutorial on the language fundamentals. The hardware-specific features of C18 as well as the functions in the C18 peripherals libraries are covered in Chapter 6.

The language elements discussed in this appendix are those of interest to the PIC 18F452 programmer. Language features that apply to conventional computer environments, such as video display, keyboard input, and printer controls, are not part of the contents of this appendix even if some of these features are supported by MPLAB C18. The reader interested in learning standard C language programming should refer to one of the many books on the subject.

The installation and testing of the MPLAB C18 Compiler is discussed in Chapter 5. The authors assume that at this point the reader has already installed the MPLAB IDE and the C18 compiler. The source files for the sample programs developed in this appendix can be found in the book's online package in the directory path named SampleCode/AppendixA.

Because the discussion in this appendix takes place in the context of microcontroller hardware, the reader should have covered the material up to and including book Chapter 5.

A.1.1 About Programming

In the present context, a program can be defined as a sequential set of instructions designed to perform a specific task. The set of instructions that must be followed to start up a particular model of automobile could be described as the start-up program for that vehicle. On the other hand, a computer program is a set of logical instructions that makes the computer perform a specific function.

For example, you can write a computer program to calculate the interest that accrues when you invest a given amount of money, at a certain interest rate, for a specific period of time. Another program could be used to tell a robot when it is time to recharge its batteries. Yet another one could help a physician diagnose a childhood disease by examining the patient's symptoms. In all these cases, the program consists of a set of instructions that perform conditional tests, follow a predictable path, and reach an also predictable result. A set of haphazard instructions that leads to no predictable end is not considered a program.

A.1.2 Communicating with an Alien Intelligence

It can be said that a computer program is a way of communicating with an alien intelligence. A computer is a machine built of metal, silicon, and other composite materials: it has no knowledge of the world and no common sense. In this sense a computer is nothing more than a tin can. If 50 years ago someone had found you attempting to communicate and give orders to a tin can, you would have probably been committed to a mental institution. Today we go unnoticed as we attempt to communicate with a hardware device.

Our main challenge is that the tin can never "knows what you mean." A human intelligence has accumulated considerable knowledge of the world and of society at large. The set of instructions for a human to get us a can of soda from a vending machine can be rather simple:

"Joe, here is fifty cents, would you please get me a Pepsi?"

Joe, who has knowledge of the world, understands that he must walk out of the room, open the necessary doors and walk up and down stairs, reach the vending machine, wait in line if someone if using it, then place the coins in the adequate slot, punch the Pepsi button, retrieve the can of soda, and bring it back to me, again opening doors and walking up and down stairs as necessary. Joe has knowledge of doors, of stairs, of money, of waiting in line, of vending machines, and of thousands of other worldly things and social conventions that are necessary to perform this simple chore.

The machine, on the other hand, has no previous knowledge, does not understand social conventions, and has no experience with doors, stairs, people standing in line, or vending machine operation. If we forget to tell the robot to open the door, it will crash through and leave a hole shaped like its outline. If we forget to tell it to wait in line if someone else is using the vending machine, then the robot may just walk over the current customer in its effort to put the coins in the slot. The tin can has no experience, no social manners, and no common sense. Giving instructions to a machine is different and much more complicated than giving instructions to an intelligent being.

This is what computer programming is about. It is sometimes considered difficult to learn, not so much because it is complicated, but because it is something to which we are not accustomed. Learning programming requires learning the grammar and syntax of a programming language, but, perhaps more importantly, it requires learning to communicate with and issue commands to a tin can. A task indeed!

A.1.3 Flowcharting

Computer scientists have come up with tools and techniques to help us develop programs. One of the simplest and most useful of these tools is the flowchart. A flowchart, like the word implies, is a graphical representation of the flow of a program. In other words, a flowchart is a graph of the tests, options, and actions that a program must perform in order to achieve a specific logical task.

Flowcharting is useful because computer programs must leave no loose ends and presume no reasonable behavior. You cannot tell a computer, "Well...you know what I mean!" or assume that a certain operation is so obvious that it need not be explicitly stated. The programmer uses a flowchart to make certain that each processing step is clearly defined and that the operations are performed in the required sequence.

Flowcharts use symbols and special annotations to describe the specific steps in program flow. The most common ones are shown in Figure A.1.

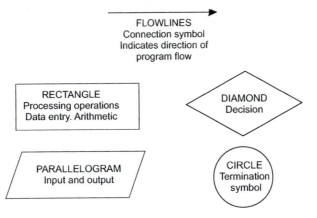

Figure A.1 *Flowcharting symbols.*

Suppose you needed to develop a program to determine when a domestic robot needs to recharge its own batteries. Assume that the robot contains a meter that measures the percent of full charge in its batteries, as well as a clock that indicates the time of day. The program is to be based on the following rules:

1. The robot should recharge itself after 5:00 PM.

2. The robot should not recharge itself if the batteries are more than 80% full.

The logic for recharging the robot batteries requires first reading the internal clock to determine if it is after 5:00 PM. If so, then it will read the robot's battery meter to determine if the batteries are less than 80% full. If both tests are true, then the robot is instructed to plug itself into a wall outlet and recharge. If not, it is instructed to continue working. The logic can be expressed in a flowchart, as shown in Figure A.2.

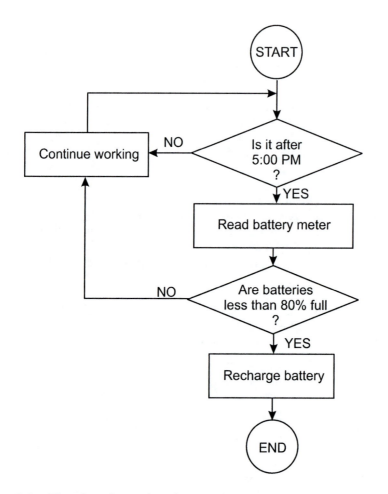

Figure A.2 *Flowchart for recharging a robot battery.*

Note in the flowchart of Figure A.2 that the diamond symbols represent program decisions. These decisions are based on elementary logic, which requires that there must be two, but not more than two choices. The possible answers are labeled YES and NO in the flowchart. Decisions are the crucial points in the logic. A program that requires no decisions is probably based on such simple logic that a flowchart would be unnecessary. For instance, a program that consists of several processing steps that are always performed in the same sequence does not require a flowchart.

The logic in computer programs often becomes complicated, or contains subtle points that can be misinterpreted or overlooked. Even simple programming problems usually benefit from a flowchart. The logic flowcharted in Figure A.2 is based on recharging the batteries if it is after 5:00 PM "and" if the battery meter reads less than 80%. In this case both conditions must be true for the action of recharging the battery to take place. An alternative set of rules could state that the robot must recharge itself if it is after 5:00 PM "or" if the battery is less than 80% full. In this case. either condition determines that the robot recharges. The flowchart to represent this logic must be modified, as shown in Figure A.3.

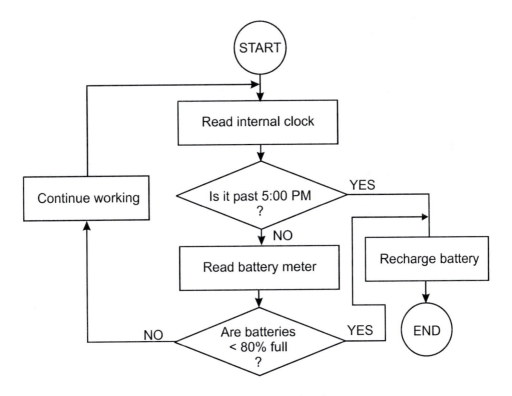

Figure A.3 *Alternative logic for recharging a robot battery.*

Now suppose that there are critical functions that you do not want the domestic robot to interrupt, even if it is after 5:00 PM or if the battery is less than 80% charged. For example, if the robot is walking the dog, you may not want it to let go of the leash and go plug itself into the wall outlet. In this case, you would have to modify the flowchart and insert an additional test so that recharging does not take place if the robot is currently performing a critical activity. Furthermore, you may decide that a very low battery could damage the machine; therefore, if the meter shows less than 20% full charge, the robot should plug itself into the outlet, no matter what. Here again, the program logic and the flowchart would have to be modified to express the new set of conditions that determines which program action takes place. It is easy to see how program logic can easily get complicated and why flowcharts and other logic analysis tools are so important to the programmer.

A.1.4 C Language Rules

The C language follows a few simple rules of syntax:

- Upper- and lower-case letters are different symbols in C. When typing C code, you must be careful to follow the required capitalization, for example, Main and main refer to different program elements.

- In general, C ignores white space. White space characters are those that do not appear on the screen, such as blank spaces, tabs, and line-end codes. C programmers use white space to make the code more pleasant and readable but these characters are ignored by the compiler..

- C uses braces {} as grouping symbols. They mark the beginning and the end of a program section. A C program must have an equal number of left and right braces. The part of a C program located between braces is called a block.

- Every C statement ends in the ; symbol. A statement is a program element that generates a processing action. Not every C expression is a statement.

Comments

Computer programs usually contain text that clarifies or explains the code. These are called comments. Comments must be preceded by a special symbol so that the text is ignored by the compiler. In C there are two ways of inserting comments into your code:

- The // symbol creates a comment that extends to the end of the line, for example,

```
// This is a single-line comment
```

The // symbol can appear anywhere in a program line.

- The /* and */ symbols are used to delimit a comment that can span over more than one line, for example:

```
/* This is a
   multiple line
   comment  */
```

Some programmers object to multiple line comments because comments that span several lines force the reader to look up or down in the text for the presence of comment symbols. These programmers consider that a much clearer style is to use the double slash symbol to mark each comment line. In this book we have used multiple-line comments in text blocks and labels. For all comments inside code, we have used the single-line comment (//) symbol.

Program Header

Programs often begin with several commented lines, sometimes called the program header, that contain general information about the code that follows. The following elements are usually found in the program header:

- The program name

- The name of the author or authors

- A copyright notice, if one is appropriate

- The date of program creation and modifications

- A description of the program's purpose and the functions that it performs

- A history of the program changes and updates

- A listing of the tools used in developing the program

- A description of the software and hardware environment required to run the program

Programmers usually create their own program headers, which they paste into all their sources.

Programming Templates

Programming templates contain not only the program header, but also a skeleton that compiles correctly in the target environment. The following is a programming template for C language applications compatible with the PIC 18F452 and the MPLAB C18 compiler in the MPLAB IDE.

```
/* Project name:
   Source files:
   Date:
   Copyright 201x by
   Processor: PIC 18F452
   Environment:    MPLAB IDE Version 8.86
                        MPLAB C-18 Compiler

   TEST CIRCUIT:

   PROGRAM DESCRIPTION:

*/

#include <p18f452.h>

// DATA VARIABLES

#pragma config WDT = OFF
#pragma config OSC = HS
#pragma config LVP = OFF
#pragma config DEBUG = OFF

// Function prototypes


/* =======================================================
                       main program
   ======================================================= */
void main(void)
{
    // Initalize direction registers
    Return;
}
```

To save space, we often eliminate or compress the header from the source listings in this book.

A.2 Structure of a C Program

A C language source file consists of a series of symbolic and explicit program elements (sometimes called expressions) that can be interpreted by the compiler. Some of these elements would be undecipherable to a person unfamiliar with the C programming language. Other elements of the program are quite understandable even to someone totally unfamiliar with C. In the following sections we present several MPLAB C18 programs of increasing complexity. The programs are then analyzed in detail. However, this analysis is limited mostly to the program's language features. The devices and hardware are covered elsewhere in the book.

A.2.1 Sample Program C_LEDs_ON

The following C language source fragment from the program C_LEDs_ON.c compiles into an executable that turns on the odd numbered LEDs wired to Port C.

```
/**************************************************************
                    main program
**************************************************************/
void main (void)
{
    // Initalize direction registers
    TRISC = 0;
    PORTC = 0;
    // Endless loop to turn on four LEDs
    while (1)
    PORTC = 0b10101010;
    ;
}
```

In the following paragraphs we attempt to dissect the above program and analyze its component elements. This walk-through is intended to provide a preliminary knowledge of C. Most of these concepts will be explained later in the book.

The lines

```
/**************************************************************
                    main program
**************************************************************/
```

are multiple-line comments. The symbols /* are used in C to mark the start of a text area that is ignored by the compiler. The symbols */ indicate the end of the comment. Comments are used to document or explain program operation. The actual contents of a comment are ignored by the program.

The program lines

```
    // Initalize direction registers
```

and

```
    // Endless loop to turn on four LEDs
```

are single-line comments.

Identifiers

An identifier is a name or phrase in a C language program the words TRISC and PORTC are identifiers. A C language identifier consists of one or more letters, numbers, or the underscore symbol. Upper and lower case letters are considered as different symbols by C language. For example, the names PORTC and PortC are different identifiers. An identifier cannot start with a digit. For example, the identifier 1ABC is not legal in C.

The ANSI C standard requires that the compiler recognize at least the first 31 characters in an identifier. The following are legal identifiers:

```
personal_name
PORTA
PI
y_121
XY_value_128
```

In the sample program C_LEDs_ON listed previously, the identifiers TRISC and PORTC refer to hardware registers in the 18F452 microcontroller. TRISC is the direction control register for Port C, and PORTC is the register itself. These registers are defined in the include file p18f452.h, which is referenced in the following #include statement:

```
#include <p18f452.h>
```

The program line

```
TRISC = 0;
```

defines Port C for output, and the program line

```
PORTC = 0;
```

Initializes all Port C lines to zero.

Reserved Words

A C identifier cannot be a word used for other purposes by the language. These special words, called reserved words, are listed in Table A.1.

Table A.1
C Language Reserved Words

asm	default	float	register	switch
auto	do	for	return	typedef
break	double	fortran	short	union
case	else	goto	signed	unsigned
char	entry	if	sizeof	void
const	enum	int	static	volatile
continue	extern	long	struct	while

main() Function

The program lines

```
void main(void)
{
```

indicate a C language function named main(). A function is a unit of program execution. Every C language program must contain one function named main(). All sample programs in this book contain a main() function. Program execution always begins at the main function. The programmer can create other functions and give them names following the guidelines for C language identifiers mentioned above. If a program has more than one function, execution will always start at the one named main(), regardless of where it is located. The main function starts at the opening roster ({) and ends at the closing roster (}). The function body consists of program lines between these rosters.

The data type, in this case void, that precedes the function name indicates the type of the value returned by the function. In the case of main(), the type void indicates that the function returns no value. The data types or types enclosed within the function's parenthesis define the arguments that are passed to the function. In the case of main(), the term void indicates that the function receives no parameters as arguments.

A.2.2 Sample Program C_LEDs_Flash

The program C_LEDs_Flash, in the book's software resource, is a demonstration to flash red and green LEDs connected to Port C pins. The code in the source file is as follows:

```
// INCLUDED CODE
#include <p18f452.h>

// DATA VARIABLES AND CONSTANTS
unsigned int count;
#define MAX_COUNT 16000

#pragma config WDT = OFF
#pragma config OSC = HS
#pragma config LVP = OFF
#pragma config DEBUG = OFF

// Function prototypes
void FlashLEDs(void);

/*************************************************************
                     main program
*************************************************************/
void main(void)
{
    /* Initalize direction registers */
    TRISC = 0;
    PORTC = 0;
    /* Endless loop to flash LEDs */
    while (1)
    FlashLEDs()
    ;
}

/*************************************************************
                     local functions
*************************************************************/
void FlashLEDs()
{
    count = 0;
    PORTC = 0x0f;
```

```
            while (count <= MAX_COUNT)
            {
                    count++;
            }
    count = 0;
    PORTC = 0xf0;
            while (count <= MAX_COUNT)
            {
                    count++;
            }
    return;
}
```

Expressions and Statements

Expressions are the building blocks of high level computer languages. An expression is a character or a sequence of characters (which can include one or more operators) that result in a unique value or program action. An expression can be composed of numeric or alphanumeric characters. The following are valid C language expressions:

```
x_variable
3.1415
goto LABEL
Larray[]
"Legal Department"
4
z = (x * y)/2
```

This leads us to conclude that a variable, a number, a string of characters, a program control keyword, a function call, and several numbers or variables connected by operators are all considered expressions.

A statement is the fundamental organizational element of a computer language. In C language, a statement consists of one or more expressions. All statements must end in the semicolon (;) symbol, which is called the statement terminator. In fact, in C language, the semicolon symbol converts an expression into a statement. The following valid C language statements were derived from the expressions listed above:

```
x_variable =   4;
goto LABEL;
char Larray[] = "Legal department";
z = (x * y)/2;
```

Note in the programs listed earlier in this appendix that not all program lines constitute statements. For example, the lines containing the #define directive and the main() function are not C language statements and therefore do not end in a semicolon. Also, note that braces are not followed by the semicolon.

Variables

The program line

```
unsigned int count;
```

is a variable declaration. In computer terminology, a variable is a memory structure for holding program data. In C language, all variables must meet the following requirements:

1. A variable is assigned a variable name, which must meet the requirements for identifiers listed previously in this chapter.

2. A variable must be of a certain variable type.

3. A variable must be declared before it is used in a C language program. The variable declaration must include the variable name and the assigned data type. The declaration can optionally initialize the variable to a value.

The variable type defines the category of information stored in the variable. In this sense, a C variable can be an integer, a character, a floating point number, or a character string, among others. Table A.2 lists the integer data types supported by MPLAB C18.

Table A.2
MPLAB C18 Integer Data Types

Type	Size	Minimum	Maximum
char[1,2]	8 bits	−128	127
signed char	8 bits	−128	127
unsigned char	8 bits	0	255
int	16 bits	−32,768	32,767
unsigned int	16 bits	0	65,535
short	16 bits	−32,768	32,767
unsigned short	16 bits	0	65,535
short long	24 bits	−8,388,608	8,388,607
unsigned short long	24 bits	0	16,777,215
long	32 bits	−2,147,483,648	2,147,483,647
unsigned long	32 bits	0	4,294,967,295

Note 1: A plain *char* is signed by default.

2: A plain *char* may be unsigned by default via the -k command-line option.

MPLAB C18 also supports two floating point. These are shown in Table A.3

Table A.3
MPLAB C18 Floating Point Data Types

Type	Size	Minimum Exponent	Maximum Exponent	Minimum Normalized	Maximum Normalized
float	32 bits	−126	128	$2^{-126} \approx 1.17549435e-38$	$2^{128} * (2-2^{-15}) \approx 6.80564693e+38$
double	32 bits	−126	128	$2^{-126} \approx 1.17549435e-38$	$2^{128} * (2-2^{-15}) \approx 6.80564693e+38$

Scope and Lifetime of a Variable

A variable assumes certain attributes at the time it is declared. The two principal ones are called the variable's scope and lifetime. The term scope refers to the part of a program over which the variable is recognized.

Regarding scope, MPLAB C18 language variables are classified as extern, register, static, auto, typedef, and overlay. Other programming languages use the term global to represent external variables and the term local to represent automatic variables.

Variables of type extern are declared outside all functions, often at the beginning of the program. The scope (also called the visibility) of an external variable starts at the point it is declared and extends to the end of the program. For this reason if an external variable is declared between two functions it will be visible to the second function, but not to the first one, as in this example:

```
void main()
{
    // Cannot see alpha.
    .
    .
{

int alpha;

user_1()
{
    // Can see and access alpha
    .
    .
}
```

In this case, the variable alpha is visible and can be referenced by the user_1() function but not by main().

Variables of type auto (previously named automatic) are declared inside a function body (after the opening brace). The name auto is related to the fact that these variables automatically disappear when the function concludes its execution. For this reason, the scope of automatic variables is limited to the function in which they are declared, as in the following example:

```
void main()
{
        .
        .
{

user_1()
{
    int alpha;
    .
    .
}
```

In this case the variable alpha is visible to the user_1() function but not the main() function.

A variable's lifetime refers to the time span of program execution over which the variable retains its value. The lifetime of automatic variables is the execution time of the function that contains it. In other words, automatic variables are created when the function in which they are located executes, and disappear when the function concludes. External variables, on the other hand, have a lifetime that extends through the course of program execution. This means that an external variable retains its value at all times.

The static keyword is used in C language to modify the scope and lifetime of variables. For instance, a static variable declared inside a function has the same scope as an automatic variable and the same lifetime as an external variable. For example,

```
void main()
{
      .
      .
{

user_1()
{
   static int alpha;
      .
      .
}
```

In this case, the internal static variable alpha is visible only inside the function user_1() but the value of alpha is preserved after the function concludes. In this manner, when execution returns to the user_1() function, the variable alpha will still hold its previous value.

Constants

Using mathematical terminology we often classify computer data as variables and constants. In C language, variables are assigned names and stored in a memory structure determined by the variable type, as previously discussed. The contents of a variable can be changed anywhere in the program. For this reason, a variable can be visualized as a labeled container, defined by the programmer, for storing a data object. Constants, on the other hand, represent values that do not change in the course of program execution. In the program C_LEDs_Flash.c, the constant MAX_COUNT is defined in the following line:

```
#define MAX_COUNT 16000
```

In this case, the C language #define directive is used to give name and value to a constant that is used later by the program. Notice that because #define is a directive, the program line does not constitute a statement and does not end in the ; symbol. Also that the = sign is not used in assigning a value to the constant.

Although the C language allows entering values directly in statements, the use of undefined constants is generally considered a bad programming practice, because hard-coded constants, introduced unexpectedly, decrease the readability of the code and make coding errors more likely.

Local Functions

Functions can be used locally as subprograms. This allows organizing and simplifying code and provides a simple mechanism for reusing routines. Any function in a C program that is not main() is defined as a local function.

In all functions, the C language requires that the function name be followed by a left parenthesis symbol. There cannot be a character or symbol between the last character of the function name and the left parenthesis. The parentheses following the function name are used to enclose data optionally passed to the function. The data items enclosed between the parentheses symbols are called the function arguments. Inside the function, these arguments are referred to as parameters.

Some functions have no arguments, as is the case with the main(void) function in the sample programs listed previously. The word void enclosed in the function's parenthesis indicates that the function receives no parameters as arguments. Alternatively, the term void can be assumed and the function can be coded as

```
void main()
```

The function FlashLEDs() in the sample program C_LEDs_Flash.c listed previously is coded as follows:.

```
void FlashLEDs()
{
    count = 0;
    PORTC = 0x0f;
        while (count <= MAX_COUNT)
        {
            count++;
        }
    count = 0;
    PORTC = 0xf0;
        while (count <= MAX_COUNT)
        {
            count++;
        }
    return;
}
```

Functions are discussed in detail later in this appendix. They are introduced intuitively at this point as a program subroutine. The FlashLEDs() function uses the variable count and the constant MAX_COUNT to introduce two delays. In the first delay the red LEDs on Port C are turned on and the green LEDs are turned off. During the second delay, the red LEDs are off and the green LEDs are on. The while statement that keeps the LEDs turned on and off is discussed later in this appendix. Notice that, like main(), the function FlashLEDs() returns no value and receives no arguments by the caller.

A.2.3 Coding Style

C language is flexible in the use of white space (tabs, spaces, and other inactive characters) and in the positioning of separators and delimiters so that the programmer can format the code according to personal preference. For example, in the programs listed in this appendix, we have placed the left brace delimiter on a separate line from the function call and have used tabs to indent program lines, as follows:

```
void main(void)
{
    // Initialize direction registers
    TRISC = 0;
    PORTC = 0;
    // Endless loop to flash LEDs
    while (1)
    FlashLEDs()
    ;
}
```

Other programmers prefer to type the left brace delimiter in the same line as the main function and maintain the program lines flush with the left margin, in this manner:

```
void main(void) {
// Initialize direction registers
TRISC = 0;
PORTC = 0;
// Endless loop to flash LEDs
while (1)
FlashLEDs()
;
}
```

Another reason for variations in the use of white space is that some programmers use spaces, tabs, and comment lines to embellish the source; enhance the appearance; or to mark the beginning of functions, routines, or other locations in the program. For example,

```
/****************************************************************
                    local functions
****************************************************************/
```

These comment banners are useful when trying to find a program function or area during editing operations, especially in large programs. On the other hand, some programmers prefer a more sober style and avoid any unnecessary text in their code. Most personal coding styles are acceptable, as long as they do not compromise the readability of the code or the basic principles of clear, understandable programs.

A.3 C Language Data

One of the main functions and a major advantage of high-level programming languages, such as C language, is that they simplify the classification, manipulation, and storage of computer data. Note that the word information is often used as a synonym for the word data, although, strictly speaking, the term data refers to the raw facts and figures and the term information to these facts after they have been processed and refined.

Data is a generic designation that applies to many types of objects. Computer scientists often speak of scalar data types to refer to those that encode individual objects and to structured data types when referring to those that encode a collection of individual objects. But, more commonly, we speak of numeric, alphanumeric, date, and logical data types. Numeric data is that with which we can perform mathematical operations, such as basic arithmetic, exponentiation, or the calculation of transcendental functions. Alphanumeric data are letters and numbers considered as language symbols. The phrase "I am a 1st class dumb computer" is a collection of alphanumeric characters usually described as a string.

Programming languages provide a means for storing and manipulating data objects of several types. Once a data item is assigned to a particular type, its processing will be performed according to the rules for that type. For example, data objects assigned to alphanumeric types cannot be manipulated arithmetically, while numeric data cannot be separated into its individual symbols. The programmer must be careful to assign each data object to the data type that corresponds with the object's nature, rather than with its appearance. A telephone number, for instance, is typically considered an alphanumeric data type, as we usually have no need for performing arithmetic operations on the digits of a telephone number. In C, all data objects are stored in the computer's memory as an integral number of byte size units. In conventional computers, each byte is made up of eight individual cells, called bits.

A.3.1 Numeric Data

Numeric data is that with which we can perform mathematical operations. In other words, numeric data consists of number symbols used to represent quantities. This definition excludes the use of numbers as designators, for example, a telephone or a social security number. It is difficult to conceive a need for performing arithmetic operations on telephone or social security numbers.

We have mentioned that numeric data can appear in the form of variables or constants. C language numeric constants can be predefined by means of the #define directive or they can be entered explicitly in the operations. Numeric variables are classified into two data types, integral types, and floating point types. The floating point types are sometimes called reals.

Each data type corresponds to a specific category of numbers; for example, the integer data type allows representing whole numbers while the real data type allows representing fractional numbers. For each data type C provides several type-specifiers that further determine the characteristic and range of representable values. The numeric data types supported by MPLAB C18 are shown in Table A.2 and Table A.3.

A.3.2 Alphanumeric Data

Alphanumeric data refers to items that serve as textual designators and that are not the subject of conventional arithmetic. For example, the phrase "Enter radius: " is an alphanumeric text string. By the same token, the letters of the alphabet and other non-numeric symbols are often used as designators. We have also mentioned that the programmer can designate one or more numeric symbols as alphanumeric data if these are used for nonmathematical purposes. Such is often the case with telephone numbers, numbers used in street addresses, zip codes, social security numbers, and others. In C, alphanumeric data belongs to the integral data type and requires the char specifier. This means that data items defined using the char specifier are considered a number or a character, depending on how the item is defined in the program. For example, the variable declaration

```
char num_var = 65;
```

defines a numeric variable named num_var and initializes it to a value of 65. On the other hand, the declaration

```
char alpha_var = 'A';
```

defines an alphanumeric variable named alpha_var and initializes it to the alphanumeric character "A." Note that in this case the single quote symbols are used to enclose the alphanumeric character. A string is a contiguous set of alphanumeric characters. In practice, the set can consist of a single character. C strings can be defined as constants or as variables. A string constant can be created with the #define statement, for example,

```
#define USA "United States of America"
```

This directive equates the name USA with the string in quotes. At compile time, the name USA is substituted with the defined string.

A.3.3 Arrays of Alphanumeric Data

Most programming languages, C included, allow the grouping of several data items of the same type into an individualized data structure called an array. In C, a string is defined as an array of char type elements and a string variable (array) is not considered an independent data type but a collection of char type characters.

The most important characteristic that distinguishes an array from other data structures is that in an array, all the elements must belong to the same data type. Other C language constructs (for example, a structure) can contain elements of different data types. These are discussed later in this appendix. For example, in the line

```
char college_addr[] = {"1211 N.W. Bypass"};
```

the string variable named college_addr is defined as being of char type. If the declaration takes place outside a function (typically before main()), the variable is external and therefore visible to all functions in the module as well as to other modules. The following ASCII graph shows the declaration and initialization of an array variable:

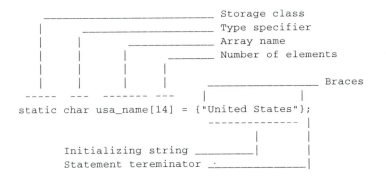

```
        _____ Storage class
    |       _____ Type specifier
    |      |      _____ Array name
    |      |     |      _____ Number of elements
    |      |     |     |
    |      |     |     |
    |      |     |     |              _____ Braces
 _____   ___  _____ ___         |                |
static char usa_name[14] = {"United States"};
                            -------------- |
                                 |         |
    Initializing string _____|         |
    Statement tereminator _____|
```

Note that in the preceding code sample the bracket symbols are used to identify an array. In this case, the number 14, enclosed in the brackets, is a count of the number of letters in the initilization string ("United States") plus one NULL terminator that is automatically added by the compiler at the end of the array. This character count can be omitted because the length of the array is automatically calculated at compile time. Also optional in string arrays are the brace symbols enclosing the initialization string.

A.3.4 Arrays of Numeric Data

An array can also contain numeric data. For example, the array

```
int nums_array1[] = { 50, 60, 700 };
```

contains the following numeric values

```
nums_array[0]  =   50
nums_array[1]  =   60
nums_array[2]  =   700
```

Note that the index number used in addressing the first array element is [0]. Also that note the index can be represented by a program variable, as in the following program to display all the elements of a numerical array:

```
int list1[] = { 10, 20, 30, 4000 };

void main() {
   int count;
   count = list1[0];
   . . .
```

In the previous code fragment, the variable count is assigned the value of the first element of the array list1[], which is 10.

A.4 Indirection

A C language variable can be visualized as a labeled container in the system's memory space. The label associated with the variable is the identifier that was assigned as a name at the time of variable declaration. For example, in the variable declaration

```
float radius = 7;
```

the label for the variable is the variable name "radius."

The contents of a variable is the string or numeric value that it presently holds. For example, after the initialization line of the previous paragraph, the variable named radius will have a content (or value) of 7. While a program is executing, the contents of variables can change due to user input or as the result of processing operations. For example, if the variable num1 has a contents of 4, the statement

```
num1 = num1 * 2;
```

will change its contents to 8.

The third element of a C language variable is its address, which is a representation of the variable's location in the system memory space. It is interesting to note that while the programmer assigns the variable its name and can initialize and change its contents, the address of the variable is determined by C language and system software (compiler, linker, and loader).

A.4.1 Storage of C Language Variables

The contents of a memory cell can be interpreted as a numeric value, part of a numeric representation, or as an ASCII encoding of an alphanumeric character. Also, memory cells can be grouped to hold integer or floating point numbers that exceed the range of a single byte (see Table A.2) or as a string of alphanumeric characters in the form of a string constant or of an array. For example, an unsigned integer (range 0 to 65,535) requires 2 bytes of storage. Therefore, the C statements

```
unsigned num1;
char let1;
```

create an unsigned integer variable named num1 and reserve 2 bytes of memory for this variable, while for the char variable let1, one byte of storage is allocated. Later in the program the statements

```
num1 = 44556;
let1 = 'N';
```

determine that the value 44556 is stored in the 2-byte space reserved for the variable named num1 and that the ASCII encoding for the letter N is stored in the 1-byte space reserved for the variable let1.

A.4.2 *Address of* Operator

In C the programmer can obtain the address of a variable by means of the address of operator, which is the & symbol. Because the address of a variable is a constant assigned by the system software, it cannot be changed by the programmer. For this reason, the statement

```
&num1 = 0x8600;
```

is illegal.

A.4.3 Indirection Operator

In C a pointer is a variable that holds the address of another variable. This address is obtained by means of the address of operator (& symbol) mentioned in the previous section. Pointers are special variables, and are managed differently than conventional ones. In the first place, pointer variables are declared using the C indirection operator, the * symbol. For example, the statement

```
char *add1;
```

declares a pointer variable named add1. It is important to note that in the case of pointer variables, the type does not refer to the type of the pointer, but to the type of variable whose address the pointer will store. In other words, after this declaration, we will have created a variable named add1 that can be used as a pointer to any variable of int type. The actual operation of assigning an address to a pointer variable is performed by means of the address of operator, as in the following statement

```
add1 = &num1;
```

Now the pointer variable add1 holds the address of the variable num1.

Figure A.4 is a screen snapshot of the project C_AddOf in this book's software package.

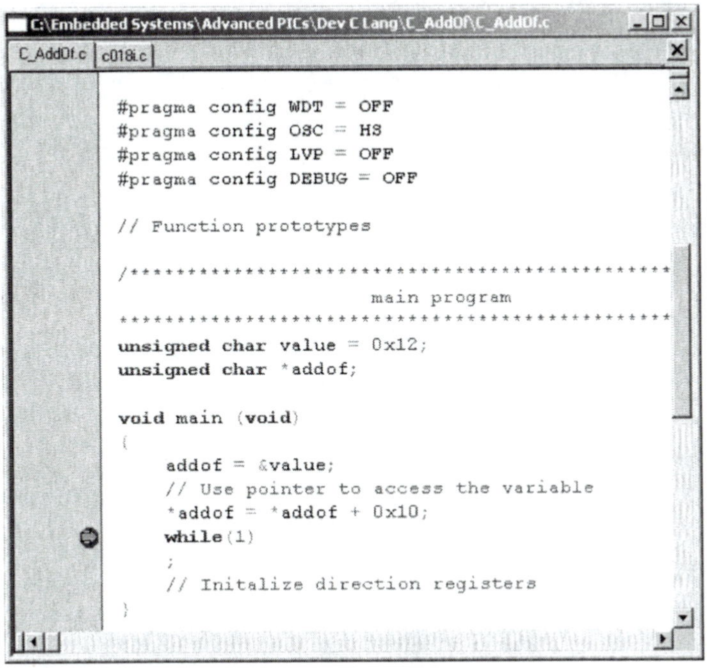

```
#pragma config WDT = OFF
#pragma config OSC = HS
#pragma config LVP = OFF
#pragma config DEBUG = OFF

// Function prototypes

/*************************************************
                    main program
*************************************************
unsigned char value = 0x12;
unsigned char *addof;

void main (void)
{
    addof = &value;
    // Use pointer to access the variable
    *addof = *addof + 0x10;
    while(1)
    ;
    // Initalize direction registers
}
```

Figure A.4 *Screen snapshot of a debug session of the program C_AddOff.c.*

Note that there are two global variable declarations in the program C_AddOf.c. The first variable, named value, of type unsigned char, is initialized to a value of 12. The second variable is a pointer to type unsigned char. In main() the & operator is used to obtain the address of the variable value that is stored in the pointer variable addof. In the example in Figure A.4 a breakpoint is inserted in the while() loop and the program is run to the breakpoint. When the cursor is placed over the pointer variable addof, the address of the variable value is displayed (0x008C). In Figure 2.10 we can see that address 0x0086 is located in Bank 0 of the 18F452 memory space.

Pointer variables would be of limited use if they served only as shorthand for the expression &num1. A powerful feature of C is that it also allows the use of pointers to access, indirectly, the contents of a variable. For example, once the pointer variable addof in Figure A.4 holds the address of the variable value, we can add 0x10 to the contents of value using the following statement

```
*addof = *addof + 0x10;
```

Observe that the * symbol preceding a pointer variable indicates the contents of the variable whose address the pointer holds, value in this case.

A.4.4 Pointers to Array Variables

Assuming that the pointer variable addof holds the address of the variable value, the following statements will perform identical operations:

```
value = value + 0x10;
*addof = *addof + 0x10;
```

In this case there is no advantage to changing the value of a variable by using a pointer to its contents rather than by using the variable name. However, pointer variables become particularly useful in addressing array elements. Consider the array

```
int array1[] = { 10, 20, 30, 4000, 5000 };
```

which creates five integer variables. These variables can be accessed by hard coding the offset in the array brackets or by representing the offset with another variable. In this manner, we could say

```
int num1;
.
.
num1 = array1[3];
```

or

```
count = 3;
num1 = array1[count];
```

In either case, we would assign the value 4000 (stored in the fourth array element) to the variable num1.

   Another way of accessing array elements is by creating a pointer variable to hold the address of the first element of the array. For example, we first declare a pointer variable to int type

```
int *add1;
```

then we can then set the pointer variable to hold the address of the first array element

```
add1 = &array1[0];
```

   As is the case with conventional pointer variables, an array pointer can be used to gain access to the contents of a variable by preceding it with the indirection operator symbol (*). For example, after initializing add1 with the address of array1[0] we can change the value of the first array element by coding

```
*add1 = 100;
```

A.4.5 Pointer Arithmetic

We have seen that it is possible in C language to use pointers to gain access to data elements within an array. But we must proceed carefully, because C follows special rules regarding arithmetic with pointer variables.

   In the previous paragraphs we created a numerical array of type int, defined a pointer variable, and initialized it with the address of the first array element. At this point we were able to access the first array element by initializing the pointer variable to its address. But how do we access the next element of the array by means of this pointer? Because each element of a int data type requires 2 bytes of storage, would we add 2 to the value of the pointer to access the next element in an array of int type? By the same argument, if the array were of type float, would we add 4 to access each successive elements? In reality accessing array elements by means of a pointer is simplified by the fact that C automatically takes into account the size of the array elements. For example, the statements

```
add1++;
```
or
```
add1 = add1 + 1;
```

result in bumping the pointer variable to the next array element, whatever its size. In other words, C language pointer arithmetic is scaled to the size of the elements in the array. The following code fragment indexes through the elements of an array of type float by adding one to the array pointer variable

```
    float array2[] = { 10.1, 2022, 30.44, 4.01 };
void main()
{
    float anum;    float *add2;
    add2 = &array2[0];
    add2 = add2 + 1;        // Retrieves the address of 2022
    anum = *add2;      // anum = 2022
    add2 = add2 + 1;        // Retrieves the address of 30.44
```

```
    anum = *add2;       // anum = 30.44
    add2 = add2 + 1;    // Retrieves the address of 4.01
    anum = *add2;       // anum = 4.01
}
```

A.5 C Language Operators

Operators are the symbols and special characters used in a programming language to change the value of one or more program elements. The program elements that are changed by an operator are called the operand. We use the + symbol to perform the sum of two operands, as in the following example:

```
int val1 = 7;
int val2 = 3;
int val3 = val1 + val2;
```

In this code fragment, the value of integer variable val3 is found by adding the values of variables val1 and val2. Note that the + symbol is also used in appending strings. When used in this manner, it is called the concatenation operator. String concatenation is discussed later in this appendix.

The fundamental C language operators can be functionally classified as follows:

- Simple assignment
- Arithmetic
- Concatenation
- Increment and decrement
- Logical
- Bitwise
- Compound assignment

We first discuss the assignment, arithmetic, concatenation, increment, and decrement operators. Later in this appendix we deal with the logical and bitwise operators.

According to the number of operands C language operators are classified as follows:

- unary
- binary
- ternary

The one C language ternary operator (?:) is discussed in the context of decision constructs later in this appendix.

A.5.1 Operator Action

C language operators are used in expressions that produce program actions. For example, if a, b, and c are variables, the expression

```
c = a + b;
```

uses the = and the + operators to assign to the variable c the value that results from adding the variables a and b. The operators in this expression are the = (assignment) and + (addition) symbols.

C language operators must be used as elements of expressions; they are meaningless when used by themselves. In this sense, the term

```
-a;
```

is a trivial expression that does not change the value of the variable. On the other hand the expression

```
b = -b;
```

assigns a negative value to the variable b.

A.5 2 Assignment Operator

While learning a programming language, it is important to keep in mind that the language expressions are not usually valid mathematically. In the expression

```
a = a + 2;
```

we observe that it is mathematically absurd to state that the value of the variable a is the same as the value that result from adding 2 to the variable a. The reason for this apparent absurdity is that C's = operator does not represent an equality but rather a simple assignment. The result of the statement

```
b = b - 4;
```

is that the variable b is "assigned" the value that results from subtracting 4 from its own value. In other words, b becomes b – 4. It is important to note that this use of the = sign in C is limited to assigning a value to a storage location referenced by a program element. For this reason, a C expression containing the = sign cannot be interpreted algebraically.

There are other differences between a C expression and an algebraic equation. In elementary algebra we learn to solve an equation by isolating a variable on the left-hand side of the = sign, as follows:

```
2x = y
x = y/2
```

However, in C language, the statement line

```
2 * x = y;
```

generates an error. This is due to the fact that programming languages, C included, are not designed to perform even the simplest algebraic manipulations.

If we look at a C language expression that uses the assignment operator, we notice one part to the left of the = sign and another one to the right. In any programming language, the part to the left of the = sign is called the lvalue (short for left value) and the one to the right is called the rvalue (short for right value). Therefore, an lvalue is an expression that can be used to the left of the = sign. In a C assignment statement, the lvalue must represent a single storage location. In other words, the element to the left of the = sign must be a variable. In this manner, if x and y are variables, the expression

```
x = 2 * y;
```

is valid. However, the expression

```
y + 2 = x;
```

is not valid because in this case, the lvalue is not a single storage location but an expression in itself. By the same token, an expression without an rvalue is illegal; for example,

```
y = ;
```

An assignment expression without an lvalue is also illegal, such as

```
= x;
```

A.5.3 Arithmetic Operators

C language arithmetic operators are used to perform simple calculations. Some of C's arithmetic operators coincide with the familiar mathematical symbols. Such is the case with the + and – operators which indicate addition and subtraction. But not all conventional mathematical symbols are available in a computer keyboard. Others would be ambiguous or incompatible with the rules of the language. For example, the conventional symbol for division is not a standard keyboard character. On the other hand, using the letter x as a symbol for multiplication is impossible, because the language would be unable to differentiate between the mathematical operator and the alphanumeric character. For these reasons, C uses the / symbol to indicate division and the * to indicate multiplication.

Table A.4 lists the C arithmetic operators.

Table A.4

C Language Arithmetic Operators

OPERATOR	ACTION
+	Addition
–	Subtraction
*	Multiplication
/	Division
%	Remainder

Remainder Operator

One of the operators in Table A.4 requires additional comment. The % operator gives the remainder of a division. The % symbol is also called the modulus operator. Its action is limited to integer operands. The following code fragment shows its use:

```
int val1 = 14;
int result = val1 % 3;
```

In this case, the value of the variable result is 2 because this is the remainder of dividing 14 by 3.

The remainder of a division finds many uses in mathematics and in programming. Operations based on the reminder are sometimes said to perform "clock arithmetic." This is due to the fact that the conventional clock face is divided into 12 hours, which repeat in cycles. We can say that the modulo of a clock is 12. The hour-of-day from the present time, after any number of hours, can be easily calculated by the remainder of dividing the number of hours by 12 and adding this value to the present time.

For example, it is 4 o'clock and you wish to calculate the hour-of-day after 27 hours have elapsed. The remainder of 27/12 is 3. The hour-of-day is then 4 + 3, which is 7 o'clock. In C language, you can obtain the remainder with a single operator. The following code fragment shows the calculations:

```
int thisHour = 4;
int hoursPassed = 27;
int hourOfDay = thisHour + (hoursPassed % 12);
```

Notice that the expression

```
hoursPassed % 12
```

produces the remainder of 27/12, which is then added to the current hour to obtain the new hour-of-day.

Modular arithmetic finds many computer uses. One of them is in calculating functions that have repeating values, called periodic functions. For example, the math units of several microprocessor produces trigonometric functions in the range 0 to 45 degrees. Software must then use remainder calculations to scale the functions to any desired angle.

A.5.4 Concatenation

In C language the + operator, which is used for arithmetic addition, is also used to concatenate strings. The term "concatenation" comes from the Latin word *catena*, which means chain. To concatenate strings is to chain them together. The following code fragment shows the action of this operator:

```
// Define strings
String str1 = "con";
String str2 = "ca";
```

```
String str3 = "ten";
String str4 = "ate";
// Form a new word using string concatenation
String result = str1 + str2 + str3 + str4;
// result = "concatenate"
```

The operation of the concatenation operator can be viewed as a form of string "addition." In C, if a numeric value is added to a string, the number is first converted into a string of digits and then concatenated to the string operand, as shown in the following code fragment:

```
String str1 = "Catch ";   // Define a string
int value = 22;           // Define an int
result = str5 + value;    // Concatenate string + int
                          // result = "Catch 22"
```

Notice that concatenation requires that at least one of the operands be a string. If both operands are numeric values, then arithmetic addition takes place.

A.5.5 Increment and Decrement

Programs often have to keep count of the number of times an event, operation, or a series of events or operations has taken place. In order to keep the tally count, it is convenient to have a simple form of adding or subtracting 1 to the value of a variable. C language contains simple operators that allow this manipulation. These operators are called the increment (++) and decrement (--).

The following expressions add or subtract 1 to the value of the operand.

```
x = x + 1;    // add 1 to the value of x
y = y - 1;    // subtract 1 from the value of y
```

The increment and decrement operators can be used to achieve the same result in a more compact way, as follows:

```
x++;          // add 1 to the value of x
y--;          // subtract 1 from the value of y
```

The ++ and -- symbols can be placed before or after an expression. When the symbols are before the operand, the operator is said to be in prefix form. When it follows the operand, it is said to be in postfix form, as follows:

```
z = ++x;  // Prefix form
z = x++;  // Postfix form
```

The prefix and postfix forms result in the same value in unary statements. For example, the variable x is incremented by 1 in both of these statements:

```
x++;
++x;
```

However, when the increment or decrement operators are used in an assignment statement, the results are different if the operators are in prefix or in postfix form. In the first case (prefix form), the increment or decrement is first applied to the operand and the result assigned to the lvalue of the expression. In the postfix form, the operand is first assigned to the lvalue and then the increment or decrement is applied. The following code fragment shows both cases:

```
int x = 7;
int y;
y = ++x;        // y = 8, x = 8
y = x++;        // y = 7, x = 8
```

A.5.6 Relational Operators

Computers and digital devices make simple decisions. For example, a program can take one path of action if two variables, a and b, are equal; another path if a is greater than b, and yet another one if b is greater than a. The C language relational operators are used to evaluate if a simple relationship between operands is true or false. Table A.5 lists the C language relational operators.

Table A.5

C Language Relational Operators

OPERATOR	ACTION
<	Less than
>	Greater than
<=	Less than or equal to
>=	Greater than or equal to
==	Equal to
!=	Not equal to

Other programming languages have special operators, called Boolean operators, that can store one of two logical values: TRUE or FALSE. C language does not implement Boolean data types. Instead, it uses integer variables that evaluate to a positive number if the relation is true and to 0 if it is false. The most commonly used data type to encode Boolean values (FALSE = 0, TRUE = NOT ZERO) is an unsigned char.

The == operator is used to determine if one operand is equal to the other one. It is unrelated to the assignment operator (=), which has already been discussed. In the following examples we set the value of an unsigned char variable according to a comparison between two numeric variables, x and y.

```
boolean result;
int x = 4;
int y = 3;
result = x > y;      // Case 1 - result true
result = x < y;      // Case 2 - result false
result = x == 0;     // Case 3 - result false
result = x != 0;     // Case 4 - result true
result = x <= 4;     // Case 5 - result true
```

Note in case 3 the different action between the assignment and the relational operator. In this case the assignment operator (=) is used to assign to the variable result the Boolean true or false that results from comparing x to 0. The comparison is performed by means of the == operator. The result is false because the value of the variable x is 4 at this point in the code. One common programming mistake is to use the assignment operator in place of the relational operator, or vice versa.

A.5.7 Logical Operators

The relational operators described in the previous sections are used to evaluate whether a condition relating two operands is true or false. However, by themselves, they serve only to test simple relationships. In programming, you often need to determine complex conditional expressions. For example, to determine if a user is a teenager, you test whether the person is older than twelve years, and younger than twenty years.

The logical operators allow combining two or more conditional statements into a single expression. As is the case with relational expressions, expressions that contain logical operators return true or false. Table A.5 lists the C language logical operators.

Table A.5

C Language Logical Operators

OPERATOR	ACTION
&&	Logical AND
||	Logical OR
!	Logical NOT

For example, if a = 6, b = 2, and c = 0, then the Boolean result evaluates to either true or false, as follows:

```
unsigned char result;
int a = 6;
int b = 2;
int c = 0;
result = a > b && c == 0;    // Case 1 - result is true
result = a > b && c != 0;    // Case 2 - result is false
result = a == 0 || c == 0;   // Case 3 - result is true
result = a < b || c != 0;    // Case 4 - result is false
```

In case 1, the Boolean result evaluates to true because both relational elements in the statement are true. Case 4 evaluates to false because the OR connector requires that at least one of the relational elements be true and, in this case, both are false (a > b and c = 0).

The logical NOT operator is used to invert the value of a Boolean variable or to test for the opposite. For example:

```
boolean result;
boolean tf  = false;
result = (tf == !true);  // result is true
```

The preceding code fragment evaluates to true because !true is false and tf is false. The principal use of conditional expressions is in making program decisions.

A.5.8 Bitwise Operators

Computers store data in individual electronic cells that are in one of two states, sometimes callled ON and OFF. These two states are represented by the binary digits 1 and 0. In practical programming, you often disregard this fact and write code that deals with numbers, characters, Boolean values, and strings. Storing a number in a variable of type double, or a name in a String object, does not usually require dealing with individual bits. However, there are times when the code needs to know the state of one or more data bits, or needs to change individual bits or groups of bits. Bit manipulations are quite frequent in programming embedded hardware.

One reason for manipulating individual bits or bit fields is based in economics. Suppose an operating system program is needed to keep track of the input and output devices. In this case program code may need to determine and store the following information:

- The number of output devices (range 0 to 3)

- If there is a pushbutton switch (yes or no)

- The number of available lines in Port C (range 0 to 7)

- If the device is equipped with EEPROM memory

One way to store this information would be in conventional variables. You could declare the following variable types:

```
unsigned char outputDevices;
unsigned char hasPB;
unsigned char portCLines;
unsigned char hasEEPROM;
```

One objection to storing each value in individual variables is the wasted memory. When we devote a char variable for storing the number of output devise in the system, we are wasting considerable storage space. A char variable consists of one memory byte. This means that you can store 256 combinations. However, in this particular example, the maximum number of output devices is 3.

A more economical alternative, memory wise, would be to devote to each item the minimum amount of storage necessary for encoding all possible states. In the case of the number of output devices, you could do this with just two bits. Because two bits allow representing values from 0 to 3, which is sufficient for this data element. By the same token, a single bit would serve to record if a mouse is present or not. The convention in this case, followed by the C language, is that a binary 1 represents YES and a binary 0 represents NO. The number of Port C lines (range 0 to 7) could be encoded in a three-bit field, while another single bit would record the presence or absence of EEPROM memory. The total storage would be as follows:

```
Output devices              2 bits
Presence of PB switch       1 bit
Number of Port C lines      3 bits
Presence of EEPROM memory   1
```

The total storage required is 7 bits. Figure A.5 shows how the individual bits of a byte variable can be assigned to store this information.

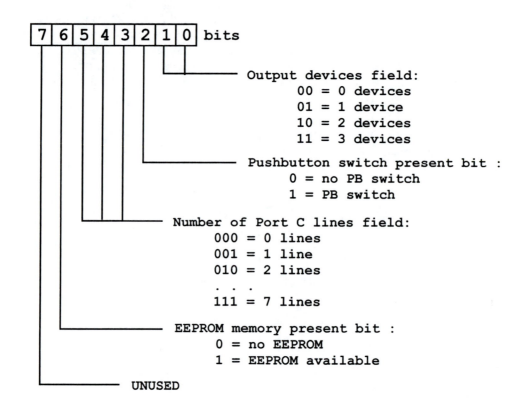

Figure A.5 *Example of bitmapped data.*

The operation of assigning individual bits and bit fields is called bit-mapping. Another advantage of bitmapped data is that several items of information can be encoded in a single storage element. Because bitmapped data is more compact, it is easier to pass and retrieve information. For example, you could devote a char variable to store the bitmapped data in Figure A.5. The variable could be defined as follows:

```
unsigned char systemDevices;
```

In order to manipulate bitmapped data you must be able to access individual bits and bit fields. This is the function of the C language bitwise operators. The operators are listed in Table A.6.

Table A.6

C Language Bitwise Operators

OPERATOR	ACTION
&	bitwise AND
|	bitwise OR
^	bitwise XOR
~	bitwise NOT
<<	bitwise left-shift
>>	bitwise right-shift

In Table A.6, the operators &, |, ^, and ~ perform bitwise functions on individual bits. The convention that a binary 1 corresponds with logical true, and a binary 0 with false, allows using binary numbers to show the results of a logical or bitwise operation. For example

```
1 AND 0 = 0
1 AND 1 = 1
1 OR 0 = 1
NOT 1 = 0
```

A table that lists all possible results of a bitwise or logical operation is called a truth table. Table A.7 has the truth tables for AND, OR, XOR, and NOT. The tables are valid for both logical and the bitwise operations.

Table A.7

Logical Truth Tables

```
    AND  |          OR   |          XOR  |          NOT  |
  ---------        ---------        ---------        --------
   0  0 | 0         0  0 | 0         0  0 | 0          0 | 1
   0  1 | 0         0  1 | 1         0  1 | 1          1 | 0
   1  0 | 0         1  0 | 1         1  0 | 1
   1  1 | 1         1  1 | 1         1  1 | 0
```

When using logical and bitwise operators, you must keep in mind that although AND, OR, and NOT perform similar functions, the logical operators do not change the actual contents of the variables. The bitwise operators, on the other hand, manipulate bit data. Thus, the result of a bitwise operation is often a value different from the previous one.

AND Operator

The bitwise AND operator (&) performs a Boolean AND of the two operands. The rule for the AND operation is that a bit in the result is set only if the corresponding bits are set in both operands. This action is shown in Table A.7.

The AND operator is frequently used to clear one or more bits, or to preserve one or more bits in the operand. This action is consistent with the fact that ANDing with a 0 bit clears the result bit, and ANDing with a 1 bit preserves the original value of the corresponding bit in the other operand.

A specific bit pattern used to manipulate bits or bit fields is sometimes called a mask. An AND mask can be described as a filter that passes the operand bits that correspond to a 1-bit in the mask, and clears the operand bits that correspond to 0-bits. Figure A.6 shows action of ANDing with a mask.

```
                              0101 1111 ◄────── operand
         bitwise AND          1111 0000 ◄────── mask
                              ─────────
                              0101 0000 ◄────── result
```

Figure A.6 *Action of the AND mask.*

A program can use the action of the bitwise AND operator to test the state of one or more bits in an operand. For example, a program can AND the contents of the Port C register with a value in which the high-order bit is set, as follows:

```
    1 0 0 0 0 0 0 0
```

When ANDing the Port C register with a mask in which only the high-order bit is set, we can assume that the seven low-order bits of the result will be zero. Recall that ANDing with a 0-bit always produces zero. Also recall that the value of the high-order bit of the result will be the same as the corresponding bit in the other operand, as shown in Figure A.7.

```
                         x x x x x x x x ◄────── Port C
         bitwise AND     1 0 0 0 0 0 0 0 ◄────── mask
                         ───────────────
                        │?│0 0 0 0 0 0 0 ◄────── result
                         └──── bit tested
```

Figure A.7 *AND-testing a single bit.*

In Figure A.7 the high-order bit of the result can be either 0 or 1. Because the seven low-order bits are always zero, you can conclude that the result will be non-zero if bit 7 of the operand is 1. If the result is zero, then bit 7 of the operand is zero.

OR Operator

The bitwise OR operator (|) performs the Boolean inclusive OR of two operands. The outcome is that a bit in the result is set if at least one of the corresponding bits in the operand is also set, as shown by the truth table in Table A.7. A frequent use for the OR operator is to selectively set bits in an operand. The action can be described by saying that ORing with a 1-bit always sets the result bit, whereas ORing with a 0-bit preserves the value of the corresponding bit in the other operand. For example, to make sure that bits 5 and 6 of an operand are set, we can OR it with a mask in which these bits are 1. This is shown in Figure A.8.

```
                            0101 0101  ◄──────  operand
        bitwise OR          1111 0000  ◄──────  mask
                            ─────────────
                            1111 0101  ◄──────  result
```

Figure A.8 *Action of the OR mask.*

Because bits 4, 5, 6, and 7 in the mask are set, the OR operation guarantees that these bits will be set in the result, independently of their value in the first operand.

XOR Operator

The bitwise XOR operator (^) performs the Boolean exclusive OR of the two operands. This action is described by stating that a bit in the result is set if the corresponding bits in the operands have opposite values. If the bits have the same value, that is, if both bits are 1 or both bits are 0, the result bit is zero. The action of the XOR operation corresponds to the truth table of Table A.7.

It is interesting to note that XORing a value with itself always generates a zero result, because all bits will necessarily have the same value. On the other hand, XORing with a 1-bit inverts the value of the other operand, because 0 XOR 1 = 1 and 1 XOR 1 = 0 (see Table A.7). By properly selecting an XOR mask the programmer can control which bits of the operand are inverted and which are preserved. To invert the two high-order bits of an operand, you XOR with a mask in which these bits are set. If the remaining bits are clear in the mask, then the original value of these bits will be preserved in the result, as in shown in Figure A.9.

```
                            0101 0101  ◄──────  operand
        bitwise XOR         1111 0000  ◄──────  mask
                            ─────────────
                            1010 0101  ◄──────  result
```

Figure A.9 *Action of the XOR mask.*

NOT Operator

The bitwise NOT operator (~) inverts all bits of a single operand. In other words, it converts all 1-bits to 0 and all 0-bits to 1. This action corresponds to the Boolean NOT function, as shown in Table A.7. Figure A.10 shows the result of a NOT operation.

```
        bitwise NOT         0101 0011  ◄──────  operand
                            ─────────────
                            1010 1100  ◄──────  result
```

Figure A.10 *Action of the NOT operator.*

Shift-Left and Shift-Right Operators

The C language shift-left (<<) and shift- right (>>) operators are used to move operand bits to the right or to the left. Both operators require an operand that specifies the number of bits to be shifted. The following expression shifts left, by 2 bit positions, all the bits in the variable bitPattern:

```
unsigned char bitPattern = 127;
bitPattern = bitPattern << 2;
```

The action of a left shift by a 1-bit position can be seen in Figure A.11.

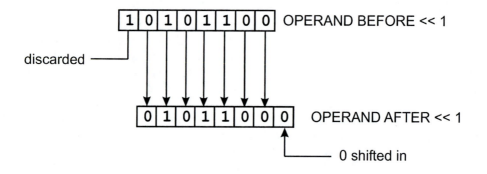

Figure A.11 *Action of the << operator.*

The operation of the left-shift, as shown in Figure A.11, determines that the most significant bit is discarded. This could spell trouble when the operand is a signed number, because in signed representations the high-order bit encodes the sign of the number. Therefore, discarding the high-order bit can change the sign of the value.

When applying the right-shift operator the low-order bit is discarded and a zero is entered for the high-order bit. The action is shown in Figure A.12.

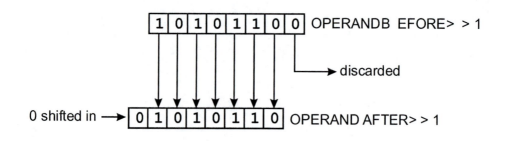

Figure A.12 *Action of the >> operator.*

The use of the right-shift operator on a negative integer number results in converting it to a positive number by replacing the leading 1-bit with a 0-bit. The ANSI Standard allows implementations to optionally handle this case by propagating the sign bit to the leftmost bit position. However, MPLAB C18 does not implement this option and does not propagate the sign bit on a right shift.

A.5.9 Compound Assignment Operators

C language contains several compound operators that were designed to make code more compact. The compound assignment operators consist of a combination of the simple assignment operator (=) with an arithmetic or bitwise operator. For example, to add 5 to the variable y, we can code

```
y = y + 5;
```

Alternatively, we can combine the addition operator (+) with the simple assignment operator (=) as follows

```
y += 5;
```

In either case, the final value of y is its initial value plus 5, but the latter form reduces the size of the program. Table A.8 lists the compound assignment operators.

Table A.8

C Language Compound Assignment Operators

OPERATOR	ACTION
+=	Addition assignment
−=	Subtraction assignment
*=	Multiplication assignment
/=	Division assignment
%=	Remainder assignment
&=	Bitwise AND assignment
|=	Bitwise OR assignment
^=	Bitwise XOR assignment
<<=	Left-shift assignment
>>=	Right-shift assignment

The programmer can remember that in compound assignments, the = sign always comes last. Notice that the compound assignment is not available for the NOT (~) bitwise unary operator or for the unary increment (++) or decrement (--) operators. The reason is that unary (one element) statements do not require simple assignment; therefore, the compound assignment form is meaningless.

A.5.10 Operator Hierarchy

Programming languages have hierarchy rules that determine the order in which each element in an expression is evaluated. For example, the expression

```
int value = 8 + 4 / 2;
```

evaluates to 6 because the C language addition operator has higher precedence than the multiplication operator. In this case, the compiler first calculates $8 + 4 = 12$ and then performs $12 / 2 = 6$. If the division operation were performed first, then the variable value evaluates to 10. Table A.9 lists the precedence of the C language operators.

Table A.9

Precedence of C Language Operators

OPERATOR	PRECEDENCE
. [] ()	Highest
++ --	
sizeof	
~	
!	
+ - (addition/subtraction)	
& (address of)	
* (indirection)	
* / (multiplication/division)	
% (remainder)	
<< >>	
< > <= >= (relational)	
== != (equality/inequality)	
& (bitwise AND)	
^	
|	
&& (logical AND)	
||	
?: (conditional assignment)	
= += -= *=	
/= %= <<= >>=	
&= ^= !=	Lowest

Associativity Rules

In some cases, an expression can contain several operators with the same precedence level. When operators have the same precedence, the order of evaluation is determined by the associativity rules of the language. Associativity can be left-to-right or right-to-left. In most programming languages, including C, the basic rule of associativity for arithmetic operators is left-to-right. This is consistent with the way we read in English and the Western European languages.

In C language, the associativity rules can be sumarized as follows:

- Binary and ternary operators are left-associative except the conditional and assignment operators, which are right-associative.

- The unary and postfix operators are described as right-associative because the expression * x ++ is interpreted as * (x++) rather than as (*x) ++.

- All others are right-associative.

Because of these variations in the associativity rules, the programmer must exercise care in evaluating some expressions. For example

```
int a = 0;
int b = 4;
a = b = 7;
```

If the expression a = b = 7 is evaluated left-to-right, then the resulting value of variable a is 4, and the value of b is 7. However, if it is evaluated right-to-left, then the value of both variables is 7. Because the assignment operator has right-to-left associativity, the value of b is 7 and the value of a is also 7.

A.6 Directing Program Flow

The main difference between a digital computing device and a calculating machine is that the computer can make simple decisions. Programs are able to process information logically because of this decision-making ability. The result of a program decision is to direct program execution in one direction or another one, that is, to change program flow. One of the most important tasks performed by the programmer is the implementation of the program's processing logic. This implementation is by means of the language's decision constructs.

A.6.1 Decisions Constructs

Making a program decision requires several language elements. Suppose an application must determine if the variable a is larger than b. If so, the program must take one course of action. If both variables are equal, or if b is larger than a, then another course of action is necessary. In other words, the program has to make a comparison, examine the results, and take the corresponding action in each case. All of this cannot be accomplished with a single operator or keyword, but requires one or more expressions contained in one or more statements. This is why we refer to decision statements and decision constructs, and not to decision operators. In programming, a construct can be described as one or more expressions, contained in one or more statements, all of which serve to accomplish a specific purpose.

C language contains several high-level operators and keywords that can be used in constructs that make possible selection between several processing options. The major decision-making mechanisms are called the if and the switch constructs. The conditional operator (?:), which is the only C language operator that contains three operands, is also used in decision-making constructs.

if Construct

The C language if construct consists of three elements:

1. The if keyword

2. A test expression, called a conditional clause

3. One or more statements that execute if the test expression is true

The sample program named C_PBFlash.c in this book's online software resource, reads the state of a pushbutton and takes action if the switch is closed. The following code fragment from the sample program implements the decision:

```
while(1)
{
    if(!(PORTB & 0b00010000))
        FlashRED();
}
```

In the program C_PBFlash.c, the action of flashing the red LEDs is implemented in the function FlashRED() not listed. The decision is made in the statement:

```
    if(!(PORTB & 0b00010000))
```

The test consists of performing a bitwise AND operation (&) with the Port B lines and a mask in which the bit to be tested (bit 4) is set and all other bits are zero. In this case the pushbutton switch is wired to Port B line 4. The pushbutton switch is wired active low so it is necessary to perform a bitwise NOT (!) on the result of the bitwise AND operation.

Statement Blocks

The simple form of the if construct consists of a single statement that executes if the conditional expression is true. The sample code listed previously uses a simple if construct. But code will often need to perform more than one operation according to the result of a decision. C language provides a simple way of grouping several statements so that they are treated as a unit. The grouping is performed by means of curly braces ({}) or roster symbol. The statements enclosed within two rosters form a compound statement, also called a statement block.

You can use statement blocking to modify the code listed previously so that a second function, named Beep(), executes if the if test reports true. The code would be as follows:

```
while(1)
{
    if(!(PORTB & 0b00010000)) {
        FlashRED();
            Beep();
    }
}
```

The brace symbols ({ and }) are used to associate more than one statement with the related if. In this example, both functions, FlashRED() and Beep(), execute if the conditional clause evaluates to true and both are skipped if it evaluates to false.

Nested if Construct

Several if statements can be nested in a single construct. The result is that the execution of a statement or statement group is determined not by a single condition, but to two or more. For example, the C_PBFlashX2.c program, in the book's software resource, is a demonstration program to monitor pushbutton switches No. 1 and 2 on DemoBoard 18F452A (or equivalent circuit). If both switches are down, then four red LEDs wired to Port C lines 0 to 3 are flashed. The following code fragment from the program C_PBFlashX2.c implements the decision:

```
void main(void)
{
    // Initialize direction registers
    TRISB = 0b00110000;// Port B, lines 4/5, set for input
//                ||_____ Pushbutton # 1
//                |_____ Pushbutton # 2
    TRISC = 0;              // Port C set for output
    PORTC = 0;              // Clear all Port C lines

    while(1)
    {
        if(!PORTBbits.RB4) {
            if(!PORTBbits.RB5) {
                FlashRED();
            }
        }
    }
}
```

In the preceding code fragment, the if statement that tests for pushbutton # 2 is nested inside the if statement that tests pushbutton # 1. In this case, the inner if statement is never reached if the outer one evaluates to false. Although white space has no effect on the code, text line indentation does help visualize logical flow. Figure A.13 is a flowchart of a nested if construct.

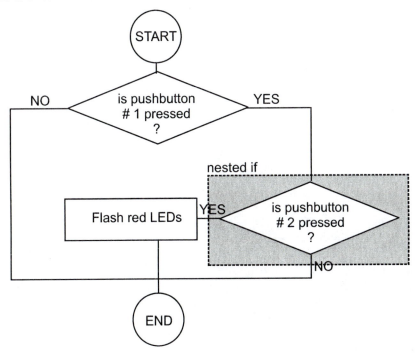

Figure A.13 *Flowchart of a nested if construct.*

Notice that in this example that we have used a C18 library macro to identify bit 5 in Port B, as follows:

```
PORTBbits.RB5
```

The macros that define pin assignments are hardware-specific facilities that are part of the MPLAB C18 compiler. The functions in the Software Peripherals Library are discussed in Chapter 6.

else Construct

An if construct executes a statement, or a statement block, if the conditional clause evaluates to true, but no alternative action is taken if the expression is false. The if-else construct allows C language to provide an alternative processing options for the case in which the conditional clause is false. The else construct is sometimes called the if-else construct.

Consider the following program fragment from the program C_PBFlash2.c in the book's software resource:

```
void main(void)
{
    // Initialize direction registers
    TRISB = 0b00001000;      // Port B, line 4, set for input
//                       |
//                       |_____ Pushbutton # 1
    TRISC = 0;               // Port C set for output
    PORTC = 0;               // Clear all Port C lines

    while(1)
    {
        if(!PORTBbits.RB4)
            FlashRED();
        else
            FlashGREEN();
    }
}
```

In this example, the red LEDs are flashed if bit 4 of Port B is clear; otherwise the green LEDs are flashed.

It is customary among C programmers to align the if and else keywords in the if-else construct. This is another example of the use of white space to clarify program flow. As in the case of the if clause, the else clause can also contain a statement block delimited by rosters. A statement block is necessary if more than one statement is to execute on the else branch. Figure A.14 is a flowchart of the preceding if-else construct.

Dangling else Case

Because the else statement is optional, it is possible to have several nested if constructs, not all of which have a corresponding else clause. This case is sometimes called the dangling else case. A dangling else statement can give rise to uncertainty about the pairing of the if and else clauses.

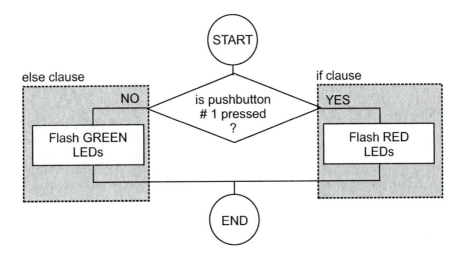

Figure A.14 *Flowchart of an if-else construct.*

The dangling else case typically takes place when there are two adjacent if statements, as in the case of the following fragment with two if statements and a single else clause:

```
if(a != 0 )
    if(a < 10)
            Action(1);
    else
            Action2();
```

In the preceding code fragment, the path of execution is different if the else clause is paired with the inner if statement, or with the outer one. If the else clause is paired with the first if statement, it will execute if the value of variable a is not zero. However, if the else clause is paired with the second if statement, it will execute if the value of the a variable is greater than or equal to ten.

The general rule used by the C language in solving the dangling else problem is that each else statement is paired with the closest if statement that does not have an else, and that is located in the same block. Indentation in the preceding code fragment helps you see that the else statement is linked to the inner if.

Because if-else pairing relates to program blocks, rosters can be used to force a particular if-else association; for example,

```
if(a != 0 ) {
    if(a < 10)
            Action(1);
    }
else
        Action2();
```

In the preceding code fragment the closest if statement without an else, and located in the same block, is the statement

```
if(a != 0)
```

Notice how the indentation serves to show this association.

else-if Clause

You have seen how the dangling else problem can the cause unpredicted association of an else clause with an if statement. The relationship between two consecutive if statements can also cause problems. The flowchart in Figure A.13 shows the case of a cascaded if construct. In this case, the second if statement is conditioned to the first one being true. However, if the second if statement is not nested in the first one, then its evaluation is independent of the result of the first if. The following variation in the coding shows this case:

```
while(1)
{
    if(!PORTBbits.RB4)
        FlashGREEN();
    if(!PORTBbits.RB5)
        FlashRED();
}
```

In this case the two conditions are tested independently. If pin RB4 is clear then the green LEDs are flashed. If pin RB5 is clear, then the red LEDs are flashed. In other words, the second if statement is unrelated to the first one, therefore, the second statement is always evaluated.

The else-if construct allows subordinating the second if statement to the case of the first one evaluating to false. All you have to do is to nest the second if within the else clause of the first one; for example,

```
int age;
. . .
if(age == 12)
  Print(12);
    else if(age == 13)
      Print(13);
        else if(age == 14)
          Print(14);
else
   Print(0);
```

In the preceding code fragment, the last if statement executes only if all the preceding if statements have evaluated to false. If one of the if statements evaluates to true, then the rest of the construct is skipped. In some cases, several logical variations of the consecutive if statements may produce the desired results, while in other cases it may not. Flowcharting is an effective way of resolving doubts regarding program logic.

The else-if is a mere coding convenience. The same action results if the else and the if clause are separated. For example,

```
if(age == 12)
    Print(12);
    else
         if(age == 13)
         Print(13);
    . . .
```

switch Construct

It is a common programming technique to use the value of an integer variable to direct execution to one of several processing routines. You have probably seen programs in which the user selects among several processing options by entering a numeric value. The software then makes a processing selection based on this value.

The C language switch construct provides an alternative mechanism for selecting among multiple options. The switch consists of the following elements:

1. The switch keyword.

2. A controlling expression enclosed in parentheses; must be of integer type.

3. One or more case statements followed by an integer or character constant, or an expression that evaluates to a constant. Each case statement terminates in a colon symbol.

4. An optional break statement at the end of each case block. When the break is encountered, all other case statements are skipped.

5. An optional default statement. The default case receives control if none of the other case statements have executed.

The switch construct provides a simple alternative to a complicated if, else-if, or else chain. The general form of the switch statement is as follows:

```
switch (expression)
{
    case value1:
       statement;
       statement;
       . . .
       [break;]
    case value2:
       statement;
       statement;
       . . .
       [break;]
       ...
    [default:]
       statement;
       statement;
       . . .
       [break;]
}
```

Note that the preceding example uses a nonexistent computer language, called pseudocode. Pseudocode shows the fundamental logic of a programming construct without complying with the formal requirements of any particular programming language. There are no strict rules to pseudocode; the syntax is left to the coder's imagination. The preceding pseudocode listing combines elements of the C language with symbols that are not part of C. For example, the ... characters (called ellipses) indicate that other program elements could follow at this point while the bracket symbols are used to signal optional components. Neither the ellipses nor the brackets are part of the C language as used in this pseudocode.

The controlling expression of a switch construct follows the switch keyword and is enclosed in parentheses. The expression, usually a variable, must evaluate to an integer type. It is possible to have a controlling expression with more than one variable, one that contains literal values, or perform integer arithmetic within the controlling expression. Each case statement marks a position in the switch construct. If the case statement is true, execution continues at the code line that follows the case keyword. The case keyword is followed by an integer or character constant, or an expression that evaluates to an integer or character constant. The case constant is enclosed in single quotation symbols (tic marks) if the control statement is an ASCII character.

The following code fragment from the sample program C_Switch_Demo.c in the book's software resource, shows a case construct. The program uses the C language switch construct to direct execution according to the setting of four DIP switches wired to Port A lines 2 to 5 in Demo Board 18F452-A or equivalent circuit. LEDs wired to Port C lines 0 to 7 are flashed as follows:

```
DIP switch              x = LEDs flashed
1 - PORTA 2             0000 00xx
2 - PORTA 3             0000 xx00
3 - PORTA 4             00xx 0000
4 - PORTA 5             xx00 0000
No DIP closed           0x0x 0x0x
```

Operating code is as follows:

```
void main(void)
{
    unsigned char DIPs = 0;

    // Init Port A for digital operation
    PORTA = 0;                  // Clear port
    LATA = 0;               // and latch register
    // ADCON1 is the configuration register for the A/D
    // functions in Port A. A value of 0b011x sets all
    // lines for digital operation
    ADCON1 = 0b00000110;            // Digital mode all Port A lines
    // Initialize direction registers
    TRISA = 0b00111100;// Port A lines 2-5 set for input
                        // to activate DIP switches
    TRISC = 0;                  // Port C set for output
    PORTC = 0;                  // Clear all Port C lines
```

```
    while(1)
    {
        DIPs = (PORTA >> 2);     // DIPs are active low
        switch(DIPs)
        {
            case 0b00000001:    // DIP # 1 closed
                FlashLED(0b00000011);
                break;
            case 0b00000010:    // DIP # 2 closed
                FlashLED(0b00001100);
                break;
            case 0b00000100:    // DIP # 3 closed
                FlashLED(0b00110000);
                break;
            case 0b00001000:    // DIP # 4 closed
                FlashLED(0b11000000);
                break;
            default:
                FlashLED(0b01010101);
                break;
        }
    }
}
/****************************************************************
                    local functions
****************************************************************/
void FlashLED(unsigned char pattern)
{
    PORTC = pattern;
    Delay1KTCYx(200);
    PORTC = 0x00;
    Delay1KTCYx(200);
    return;
}
```

Because the case constant is a numeric type (unsigned char), the case constants do not require tic marks. The following code fragment shows a case construct in which the switch variable is of type int.

```
char letter = 'A';
. . .
switch (letter)
{
   case 'A':
       ...
       break;
   case 'B':
       ...
       break;
. . .
}
```

The break keyword is optional, but if it is not present at the end of a case block, then the following case block or the default block executes. In other words, execution in a switch construct continues until a break keyword or the end of the construct is encountered. When a break keyword is encountered, execution is immediately directed to the end of the switch construct. A break statement is not required on the last block (case or default statement), although it is sometimes included for readability.

The blocks of execution within a switch construct are enclosed in rosters; however, the case and the default keywords automatically block the statements that follow. Rosters are not necessary to indicate the first-level execution block within a case or default statement.

Conditional Expressions

C language expressions usually contain a single operand; however, there is one ternary operator that uses two operands. The ternary operator is also called the conditional operator. A conditional expression is used to substitute a simple if-else construct. The syntax of a conditional statement can be sketched as follows:

```
expTest ? expTrue : expFalse;
```

In the above pseudocode expTest, expTrue, and expFalse are C language expressions. During execution, expTest is first tested. If it is true, then expTrue executes. If it is false, then expFalse executes. For example, assign the value of the smaller of two integer variables (named a and b) to a third variable named min. Conventional code could be as follows:

```
int a, b, min;
...

if (a < b)
   min = a;
else
   min = b;
```

With the conditional operator, the code can be shortened and simplified, as follows:

```
min = (a < b) ? a : b;
```

In the above statement, the conditional expression is formed by the elements to the right of the assignment operator (=). There are three elements in the rvalue:

1. The expression (a < b), which evaluates either to logical true or false.

2. The expression ? a determines the value assigned to the lvalue if the expression (a < b) is true.

3. The expression : b determines the value assigned to the lvalue if the expression (a < b) is false.

The lvalue is the element to the left of the equal sign in an assignment expression. The rvalue is the element to the right of the equal sign.

A.7 Loops and Program Flow Control

Often, computer programs must repeat the same task a number of times. Think of a burglar alarm system in which the software must disable the alarm whenever the user enters a specific control code followed by a password. A reasonable design for this program would be a routine that continually monitors the keypad for input and, on detecting a disabling command, prompts the user for a password and if correct, proceeds to turn off the alarms. The repeated processing takes place in a program construct usually called a loop. In this section we discuss three C language loop constructs: the for loop, the while loop, and the do-while loop.

A.7.1 Loops and Iterations

Loops do not offer functionality that is not otherwise available in a programming language. Loops just save coding effort and make programs more logical and efficient. In many cases, coding would be virtually impossible without loops. Imagine an application that estimates the tax liability for each resident of the state of Minnesota. Without loops, you may have to spend the rest of your life writing the code.

In talking about loops it is convenient to have a word that represents one entire trip through the processing routine. We call this an iteration. To iterate means to do something repeatedly. Each transition through the statement or group of statements in the loop is an iteration. Thus, when talking about a program loop that repeats a group of statements three times, we speak of the first, the second, and the third iteration.

A.7.2 Elements of a Program Loop

A loop always involves three steps:

1. The initialization step is used to prime the loop variables to an initial state.

2. The processing step performs the processing. This is the portion of the code that is repeated during each iteration.

3. The testing step evaluates the variables or conditions that determine the continuation of the loop. If they are met, the loop continues. If not, the loop ends.

A loop structure can be used to calculate the factorial. The factorial is the product of all the whole numbers that are equal to or less than the number. For example, factorial 5 (written 5!) is

```
5! = 5 * 4 * 3 * 2 * 1 = 120
```

In coding a routine to calculate the factorial, you can use one variable to hold the accumulated product and another one to hold the current factor. The first variable could be named facProd and the second one curFactor. The loop to calculate facProd can be as follows:

1. Initialize the variables facProd to the number whose factorial is to be calculated and the variable curFactor to this number minus one. For example, to calculate 5!, you make facProd = 5 and curFactor = 4.

2. During each iteration, calculate the new value of facProd by making facProd = curFactor times facProd. Subtract one from curFactor.

3. If curFactor is greater than 1, repeat step 2; if not, terminate the loop.

Figure A.15 is a flowchart of the logic used in the factorial calculation described above.

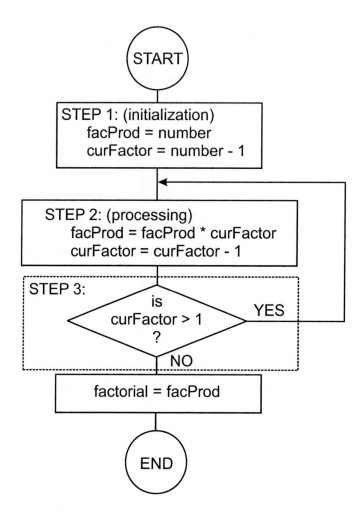

Figure A.15 *Factorial flowchart.*

Notice that in the factorial calculation we test for a factor greater than 1 to terminate the loop. This eliminates multiplying by 1, which is a trivial operation.

A.7.3 for Loop

The for loop is the simplest iterative construct of C language. The for loop repeats the execution of a program statement or statement block a fixed number of times. A typical for loop consists of the following steps:

1. An initialization step that assigns an initial value to the loop variable.

2. One or more processing statements. It is in this step where the calculations take place and the loop variable is updated.

3. A test expression that determines the termination of the loop.

The general form of the for loop instruction is shown in Figure A.16.

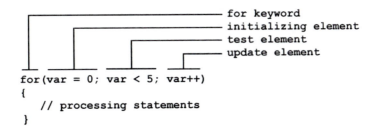

Figure A.16 *Elements of the for loop construct.*

We can use the for loop in the following code fragment for calculating the factorial according to the flowchart in Figure A.15.

```
int number = 5;            // Factorial to be calculated
int facProd, curFactor;    // Local variables
// Initialization step
facProd = number;          // Initialize operational variable
for (curFactor = number - 1; curFactor > 1; curFactor--)
// Processing step
      facProd = curFactor * facProd;
// Done
// The variable facProd now holds the factorial
```

Notice that the expression

```
for(curFactor = number - 1; curFactor > 1; curFactor --)
```

contains the loop expression and that it includes elements from Steps 1, 2, and 3. The first statement (curFactor = number - 1) sets the initial value of the loop variable. The second statement (curFactor > 1) contains the test condition and determines if the loop continues or if it ends. The third statement (curFactor --) diminishes the loop variable by 1 during each iteration.

Also notice that while the for loop expression does not end in a semicolon, it does contain semicolon symbols. In this case, the semicolon symbol is used to separate the initialization, test, and update elements of the loop. This action of the semicolon symbol allows the use of multiple statements in each element of the loop expression, as in the following case:

```
unsigned int x;
unsigned int y;
for(x = 0, y = 5; x < 5; x++, y--)
    Print("x is: " + x, " y is: " + y);
```

In the preceding code, the semicolons serve to delimit the initialization, continuation, and update phases of the for loop. The initialization stage (x = 0, y = 5) sets the variables to their initial values. The continuation stage (x < 5) tests the condition during which the loop continues. The update stage (x++, y--) increments x and decrements y during each iteration. The comma operator is used to separate the components in each loop phase.

The middle phase in the for loop statement, called the test expression, is evaluated during each loop iteration. If this statement is false, then the loop terminates immediately. Otherwise, the loop continues. For the loop to execute the first time, the test expression must initially evaluate to true. Notice that the test expression determines the condition under which the loop executes, rather than its termination. For example,

```
for(x = 0; x == 5; x++)
    Print(x);
```

The Print() statement in the preceding loop will not execute because the test expression x == 5 is initially false. The following loop, on the other hand, executes endlessly because the terminating condition is assigned a new value during each iteration.

```
int x;
for (x = 0; x = 5; x++)
     Print(x);
```

In the preceding loop the middle element should have been x = = 5. The statement x = 5 assigns a value to x and always evaluates true. It is a very common mistake to use the assignment operator (=) in place of the comparison operator (= =).

It is also possible for the test element of a for loop to contain a complex logical expression. In this case, the entire expression is evaluated to determine if the condition is met. For example,

```
int x, y;
for(x = 0, y = 5; (x < 3 || y > 1); x++, y--)
```

The test expression

```
(x < 3 || y > 1)
```

evaluates to true if either x is less than 3 or if y is greater than 1. The values that determine the end of the loop are reached when the variable x = 4 or when the variable y = 1.

Compound Statement in Loops

We have seen that the roster symbols ({ and }) are used in C language to group several statements into a single block. Statement blocks are used in loop constructs to make possible performing more than one processing operation.

while Loop

The C language while loop repeats a statement or statement block "while" a certain condition evaluates to true. Like the for loop, the while loop requires initialization, processing, and testing steps. The difference is that in the for loop, the initialization and testing steps are part of the loop itself, but in the while loop these steps are located outside the loop body.

The program C_LEDs_Flash.c in the book's software resource contains a function with a while loop to implement a do-nothing time delay. The code for the function named FlashLEDs() is a follows:

```
void FlashLEDs()
{
// Function uses a while loop to implement a do-nothing delay
    count = 0;
    PORTC = 0x0f;
        while (count <= MAX_COUNT)
        {
            count++;
        }
    count = 0;
    PORTC = 0xf0;
        while (count <= MAX_COUNT)
        {
            count++;
        }
    return;
}
```

Figure A.17 shows the elements of the while loop.

```
loopVar = 0;  ◄─────────────── External initialization
while(loopVar != 10) ◄──────── Loop continuation test
{
   // Processing statements
   loopVar++; ◄─────────────── Loop variable update
}
```

Figure A.17 *Elements of the while loop construct.*

do-while Loop

A feature of the while loop is that if the test condition is initially false, the loop never executes. For example, the while loop in the following code fragment will not execute the statement body because the variable x evaluates to 0 before the first iteration:

```
int x = 0;
. . .
while (x != 0)
. . .;
```

In the do-while loop, the test expression is evaluated after the loop executes. This ensures that the loop executes at least once, as in the following example:

```
int x = 0;
do
    while (x != 0);
```

In the case of the do-while loop, the first iteration always executes because the test is not performed until after the loop body executes. Figure A.18 shows the elements of the do-while loop.

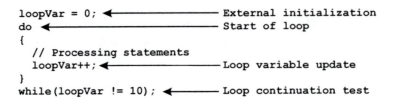

Figure A.18 *Elements of the do-while loop construct.*

In many cases, the processing performed by the do-while loop is identical to the one performed by the while loop. Note that the test expression in a do-while loop terminates in a semicolon symbol. Because a while statement does not contain a semicolon, it is a frequent programming mistake to omit it in the do-while loop.

A.8 Breaking the Flow

Some computer programs perform simple functions execute linearly, that is, they start at the first instruction or statement and continue, in order, until they reach the last one. This is the preferred scheme because these programs are easy to follow and understand. However, it sometimes happens that a simple execution pattern is not possible because the code must abruptly change the order of execution to another location. C provides four statements that allow abrupt changes in the order of execution of a program. These are named goto, break, continue, and return. The first three (goto, break, and continue) are discussed in the sections that follow. The return statement is discussed in the context of functions, later in this Appendix.

A.8.1 goto Statement

The C language goto statement transfers execution unconditionally to a specific location in the program. The destination location, at which execution resumes, is marked by a name followed by a colon symbol. This combination of name and colon symbol is called a label. For example, the following code fragment tests the variable named "divisor" for a zero value and, if zero, transfers execution to a program branch labeled DIVISION_ERROR:

```
if (divisor == 0)
     goto DIVISION_ERROR;
 .
 .
 .


DIVISION_ERROR:
 .
 .
 .
```

   The abuse of the goto statement can lead to programs in which execution jumps around the code in an inconsistent and hard-to-follow manner. Programmers sometimes say that this coding style generates spaghetti code, in reference to a program logic that is as difficult to unravel as a bowl of uncut spaghetti. Because the goto statement can always be substituted by other C language structures, its use is optional and discretionary. On the other hand, most authors agree that the goto statement provides a convenient way of directly exiting nested loops, especially in the case of errors or other unforseen conditions, and that its use is legitimate in these or similar circumstances. Sometimes in order to avoid the use of a goto statement, we must produce code that is even more difficult to follow and understand.

A.8.2 break Statement

The C language break statement provides a way for exiting a switch construct or the currently executing level of a loop. The break statement cannot be used outside a switch or loop structure. The sample program named C_Break_Demo.c in the book's software resource is a demonstration program that shows the action of a C language break statement by interrupting the/ execution of an if statement block. The program reads the state of the four DIP switches wired to Port A. If DIP switches number 2 and 3 are closed, a break statement executes and the if construct is aborted. Otherwise, the status of the four DIP switches is echoed in the green LEDs of Demo Board 18F452-A or equivalent circuit. Code is as follows:

```c
//********************************************************************
//                         main program
//********************************************************************
void main(void)
{
    unsigned char DIPs = 0;

    // Init Port A for digital operation
    PORTA = 0;                 // Clear port
    LATA = 0;                  // and latch register
    // ADCON1 is the configuration register for the A/D
    // functions in Port A. A value of 0b011x sets all
    // lines for digital operation
    ADCON1 = 0b00000110;         // Digital mode all Port A lines
    // Initialize direction registers
    TRISA = 0b00111100;// Port A lines 2-5 set for input
                       // to activate DIP switches
    TRISC = 0;                 // Port C set for output
    PORTC = 0;                 // Clear all Port C lines

    while(1)
    {
        DIPs = (PORTA >> 2);         // DIPs are active low
        if(DIPs == 0b00000110){ // Test DIPs 2 and 3
            FlashLED(0x00);
            break;                 // Exit if statement
        }
        else
            FlashLED(DIPs);
    }
}

//********************************************************************
//                       local functions
//********************************************************************
void FlashLED(unsigned char pattern)
{
    PORTC = pattern;
    Delay1KTCYx(200);
    PORTC = 0x00;
```

```
        Delay1KTCYx(200);
        return;
}


In the case of a nested loop, the break statement exits the loop level at
which it is located, as shown in the following code fragment.

main()
{
    unsigned int number = 1;
    char letter;
    ...
    while (number < 10 ) {
        Print(number);
        number ++;
            for (letter = 'A'; letter < 'G'; letter ++) {
                Print(letter);
                    if (letter == 'C')
                        break;
            }
    }
}
```

The code listed above contains two loops. The first one, a while loop, displays a
count of the numbers from 1 to 9. The second one, or inner loop, displays the capital
letters A to F, but the if test and associated break statement interrupt execution of
this nested loop as soon as the letter C is reached. In this example, because the
break statement acts on the current loop level only, the outer loop resumes counting numbers until the value 10 is reached. In this program we have used a function
named Print() to demonstrate flow of execution. Such a function does not exist in C
language and is a nor in the sample code and should be considered as pseudocode.

The break statement is also used in the switch construct, described previously. In
this case, the break statement forces an exit out of the currently executing case
block.

A.8.3 continue Statement

While the break statement can be used with either a switch or a loop construct, the
continue statement operates only in loops. The purpose of the continue statement is
to bypass all not yet executed statements in the loop and return to the beginning of the
loop construct. Continue can be used with for, while, and do loops. For example, the
following program contains a for loop to display the letters A to D; however, the continue statement serves to bypass the letter C, which is not displayed.

```
main()
{
    unsigned char letter;
      .
      .
      .
        for (letter = 'A'; letter < 'E'; letter ++) {
            if (letter == 'C')
                continue;
```

```
        Print(letter);
    }
}
```

Like the goto statement previously discussed, the abuse of the continue statement often leads to code that is difficult to follow and decipher.

A.9 Functions and Structured Programming

Early-day programmers discovered that their code often repeated identical operations. For example, a geometrical application that contains a routine to calculate the area of a circle may have the code repeated several times. This leads to wasted coding effort and to unnecessary program size. A possible solution is to create an independent program element that receives the radius parameter from the main program, proceeds to calculate the area, and then returns the result to the main code. The construction, often called a subroutine, can be reused as often as necessary by the main program with considerable savings in programming effort and in program storage.

In C language subroutines are called functions. Except for the special procedure named main(), a C language program need not contain other functions. In fact, any C programming operation, no matter how complicated or sophisticated, can always be performed without using functions.

We have already seen some functions; for example, we know that all C language programs must contain a function called main() and that execution always starts at this function, no matter where it appears in the source file. We have also encountered special hardware library functions that are part of MPLAB C18 and that are covered in the book's main text. In addition, we have used local functions in some of our sample programs without a rigorous description of their structure and application. In the sections that follow we correct this situation and discuss the details of C language functions. These programmer-created functions simplify the task of designing and coding, improve a program's readability and cohesiveness, and make it easier to understand and maintain the code.

In order to distinguish this type of function from the main() function and from the functions available in C language and hardware libraries, we will refer to them as programmer declared functions, or simply as functions. Any future reference to the main() function or to library functions will be clearly spelled out.

A.9.1 Modular Construction

Modular program construction is one of the core principles of a methodology called structured programming. The idea behind modular construction is to break a processing task into smaller units that are easier to analyze, code, and maintain. A well-designed program is divided into a few fundamental functions that constitute its principal modules. Because functions can contain other embedded functions, the process of dividing a program into individual modules can continue until processing reaches the simplest possible stage. In this manner the main modules may contain second-level modules, and the second-level modules may contain third-level modules, and so on.

A properly designed function should perform a specific and well-defined set of processing operations. Each function contains a single entry and a single exit point. The individual functions should be kept within a manageable size. In practice, a well conceived function rarely exceeds two or three pages of code. Except where program performance is an issue, it is better programming practice to divide processing into several smaller functions than to create a single, more complicated one.

A.9.2 Structure of a Function

A programmer-defined function in C language can be described as a collection of declarations and statements, grouped under a function name, and designed to perform a specific task. Every function contains three clearly identifiable elements:

1. The function prototype (also called the declaration)
2. The function definition (or the function itself)
3. The function call

Function Prototype

We have seen that a characteristic of C is that all variables must be declared, specifying name and data type, before they can be used, however, the original version of C (defined by Kernighan and Ritchie in their 1978 book *The C Programming Language*) did not require the pre-declaration of functions. Later on, the C ANSI Committee decided that this practice was inconsistent with the language's handling of variables and tended to obscure the code. Therefore, the ANSI standard for C language introduced the pre-declaration of functions. This operation is called function prototyping.

A function prototype is a pre-declaration of the function's name, the data type of the returned value, and the number and data types of the arguments passed to it. The prototype must appear before the main() function. Figure A.19 shows the elements of a function prototype.

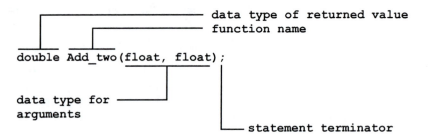

Figure A.19 *Elements of a function prototype.*

In the program C_LEDs_Flash.c listed previously, the local function FlashLEDs() is prototyped as follows:

```
// Function prototypes
void FlashLEDs(unsigned char);
```

Because a function prototype is required before the function is called by code, it is possible to avoid it by placing a function's declaration in a location in the code that precedes the function call. The one objection to this way of avoiding function prototypes is that it forces the programmer to place functions before their call. The result is a code listing that starts with functions other than main().

Function Definition

The function itself appears in the source file either after its prototype or before it is referenced. There are two sections of the function definition that can be clearly identified: the function declaration section (sometimes called the header) and the function body section. The declaration section, which is reminiscent of the prototype, contains the function name, the return type, and the data type of the parameters passed to the function. The function body section, which is enclosed in braces, contains the declarations and statements required for performing the function's processing operations. Figure A.20 shows the declaration section of a function definition.

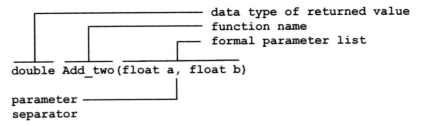

Figure A.20 *Declaration section of a function definition.*

Note that, unlike the prototype, the declaration section of the function definition does not end in a semicolon symbol. Also note that the formal parameter list includes the parameter names and their data types, while in the prototype, the names are not referenced. This listing of parameter names and types in the function definition serves as a declaration for these variables.

The body section of the function definition is where the actual processing takes place. It can include all the program elements of the C language, such as variable declarations, keywords, statements, and any legal program element. There is no structural difference between the body of a programmer-defined function and the body of the main() function.

The return keyword, discussed later in this chapter, ends the execution of a function. Return can include a variable, constant, expression or statement, conventionally enclosed in parentheses, which represents the value returned by the function. The only restriction is that the type of the actual returned value must conform to the one listed in the declaration and in the prototype.

Function Call

A programmer defined function receives control when the function's name is referenced in a program statement. This reference, or function call, appears within the main() or another programmer defined function. Furthermore, C allows a function to call itself; this operation is called recursion. The following trivial program requests from the user the radius of a circle and uses a function called Circum to calculate its circumference.

```
#define PI 3.14159

float GetCircum(float);
float circumference;

main()
{
    float radius = 12.44;

    circumference = GetCircum(radius);
}

float GetCircum(float r)
{
    return((r + r) * PI);
}
```

Return Keyword

The return keyword is used in C language to end the execution of a function and return control to the line following the function call. We have seen that a return statement can contain an expression, conventionally enclosed in parenthesis, that represents the value returned to the caller. In the function GetCircun() listed above, the return statement

```
    return((r + r) * PI);
```

also performs the circumference calculation. The function could have been coded so that the circumference is first stored in a local variable; for example,

```
float GetCircum(float r)
{
    float cir;
    cir = (r + r) * PI;
    return (cir);
}
```

Because the parenthesis symbols are optional, the return statement in the previous code sample could have been coded in the form

```
    return cir;
```

If no value is associated with the return statement, then the returned value is undefined. If no return statement appears in a function body, then the function concludes at its last statement, that is, when it encounters the closing brace (}). In this case, programmers sometimes say that the function "fell off the edge." Functions that return no values should be prototyped and declared with a return value of type void. The return statement can also be used to return a constant to the calling program; for example,

```
    return (6);
```

A return statement can appear anywhere in the function body. A function can contain more than one return statement, and each one can return a different value to the caller; for example,

```
if ( program_error > 0 )
    return (1);
else
    return (0);
```

In conclusion, a C function can contain no return statement, in which case execution concludes at the closing brace (function falls of the edge) and the returned value is undefined. Alternatively, a function can contain one or more return statements. A return statement can appear in the form

```
return;
```

In this case, no value is returned to the caller. Or a return statement can return a constant, optionally enclosed in parentheses

```
return 0;
```

or

```
return (1);
```

Finally, a return statement can return a variable or include an expression:

```
return (error_code);
return((r + r) * PI);
return 2 * radius;
```

If a function returns no value, it should be prototyped and declared of void type. In this case, the return statement, if used, does not contain an expression.

Matching Arguments and Parameters

The prototype and the declaration of a function must include (in parentheses) a list of the variables whose value will be passed to the function, called the formal parameter list. In Figure A.20, the formal parameter list includes the type float variables a and b. The function call contains, also in parentheses, the names of the variables whose value is passed to the function. Note that the term function argument relates to the variables referenced in the function call, while the term function parameter refers to the variables listed in the function declaration. In other words, a value passed to a function is an argument from the viewpoint of the caller and a parameter from the viewpoint of the function itself. The programmer should remember that data is passed to a function in the order in which the arguments are referenced in the call and the parameters listed in the function declaration. In other words, the first argument in the function call corresponds to the first parameter in the header section of the function declaration. The second argument referenced in the call is assigned to the second parameter in the declaration, and so on.

A.10 Visibility of Function Arguments

The value returned by a function is associated with the function name and not with the variable or variables that appear in the return statement. For this reason, the variable referenced in a functions return statement can be an automatic variable because it is not used to return a value to main(). Also observe that the function returns a single value to the caller. There are several ways of getting around this limitation, some of are discussed in the following paragraphs.

A.10.1 Using External Variables

Functions can receive and return values. The function definition

```
float GetCircum(float radius);
```

describes a function that receives and returns arguments of type float. The argument passed to the function, in this case the radius of the circle, is not the variable itself but a copy of its value at the time of the function call. This method, usually described as passing by value, allows the function to use this temporary copy of a variable's value but not to change the variable itself. In this manner, the GetCircum() function cannot change the value of the variable named radius, which is visible only inside main(). The same limitation applies to arguments returned by a function.

The pass by value feature of C provides a way for protecting variables from unexpected or undesired changes. This action is due to the fact that the scope of an automatic variable is the function in which it was declared. On the other hand, external variables are visible to all functions that appear later in the code. This mechanism provides an alternative way of passing data between functions. For example, the program to calculate the circumference of a circle could also be coded using the external variables, as in the following sample:

```
#define PI 3.14159


void GetCircum(void);
float radius, cir;              /* <== external variables /*

main()
{
    radius = 12.33;
    GetCircum();
    // Now the global variable cir holds the circumference
    return(0);
}

void GetCircum(void)
{
    cir = (radius + radius) * PI;
    return;
}
```

In this code sample the variables radius and cir are declared outside the main() function. Therefore they are visible to main() and to GetCircum(). Because both functions are able to access data directly from storage, no values are passed to Circum() in the call. By the same token, the results of the circumference calculations are recovered by main() in the external variable named cir. Because no parameters are passed to or returned by the function Circum(), it is prototyped and declared using the void keyword. In this manner, the format void function() is used when a function returns no arguments. While the format function(void) serves as a notice to the compiler that the function is passed no arguments. By extension, the format void function(void) represents a function that neither receives nor returns arguments to or from the caller.

A.10.2 Passing Data by Reference

The mechanism of a C language function allows passing to it multiple arguments but a single one can be returned to the caller. We have seen how it is possible to overcome this limitation using external variables that are visible both to the calling and to the called functions. Another way in which a function can change the value of one or more of the caller's variables is by passing to the function the address of these variables. In this case, we can say that the default mode of operation (passing function arguments by value) is circumvented and the function arguments are passed by reference.

Pointers and Functions

We have seen that a C language program can obtain the address of a variable (the & symbol is the address of operator) and how the * symbol (indirection operator) is used to create variables that hold the address of another variable. It is by means of these operators that a calling routine can pass to a function the address of a variable, and the called function can access the contents of an external variable. The process is shown in the following code fragment:

```
#define PI 3.14159

void CircCalc(float *, float *, float *);

main()
{
    float radius = 12.33;
    float diameter, circumference;

    CircCalc(&radius, &diameter, &circumference);
    // Data for the diameter and circumference is now
    // stored in the corresponding local variables
    return(0);
}

void CircCalc(float *rad, float *dia, float *circ)
{
    *dia = *rad + *rad;
    *circum = *dia * PI;
    return;
}
```

In this case, note that the function CircCalc() is prototyped and declared of void return type. This is due to the fact that the function will access the caller's variable directly and, therefore, will not return a value in the conventional way. Also note that the variables named radius, diameter, and circumference are declared inside main(), which would normally make them invisible to CircCalc(). However, the function call statement

```
CircCalc(&radius, &diameter, &circumference);
```

passes to the function the address of the variables radius, diameter, and circumference, rather than a copy of their current value. The fact that the function CircCalc() receives pointers, rather than values, can be seen in the function prototype as well as in the declaration:

```
void CircCalc(float *rad, float *dia, float *circum)
```

In this case *rad, *dia, and *circum are pointer variables that hold the addresses of the variable named radius, diameter, and circumference respectively.

Passing Array Variables

We have seen how a pointer variable is used to gain access to the elements of an array, and, in previous paragraphs, we discussed how a function can receive the address of a variable encoded in a pointer. The following code fragment demonstrates some simple manipulations in passing an array to a function.

```
void ShowArray(char *);

main()
{
static char USA_name[] = "United States of America";

    ShowArray(USA_name);
    return(0);
}

void ShowArray(char *USA_name_ptr)
{
    unsigned int counter = 0;

    while(USA_name_ptr[counter]) {
        Print(*USA_name_ptr[counter]);
    return;
}
```

Note that arrays must be passed by reference if the function is to have access to its elements. For this reason, the function declaration

```
void Show_array(char *USA_name_ptr)
```

contains a pointer variable to an array of char type. This pointer, named USA_name_ptr, allows the function to access the elements of an array declared in main(). In the previous code listing, the function ShowArray() first displays the array using a pointer.

A.10.3 Function-Like Macros

We have seen how the #define directive can be used to associate an identifier with a constant; for example,

```
#define PI 3.1415927
```

which thereafter assigns the constant value 3.1415927 to the identifier PI. By the same principle the #define directive can be used to assign a string value to an identifier; for example, the program line

```
#define MSU "Minnesota State University\n"
```

assigns the string enclosed in double quotation marks to the identifier MSU. Thereafter, any reference to MSU will be replaced with the string "Minnesota State University\n". Note that it is conventional practice in C to use uppercase letters for identifiers associated with the #define directive.

Macro Argument

In addition to the literal replacement of a numeric or string value for an identifier, C allows the use of an argument as the replacement element of a #define directive. This construction is usually called a macro. The most important element in the macro concept is the substitution or replacement that takes place when the macro is encountered during compilation. This substitution, sometimes called the macro expansion, is handled by a section of the compiler called the preprocessor.

The simplest version of a macro argument is a macro formula; for example,

```
#define PI 3.1415927
#define DOUBLE_PI  2 * PI
```

In this case, the macro formula DOUBLE_PI generates the product of the constant 2 times 3.1415927, which is represented by the identifier PI defined in another macro.

Finally, a macro can contain an argument that is replaced by a variable on expansion, as in the following code fragment:

```
#define PI 3.1415926
#define CIRC(x) ((x + x) * PI)
```

In this example, the macro named CIRC calculates the circumference of any variable (represented by x in the macro definition). Regarding the macro argument, it is important to note that the replaced variable, which can be any legal C identifier, is enclosed in parentheses following the macro name. Also note that there can be no space between the macro name and the argument's left parenthesis. This use of spaces as separators in the macro definition creates a lexical peculiarity that sometimes confuses the novice programmer. A final point to keep in mind is that the macro definition does not end in a semicolon.

The use of parentheses in the argument makes macros somewhat reminiscent of C functions, to the point that some authors speak of function-like macros. However, there are differences between a macro and a function. Perhaps the most important one is that the macro expansion is the replacement of the identifier (the macro name) by its equivalent expression. Therefore, the code generated during a macro expansion is in line, while a function call requires that execution be transferred to the function body and then returned to the line following the call. The call/return operation associated with functions brings about an overhead in execution time. This determines that if identical calculations are encoded as a macro and as a function, the function will take slightly longer to execute than the equivalent macro. On the other hand, we have seen that the macro is expanded every time that the macro name is referenced in the code, whereas a function appears only once. This means that a macro that is referenced more than once will add more to the size of the executable file than an equivalent function.

A.11 Structures, Bit Fields, and Unions

In digital systems, a record is defined as a set of related data items. A typical example is the record for an individual employee in an organization's database. Typically, an employee record holds the employee's name, address, Social Security number, mailing address, base salary or wage, number of dependents, and other pertinent information. To store this information in a computer program, we will have to resort to several different data types. For example, string items, such as the employee's name, Social Security number, and mailing address, would go into arrays of type char. Decimal data items, such as salary or wage, would be stored in variables of type real. Finally, integral data items, such as the number of dependents, would probably be encoded using the unsigned char or int data types.

We have seen that C language arrays are a collection of data items of the same type (see Section 3.1). For this reason, a data record that consists of data items of different type cannot be stored in a single array. A C language structure is a data structure that can contain one or more variables of the same or of different types. The structure allows treating a group of related items as a unit, such as the employee record previously mentioned, thus helping the programmer organize and manage the more complicated data processing operations.

Although C language structures share some of the properties of functions and of arrays, they constitute a distinct programming concept. In fact, C structures contain lexical elements that have become characteristic of the C language.

A.11.1 structure Declaration

Like other elements of the C language, structures must be pre-declared. This operation usually requires two distinct steps: the structure type declaration and the structure variable declaration.

structure type Declaration

In the structure type declaration, the programmer defines the structure name, sometimes called the structure tag, and lists one or more associated variables, called the structure members. These members can be of different data types. Figure A.21 shows the elements of a structure type declaration.

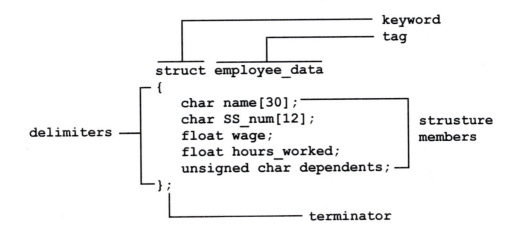

Figure A.21 *Elements of a structure type declaration.*

Note the following characteristics of a structure type declaration.

1. Starts with the keyword struct.

2. The identifier following the struct keyword is the structure tag. The tag defines a structure type.

3. The elements of the structure are enclosed in braces, in a similar manner as the body of a function.

4. Ends in a semicolon.

Because the structure is a pattern of data objects to be associated in storage, the structure tag is not a variable name, but a type name. In other words, a structure is a programmer-defined data type. The type declaration can be visualized as a description of a template to be used in grouping several data objects, which can be of different type. The type declaration, which is reminiscent of the function prototype, serves as a mere description of a structure and therefore assigns no physical storage space; it creates a structure template.

structure variable Declaration

Once the structure type declaration has defined the tag and data format for a structure, the program can allocate storage space for one or more structures by means of the structure variable declaration. For example, once the type of structure for employee data (see Figure A.21) has been declared, the program can assign storage space for five employees, as follows:

```
struct employee_data shop_foreman;
struct employee_data machinist_1;
struct employee_data machinist_2;
struct employee_data welder_1;
struct employee_data welder_2;
```

Hereafter, the program will have reserved storage for five structures, of the type declared in Figure A.21. In this example, each structure requires the following memory space, including the string terminator codes:

```
MEMBER                                    BYTES
char name [30]                              31
char SS_num [12]                            13
float wage                                   4
float hours_worked                           4
unsigned char dependents                     1
                                     --------------
              total storage            53 bytes
```

Note that each structure variable declaration statement performs a similar operation as a variable declaration; that is, it reserves storage for a data item and assigns to it a particular identifier (name).

C language also allows declaring a structure variable at the same time as the type declaration; for example,

```
struct employee_data
    {
    char name [30];
    char SS_num [12];
    float wage;
    float hours_worked;
    unsigned int dependents;
    } welder_1, welder_2;
```

In this case, the variables welder_1 and welder_2, of the structure tag employee_data, are declared at the time of the type declaration.

A.11.2 Accessing Structure Elements

We have seen that individual array elements are accessed by means of a subscript that encodes the relative position of each item in the array. In accessing a particular member of a structure, we must take into account both the type declaration and the variable declaration. C language provides a membership operator symbol, represented by a dot (.), which is also called the dot operator or the member of operator. By means of the membership operator, we can connect a structure member (defined in the structure type declaration) and a variable name, as in the statement

```
machinist_1.wage = 14.55;
```

This statement creates an addressable variable by relating the structure variable "machinist_1" with the structure element "wage." The following code fragment demonstrates structure type and variable declarations and the use of the membership operator:

```
main()
{
    float salary;

    /* Structure type declaration */
    struct employee_data
        {
        char name [30];
        char SS_num [12];
        float wage;
        float hours_worked;
        unsigned int dependents;
        };

        /* Structure variable declaration */
        struct employee_data machinist_1;
        struct employee_data machinist_2;

    /* Use of the membership operator */
    machinist_1.wage = 14.55;
    machinist_2.wage = 16.00;
    machinist_1.hours_worked = 12;
    machinist_2.hours_worked = 40;

    ...
```

Initializing Structure Variables

Like conventional variables, structure variables can be initialized at the time they are declared. The initialization of a structure variable, which is reminiscent of the initialization of an array variable, can be seen in the following statement:

```
static struct employee_data machinist_1 =
    { "Joe Smith", "234 43 274", 14.55, 40.2, 5 };
```

Note that as is the case with arrays (see Section 3.1) a structure variable must be declared static if it is to be initialized inside a function. By the same token, function variables declared external do not require the static keyword. An alternative style for initializing function variables is to place each item on a separate line, as follows:

```
static struct employee_data machinist_2 =
    {
    "Jim Jones",
    "200 12 345",
    16.00,
    10.5,
    2
    };
```

Manipulating a Bit Field

We have seen how the bitwise operators can be used to manipulate the individual bits within an integral data type. These operations can be further simplified using the #define directive to isolate individual bits or fields. For example, because each binary digit corresponds to a power of 2, we can mask the individual bits as follows:

```
#define   BIT0_MASK   01
#define   BIT1_MASK   02
#define   BIT2_MASK   04
#define   BIT3_MASK   08
#define   BIT4_MASK   16
#define   BIT5_MASK   32
#define   BIT6_MASK   64
#define   BIT7_MASK   128
```

Thereafter, a program can use these bit masks in conjunction with the bitwise operators AND, OR, XOR, and NOT (see Section 5.2) to test and change the individual bits in an integral variable.

In addition to the bitwise operators, C allows the definition of individual bits within a structure composed of one or more members of an integral data type. This specialized type of structure is called a bit field. The following is a bit field structure declaration

```
struct low_bits
{
    unsigned int bit0     : 1;
    unsigned int bit1     : 1;
    unsigned int bit2     : 1;
    unsigned int bit3     : 1;
    unsigned int bits4_7 : 4;
    unsigned int padding : 8;
} mask_1;
```

Notice that the colon symbol is used in the type declaration of a bit field structure to assign a dimension to each field. In the above declaration for the tag low_bits, the first four fields, named bit0, bit1, bit2, and bit3, are declared to be 1 bit wide, while the field named bits4_7 is declared to be 4 bits wide, and the field named padding is declared to be 8 bits wide.

We have seen that a structure variable can be created at the time of the type declaration. In the previous bit field, we created a variable named mask_1. The fields in this variable can now be accessed by means of the membership operator (.) in the manner explained previously. For example, we can initialize bits 0 to 3 to a value of one and the remaining bits to zero with the statements

```
mask_1.bit0 = mask_1.bit1 = mask_1.bit2 = mask_1.bit3 = 1;
mask_1.bits4_7 = mask_1.padding = 0;
```

However, C does not allow direct access to a bit field variable. In other words, a C statement cannot reference the variable mask_1 as it would a conventional variable. For example, the program fragment

```
unsigned int value, result;
value = 127;
result = value & mask_1;
```

is illegal because the bitwise AND (&) operator cannot be applied to a structure variable. The only operations that can be performed on structures are to obtain the structure's address by means of the address of (&) operator and to access the structure members. Therefore, a structure variable such as mask_1 cannot be accessed as a unit. However, because there are no restrictions regarding pointers to structures we can use the address of and the indirection operators to gain access to a bit field variable. The following code fragment shows the necessary manipulations:

```
main()
{

    struct
    {
    unsigned int bit0          : 1;
    unsigned int bit1          : 1;
    unsigned int bits2_6       : 5;
    unsigned int bit7          : 1;
    unsigned int padding       : 8;
    } mask;

    unsigned char value, result;
    unsigned int *maskptr;
                                    /* Create a pointer to an  */
                                    /* unsigned int variable    */

    mask.bit0 = mask.bit1 = mask.bit7 = 1;
    mask.bits2_6 = mask.padding = 0;

    maskptr = (unsigned int *)&mask;
                                    /* Initialize the pointer   */
                                    /* to the address of the    */
                                    /* structure variable       */

    Print("\nMask is: %u", mask);

    Print("bit0 is: ", mask.bit0);
    Print("bit1 is: ", mask.bit1);
    Print("bits2_6 is: ", mask.bits2_6);
    Print("bit7 is: ", mask.bit7);

    value = 127;
    result = value & *maskptr;
                                    /* Now it is possible to    */
                                    /* access the entire bit    */
                                    /* field through the        */
                                    /* contents of the pointer */

    Print("Contents of maskptr variable is: ", *maskptr);
    Print("Result of ANDing 127 with mask is: ", result);
```

```
    return(0);
}
```

Type Casting

The preceding code fragment contains the statement

```
    maskptr = (unsigned int *)&mask;
```

that assigns the address of the structure variable named mask to the pointer variable named maskptr. Note that the statement assigns the address of an array (&mask) to a pointer variable of type unsigned int (maskptr). This maneuver is necessary because we are trying to access the structure as a unit. The element enclosed in parentheses, called a type cast, indicates to the compiler that we are forcing a type conversion. In this case, the forced conversion consists of assigning the address of a structure to an unsigned int pointer.

If the assignment statement did not include a type cast (in parentheses), the compiler would generate a warning message indicating that an indirection operation referred to values of different type, however, the program would compile and run. In many cases, the type cast serves the additional purpose of documenting and clarifying the programmer's intentions.

A.11.3 Unions

A union is a variation on the concept of structure in which the member variables can hold different data types at different times. The union type declaration consists of a list of data types matched with the variables names in a similar manner to that of a structure type declaration. This means that the union variable declaration is similar to a structure variable declaration except that the structure declaration causes the compiler to reserve storage for all the members listed, whereas in the union declaration the compiler reserves space for the largest member in the aggregate. In other words, the members of a structure variable have their own memory space while the members of a union variable share a common memory space. This means that in relation to unions, the programmer must keep track of the current occupant of the common memory space. The following code fragment demonstrates the use of a union to hold three different data items of type double, float, and int:

```
    union vari_data
    {
    double num_type1;
    float num_type2;
    int num_type3;
    };

main()
{

union vari_data first_set;    /* Declare the union variable   */
                              /* named first_set              */

first_set.num_type3 = 12;     /* Store an integer in union    */
...
```

```
first_set.num_type2 = 22.44; /* Store a float in union     */
...
first_set.num_type1 = 0.023; /* Store a double in union    */
...
```

A.11.4 Structures and Functions

Because of the limitation that structures cannot be accessed as a unit, a C program cannot pass a structure to a function directly. Nevertheless, as in the case of bit fields, we can get around this limitation by storing the address of the structure in a pointer variable and passing this address to the function. Keep in mind that the value of a structure variable (the structure member) can be passed as a parameter.

Pointers to Structures

The declaration, initialization, and passing of structure pointers is quite similar to that of conventional pointer variables. Nevertheless, the handling of structure pointers requires special symbols and operators. In the first place, the pointer declaration must specify that the object of the pointer is a structure. For example, the statement

```
struct triangle *tr_ptr;
```

creates a pointer to a structure of the tag triangle. But the structure pointer must be initialized to a particular structure, not to a template. For example, if triangle is the tag for a structure template, the statement

```
tr_ptr = &triangle;
```

is illegal. However, the statement

```
struct triangle tr1;
```

creates a structure variable named tr1, according to the tag triangle. Thereafter, we can initialize a pointer to the structure tr1 as follows:

```
tr_ptr = &tr1;
```

Pointer Member Operator

Accessing individual structure members by means of a pointer requires a special symbol called the pointer member operator. This operator symbol consists of a dash and an angle bracket combined to simulate an arrow (–>). For instance, if one member of the structure tr1 is named side_a, we can use the pointer member operator as follows

```
tr_ptr -> side_a
```

Note that the pointer member operator (->) is combined with the structure pointer in a similar manner as the membership operator (.) is combined with the structure name.

Passing Structures to Functions

Structures, like other variables, can be declared external in order to make them visible to all functions in a program. Local structures (declared inside a function) can be passed to other functions by means of structure pointers, which then can use the pointer member operator to gain access to the data stored in the structure. The following code fragment uses a function named perimeter() to calculate the perimeter of a triangle whose sides are stored in a structure.

```
struct triangle
{
float side_a;
float side_b;
float side_c;
};

void perimeter(struct triangle *);

main()
{
    struct triangle tr1;        /* Declaring structure variables */
    struct triangle *tr_ptr;

    tr1.side_a = 12.5;          /* Initializing variables      */
    tr1.side_b = 7.77;
    tr1.side_c = 22.5;
    tr_ptr = &tr1;

    perimeter(tr_ptr);          /* Calling the function        */
    return(0);
}

void perimeter(struct triangle *tr1_ptr)
{
    float a, b, c, p;
    a = tr1_ptr  > side_a;
    b = tr1_ptr  > side_b;
    c = tr1_ptr  > side_c;
    p = a + b + c;
    Print("Perimeter is: ", p);
    return;
}
```

In the preceding code fragment, note the following points:

1. The structure declaration is external. This makes the function tag visible to all functions.

2. The prototype for the function named perimeter declares a structure pointer variable as the argument passed to the function. This is done in the statement

   ```
   void perimeter(struct triangle *);
   ```

   Also note that the name of the structure variable need not appear in the prototype.

3. The actual structure is declared in main(), as is the structure pointer variable. These declarations take place in the statements

```
struct triangle tr1;
struct triangle *tr_ptr;
```

4. Initialization of structure variables (members) is performed in the conventional manner by means of the membership operator.

```
tr1.side_a = 12.5;
tr1.side_b = 7.77;
tr1.side_c = 22.5;
```

The pointer variable is initialized using the address of operator.

```
tr_ptr = &tr1;
```

5. The function named perimeter() is declared to receive a structure pointer in the line

```
void perimeter(struct triangle *tr1_ptr)
```

The pointer name (tr1_ptr) is a local variable and therefore need not coincide with the one referenced in main.

6. The function gains access to the data stored in the structure members in the statements

```
a = tr1_ptr -> side_a;
b = tr1_ptr -> side_b;
c = tr1_ptr -> side_c;
```

Access to structure data through a structure pointer variable (tr1_ptr) requires the use of the pointer member operator.

A.11.5 Structures and Unions in MPLAB C18

MPLAB C18 supports structures, bit fields, and unions in compliance with the requirements of the ANSI standard. In addition, the MPLAB C18 contains extensions to the standard that facilitate C programming in embedded systems. One such extension is the support for anonymous structures inside of unions. An anonymous structure has the form

```
struct { member-list };
```

It is designed to define an unnamed object, although the names of the members of an anonymous structure must be distinct from other names in the scope in which the structure is declared.

For example, the processor-specific header file p18f452.h contains several anonymous structures inside unions that help access the bits, bit fields, and hardware registers of the 18F452 device. The union named PORTAbits is defined as follows:

```
extern volatile near unsigned char        PORTA;
extern volatile near union {
  struct {
    unsigned RA0:1;
    unsigned RA1:1;
    unsigned RA2:1;
    unsigned RA3:1;
```

```
    unsigned RA4:1;
    unsigned RA5:1;
    unsigned RA6:1;
  };
  struct {
    unsigned AN0:1;
    unsigned AN1:1;
    unsigned AN2:1;
    unsigned AN3:1;
    unsigned :1;
    unsigned AN4:1;
    unsigned OSC2:1;
  };
  struct {
    unsigned :2;
    unsigned VREFM:1;
    unsigned VREFP:1;
    unsigned T0CKI:1;
    unsigned SS:1;
    unsigned CLK0:1;
  };
  struct {
    unsigned :5;
    unsigned LVDIN:1;
  };
} PORTAbits;
```

Because of this union declaration and its anonymous structures, a program can test if bit 3 of Port A is set with the expression

```
    if(PORTAbits.RA3)
```

In this expression we have used the union PORTAbits and the member RA3 of an anonymous structure defined within this union. In the program named C_Bitfield_Demo.c, in the book's software resource, we have used define statements to extend the PORTAbits union defined in the header file p18f452.h. This allows us to include the bitfield in the Demo Board 18F452-A. Alternatively, the header file can be edited with an additional structure that defines these bits. Code is as follows:

```
#define SW1 RA2
#define SW2 RA3
#define SW3 RA4
#define SW4 RA5
#define DIPSW PORTAbits

...

while(1)
    {
    pattern = 0;
    // Cascaded if statements using the new bitfields and the
    // alias defined for the PORTAbits union
    if(DIPSW.SW1)
        pattern |= 0b00000001;
    if(DIPSW.SW2)
        pattern |= 0b00000010;
```

```
if(DIPSW.SW3)
    pattern |= 0b00000100;
if(DIPSW.SW4)
    pattern |= 0b00001000;
FlashLED(pattern);
```

Appendix B

Debugging 18F Devices

B.1 Art of Debugging

Detecting and correcting program defects are an important part of program development. In an ideal world, a program could be designed and constructed following principles that would preclude defects and ensure that the resulting code correctly performs all the functions required of it and would do so with minimal complexity. A field of software engineering sometimes called scientific programming or formal specifications has explored the possibility of using predicate calculus to develop programs following a methodology that precludes errors and ensures the code itself is proof that it solves the intended tasks. The work of Edsger W. Dijkstra, David Gries, and Edward Cohen has investigated this methodology but, unfortunately, so far it has not been found to be useful in practical programming. Until this or another rigorous program design and coding methodology succeeds, we will continue to be stuck with program bugs, because they are a natural consequence of trial-and-error programming.

The previous statement should not be interpreted to mean that programming is, by necessity, a haphazard process. Good program design and coding principles, based on solid software engineering, generate code that may not be logically perfect or mathematically provable but is certainly better structured and more effective. In this sense, that art of programming is more thinking than writing code. Most programmers spend less time in design considerations than would be required for a solid program structure. Knowing that embedded systems are often coded by engineers rather than computer scientists.

Furthermore, throughout this book we have attempted to demonstrate and emphasize the advantages of well-commented and well-structured code. The typical programming student's attitude that states, "I will now write the code and later come back and add the comments" is guaranteed to fail. The main reason for program comments is to leave a record of the thoughts that were on the programmer's mind at coding time. Postponing this task is a self-defeating proposition. Most professional programmers will agree that a program is as good as its comments, and the most valuable attribute of an excellent programmer is the ability to write solid code decorated with clear and elegant comments. Debugging badly commented or carelessly designed code is usually a futile effort. In many such cases, we are forced to conclude that it will be simpler and easier to rewrite the code from scratch than to try to fix a messy and disorganized piece of bad programming.

B.1.1 Preliminary Debugging

In all the examples discussed in this appendix we have assumed a program that assembles or compiles correctly. This preliminary step usually consists of finding errors in syntax or fundamental program structure that make the assembly or the compile process fail. The diagnostics provided by the development software are helpful in locating and correcting the syntax or construction errors.

One of the possible causes of errors at the development stage is the incorrect selection of development tools. For example, assembly language programs can be developed in absolute or relocatable code, and each type requires a particular development tool set. Attempting to build a relocatable program using an assembly file for absolute code will generate errors that relate to an incompatible development environment. The use of coding templates for each type of program helps to prevent these errors.

Another frequent source of preliminary or developmental errors is the incorrect selection of the MCU device. If the processor or its #include file were defined in the source, then the development environment returns a "header file mismatch" error. Otherwise, the individual errors are listed by offending source line. This can result in a long list of errors all caused by a single flaw. Unfortunately, there is no way of ignoring the setting of the device configuration selected in the MPLAB Configure menu. This is due to the fact that the MPLAB environment is modified to match the individual processor selected.

Once the code assembles and compiles without error and an executable file (.hex format) is generated, debugging can proceed to the next step.

B.1.2 Debugging the Logic

A program of any complexity is an exercise in intuitive logic, if not in formal logic. Programmers often marvel at the basic logical errors that flaws our thinking. Sometimes we refer to these elementary errors in reasoning as "stupid" or "dumb" mistakes. Detecting and correcting the more subtle errors in our logical thinking can be a formidable task. The primary, perhaps the only practical tools at this stage are flowcharting and code step analysis. Building a flowchart based on the actual code often reveals elementary errors in reasoning. Single-stepping execution of the offending code with the help of a debugger can be used to make certain that processing actually follows the necessary logical steps. In other words, logic flaws can consist of errors in the logical thinking itself, or in the code not following the predefined logic.

B.2 Software Debugging

Whatever the reason that a program does not perform as expected, it is convenient to have schemes and tools that allow for inspecting and testing the code. These range in complexity from simple devices and techniques developed by the programmer to more-or-less sophisticated systems that allow inspecting the code, variables, and data at execution time. In the present section we discuss software debugging, that is, locating and fixing flaws and defects in the program code. In Section B.3 we discuss tools and techniques for debugging both the hardware and the software.

B.2.1 Debugger-Less Debugging

Even without any debugging tools it is often possible to detect a program error by skillful trickery. For example, a program fails by apparently hanging up before its termination but the programmer cannot tell at which of it various stages this is happening. If the hardware contains several LEDs or if these can be attached to the breadboard circuit, the program can be modified to make calls to light up each one of the LEDs at different stages of execution. The programmer can then run the code and observe which is the last LED that lights up in order to tell the stage at which the program is failing. Similarly, the fail point can be echoed with a numeric code on a seven-segment LED or a message displayed on an LCD device if these are available in the hardware.

Another primitive debugging technique that can be used if LEDs or other output devices are unavailable is storing an error code in EEPROM memory. In Chapter 10 we developed simple EEPROM write routines in both C and assembly language that can be used for this purpose. Most programmers provide utilities to read and display EEPROM memory. The offending program can be modified to store a specific code in an EEPROM memory address. The microcontroller can be placed on the target board, the program executed, and then the MCU removed and replaced in the programming board in order to inspect the code stored in its EEPROM.

Either one of these methods is not suitable for every occasion but there are cases in which we must resort to them due to the lack of a code or hardware debugger.

B.2.2 Code Image Debugging

The most elementary debugging tool is a program that simulates program execution by providing a software model of the circuit being examined. MPLAB SIM, which is furnished with the MPLAB development environment, is one such simulator-debugger for the PIC MCUs.

MPLAB SIM simulates the hardware at the register level, not the pin level. With the simulator, RB0 represents the value in bit0 of PORTB, not the value on the pin RB0. This is due to the fact that the simulator provides only a software model, not the actual device hardware. In many cases, the binary values stored at the register bit level and the pin level are the same. However, there are cases in which the pin and port bit levels are different. For example, the ADC comparator requires that port register read '0', which may not be the value in the actual pin. Additionally, device I/O pins are often multiplexed with those of other functions or peripherals. In this case, the simulator recognizes the pin names defined in the corresponding .inc file, although most multiplexed bit/pin names may be used interchangeably.

MPLAB SIM allows the following operations:

- Modify object code and reexecute it
- Inject external stimuli to provide the simulation of hardware signals
- Set pin and register values
- Trace the execution of the program
- Detect "dead" code areas
- Extract program data for inspection

B.2.3 MPLAB SIM Features

The MPLAB SIM debugger provides two basic modes of code execution: run and step. The step mode includes a variation called animate.

Run Mode

In Run mode program code is executed until a breakpoint is encountered or until the Halt command is received. Status information is updated when program execution is halted.

Run mode is entered either by clicking the Run button on the Debug toolbar (see Figure B-1 later in this appendix), by selecting Debugger>Run from the menu bar, or by pressing <F9> on the keyboard. The Run mode is halted using a breakpoint or other type of break, for example, "Halt on Trace Buffer Full". The Run mode is manually halted by either clicking the Halt button on the Debug toolbar, selecting Debugger>Halt from the menu bar, or pressing <F5> on the keyboard.

Step Mode

There are several types of Step modes: Step Into, Step Over, and Step Out. The Step modes are discussed in Section B.6.2.

Animate

The Animate variant of the Step mode causes the debugger to actually execute single steps while running, updating the values of the registers as it runs. Animate executes the program slower than the Run function. This allows viewing the register values as they change in a Special Function Register window or in the Watch window. To halt Animate, use the menu option Debugger>Halt instead of the toolbar Halt or <F5>.

Mode Differences

Functions and features that work in Run mode do not always work when in Step or Animate mode. For example:

- Peripheral modules do not work, or do not work as expected, when code is stepped.
- Some interrupts do not work when code is stepped. Because some peripherals are not running, their interrupts will not occur.

- Stepping involves executing one line of code and then halting the program. Because open MPLAB IDE windows are updated on Halt, stepping may be slow. It is possible to minimize as many windows as possible to improve execution speed.

Build Configurations

MPLAB IDE provides two options for building a project: Debug and Release. When the Quickbuild option is selected for assembly language programs, only the Release option is available. In programs contained in an MPLAB project, a Build Option drop-down box is visible in the Project Manager toolbar. The Build Option can also be selected by the Build Configuration command in the Project menu.

Programs assembled in the Release configuration with the Quickbuild option can be debugged using MPLAB SIM, but not with hardware debuggers.

Setting Breakpoints

Breakpoints are debugger controls that produce a conditional program halt. During the halt program, execution can be evaluated by observing memory, register, or variable. Breakpoints can be set in the file (editor) window, in the program memory window, or in the disassembly window, as follows:

- Double-clicking on the line of code where you want the breakpoint. Double-click again to remove the breakpoint. For this to work, the "Double Click Toggles Breakpoint" must have been selected in the Editor>Properties dialog.

- Double-clicking in the window gutter, in the next line of code where the breakpoint is wanted. Double-clicking again removes the breakpoint.

- Placing the cursor over the line of code where the breakpoint is desired then, right- clicking to pop up a menu and select "Set Breakpoint." Once a breakpoint is set, "Set Breakpoint" will change to "Remove Breakpoint" and "Disable breakpoint." Other options on the pop-up menu under Breakpoints are for deleting, enabling, or disabling all breakpoints.

- Opening the Breakpoint dialog (Debugger>Breakpoints) to set, delete, enable, or disable breakpoints. A debug tool must have been selected before this option is available.

B.2.4 PIC 18 Special Simulations

The conditions listed in the following subsections refer to the simulation of PIC 18 devices.

Reset Conditions

MPLAB SIM supports all Reset conditions. The condition causing the reset can be determined by the setting in the RCON register, such as the Time-Out (TO) and Power-Down (PD) bits. However, the device cannot be reset by toggling the MCLR line using stimulus control. Stimulus features are described later in this appendix.

Sleep

If a sleep instruction is in the code stream, the simulator will appear "asleep" until a wake-up from sleep condition occurs. If the Watchdog Timer has been enabled, it will wake up the processor from sleep when it times out, depending on the pre/postscaler setting.

Watchdog Timer

The Watchdog Timer is fully simulated in the simulator. The period is determined by the pre/postscaler Configuration bits WDTPS0:2. The WDT is disabled by clearing the WDTEN bit unless it is enabled by the SWDTEN bit of the WDTCON register. Setting the Configuration bit WDTEN to 1 will enable the WDT regardless of the value of the SWDTEN bit.

A WDT time-out is simulated when WDT is enabled, proper pre/postscaler is set, and WDT actually overflows. On WDT time-out, the simulator will halt or Reset, depending on the selection on the Break Options tab of the Settings dialog (Debugger>Settings).

Special Registers

To aid in debugging PIC 18 devices, certain items that are normally not observable are defined as "special" registers. For example, because prescalers cannot be declared in user code as "registers," the following special symbols are available as Special Function Registers:

```
T0PRE (Prescaler for Timer 0)
WDTPRE (Prescaler for WDT)
```

B.2.5 PIC 18 Peripherals

The following peripherals are supported by MPLAB SIM:

- Timers
- CCP/ECCP
- PWM
- Comparators
- A/D Converter
- USART
- EEPROM Data Memory
- Remapping of I/O pins
- OSC Control of IO
- IO Ports

Timers are supported, except those that use an external crystal. Timer interrupts on overflow and wake-up from sleep are also supported. Comparator modes that do not use Vref are simulated. Comparator pins cannot be toggled. The A/D Converter module is simulated in all registers, timing functions, and interrupts. Here again, simulation is at the register level. USART and UART functions are functionally supported with certain limitations. EEPROM data memory is fully simulated. Remapping and the lock/unlock functions of I/O pins are supported. I/O ports are supported at the register level for input/output/ interrupts, and changes.

B.2.6 MPLAB SIM Controls

The basic controls are available in an MPLAB SIM control window displayed when the debugger is selected or when the Debug toolbar is displayed from the View command. The window is shown in Figure B.1.

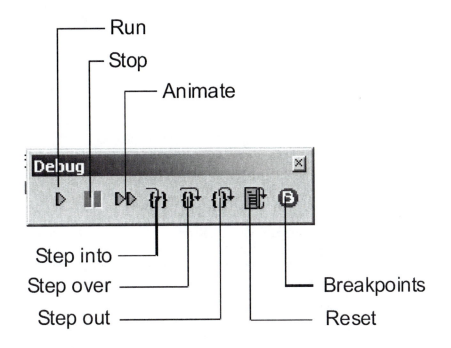

Figure B.1 *MPLAB SIM control window.*

Once MPLAB SIM is selected with the Debugger>>Select tool command, the MPLAB SIM control window is displayed in the MPLAB commands line. The window can be moved to another area of the desktop as the window shown in Figure B.1.

The Run button in Figure B.2 will start program execution for the first time at address 0x000. If a breakpoint stopped execution, then the Run button will resume it at the following instruction code. Execution continues until a breakpoint is reached, the Stop button is pressed, or the program terminates.

The Animate button starts a program animation. Animation is a method of execution that allows selecting the step time for each instruction. The rate is set in the Simulator Settings dialog and the Animation/Realtime Updates tab. The slider allows selecting an animation range from 0 to 5 sec. The updates refer to how often the Watch window is updated.

The three Step buttons refer to single-stepping execution. In assembly language, single stepping refers to executing one machine instruction at a time. In a C language program single stepping refers to one line of a high-level program statement. The Step into button moves the program counter to the next instruction to be executed, whether it is the next one in line or an instruction inside a procedure call.

The Step over button allows bypassing subroutines. When a call is encountered in the code, the Step over command causes the breakpoint to be set at the instruction following the call. The Step out button stops execution at the return address of the current subroutine. This allows breaking out of the subroutine currently executing.

The Reset button is used to restart the debugging section by returning execution to the program's first line of code. The Reset command does not erase existing breakpoints or debugger setup switches and parameters.

Finally, the button labeled B displays a dialog box that allows setting a breakpoint at an address or line number, including the source file path, removing a single or all breakpoints, enabling and disabling individual breakpoints, and displays the limit number of breakpoints and the number of available breakpoints. Most conveniently, a breakpoint can be set or removed by double-clicking the desired code line in the editor.

B.2.7 Viewing Commands

Several commands in the MPLAB View menu allow inspecting program elements during debugging. Although perhaps the simplest and most useful way of viewing registers and program components is by placing the cursor over the desired name and MPLAB displays a flag showing its contents. The view debugger-related view commands available in MPLAB SIM are the following:

- Disassembly Listing
- EEPROM
- File Registers
- Hardware Stack
- Locals
- Program Memory
- Special Function Registers
- Watch
- Simulator Trace
- Simulator Logic Analyzer

In the subsections that follow we discuss the View menu commands whose contents are not obvious. The Simulator and Tracing functions are discussed in a separate section.

Dissasembly Listing

This view command displays a disassembly of the code being debugged showing addresses and instruction opcodes. Figure B.2 shows a few lines of a disassembly listing of the sample program LedFlash_PB_F18.asm.

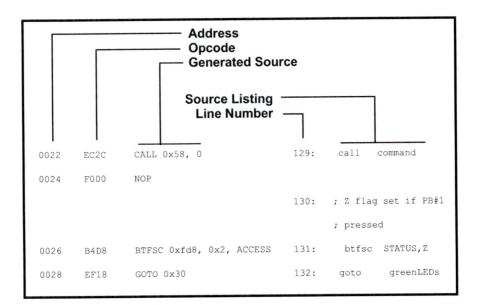

Figure B.2 *Disassembly listing.*

In Figure B.2 the right-hand column of the listing shows the original source code. The second line, right-to-left, is the consecutive line number. The next column, labeled Generated Source, contains the source generated by the assembler or compiler. Notice that this column shows the instruction mnemonics in capital letters and lists the actual address of elements represented by program names in the original source. Also note that implicit operands are displayed in this column, as is the memory bank (ACCESS) used by the instruction. The columns labeled Address and Opcode require no explanation.

File Registers

This command displays the address and contents of the file registers in the format of a conventional memory dump. Figure B.3 shows a File Registers screen.

File Registers																	_ □ ✕
Address	00	01	02	03	04	05	06	07	08	09	0A	0B	0C	0D	0E	0F	ASCII
F60	--	--	--	--	--	--	--	--	--	--	--	--	--	--	--	--	-------- --------
F70	--	--	--	--	--	--	--	--	--	--	--	--	--	--	--	--	-------- --------
F80	00	00	02	80	00	--	--	--	--	00	00	02	80	04	--	--	.....--- ......--
F90	--	--	7F	10	00	FF	07	--	--	--	--	--	--	00	00	FF	--......- -----...
FA0	00	00	1F	--	--	--	00	00	00	00	--	00	02	00	00	00	...--.. ..-....-
FB0	--	00	00	00	--	--	--	--	--	--	00	00	00	00	00	00	-...---- --.....
FC0	--	00	00	00	00	00	00	00	00	00	00	FF	00	00	00	00	-........ ....
FD0	0C	00	05	00	--	FF	4A	00	00	00	00	--	--	--	--	--	....-.J. ...-----
FE0	00	00	00	--	--	--	--	--	10	00	00	--	--	--	--	--	...----- ...-----
FF0	C0	F5	00	00	00	00	00	00	00	22	00	00	00	00	00	00	...."......

Hex	Symbolic

Figure B.3 *File Registers screen.*

Hardware Stack

The processor's hardware stack is mapped into the simulator's data memory space. All 32 stack locations can be viewed in a display window, as shown in Figure B.4.

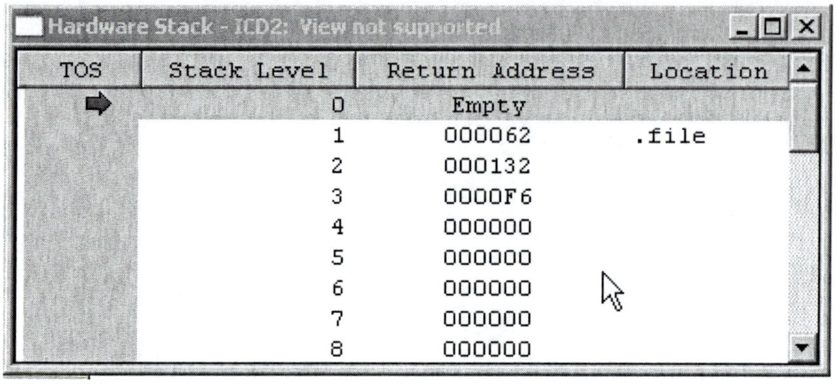

Figure B.4 *Hardware Stack window.*

Locals

The Locals window allows monitoring automatic variables that have local scope. This screen is used in projects coded in high-level languages such as C.

Program Memory

The Program Memory command in the View menu displays a window that shows the contents of all locations in the program memory space of the current processor. The default program memory screen is shown in Figure B.5.

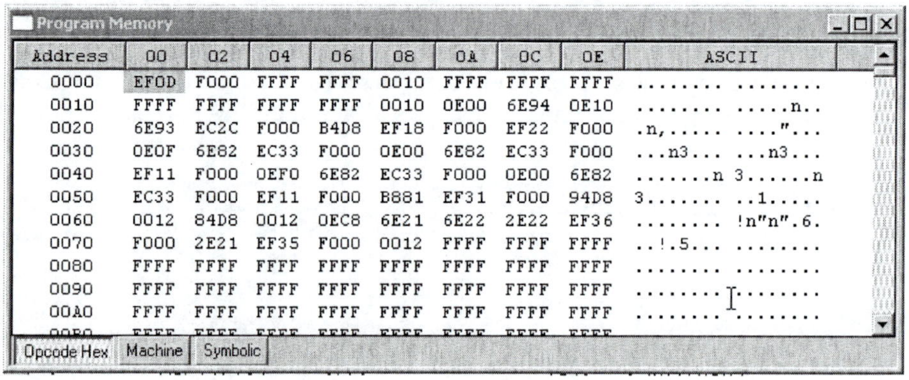

Figure B.5 *Program Memory window.*

The Status line of the Program Memory screen in Figure B.5 shows the three display modes available. Opcode Hex is the default mode and displays the data in a conventional screen dump format. The Machine display mode shows data as it appears in the Generated Source line of a disassembly listing (see Figure B.2). The Symbolic display mode shows the data in the user's source code format.

The various subcommands available in the Program Memory command are listed by pressing the right mouse button in the window's display area. Figure B.6 shows the commands menu.

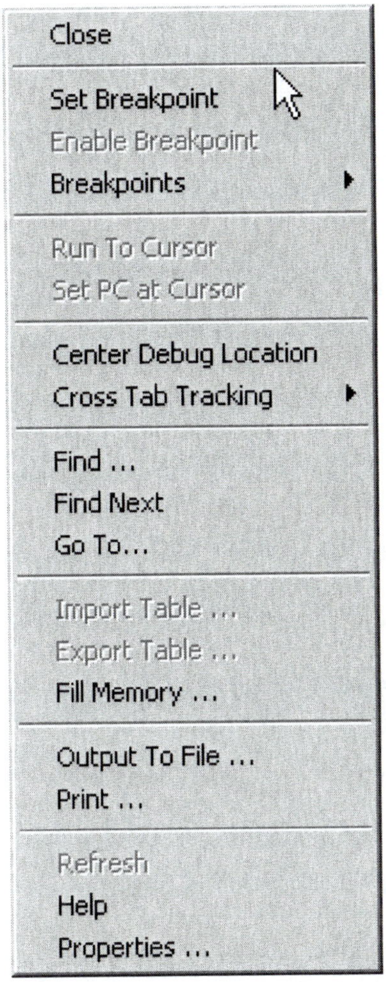

| Close |
| Set Breakpoint |
| Enable Breakpoint |
| Breakpoints ▶ |
| Run To Cursor |
| Set PC at Cursor |
| Center Debug Location |
| Cross Tab Tracking ▶ |
| Find ... |
| Find Next |
| Go To... |
| Import Table ... |
| Export Table ... |
| Fill Memory ... |
| Output To File ... |
| Print ... |
| Refresh |
| Help |
| Properties ... |

Figure B.6 *Commands in the Program Memory window.*

Special Function Registers

The Special Function Registers window displays the contents of the SFRs available in the selected processor. The contents of this window are the same as the File Registers window mentioned previously (see Figure B.3), however; the format is more convenient because each SFR name is included and several number formats are available. If a data memory register is not physically implemented on a particular device, it may not appear in the SFR display window. However, some tools such as simulators sometimes allow viewing registers that do not exist on the actual device, such as prescalers. Figure B.7 is a screen snapshot of the Special Function Registers window.

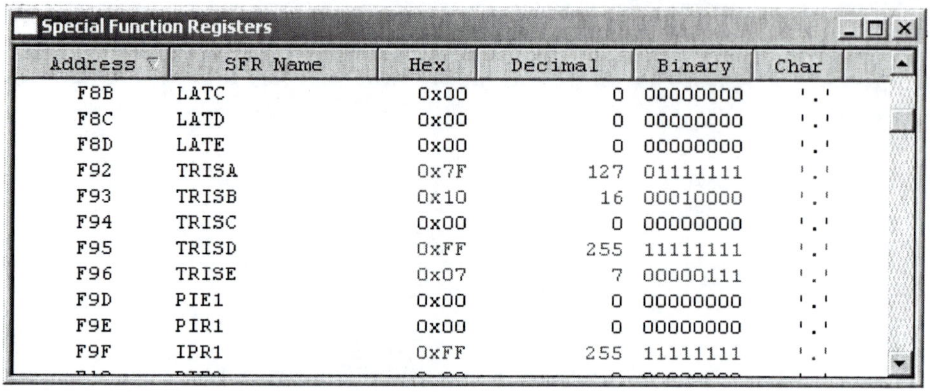

Figure B.7 *Special Function Registers window.*

While debugging, the registers that have changed during program execution are shown in red characters in the FSR window. In Figure B.7 these are shown in gray characters.

Watch

The Watch window is one of the most useful commands in the View menu while debugging. The window's Status line allows selecting up to four watch windows. The Watch window allows selecting the variables and SFRs that are meaningful to your application or debugging session. Figure B.8 is a screen snapshot of a Watch window.

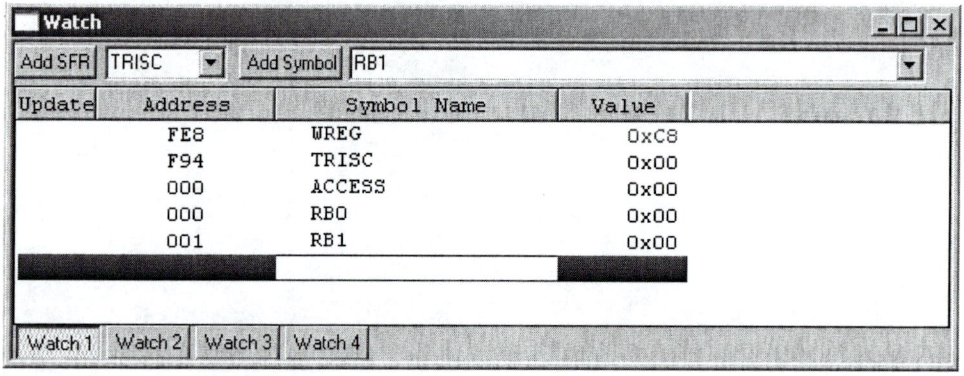

Figure B.8 *Watch window snapshot.*

The Update column is used to set the data capture or the runtime watch function of different debugger. The Address column displays the hex location of the symbol or SFR. You may enter a specific address to watch by clicking the Address column of the first available row. Addresses preceded by the debugger with the letter "P" indicate that the symbol is defined in program memory.

To add a SFR or Symbol to the Watch window click the corresponding triangle symbol to expand the window, select the desired element, and then click the Add

SFR or Add Symbol button. The value in the Watch window can be changed by double clicking the existing value box. Binary information in hex, decimal, binary, or
char can be displayed by right clicking the corresponding column header. Drag and
drop can be used to re-arrange the Watch variables.

The subcommands available in the Watch window are displayed by right-clicking
the window area. Figure B.9 is a screen snapshot of the resulting menu.

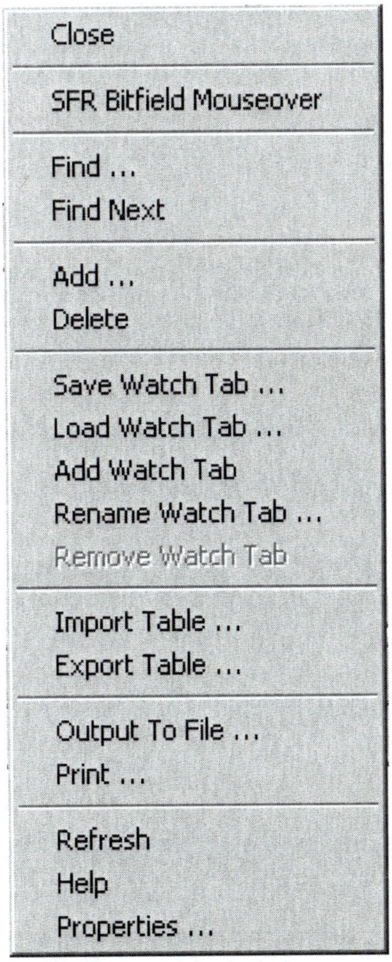

Figure B.9 *Watch window subcommand menu.*

While debugging, the contents of the Watch window are updated on a program
halt. Watch window columns can be resized by dragging the line that separates two
columns. A column of the Watch window can be made invisible by right clicking the
column header, selecting more from the menu, and then clicking the Hide button in
the displayed dialog box. Columns made invisible can be redisplayed by similarly
clicking on the Show button. Columns can be reordered by dragging and dripping
the column header.

Watch Window in C language

The following features of the Watch window are available to C language code:

- The Bitfield value Mouseover is made available by using the right mouse button in an active Watch window menu.

- A pointer to an intrinsic type or structure can be made visible by entering the pointer variable name (for example, *thisPtr) in the Symbol Name field.

- A structure member name is made visible by entering its name in the Symbol Name field, for example, porta.ra2. The variable must be in the form struct.membername.

- A structure pointer member name is made visible by entering the name in the Symbol Name field, for example, porta>pin2. The variable must be in the form struct->membername.

The Watch window displays 16 bits for the PIC18 MCU compiler type short long int.

B.2.8 Simulator and Tracing

MPLAB SIM supports a Trace function and its corresponding window, which is available from the View menu. The Trace window is used to monitor processor operation. Trace is enabled in the Simulator Settings dialog box as shown in Figure B.10.

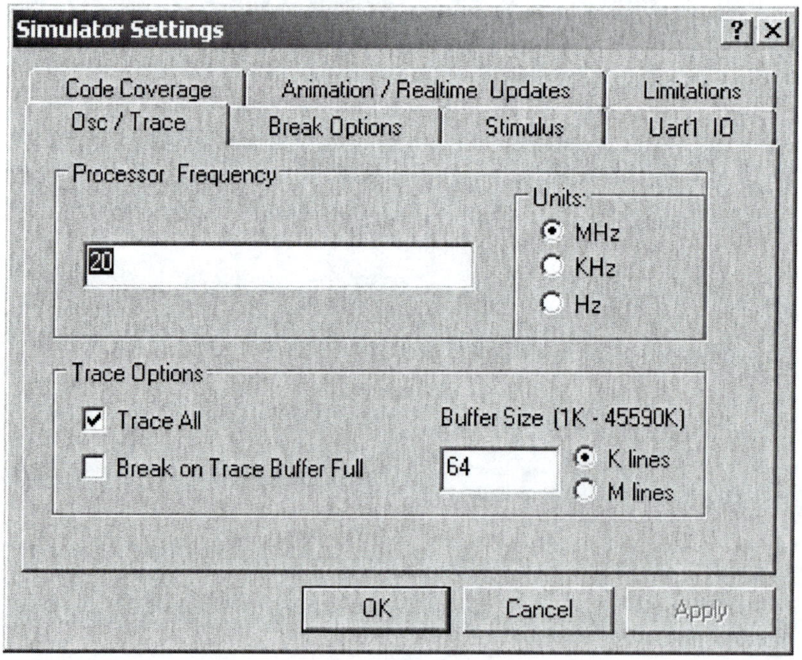

Figure B.10 *Trace options in Simulator Settings dialog.*

The Trace window is displayed by selecting the Simulator Trace option in the View menu. Figure B.11 is a screen snapshot of a Trace window.

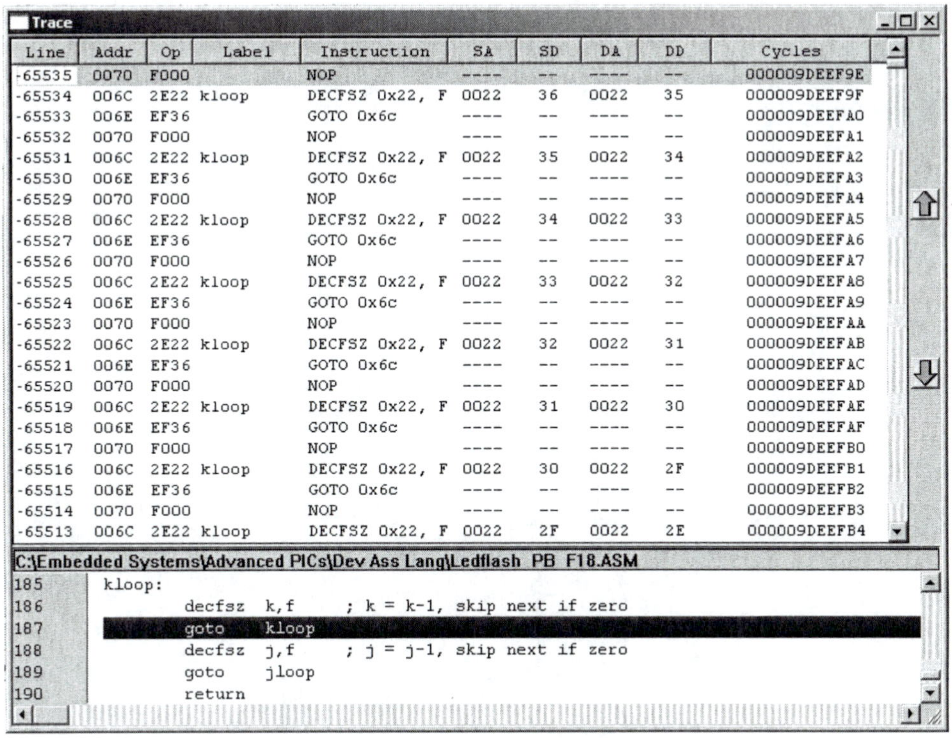

Figure B.11 *Screen snapshot of Trace window.*

The image in Figure B.11 shows tracing execution through an endless loop in the sample program. Some of the columns columns have obvious contents. The first column, labeled Line, refers to the line number, which is the cycle's relative position. The column labeled SA displays the address of the source data. The column labeled SD is the value of the source data, in this case the iteration number in the loop being traced. Column DA shows the address of the destination data and column DD the value of the destination data. The column labeled Cycles is a sequential count of machine cycles or seconds.

Setting Up a Trace

Up to 32,767 instruction cycles can be displayed in the Trace buffer. Setting up a trace is done from the Editor screen. Right-clicking a source code line displays a menu whose first four entries relate to the Trace function. The two trace modes are labeled filter-in and filter-out, These modes are mutually exclusive and setting one clears the other one.

To setup a trace, use the editor cursor to highlight a line to be traced, and then select the Add Filter-in Trace or Add Filter-out Trace options from the menu. Lines selected for a trace are marked with a rectangle symbol on the editor's right margin. In the Filter-in option, the code selected is the one traced by the debugger. In the Filter-out mode, the code selected is excluded from the trace. Traces are removed from the editor menu by selecting Remove Filter Trace or Remove All Filter Traces. The Remove Filter Trace option is useful in removing one or more lines from the code being traced.

Trace Menu

Tracing code is an advanced debugger function most useful in detecting program hang-ups. The Trace subcommands are displayed on a menu by right-clicking on the Trace window. Figure B.12 shows this menu.

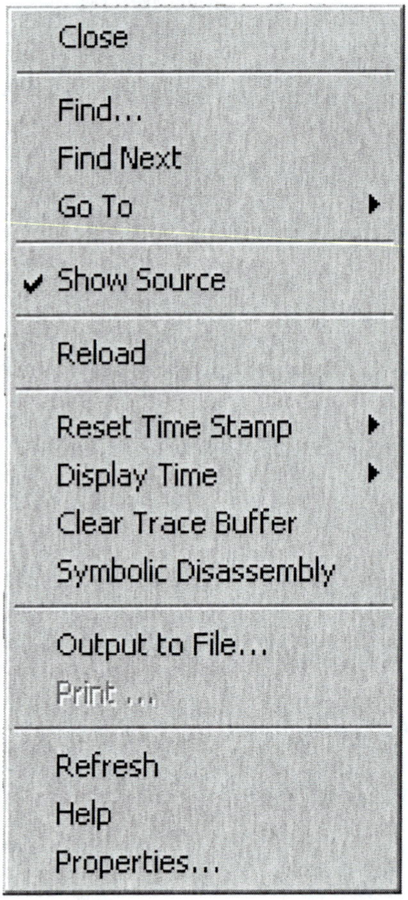

Figure B.12 *Trace subcommands.*

The Close subcommand closes the Trace window. The Find command opens the Find dialog. The Find subcommand allows entering a string to be located in the Trace window. The Find dialog allows entering a string or selecting from a list of names referenced in the code. The search can be directed to match the whole world and to match or ignore the case. The search direction can be up or down in the code. Use the What field to enter a string of text you want to find, or select text from the drop-down list. Figure B.13 shows the Find dialog box.

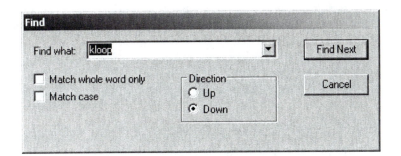

Figure B.13 *Trace/Find dialog.*

The Find Next subcommand locates the next instance of the string defined with the Find subcommand. The <F3> key serves as a shortcut to this subcommand. The Go To subcommand jumps to the item selected in the popup, which can be one of the following:

- Trigger (jump to the trigger location)
- Top (jump to the top of the window)
- Bottom (jump to the bottom of the window)
- Go To Trace Line (go to the specified trace line)
- Go To Source Line (open a File window and go to the selected source line)

The Show Source subcommand is used to show or hide the source code listing on the bottom of the window. The window bar dividing the trace and source code may be dragged to resize each portion. The Reload reloads the trace window with the contents of the trace buffer. The Reset Time Stamp subcommand reset, the time stamp conditionally on processor Reset, on run, or manually — or forces an immediate reset by selecting Reset Now.

The Display Time subcommand displays the time stamp as a cycle count. Time can be displayed in elapsed seconds or in engineering format. The subcommand Clear Trace Buffer clears the contents of the Trace buffer and window. The subcommand Symbolic Disassembly uses names for SFRs and symbols instead of numeric addresses.

The subcommand Output to File exports the contents of the trace memory window to a file. A dialog box allows entering a filename and selecting save options. The Refresh subcommand refreshes the viewable contents of the window. The Properties subcommand displays a dialog that allows selecting the columns displayed as well as the pixel width of each column.

B.2.9 Stimulus

Software debuggers such as MPLAB SIM have the limitation that the action of hardware devices cannot be modeled or traced. This means that if a program fails while interacting with a hardware component, it will be difficult to detect the reason for the failure by simple code tracing. For example, suppose a program that monitors the action of a pushbutton switch, wired active high, to Port C, line 2. Now suppose that while executing the program, pressing the pushbutton switch fails to produce the expected action. In this case, debugging the code by means of conventional tracing cannot help.

The Stimulus simulator that is part of MPLAB SIM provides assistance in these cases by allowing the simulation of hardware signals. For example, we can set up a Stimulus window that allows us to produce a change in the state of a specific pin in a port register. Following the previous example we could define pin 2 of PORTC so that it reports low but becomes high when the button labeled <Fire> is pressed. Now we can run the program under the debugger, insert a breakpoint in the code that is expected to execute when the pushbutton is pressed, and click on the <Fire> button to change the state of pin 2, PORTC, to low. In this case, execution reaching the breakpoint indicates that the code is responding as expected.

It is a fact that the simulation provided by Stimulus, although powerful and useful, does not entirely substitute a hardware debugger. If in the previous example the pushbutton switch is defective, or has been wired improperly, the simulation cannot detect the flaw. In other words, a software simulation assumes that the hardware is operating correctly. A hardware debugger, such as MPLAB ICD2 discussed later in the appendix, allows us to actually monitor the signal on the board and follow execution when the physical device is active.

Another limitation of software simulators is that some devices are quite difficult to model. For example, a typical Liquid Crystal Display requires three control lines and four or eight data lines. Trying to use a software simulator to find a defect in LCD code could be a fruitless endeavor.

Stimulus Basics

Stimulus provides a simulation of hardware signals that are received by the device. For example, a change in the level of a signal or a pulse to an I/O pin of a port. Stimulus can also be used to change the values in an SFR (Special Function Register) or any other element in data memory.

Stimulus may need to happen at a certain instruction cycle or a specific time during the simulation. It is also possible that a stimulus may need to occur when a condition is satisfied, for example, when execution has reached a certain instruction. Basically, there are two types of stimulus:

- Asynchronous stimuli consist of a one-time change to the state of an I/O pin or to RCREG used in USART operation. The action is triggered by "firing" a button in the Stimulus window.

- Synchronous stimuli consist of a predefined series of signal/data changes to an I/O pin, SFR, or GPR (as would be the case with a clock cycle).

The Stimulus dialog allows defining when, what, and how external stimuli are to happen. The dialog is used to create both asynchronous and synchronous stimuli on a stimulus workbook. A more advanced Stimulus Control Language (SCL) can be used to create a file for custom stimuli. The elements of the basic mode stimulus are shown in Figure B.14.

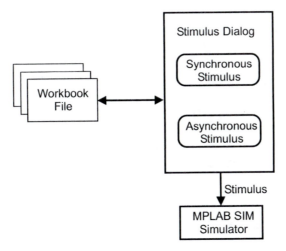

Figure B.14 *Basic mode stimulus.*

The Stimulus settings are saved to a Workbook file. Both synchronous and asynchronous stimuli become available when the Stimulus dialog window is open. More advanced use of the Stimulus allows exporting the synchronous stimulus to an SCL file.

Using Stimulus

Select Debugger>Stimulus>New Workbook to start the Stimulus dialog. To open an existing workbook, select Debugger>Stimulus>Open Workbook. The Stimulus window must be open for Stimulus to be active. The Stimulus dialog is shown in Figure B.15.

Fire	Pin / SFR	Action	Width	Units	Comments / Message
>	AN0	Pulse High	1	cyc	This pin is wired to the LCD device
>	TOCKI	Set High			
>	RD7	Toggle			This pin is wired to a pushbutton switch
>	RCREG	Direct Message			"This is a test"

Tabs: Asynch | Pin / Register Actions | Advanced Pin / Register | Clock Stimulus | Register Injection | Register Trace

Buttons: Advanced... | Apply | Remove | Delete Row | Save | Exit | Help

Figure B.15 *Stimulus dialog window.*

To set up a an asynchronous simulation, click on the Asynch tab. A simulation is initialized by clicking the various elements in a row. A row is removed by selecting it and then clicking the Delete Row button. There are two types of asynchronous simulations: regular simulations are based on activating pins and SFRs while message-based stimuli are used mainly in programs that used USART or UART operations.

Asynch Tab

The row labeled <Fire> triggers the stimulus in that row. A stimulus must be set up before the <Fire> command can be used.

Selecting the Pin/SFR column and then clicking the expansion button displays a list of elements to which the stimulus can be applied. The drop-down list includes pins and Special Function Registers that can be triggered.

The <Action> button is functional only after the PIN/SFR has been selected. The available options in the drop-down list include "Set High," "Set Low," "Toggle," "Pulse High," or "Pulse Low."

If "Pulse" is chosen as an Action, then the Width column provides a way to specify the number of units for the pulse, and the column labeled Units allows selecting instruction cycles, nanoseconds, microseconds, milliseconds, or seconds.

With regular stimulus, the comments field is used to note or as a reminder, as shown in the first and third rows in Figure B.15.

Message-Based Stimulus:

Message-based stimuli are available when RCREG is selected in the Pin/SFR column. In this case, the Action column allows choosing either "File Message" or "Direct Message." The "File Message" option means that the messages to be used will be contained in a file. The "Direct Message" option means that the Comments/Message cell will be used to define a one-line message packet.

With Message-Based Stimulus, the Comments/Message column is used to specify or change the stimulus message. If it is a text message, the string must be enclosed in double quotes, as in Figure B.15.

Pin/Register Actions Tab

When the Pin/Register Actions tab is selected, the basic synchronous pin or register actions may be entered. This is the simplest time-based stimulus. Some possible uses for this tab could be:

- Initialize pin states at time zero so when a simulation is restarted, the pins will be in a predetermined state after each POR.
- I/O port pins do not change state on a reset. In this case, the simulator starts off by treating all IO pins as inputs of a zero state.
- Set a register to a value at a specific time.

- Set multiple interrupt flags at exactly the same time. This allows seeing the effect within the interrupt handler for priority.

- Create a pulse train with different periods of a pulse over time, or an irregular wave form based on run time

- Repeat a sequence of events for endurance testing.

To enter data first select the unit of time in the "Time Units" list box. The units are cycles, hours-minutes-seconds, milliseconds, microseconds, and nanoseconds. The number of units (decimal format) is then entered in the Time column. Clicking on the text that states "Click here to Add Signals" allows opening the Add/Remove Pin/Registers dialog. This dialog box is used to add or remove pins, registers, or other signals to which the stimulus will be applied. The selections will become the titles of the columns. Fill out each row, entering a trigger time in the ("Time") and value for each pin/register column. Trigger time for each row is accumulative time because the simulation began.

Check the checkbox "Repeat after X (dec)" to repeat the stimulus on the tab after the last stimulus has occurred. Specify a delay interval for repeating the stimulus. To restart at a specific time, make a selection from the "restart at: (dec)" list box. The list box selections are determined by the trigger times (in the Time column) for each row. Once the tab is filled, you may proceed to another tab or click Apply to use the stimulus. To remove a previously applied stimulus, click Remove. Figure B.16 is a screen snapshot of the Pin/Register Actions tab in the Stimulus dialog.

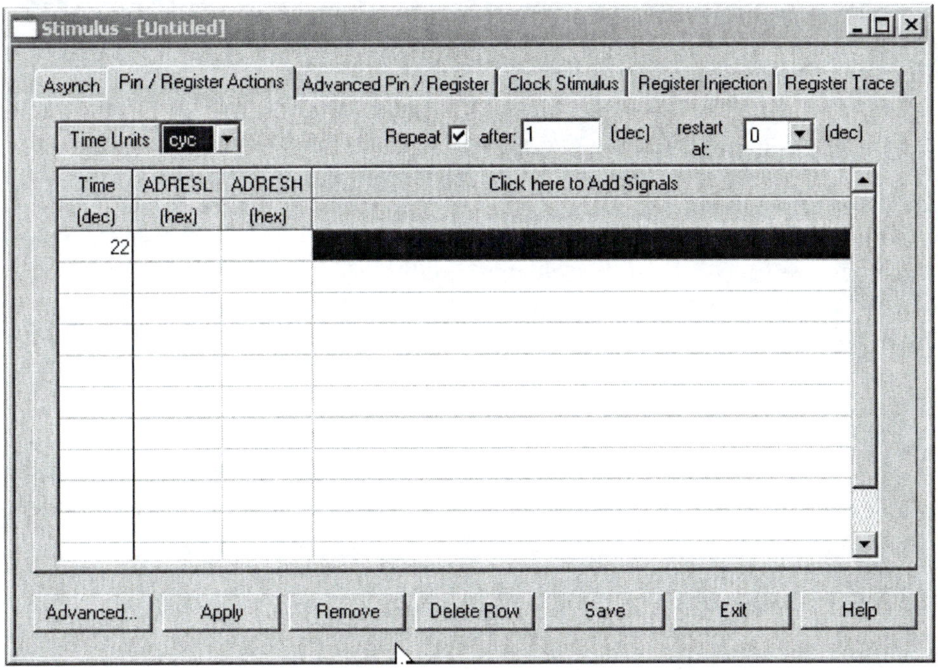

Figure B.16 *Snapshot of the Pin/Register Action tab.*

Advanced Pin/Register Tab

The Advanced Pin/Register tab in the Stimulus screen allows entering complex synchronous pin/register actions. The Advanced dialog is shown in Figure B.17.

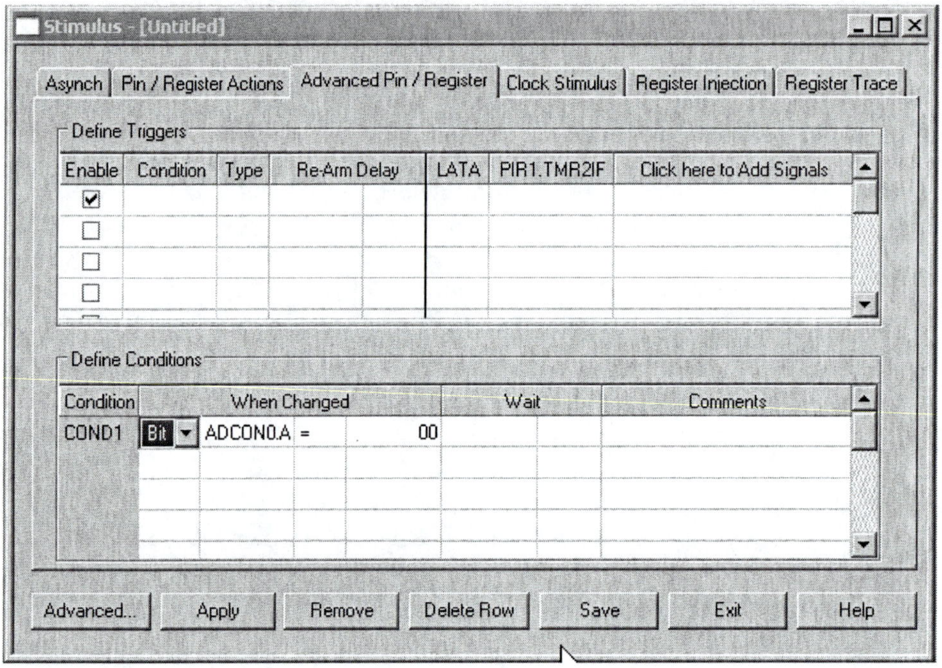

Figure B.17 *Advanced Pin/Register tab screen snapshot.*

For entering data for advanced actions, first define the conditions and then the triggers. The name for the condition being specified is automatically generated when you enter data in any other column. This name is used to identify the condition in the Condition column of the Define Triggers section of this tab. The When Changed condition defines the change condition. This condition is true when the value of the pin/register in Column 2 (its type specified in Column 1) changes to the relationship of Column 3 to the value of Column 4. The columns operate as follows:

- Column 1 is used to select the type of pin/register, either "SFR," "Bitfield," "Pin," or "All" of the above. This will filter the content of Column 2.

- Column 2 is used to select the pin/register to which the condition will apply.

- Column 3 is used to select the condition, either equal (=), not equal (!=), less than or equal (<=), greater than or equal (>=), less than (<), or greater than (>).

- Column 4 is used to enter the value for the condition.

Certain precautions must be taken when using FSR values as conditions because conditions will only occur when the SFR is updated by the user code, not the peripheral. For example, suppose the condition within the Advanced Pin Stimulus dialog is set up to trigger when TMR2 = 0x06. Then when TMR2 is incremented past 0x06, the condition is not met. However, if the following sequence is executed in user code, then the condition will occur:

```
movlw     0x06
movwf     TMR2
```

For example, a conditions can consist of a register being set equal to a value, as shown in Figure B.18.

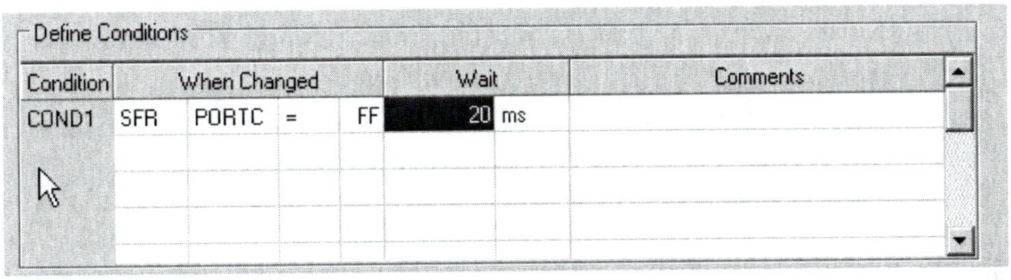

Figure B.18 *Definition of Condition dialog.*

Referring to the condition in Figure B.18, if PORTC has an initial value of 0xff or never reaches this value, the condition will never be met and the stimulus will not be applied.

Clock Stimulus Tab

The Clock Stimulus tab refers to a low or high pulse applied to a pin. The Clock Stimulus dialog is shown in Figure B.19.

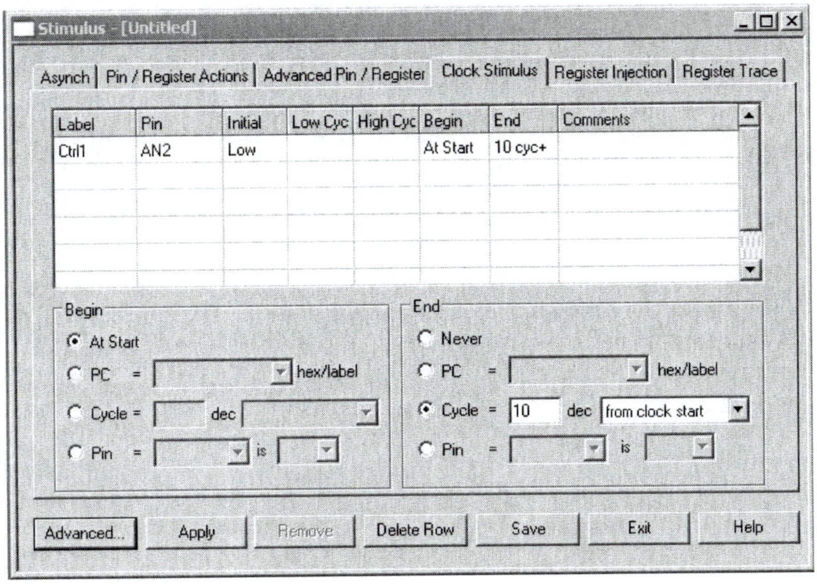

Figure B.19 *Clock Stimulus dialog.*

The following elements in the Clock Stimulus dialog define its operation:

- Label is a unique, optional name assigned to s specific clock stimulus.

- Pin defines the individual pin on which the clocked stimulus is applied.

- Initial defines the initial state of the clocked stimulus. It can be low or high.

- Low Cycles allows entering a value for the number of low cycles in a clock pulse. High Cycles allows entering a value for the number of high cycles in a clock pulse.

- Begin defines the left-side area at the lower part of the Clock Stimulus dialog (see Figure B.19). In the Begin area, the following elements can be defined: At Start (default) begins the stimulus immediately on program run. PC= begins the stimulus when the program counter equals the entered value. Cycle= begins the stimulus when the instruction cycle count equals the entered value. Pin= selects the specific pin.

- End defines the right-side area at the lower part of the Clock Stimulus dialog (see Figure B.19). In the End area the following elements can be defined: Never (default) applies the stimulus until program halt. PC= ends the stimulus when the program counter equals the entered value. Cycle= ends the stimulus when the instruction cycle count equals the entered value. Pin= ends the stimulus when the selected pin has the selected value (low or high).

- Comments adds descriptive information about the stimulus.

Register Injection Tab

Stimulus allows injecting registers with values defined in a file. Enter information for register injection here. A single byte is injected into General Purpose Registers but more than one byte may be injected for arrays. Figure B.20 is a screen snapshot of the Register Injection tab of the Stimulus dialog.

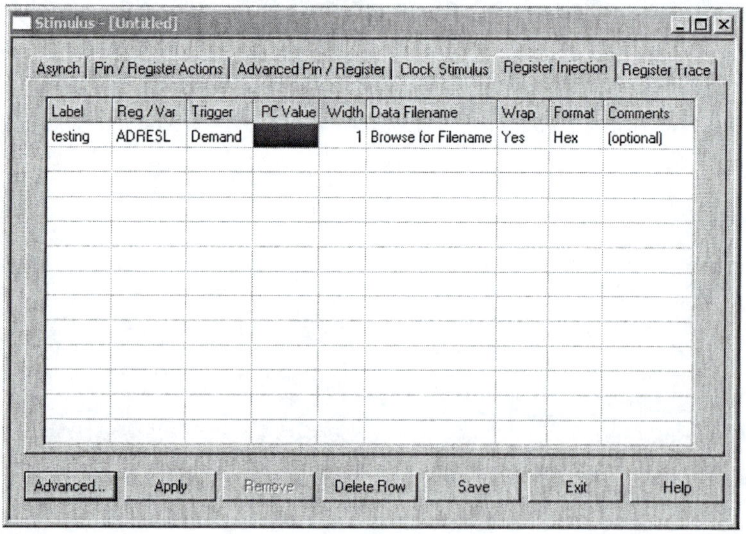

Figure B.20 *Register Injection dialog.*

The following columns (see Figure B.20) are used for Register Injection data:

- Label allows assigning an optional name to the specific Register Injection.

- Reg/Var allows selecting from a drop-down list a destination register for data injection. Listed registers include SFRs (top of list) and any GPRs used (bottom of list). The GPR registers are shown only after the program is compiled.

- Trigger selects when to trigger injection. For most registers, the Trigger is either on Demand or when the PC equals a specified value.

- PC Value is used for entering a PC value for triggering injection. The value can be an absolute address or a label in the code.

- Width refers to the number of bytes to be injected if Trigger = PC.

- Data Filename allows browsing for the injection data file.

- Wrap is Yes indicates that once all the data from the file has been injected, start again from the beginning of the file. Wrap is No indicates that once all the data from the file has been injected, the last value will continue to be used.

- Format allows selecting the format of the injection file. Format for a Regular Data File Format can be Hex (ASCII hexadecimal), Raw (interpreted as binary), SCL , or Dec (ASCII decimal). For Message-Based Data File, format can be Pkt (Hex or Raw packet format).

- Comments is used to add optional descriptive information about the register injection.

Register Trace Tab

Under certain conditions, Stimulus allows that specific registers be saved (traced) to a file during a debugger run session. Figure B.21 is a screen snapshot of the dialog screen for the Register Trace option.

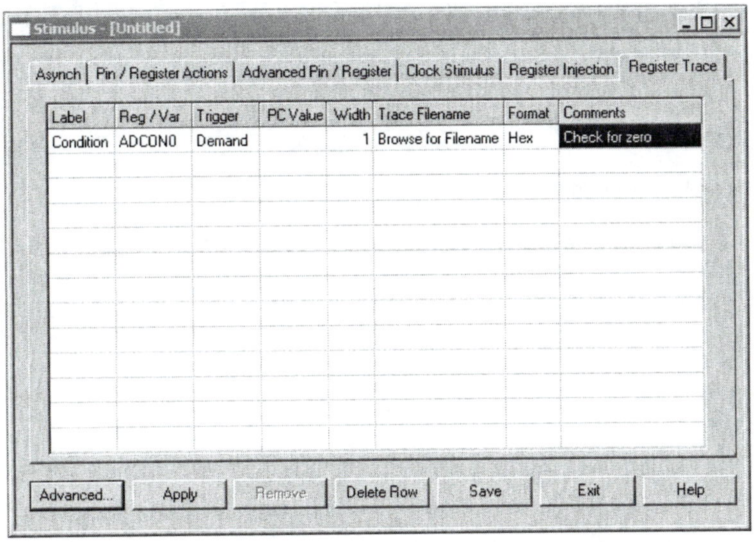

Figure B.21 *Register Trace tab.*

The following columns (see Figure B.21) are used for entering Register Trace data:

- Label allows assigning an optional name to the specific Register Injection.

- Reg/Var allows selecting from a drop-down list a destination register for data injection. Listed registers include SFRs (top of list) and any GPRs used (bottom of list). The GPR registers are shown only after the program is compiled.

- Trigger selects when to trigger the trace. For most registers, the Trigger is either on Demand or when the PC equals a specified value.

- PC Value is used for entering a PC value for triggering the trace. The value can be an absolute address or a label in the code.

- Width refers to the number of bytes to be injected if Trigger = PC.

- Trace Filename allows browsing for the location of the trace file.

- Format allows selecting the format of the injection file. Format for a Regular Data File format can be Hex (ASCII hexadecimal), Raw (interpreted as binary), SCL, or Dec (ASCII decimal).

- Comments is used to add optional descriptive information about the register injection.

B.3 Hardware Debugging

Software debuggers such as MPLAB SIM (discussed in previous sections) are sophisticated and useful tools in detecting defects and flaws in the code. However, very often code is not the sole culprit of the malfunctioning of an embedded system. In some cases, the reason for the flaw lies in a hardware or wiring defect and the engineer must resort to electronic testers, logical probes, and ultimately oscilloscopes to locate and fix the problem. A defective hardware component has been the source of more than one monumental debugging problem. In other cases, perhaps the most common, an embedded system malfunctions but we are unable to tell if the cause resides in hardware, in software, or in the interaction of both of these elements. Hardware debuggers allow testing a program as it executes in the circuit, thus providing information that cannot be obtained with software debuggers or by electronic testing alone.

B.3.1 Microchip Hardware Programmers/Debuggers

Microchip and other vendors make available hardware debuggers that range from entry-level devices to expensive professional tools used mostly in industry. Microchip refers to hardware debuggers as in-circuit devices. The following subsections describe the various hardware debuggers supplied by Microchip for the 18F PIC family at the present time.

MPLAB ICD2

An entry-level in-circuit debugger and in-circuit serial programmer, MPLAB ICD2 provides the following features.

- Real-time and single-step code execution

- Breakpoints, Register, and Variable Watch/Modify

- In-circuit debugging

- Target Vdd monitor

- Diagnostic LEDs

- RS-232 serial or USB interface to a host PC

MPLAB ICD2 is still available in the marketplace but has been superseded by MPLAB ICD3. Microchip has stated that MPLAB ICD2 will not support devices released after 2010, and is not recommended for new designs. MPLAB ICD2 is described in greater detail in Section B.3.3.

MPLAB ICD3

Follow-up version of MPLAB ICD2. The MPLAB ICD 3 is described as an in-circuit debugger system used for hardware and software development of Microchip PIC microcontrollers (MCUs) and dsPIC Digital Signal Controllers (DSCs) that are based on In-Circuit Serial Programming (ICSP) and Enhanced In-Circuit Serial Programming 2-wire serial interfaces.

The debugger system will execute code like an actual device because it uses a device with built-in emulation circuitry, instead of a special debugger chip, for emulation. All available features of a given device are accessible interactively, and can be set and modified by the MPLAB IDE interface.

MPLAB ICE 2000

An in-circuit emulator that provides emulation in real-time of instructions and data paths. Figure B.22 shows the components furnished in the MPLAB ICE 2000 package.

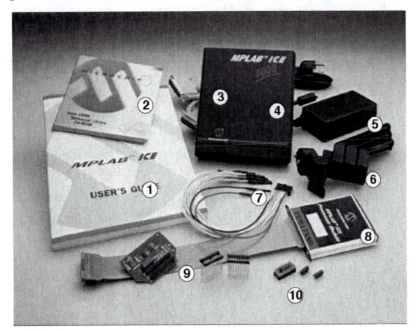

Figure B.22 *Components of MPLAB ICE 2000.*

MPLAB ICE 2000 allows the following operations:

- Debug the application on your own hardware in real-time.
- Use both hardware and software breakpoints.
- Measure timing between events.
- Set breakpoints based on internal and/or external signals.
- Monitor internal file registers.
- Emulate at full speed (up to 48 MHz).
- Select oscillator source in software.
- Program the application clock speed.
- Trace data bus activity and time stamp events.
- Set complex triggers based on program and data bus events, and external inputs.

MPLAB ICE 4000

MPLAB ICE 4000 is an In-Circuit Emulator (ICE) designed to emulate PIC18X microcontrollers and dsPIC digital signal processors. It uses an emulation processor to provide full-speed emulation and visibility into both the instruction and the data paths during execution. Figure B.24 shows the components of the MPLAB ICE 4000 system.

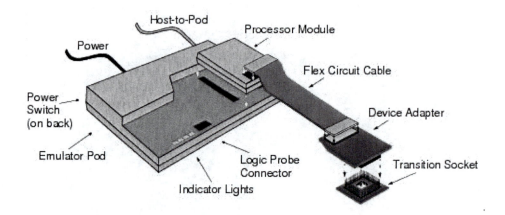

Figure B.23 *Components of MPLAB ICE 4000.*

In Figure B.23, the emulator pod connects to the PC through a USB port. The pod contains the hardware necessary to perform the common emulator functions, such as trace, break and emulate. The processor module inserts into two slots on top of the emulator pod. It contains the hardware required in order to emulate a specific device or family of devices.

The flex cable connects the device adapter to the processor module. Device adapters are interchangeable assemblies that allow the emulator to interface to a target application. It is the device adapter that has control logic for the target application to provide a clock signal and power to the processor module. The transition socket is connected to the device adapter. Transition sockets are available in various styles to allow a common device adapter to be connected to one of the supported surface-mount package styles. Logic probes can be connected to the logic probe connector on the emulator pod.

MPLAB REAL ICE

Real Ice is the high end of the Microchip debuggers described as "economical." The others are MPLAB PICkit 3, MPLAB ICD2, and MPLAB ICD3. The MPLAB REAL ICE is described as an in-circuit emulator that supports hardware and software debugging. It is compatible with Microchip MCUs and Digital Signal Controllers that support In-Circuit Serial Programming (ICSP) capabilities.

MPLAB REAL ICE executes code in a production device because these Microchip devices have built-in emulation circuitry. All available features of a given device are accessible interactively, and can be set and modified by the MPLAB IDE interface. The following are features of the emulator:

- Processors run at maximum speeds
- Debugging can be done with the device in-circuit
- No emulation load on the processor bus
- Simple interconnection
- Capability to incorporate I/O data
- Instrumented Trace (MPLAB IDE and Compiler Assisted)
- PIC32 Instruction Trace

In addition to emulator functions, the MPLAB REAL ICE in-circuit emulator system also may be used as a production programmer.

MPLAB PICkit 2 and PICkit 3

These are starter kits that can be used as programmers and debuggers, and include the following components:

- The PICkit 2 Development Programmer/Debugger
- USB cable
- PICkit Starter Kit and MPLAB IDE CD-ROMs
- A demo board with a PIC microcontroller device

The PICkit products are the most inexpensive programmer/debuggers sold by Microchip. Both kits are sold in several packages, including a Starter Kit and a Debug Express kit.

B.3.2 Using Hardware Debuggers

The hardware debuggers described in the previous section all provide considerable advantages over their software counterparts. Fortunately, hardware debuggers are easier to use than software simulators, such as MPLAB SIM described earlier in this appendix. The reason is that much of the signal, port, and device modeling and simulation provided by Stimulus or similar software is not necessary in hardware devices because the signals and devices are read directly from the hardware. For example, to test a program that reads an input device, such as a pushbutton switch, a software debugger requires that a stimulus be set up on the port line to which the switch will be connected. The programmer can then test the action of the program by "firing" the pin that represents the switch and observing the resulting code flow. The hardware debugger connected to the target board reports the status of the physical line connected to the actual switch, making the software simulation unnecessary.

Another consideration that limits the use of software emulators (such as MPLAB SIM) is that their use assumes that the hardware operates perfectly, and that the wiring and component interconnections are also correct. The emulated pushbutton switch is a virtual device that never fails to produce the expected result. In the real world, systems often fail because components are defective or have been incorrectly connected. The hardware debugger allows us to read the actual signal on the board and to detect any electronic, mechanical, or connectivity defect.

Which Hardware Debugger?

All the hardware debuggers listed and described in Section B.3.1 are usable to some degree. The high-end devices, such as ICE 2000 and ICE 4000, are designed for professional application and can be used as production tools for programming and product development. The low-end devices, such as MPLAB PIKkit 2 and 3, MPLAB ICD2 and ICD3, and MPLAB REAL ICE, are not intended as production tools and have limited functionality.

The minimal functionality required of a hardware debugger is the ability to set a breakpoint, to single-step through code, and to inspect memory and code. All of the products described previously have these capabilities. In later sections we describe the operation of the MPLAB ICD2 debugger.

ICSP

All Enhanced MCU devices can be In-Circuit Serial Programmed (ICSP) while in the end application circuit. This requires one line for clock, one line for data, and three other lines for power, ground, and the programming voltage. ICSP allows assembling a product with a blank microcontroller, which can be later programmed with the latest version of the software. The application circuit must be designed to allow all the programming signals to be directly connected to the microcontroller. Figure B.24 shows a typical circuit for ICSP operations.

In Figure B.24 notice the following features:

1. The MCLR/VPP pin is isolated from the rest of the circuit.

2. Two pins are devoted to CLOCK and DATA.

3. There is capacitance on each of the VDD, MCLR/VPP, CLOCK, and DATA pins.

4. There are a minimum and maximum operating voltages for VDD.

5. The circuit board must have an oscillator.

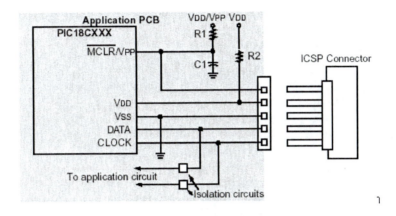

Figure B.24 *ICSP circuit elements.*

In-series debugging provided by the devices mentioned previously assumes connectivity to support ICSP

B.3.3 MPLAB ICD2 Debugger Connectivity

Microchip has described MP[LAB ICD2 and ICD3 as cost-efficient alternatives to their more expensive ICE emulators. The trade-off of the ICD products is that the developer must design products to be ICD compatible. Some of the requirements of the in-circuit debugger are

- The in-circuit debugger requires exclusive use of some hardware and software resources of the target.
- The target microcontroller must have a functioning clock and be running.
- The ICD can debug only when all the links in the system are fully functional.

An emulator provides memory and a clock, and can run code even without being connected to the target application board. During the development and debugging cycle, an ICE provides the most power to get the system fully functional, whereas an ICD may not be able to debug at all if the application does not run. On the other hand, an in-circuit debug connector can be placed on the application board and connected to an ICD even after the system has been produced, allowing easy testing, debugging, and reprogramming of the application. Even though an ICD has some drawbacks in comparison to an ICE, in this situation it has some distinct advantages:

- The ICD does not require extraction of the microcontroller from the target board in order to insert an ICE probe.
- The ICD can reprogram the firmware in the target application without any other connections or equipment.

An ICE emulates the target microcontroller by means of custom hardware. An ICD, on the other hand, uses hardware on the target microcontroller itself to reproduce some of the functions of an ICE. An ICD also employs software running on the target to perform ICE-like functions and, as a result, relies on the target microcontroller for some memory space, CPU control, stack storage, and I/O pins for communication.

Connection from Module to Target

The MPLAB ICD 2 is connected to the target board with a modular telephone connector. The six-conductor cable follows RJ 12 specifications. When wiring your own systems, it is important to note that RJ 11 cables and plugs may look identical to the RJ 12 but have four active connectors instead of six. In the ICD 2, only five of the six RJ 12 lines are used. Figure B.25 shows the pin numbers and wiring of a target board compatible with MPLAB ICD 2.

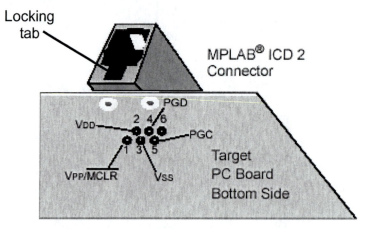

Figure B.25 *MPLAB ICD 2 connections to the target board.*

Figure B.25 shows the connections of the five pins of the modular jack RJ 12. Notice that the connector socket has the locking tab facing down, toward the board. If the RJ 12 socket in the device has the locking tab facing up, then the cable wiring must be reversed.

The actual connections to the PIC 18F series microcontroller (as well as many other MPUs that support in-circuit debugging) require the following five lines:

1. Line 1 on the connector goes to the VPP/MCLR pin on the target. In the 18F452, this is pin 1.

2. Line 2 on the connector goes to the VDD line on the target microcontroller. In the 16F452, this is the 5-volt source lines which can be either pin 11 or 32.

3. Line 3 on the connector goes to ground on the device. In the 18F452, this can be either pin 12 or pin 31.

4. Line 4 on the connector goes to the PGD pin on the target device. In the 18F452, this is pin 40.

5. Line 5 on the connector goes to the PGC pin on the target device. In the 18F452, this is pin 39.

Figure B.26 shows the actual wiring.

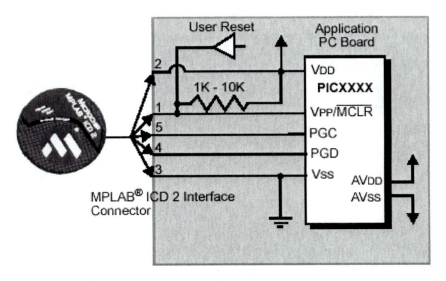

Figure B.26 *ICD 2 connection to target board.*

In the circuit of Figure B.26, pin 2 (V_{DD}) can supply a limited amount of power to the target application. Actually, only three lines are active and relevant to MPLAB ICD 2 operation: Vpp/MCLR, PGC, and PGD. If MPLAB ICD 2 does not have voltage on its V_{DD} line (pin 2 of the ICD connector), either from power being supplied to the target by MPLAB ICD 2 or from a separate target power supply, it will not operate.

Debug Mode Requirements

To use MPLAB ICD 2 in debug mode, the following requirements must be met:

1. MPLAB ICD 2 must be connected to a PC.

2. The MPLABf the target device. V_{SS} and V_{DD} are also required to be connected ICD 2 must be connected to the Vpp, PGC, and PGD pins obetween the MPLAB ICD 2 and target device.

3. The target PIC MCU must have power and a functional, running oscillator.

The target PIC MCU must have its configuration words programmed as follows:

- The oscillator bits should match the target oscillator.

- The target must not have the Watchdog Timer enabled.

- The target must not have code protection enabled.

- The target must not have table read protection enabled.

Debug Mode Preparation

The first preparatory step consists of setting MPLAB ICD 2 12 as the current debugger. This is accomplished in the Debugger>Select Tool dialog.

- When Debugger>Program is selected, the application code is programmed into the microcontroller memory via the ICSP protocol.

- A small "debug executive" program is loaded into the high area of program memory of the target PIC MCU. The application program must not use this reserved space. This space typically requires 0x120 words of program memory.

- Special "in-circuit debug" registers in the target PIC MCU are enabled. These allow the debug executive to be activated by the MPLAB ICD 2.

- The target PIC MCU is held in reset by keeping the VPP/MCLR line low.

Debug Ready State

Figure B.27 shows the conditions of the debug ready state.

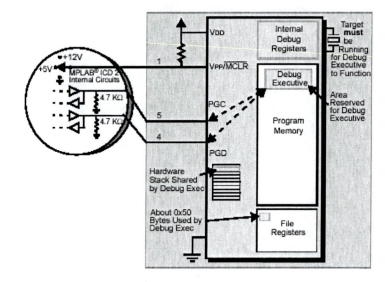

Figure B.27 *MPLAB ICD 2 debug ready state.*

Testing the MPLAB ICD 2 debug state is accomplished by executing a Build-All command on the project. The breakpoint is set early in the program code. When a breakpoint is set from the user interface of the MPLAB IDE, the address of the breakpoint is stored in the special internal debug registers of the target PIC MCU. Commands on PGC and PGD communicate directly to these registers to set the breakpoint address.

Next, the Debugger>Run function or the Run icon (forward arrow) is pressed from MPLAB IDE. This raises the VPP/MCLR line to allow the target to run; the target will start from address zero and execute until the program counter reaches the breakpoint address previously stored.

After the instruction at the breakpoint address is executed, the in-circuit debug mechanism of the target microcontroller transfers the program counter to the debug executive and the user's application is halted.

MPLAB ICD 2 communicates with the debug executive via the PGC and PGD lines. Through these lines, the debugger gets the breakpoint status information and sends it back to the MPLAB IDE. The MPLAB IDE then sends a series of queries to MPLAB ICD 2 to get information about the target microcontroller, such as file register contents and the processor's state.

The debug executive runs like an application in program memory. It uses one or two locations on the hardware stack and about fourteen file registers for its temporary variables. If the microcontroller does not run, for any reason, such as no oscillator, a faulty power supply connection, shorts on the target board, etc., then the debug executive cannot communicate with MPLAB ICD 2. In this case, MPLAB IDE will issue an error message.

Another way to get a breakpoint is to press the MPLAB IDE's Halt button. This toggles the PGC and PGD lines in such a way that the in-circuit debug mechanism of the target PIC MCU switches the program counter from the user's code in program memory to the debug executive. At this time, the target application program is effectively halted, and MPLAB IDE uses MPLAB ICD 2 communications with the debug executive to interrogate the state of the target PIC MCU.

Breadboard Debugging

Most commercial demo boards, including PICDEM 2 PLUS, dsPICDEM, and LAB X1, and this book's Demo Boards, are wired for compatibility with MPLAB ICD 2. These boards have the female RJ 12 connector and can be hooked up to ICD 2 modules as previously described. Demo boards are a great learning tool, and the connectivity to ICD 2 allows for gaining skills in hardware debugging.

However, when developing a new circuit, it would be a coincidence to find a commercial demo board that duplicates the components and connectors. Because most new systems are developed on breadboards it is quite useful to connect the breadboard circuit to the debugger hardware. Conventional female RJ 12 connectors cannot be used on a breadboard but there are several connectors on the market that can. One type contains pigtails that can be plugged into the motherboard while the other end is an RJ 12 female connector. This device is shown in Figure B.28.

Figure B.28 *Wired RJ 12 connector for breadboard use.*

The wires in the connector in Figure B.28 are color coded. Other variations are also available on the market.

B.4 MPLAB ICD 2 Tutorial

We have developed a very simple example to demonstrate debugging using the MPLAB ICD 2 debugger. The sample project is named ICD2_Tutor.mcp, and the single source file is LedFlash_Reloc.asm.

B.4.1 Circuit Hardware

The circuit used in this tutorial consists of four pushbutton switches and four LEDs. Figure B.29 shows the circuit diagram.

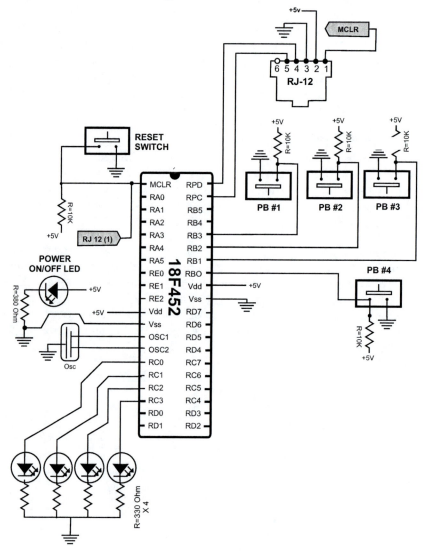

Figure B.29 *Circuit diagram for debugging tutorial.*

B.4.2 LedFlash_Reloc Program

The program LedFlash_Reloc.asm monitors action on four pushbutton switches wired to lines RB0 to RB3 in PORT B and lights the corresponding LED wired lo lines RC0 to RC3 in PORT C. The pushbutton switches and LEDs are paired as follows:

```
LED to pushbutton association:
RB3 - RB2 - RB1 - RB0 ==> Pushbuttons
 |     |     |     |
RC3 - RC2 - RC1 - RC0 ==> LEDs
```

In this tutorial, a defect in the code has been introduced intentionally to demonstrate debugging. It consists of the fact that because the pushbutton switches are wired active low, the value in PORT B lines 0 to 3 must be inverted before it is reflected in PORT C. Howevwer, if the negf instruction is used, the results are incorrect because negf generates a two's complement result. Many microprocessors and microcontrollers contain a pure NOT instruction that turns all the one bits to zero and all zero bits to one, but the instruction set of the PIC 18 family does not include this opcode. In this case, the pure NOT is obtained using xorlw with a mask of 1-digits for the field to be negated:

```
    xorlw    B'00001111'
```

This works because XORing with a 1 digit inverts the bit in the other operand.

B.4.3 Relocatable Code

MPLAB ICD 2 requires that the program be built using relocatable code. Additionally, the code must comply with the requirements and restrictions of the ICD 2 system. MPASM assembler, used with MPLINK object linker, has the ability to generate object modules and executable files. However, writing source code that assembles into an object module is slightly different from writing absolute code. MPASM assembler routines designed for absolute address assembly will require minor modifications to compile correctly into relocatable object modules.

Header Files

The standard header files (such as p18f452.inc) should be used when generating the object module.

Program Memory

In relocatable code, program memory code must be organized into logical code sections. This means that the code must be preceded by a code section declaration following the format

```
[label]   code [ROM_address]
```

In the sample program, the code sections are defined as follows:

```
Reset_Vector  code 0x000
    goto Start

    ; Start application beyond vector area

    code 0x002A
```

```
Start:
```

    If more than one code section is defined in a source file, each section must have a unique name. If the name is not specified (as in the sample program), it will be given the default name .code. Each program memory section must be contiguous within a single source file. A section may not be broken into pieces within a single source file. The physical address of the code can be fixed by supplying the optional address parameter of the code directive. Situations where this might be necessary are

- To specify reset and interrupt vectors
- To ensure that a code segment does not overlap page boundaries

Configuration Requirements

After the conventional list and include directives, the source code sets the required configuration bits as follows:

```
; ============================================================
;                    configuration bits
; ============================================================
; Configuration bits set as required for MPLAB ICD 2
    config OSC = HS              ; Assumes high-speed resonator
    config WDT = OFF             ; No watchdog timer
    config LVP = OFF             ; No low voltage protection
    config DEBUG = OFF           ; No background debugger
    config PWRT = OFF            ; Power on timer disabled
    config CP0 = OFF             ; Code protection off
    config CP1 = OFF
    config CP2 = OFF
    config CP3 = OFF
    config WRT0 = OFF            ; Write protection off
    config WRT1 = OFF
    config WRT2 = OFF
    config WRT3 = OFF
    config EBTR0 = OFF           ; Table read protection off
    config EBTR1 = OFF
    config EBTR2 = OFF
    config EBTR3 = OFF
```

    In the PIC 18 family the config directive is not preceded by one or more underscore characters, as is the case with other PICs.

RAM Allocations

Relocatable code will build without error if data is defined using the equ directives that are commonly used when programming in absolute code. But this practice is likely to generate linker errors. Additionally, variables defined with the equ directive are also not visible to the MPLAB ICD 2 debugger. MPLAB MASM supports several directives that are compatible with relocatable code and that make the variable names visible at debug time. The ones most often used are

- udata defines a section of uninitialized data. Items defined in a udata section are not initialized and can be accessed only through their names.
- udata_acs defines a section of uninitialized data that is placed in the access area. In PIC 18 devices, access RAM is always used for data defined with the udada_acs directive. Applications use this area for the data items most often used.

- udata_ovr defines a section of ovr uninitialized, overlaid data. This data section is used for variables that can be declared at the same address as other variables in the same module or in other linked modules, such as temporary variables.

- udata_shr defines a section of uninitialized, shared data. This directive is used in defining data sections for PIC12/16 devices.

- idata defines a section of initialized data. This directive forces the linker to generate a lookup table that can be used to initialize the variables in this section to the specified values. When linked with MPLAB C18 code, these locations are initialized during execution of the start-up code. The locations reserved by this section can be accessed only by the defined labels or by indirect accesses.

The following example shows the use of several RAM allocation directives:

```
    udata_acs      0x10    ; Allocated at address 0x10
j   res 1                  ; Data in access bank
temp res 1

    idata
ThisV        db    0x29         ; Initialized data
Aword        dw    0xfe01

    udata                  ; Allocated by the Linker
varx res 1                 ; One byte reserved
vary res 1                 ; Another byte
```

The location of a section may be fixed in memory by supplying the optional address, as in the udata_acs example listed previously. If more than one of a section type is specified, each one must have a unique name. If a name is not provided, the default section names are .idata, .udata, .udata_acs, .udata_shr, and .udata_ovr.

When defining initialized data in an idata section, the directives db, dw, and data can be used. The db directive defines successive bytes of data, while the dw directive defines successive words; for examp,le

```
    idata
Bytes db 1,2,3
Words dw 0x1234,0x5678
String db "This is a test",0
```

LedFlash_Reloc.asm Program

The program LedFlash_Reloc that is part of the ICD2_Tutor project is used as a simple demonstration of hardware debugging with MPLAB ICD 2. The electronic files for the project and program are found in the book's online software package.

```
; File name: LedFlash_Reloc.asm
; Project name: ICD2_Tutor.mcp
; Date: February 11, 2013
; Author: Julio Sanchez
; PIC 18F452
;
;==========================================================
; Description:
; A demonstration program for the tutorial on MPLAB ICD 2
; is presented in Appendix B. Program monitors action on the
```

```
;  four pushbutton switches and lights the corresponding
;  LED if the switch is closed.
;
;              LED to pushbutton association:
;              RB3 - RB2 - RB1 - RB0 ==> Pushbuttons
;               |     |     |     |
;              RC3 - RC2 - RC1 - RC0 ==> LEDs
;
;  Bug:
;  Because the pushbutton switches are wired active low,
;  the value in PORT B lines 0 to 3 must be inverted before
;  it is reflected in PORT C. If the negf instruction is
;  used the results are incorrect because negf generates a
;  two's complement result. Because the instruction set does
;  not include a pure NOT operator, the correct result
;  is obtained using xorlw with a mask of one-digits for
;  the field to be negated: xorlw     B'00001111'
;
;  Program uses relocatable code (required for MPLAB ICD 2
;  operation). Code includes several nop instructions to
;  facilitate inserting breakpoints
;===========================================================
;
     list p=18f452
     ; Include file, change directory if needed
     include "p18f452.inc"
;  =========================================================
;                    configuration bits
;===========================================================
;  Configuration bits set as required for MPLAB ICD 2
     config OSC = HS               ; Assumes high-speed resonator
     config WDT = OFF              ; No watchdog timer
     config LVP = OFF              ; No low voltage protection
     config DEBUG = OFF            ; No background debugger
     config PWRT = OFF          ; Power on timer disabled
     config CP0 = OFF           ; Code protection block x = 0-3
     config CP1 = OFF
     config CP2 = OFF
     config CP3 = OFF
     config WRT0 = OFF          ; Write protection block x = 0-3
     config WRT1 = OFF
     config WRT2 = OFF
     config WRT3 = OFF
     config EBTR0 = OFF         ; Table read protection block x = 0-3
     config EBTR1 = OFF
     config EBTR2 = OFF
     config EBTR3 = OFF


;===========================================================
;                    variables in PIC RAM
;===========================================================
     udata_acs
;  Declare variables at 2 memory locations in acess RAM
j    res     1
temp           res     1
;===========================================================
     ; Start at the reset vector
Reset_Vector  code 0x000
     goto Start
     ; Start application beyond vector area
```

```
       code      0x002A
Start:
       clrf      TRISC         ; PORTC all lines are output
; PORT B lines 0 to 3 to input
       movlw     B'00001111'
       movff     WREG,temp
       movwf     TRISB
       clrf      PORTC
       nop
read_PBs:
; Read pushbuttron switches
       nop
       movf      PORTB,W       ; PORTB to W
; The negf instruction fails because it produces a result
; in two's complement form. It must be replaced with the
; xorlw instruction that follows
       negf      WREG          ; Negate???
;      xorlw     B'00001111' ; XOR with mask to NOT
       movff     WREG,PORTC  ; To PORT C
       nop
       goto      read_PBs

       end
```

B.4.4 Debugging Session

Running the LedFlash_Reloc program under MPLAB ICD 2 or any other hardware debugger immediately shows that the program malfunctions as originally coded. Pushbuttons number 3 and 2 perform as expected but pushbuttons number 1 and 0 do not. Breakpointing at the end of the read_PBs: routine immediately shows that the negf instruction does not produce the desired results. Because the default contents of Port B is xxxx1111B, inverting the bits should result in xxxx0000B, which is not the case because the resulting value is xxxx0001B.

The error results from the fact that the negf instruction generates a two's complement of the target operand. The NOT instruction that is used in many microcontrollers and microprocessors produces a binary negation of the operand in which each binary digits is complemented, but there is no NOT opcode in the 18F processor family. In processors that do not provide a negate instruction (such as negf), the two's complement can be obtained by complementing all binary digits and adding one to the result. This implies that the result of a negf instruction can be converted into a pure NOT by subtracting one from the result.

Appendix C

Building Your Own Circuit Boards

Several methods have been developed for making printed circuit boards on a small scale, as would be convenient for the experimenter and prototype developer. If you look through the pages of any electronics supply catalog, you will find kits and components based on different technologies of various levels of complexity. The method we describe in this appendix is perhaps the simplest one because it does not require a photographic process.

The process consists of the following steps:

1. The circuit diagram is drawn on the PC using a general-purpose or a specialized drawing program.

2. A printout is made of the circuit drawing on photographic paper.

3. The printout is transferred to a copper-clad circuit board blank by ironing over the backside with a household clothes iron.

4. The resulting board is placed in an etching bath that eats away all the copper, except the circuit image ironed onto the board surface.

5. The board is washed of etchant, cleaned, drilled, and the components soldered to it in the conventional manner.

6. Optionally, another image can be ironed onto the backside of the board to provide component identification, logos, etc.

The following URL contains detailed information on making your own PCBs:

        Http://www.fullnet.com/u/tomg/gooteedr.htm

C.1 Drawing the Circuit Diagram

Any computer drawing program serves this purpose. We have used CorelDraw™ but there are several specialized PCB drawing programs available on the Internet. The following is a circuit board drawing used by us for a PIC flasher circuit described in the text:

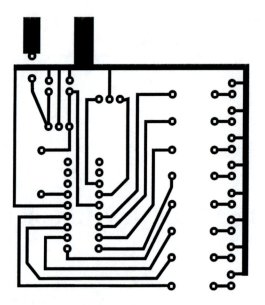

Figure C.1 PIC flasher circuit board drawing.

Note in the drawing that the circuit locations where the components are to be soldered consist of small circular pads, usually called *solder pads*. The illustration in Figure C.2 zooms into the lower corner of the drawing to show the details of the solder pads.

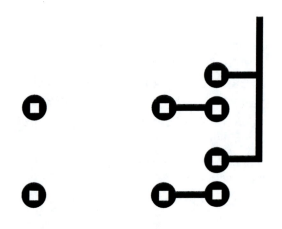

Figure C.2 Detail of circuit board pads.

Quite often it is necessary for a circuit line to cross between two standard pads. In this case, the pads can be modified so as to allow it. The modified pads are shown in Figure C.3.

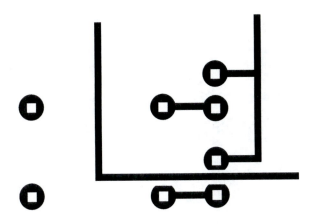

Figure C.3 Modified circuit boards pads

C.2 Printing the PCB Diagram

The circuit diagram must be printed using a laser printer. Inkjet toners do not produce an image that resist the action of the etchant. Although in our experiments we used LaserJet printers it is well documented that virtually any laser printer will work. Laser copiers have also been used successfully for creating the PCB circuit image.

With the method we are describing, the width of the traces can become an issue. The traces in the PCB image of Figure B-1 are 2 points, which is 0.027". Traces half that width and less have been used successfully with this method but as the traces become thinner the entire process becomes more critical. For most simple circuits 0.020" traces should be a useful limit. Also be careful not to touch the glossy side of the paper or the printed image with fingers.

Note that the pattern is drawn as if you were looking from the component side of the board.

C.3 Transferring the PCB Image

Users of this method state that one of the most critical elements is the paper used in printing the circuit. Pinholes in some papers can degrade the image to the point that the circuit lines (especially if they are very thin) do not etch correctly. Another problem relates to removing the ironed-on paper from the board without damaging the board surface.

Glossy, coated inkjet-printer paper works well. Even better results can be obtained with glossy photo paper. We use a common high-gloss photographic paper available from Staples® and sold under the name of "picture paper." The 30 sheets, 8-by-10 size, have the Staples® number B031420197 1713. The UPC barcode is: 7 18103 02238 5.

Transferring the image onto the board blank is done by applying heat from a common clothes iron, set on the hottest setting, onto the paper/board sandwich. In most irons, the hottest setting is labeled "linen." After going over the back of the paper several times with the hot iron, the paper becomes fused to the copper side of the blank board. The board/paper sandwich is then allowed to soak in water for about 10 minutes, after which the paper can be removed by peeling or light scrubbing with a toothbrush. It has been mentioned that Hewlett-Packard™ toner cartridges with microfine particles work better than the store-brand toner cartridges.

C.4 Etching the Board

Once the paper has been removed and the board washed, it is time to prepare the board for etching. The preliminary operations consist of rubbing the copper surface of the board with Scotchbrite® plastic abrasive pad and then scrubbing the surface with a paper towel soaked with acetone solvent.

When the board is rubbed and clean, it is time to etch the circuit. The etching solution contains ferric chloride and is available from Radio Shack™ as a solution and from Jameco Electronics™ as a powder to be mixed by the user. PCB ferric chloride etchant should be handled with rubber gloves and rubber apron because it stains the skin and utensils. Also, concentrated acid fumes from ferric chloride solution are toxic and can cause severe burns. These chemicals should be handled according to cautions and warnings posted in the containers.

The ferric chloride solution should be stored and used in a plastic or glass container, never metal. Faster etching is accomplished if the etching solution is first warmed by placing the bottle in a tub of hot water. Once the board is in the solution, face up, the container is rocked back and forth. It is also possible to aid in the copper removal by rubbing the surface with a rubber-gloved finger.

C.5 Finishing the Board

The etched board should be washed well, first in water and then in lacquer thinner or acetone; either solvent works. It is better to just rub the board surface with a paper towel soaked in the solvent. Keep in mind that most solvents are flammable and explosive, and also toxic.

After the board is clean, the mounting holes can be drilled using the solder pads as a guide. A small electric drill at high revolutions, such as a Dremmel® tool, works well for this operation. The standard drill size for the mounting holes is 0.035 inches. A #60 drill (0.040 inches) also works well. Once all the holes are drilled, the components can be mounted from the backside and soldered at the pads.

C.6 Backside Image

The component side (backside) of the PCB can be printed with an image of the components to be mounted or with logos or other text. A single-sided blank board has no copper coating on the backside so the image is just ironed on without etching. Probably the best time to print the backside image is after the board has been etched and drilled, but before mounting the components.

Because the image is to be transferred directly to the board, it must be a mirror image of the desired graphics and text. Most drawing programs contain a mirroring transformation so the backside image can be drawn using the component side as a guide, and then mirrored horizontally before ironing it on the backside of the board.

Appendix D

PIC18 Instruction Set

This appendix describes the instructions in the PIC 18 family. Not all instructions are implemented in all devices but all of them work in the18F452 PIC discussed in the text.

Table D.1

Mid-Range PIC Instruction Set

MNEMONIC	OPERAND	DESCRIPTION	CYCLES	BITS AFFECTED
		BYTE-ORIENTED OPERATIONS:		
ADDWF	f,d	Add WREG and f	1	C,DC,Z
ADDWFC	f, d, a	Add WREG and Carry bit to f	1	C, DC, Z, OV, N
ANDWF	f,d	AND WREG with f	1	Z
CLRF	f	Clear f	1	Z
COMF	f,d	Complement f	1	Z
CPFSEQ	f,a	Compare f with WREG, skip =	1-3	None
CPFSGT	f,a	Compare f with WREG, skip >	1-3	None
CPFSLt	f,a	Compare f with WREG, skip <	1-3	None
DECF	f,d	Decrement f	1	Z
DECFSZ	f,d,a	Decrement, skip if 0	1(2)	None
DCFSNZ	f,d,a	Decrement, skip if Not 0	1(2)	None
INCF	f,d	Increment f	1	Z
INCFSZ	f,d	Increment, skip if 0	1(2)	None
INFSNZ	f,d	Increment, skip if Not 0	1(2)	None
IORWF	f,d	Inclusive OR WREG and f	1	Z
MOVF	f,d	Move f	1	Z
MOVFF	fs,fd	Move word to word	2	None
MOVWF	f	Move WREG to f	1	None
MULWF	f, a	Multiply WREG with f	1	None
NEGF	f, a	Negate f	1	C,DC,Z,
RLCF	f,d,a	Rotate left through carry	1	C,Z,N
RLNCF	f,d,a	Rotate left (no carry)	1	Z,N
RRCF	f,d,a	Rotate right through carry	1	C,Z,N
RRNCF	f,d,a	Rotate right (no carry)	1	Z,N
SETF	f,a	Set f	1	None

(continues)

Table D.1

Mid-Range PIC Instruction Set (continued)

MNEMONIC	OPERAND	DESCRIPTION	CYCLES	BITS AFFECTED
SUBFWB	f,d,a	Subtract f from WREG (with borrow)	1	C,DC,Z, OV,N
SUBWF	f,d,a	Subtract WREG from f	1	C,DC,Z. OV,N
SUBWFB	f,d,a	Subtract WREG from f (with borrow)	1	C,DC,Z, OV,N
SWAPF	f,d	Swap nibbles in f	1	None
TSTFSZ	f,a	Test f, skip if 0	1	None
XORWF	f,d,a	XOR WREG with f	1	Z,N
BIT-ORIENTED OPERATIONS				
BCF	f,b	Bit clear in f	1	None
BSF	f,b	Bit set in f	1	None
BTFSC	f,b	Bit test, skip if clear	1	None
BTFSS	f,b	Bit test, skip if set	1	None
BTG	f,d,a	Toggle bit f	1	None
LITERAL AND CONTROL OPERATIONS				
ADDLW	k	Add literal and WREG	1	C,DC,Z
ANDLW	k	AND literal and WREG	1	Z
BC	n	Branch if carry	1	None
BN	n	Brach if negatve	1	None
BNC	n	Branch if not carry	1	None
BNN	n	Branch if not negative	1	None
BNOV	n	Branch if not oveerflow	1	None
BNZ	n	Brach if not zero	1	None
BOV	n	Branch if overflow	1	None
BRA	n	Branch unconditionally	1	None
BZ	n	Branch if zero	1	None
CALL	k	Call procedure	2	None
CLRWDT	-	Clear watchdog timer	1	TO,PD
DAW	-	Decimal adjust WREG	1	TO,PD
GOTO	k	Go to address	2	None
IORLW	k	Inclusive OR literal with WREG	1	Z
MOVLB	k	Move literal to BSR	1	None
MOVLW	k	Move literal to WREG	1	None
MULLW	k	Multiply literal and WREG	1	None
NOP	-	No operation	1	None
POP	-	Pop return stack	1	None
PUSH	-	Push return stack	1	None
LFSR	f,k	Move literal to FSR	1	None
RETLW	K	Return with literal in WREG	1	None
RETFIE	-	Return from interrupt	2	-
RETLW	k	Return literal in WREG	2	-
RETURN	-	Return from procedure	2	-
RCALL	n	Relative call	2	None
RESET		Software reset	1	ALL FLAGS
RETFIE	s	Return from interrupt	2	GIE,GIEH PEIE,GIEL

(continues)

Table D.1

Mid-Range PIC Instruction Set (continued)

MNEMONIC	OPERAND	DESCRIPTION	CYCLES	BITS AFFECTED
LITERAL AND CONTROL OPERATIONS				
SLEEP	-	Go into SLEEP mode	1	TO,
PDSUBLW	k	Subtract literal and WREG	1	C,DC,Z
XORLW	k	Exclusive OR literal with WREG	1	Z
DATA AND PROGRAM MEMORY OPERATIONS				
TBLRD*		Table read	2	None
TBLRD*+		Post-increment table read	2	None
TBLRD*-		Post-decrement table read	2	None
TBLRD+*		Pre-increment table read	2	None
TBLWT*		Table write	2	None
TBLWT*+		Post-increment table write	2	None
TBLWT*-		Post-decrement table write	2	None
TBLWT+*		Pre-increment table write	2	None

Legend:
  f = file register
  d = destination: 0 = WREG
                            1 = file register
  b = bit position
  k = 8-bit constant

Table D.2

Conventions Used in Instruction Descriptions

FIELD	DESCRIPTION
f	Register file address (0x00 to 0x7F)
WREG	Working register (accumulator) also w.
b	Bit address within an 8-bit file register (0 to 7)
k	Literal field, constant data, or label (may be either an 8-bit or an 11-bit value)
x	Don't care (0 or 1)
d	Destination select;
	d = 0: store result in W,
	d = 1: store result in file register f.
dest	Destination either the WREG register or the specified register location
label	Label name
TOS	Top of Stack
PC	Program Counter
PCLATH	Program Counter High Latch
GIE	Global Interrupt Enable bit
WDT	Watchdog Timer
!TO	Time-Out bit
!PD	Power-Down bit
[ ]	Optional element
[XXX]	Contents of memory location pointed at by XXX register
( )	Contents
->	Assigned to
< >	Register bit field
italics	User-defined term

file

ADDLW   Add Literal and WREG

Syntax: [label] ADDLW   k
Operands: k in range 0 to 255
Operation: (WREG) + k -> WREG
Status Affected: C, DC, Z
Description: The contents of WREG are added to the eight
                     bit literal 'k' and the result is placed in WREG.
Words: 1
Cycles: 1

Example1:

```
ADDLW   0x15
Before Instruction:          WREG = 0x10
After Instruction:           WREG = 0x25
```

Example 2:

```
ADDLW   var1
Before Instruction:          WREG = 0x10
var1 is data memory variable
var1 = 0x37
After Instruction:           WREG = 0x47
```

ADDWF   Add WREG and f

Syntax: [label] ADDWF   f,d
Operands: f in range 0 to 127
                     d = 0 / 1
Operation: (W) + (f) -> destination
Status Affected: C, DC, Z
Description: Add the contents of the WREG register
                     with register 'f'. If 'd' is 0 the result is stored
                     in WREG. If 'd' is 1, the result is stored
                     back in Register 'f'.
Words: 1
Cycles: 1

Example 1:

```
ADDWF   FSR,0
Before Instruction:
        WREG = 0x17
        FSR = 0xc2
After Instruction:
        WREG = 0xd9
        FSR = 0xc2
```

Example 2:

```
ADDWF   INDF, 1
before Instruction:
        WREG = 0x17
        FSR = 0xC2
        Contents of Address (FSR) = 0x20
After Instruction:
        WREG = 0x17
        FSR = 0xC2
        Contents of Address (FSR) = 0x37
```

ADDWFC   Add WREG and Carry bit to f

Syntax: [label] ANDLW k
Operands: $0 \le k \le 255$
Operation: (WREG).AND. (k) -> WREG
Status Affected: Z, N
Encoding: 0000 1011 kkkk kkkk
Description: The contents of WREG are ANDed with
                  the eight bit literal 'k'. Theresult is placed in
                  WREG.

Words: 1
Cycles: 1

Example 1:

```
            ANDLW 0x5F        ; And constant to W
            Before Instruction:
                  WREG = 0xA3
                  Z, N = x
            After Instruction:
                  WREG = 0x03
                  Z = 0
                  N = 0
```

Example 2:

```
            ANDLW MYREG       ; And address of MYREG
                              ; to WREG
            Before Instruction:
                  WREG = 0xA3
                  Address of MYREG † = 0x37
                  Z, N = x
                  † MYREG is a symbol for a
                  data memory location
            After Instruction:
                  WREG = 0x23
                  Z = 0
                  N = 0
```

ANDWF   AND WREG with f

Syntax: [label] ANDWF f, d, a
Operands: 0 ≤ f ≤ 255
Operation: (WREG).AND. (f) -> destination
Status Affected: Z, N
Description: The contents of WREG is ANDed with the
                           contents of Register 'f'.
                           The 'd' bit selects the destination for the
                           operation. If 'd' is 1 the result is stored back in
                           the File Register. If 'd' is 0 the result is
                           stored in WREG.
                           The 'a' bit selects which bank is accessed for
                           the operation. If a is 1, the bank specified
                           by the BSR Register is used. If a is 0, the
                           access bank is used.

Words: 1
Cycles: 1
Example 1:

```
ANDWF REG1, 1, 1 ; And WREG with REG1
Before Instruction:
        WREG = 0x17
        REG1 = 0xC2
        Z, N = x
After Instruction:
        WREG = 0x17
        REG1 = 0x02
        Z = 0, N = 0
```

Example 2:

```
ANDWF REG1, 0, 1 ; And WREG with REG1
                 ; (destination WREG)
Before Instruction:
        WREG = 0x17
        REG1 = 0xC2
        Z, N = x
After Instruction:
        WREG = 0x02
        REG1 = 0xC2
        Z = 0, N = 0
```

BC   Branch if Carry

Syntax: [label] BC n
Operands: -128 ≤ f = ≤127
Operation: If carry bit is set
                            (PC + 2) + 2n -> PC
Status Affected: None
Description: If the Carry bit is '1', then the program will
                            branch. The 2's complement number
                            (the offset) is added to the PC. Because the
                            PC will have incremented to fetch the next
                            instruction, the new address will be (PC+2)+2n.
                            This instruction is then a two-cycle instruction.
Words: 1
Cycles: 1 (2)

```
Example 1:
      HERE      BC      C_CODE
      .                 ; If C bit is not set
      .                 ; execute this code
      C_CODE
      .                 ; Else, execute this code
      .


Example 2:
      HERE BC $ + OFFSET      ; If carry bit is set,
      NO_C    GOTO    PROCESS_CODE ; branch to HERE+
                              ; OFFSET

      PLUS0 •
      PLUS1 •
      PLUS2 •
      PLUS3 •
      PLUS4 •
      PLUS5 •
      PLUS6 •
  Case 1: Before Instruction:
              PC = address HERE
              C = 0
      After Instruction:
              PC = address NO_C
  Case 2: Before Instruction:
```

```
        PC = address HERE
         C = 1
After Instruction:
        PC = address HERE + OFFSET
```

BCF

Bit Clear f

Syntax:	[label] BCF f,b
Operands:	f in range 0 to 127
	b in range 0 to 7
Operation:	0 ->f<b>
Status Affected:	None
Description:	Bit 'b' in Register 'f' is cleared.
Words:	1
Cycles:	1

Example 1:

```
BCF       reg1,7
Before Instruction: reg1 = 0xc7 (1100 0111)
After Instruction: reg1 = 0x47 (0100 0111)
```

Example 2:

```
BCF       INDF,3
Before Instruction:    WREG = 0x17
                       FSR = 0xc2
                       [FSR]= 0x2f
After Instruction:
                       WREG = 0x17
                       FSR = 0xc2
                       [FSR] = 0x27
```

BN   Branch if Negative

Syntax: [label] BN n
Operands: -128 ≤ f ≤ 127
Operation: If negative bit is 1
                   (PC + 2) + 2n -> PC
Status Affected: None
Description: If the Negative bit is 1, then the program will
                   branch. The 2's complement number is added
                   to the PC. Because the PC will have
                   incremented to fetch the next instruction, the
                   new address will be (PC+2)+2n. This
                   instruction is then a two-cycle instruction.

Example 1:

```
      HERE     BN   N_CODE   ; If N bit is not set
      NOT_N    •             ; execute this code
               •
      GOTO     MORE_CODE
      N_CODE   •             ; Else, this code will
               •             ; execute
```

```
 Case 1:  Before Instruction:
               PC = address HERE
               N = 0
          After Instruction:
               PC = address NOT_N
 Case 2: Before Instruction:
               PC = address HERE
               N    = 1
          After Instruction:
               PC = address N_CODE
```

Example 2:

```
      HERE     BN  $ + OFFSET ; If negative bit is
                              ; set
      NOT_N    GOTO    PROCESS_CODE ; branch to HERE
                              ; + OFFSET
      PLUS0  •
      PLUS1  •
      PLUS2  •
      PLUS3  •
```

```
        PLUS4  •
        PLUS5  •
        PLUS6  •
Case 1:
        Before Instruction:
            PC = address HERE
            N = 0
        After Instruction:
            PC = address NOT_N

Case 2:
        Before Instruction:
            PC = address HERE
            N = 1
        After Instruction:
            PC = address HERE + OFFSET
```

BNC   Branch if Not Carry

Syntax: [*label*] BNC n
Operands: -128 ≤ f ≤ 127
Operation: If carry bit is '0'
                         (PC + 2) + 2n . PC
Status Affected: None
Description: If the Carry bit is '0', then the program will
                         branch. The 2's complement number is added
                         to the PC. Because the PC will have
                         incremented to fetch the next instruction, the
                         new address will be (PC+2)+2n. This
                         instruction is then a two-cycle instruction.
Words: 1
Cycles: 1 (2)

Example 1:
```
        HERE    BNC NC_CODE  ; If C bit is set
        CARRY   .            ; Execute this code

                .

                GOTO    MORE_CODE
        NC_CODE .            ; Else, this code
                .            ; executes
```

Example 2:
```
        HERE    BNC $+OFFSET ; If carry bit clear
        CARRY   GOTO PROCESS_CODE
                             ; branch to HERE +
                             ; OFFSET
        PLUS0   .
        PLUS1   .
        PLUS2   .
        PLUS3   .
        PLUS4   .
        PLUS5   .
        PLUS6   .
```

```
  Case 1:    Before Instruction::
             PC = address HERE
             C = 0
```

```
                After Instruction:
                     PC = address HERE + OFFSET
Case 2:    Before Instruction;
                     PC = address HERE
                     C = 1
                After Instruction;
                PC = address CARRY
```

BNN   Branch if Not Negative

Syntax: [*label*] BNN n
Operands: -128 ≤ f = ≤ 127
Operation: If negative bit is '0'
                   (PC + 2) + 2n . PC
Status affected: None
Description: If the Negative bit is '0', then the program wil
                   branch.The 2's complement number is added
                   to the PC. Because the PC will have
                   incremented to fetch the next instruction, the
                   new address will be (PC+2)+2n. This
                   instruction is then a two-cycle instruction.
Words: 1
Cycles: 1 (2)

Example 1:
```
          HERE      BNN POS_CODE ; If N bit is set
          NEG          .              ; execute this code
                       .
                    GOTO MORE CODE
          POS_CODE
                       .
                       .
      Case 1: Before Instruction:
                  PC = address HERE
                  N = 0
              After Instruction:
                  PC = address POS_CODE
      Case 2: Before Instruction:
                  PC = address HERE
                  N = 1
              After Instruction:
                  PC = address NEG
```

Example 2:
```
          HERE    BNN $+OFFSET ; If negative
          NEG GOTO PROCESS CODE    ; If bit is
                                   ; clear, branch
                                   ; to here +
```

```
                                        ;  OFFSET
              PLUS0    .
              PLUS1    .
              PLUS2    .
              PLUS3    .
              PLUS4    .
              PLUS5    .
              PLUS6    .
   Case 1:    Before Instruction:
                  PC = address HERE
                  N = 0
              After Instruction:
                  PC = address HERE + OFFSET
   Case 2:    Before Instruction:
                  PC = address HERE
                  N = 1
              After Instruction:
                  PC = address NEG
```

BNOV   Branch if Not Overflow

Syntax: [label] BNOV n
Operands: $-128 \le f \le 127$
Operation: If overflow bit is '0'
                           $(PC + 2) + 2n . PC$
Status Affected: None
Description: If the Overflow bit is '0', then the program will
                       branch. The 2's complement number is added
                       to the PC. Because the PC will have
                       incremented to fetch the next instruction, the
                       new address will be (PC+2)+2n. This
                       instruction is then a two-cycle instruction.
Words: 1
Cycles: 1 (2)

Example 1:
```
            HERE      BNOV      NOV_CODE
            OVFL      .         ; If overflow bit is set
                      .         ; execute this code
                      GOTO      MORE_CODE
            NOV_CODE .          ; Else, this code will
                      .         ; execute
    Case 1:
            Before Instruction:
                PC = address HERE
                OV = 0
            After Instruction:
                PC = address NOV_CODE
    Case 2:
            Before Instruction:
                PC = address HERE
                OV = 1
            After Instruction:
                PC = address OVFL
```
Example 2:
```
            HERE    BNOV    $+OFFSET   ; If overflow bit
            OVFL    GOTO    PROCESS_CODE ; is clear,
                                       ; branch here
```

```
                                              ;   + OFFSET
        PLUS0      .
        PLUS1      .
        PLUS2      .
        PLUS3      .
        PLUS4      .
        PLUS5      .
        PLUS6      .
```

BNZ   **Branch if Not Zero**

Syntax: [*label*] BNZ n
Operands: -128 ≤ f ≤ 127
Operation: If zero bit is '0'
                     (PC + 2) + 2n . PC
Status Affected: None
Description: If the Zero bit is '0', then the program will
                     branch. The 2's complement number is added
                     to the PC. Because the PC will have
                     incremented to fetch the next instruction, the
                     new address will be (PC+2)+2n. This
                     instruction is then a two-cycle instruction.
Words: 1
Cycles: 1 (2)

Example 1:

```
            HERE     BNC Z_CODE    ; If Z bit is set,
            ZERO     .             ; execute this
                                   ; code
                     GOTO MORE_CODE
            Z_CODE   .
                     .
   Case 1:  Before Instruction:
                PC = address HERE
                Z = 0
            After Instruction:
                PC = address Z_CODE
   Case 2:  Before Instruction:
                PC = address HERE
                Z = 1
            After Instruction:
                PC = address ZERO
```

Example 2:

```
            HERE     BNC $ + OFFSET
                              ; If zero bit is clear
                              ; branch to here
            ZERO     GOTO PROCESS_CODE
            PLUS0    .
```

```
                    PLUS1   .
                    PLUS2   .
                    PLUS3   .
                    PLUS4   .
                    PLUS5   .
                    PLUS6   .
Case 1:             Before Instruction:
                        PC = address HERE
                        Z = 0
                    After Instruction:
                        PC = address HERE + OFFSET
Case 2:             Before Instruction:
                        PC = address HERE
                        Z = 1
                    After Instruction:
                        PC = address ZERO
```

BOV   **Branch if Overflow**

Syntax: [*label*] BOV n
Operands: -128 ≤ f ≤ 127
Operation: If overflow bit is '1'
                         (PC + 2) + 2n . PC
Status Affected: None
Description: If the Overflow bit is '1', then the program will
                     branch. The 2's complement number is added
                     to the PC. Because the PC will have
                     incremented to fetch the next instruction, the
                     new address will be (PC+2)+2n. This
                     instruction is then a two-cycle instruction.
Words: 1
Cycles: 1 (2)

Example 1:

```
            HERE    BOV OV_CODE  ; If OV bit is set
                              ; execute this code
            OVFL       .

                       .
                    GOTO    MORE_CODE
            OV_CODE .          ; Else, this code will
                       .       ; execute
    Case 1:     Before Instruction:
                    PC = address HERE
                    OV = 0
                After Instruction:
                    PC = address OVFL
    Case 2:     Before Instruction:
                    PC = address HERE
                    OV = 1
                After Instruction:
                    PC = address OV_CODE
```


Example 2:

```
            HERE    BOV $+OFFSET    ; If OV bit
                                    ; is set
```

```
                                    ; branch to  HERE +
                                    ; OFFSET
              OVFL      GOTO     PROCESS_CODE
              PLUS0     .
              PLUS1     .
              PLUS2     .
              PLUS3     .
              PLUS4     .
              PLUS5     .
              PLUS6     .
```

```
Case 1:          Before Instruction:
                 PC = address HERE
                 OV = 0
            After Instruction:
                 PC = address OVFL
Case 2:          Before Instruction:
                 PC = address HERE
                 OV = 1
            After Instruction:
                 PC = address HERE + OFFSET
```

BRA   Branch Unconditional

Syntax: [label] BRA n
Operands: $-1024 \leq f \leq 1023$
Operation: (PC + 2) + 2n . PC
Status Affected: None
Description: The 2's complement number is added to
                     the PC. Because the PC will have incremented
                     to fetch the next instruction, the new address
                     will be (PC+2)+2n. This instruction is a
                     two-cycle instruction.
Words: 1
Cycles: 2

Example:

```
        HERE        BRA  THERE    ; Branch to a
                          .       ; program memory
                          .       ; location (THERE)
                          .       ; Which must be < 1023
                          .       ; locations forward.
                    THERE  .
    Before Instruction:
        PC = address HERE
    After Instruction:
        PC = address THERE
```

BSF

Bit Set f

Syntax:	[label] BSF f,b
Operands:	f in range 0 to 127
	b in range 0 to 7
Operation:	1-> f<b>
Status Affected:	None
Description:	Bit 'b' in register 'f' is set.
Words:	1
Cycles:	1

Example 1:

```
BSF       reg1,6
Before Instruction:    reg1 = 0011 1010
After Instruction:     reg1 = 0111 1010
```

Example 2:

```
BSF       INDF,3
Before Instruction:
        WREG = 0x17
        FSR = 0xc2
        [FSR] = 0x20
After Instruction:
        WREG = 0x17
        FSR = 0xc2
        [FSR] = 0x28
```

BTFSC   Bit Test f, Skip if Clear

Syntax: [label] BTFSC   f,b
Operands: f in range 0 to 127
                       b in range 0 to 7
Operation: skip next instruction if (f<b>) = 0
Status Affected: None
Description: If bit 'b' in register 'f' is '0' then the next
                       instruction is skipped. If bit 'b' is '0' then the
                       next instruction (fetched during the current in
                       struction execution) is discarded, and a NOP
                       is executed instead, making this a 2 cycle
                       instruction.
Words: 1

Example:

```
repeat:
        btfsc   reg1,4
        goto    repeat
Case 1: Before Instruction:
                PC = $
                reg1 = xxx0 xxxx
        After Instruction:
                Because reg1<4>= 0,
                PC = $ + 2 (goto skiped)
Case 2: Before Instruction:
                PC = $
                reg1= xxx1 xxxx
        After Instruction:
                Because FLAG<4>=1,
                PC = $ + 1 (goto executed)
```

BTFSS   Bit Test f, Skip if Set

Syntax: [label] BTFSC   f,b
Operands: f in range 0 to 127
                  b in range 0 to 7
Operation: skip next instruction if (f<b>) = 1
Status Affected: None
Description: If bit 'b' in Register 'f' is '1', then the next
                  instruction is skipped. If bit 'b' is '0', then the
                  next instruction (fetched during the current
                  instruction execution) is discarded, and a NOP
                  is executed instead, making this a two-cycle
                  instruction.
Words: 1
Cycles: 1(2)

Example:

```
repeat:
        btfss   reg1,4
        goto    repeat
Case 1: Before Instruction:
                PC = $
                Reg1 = xxx1 xxxx
        After Instruction:
                Because Reg1<4>= 1,
                PC = $ + 2 (goto skiped)
Case 2: Before Instruction:
                PC = $
                Reg1 = xxx0 xxxx
        After Instruction:
                Because Reg1<4>=0,
                PC = $ + 1 (goto executed)
```

BTG   Bit Toggle f

Syntax: [label] BTG f, b, a
Operands: $0 \leq f \leq 255$
                     $0 =< b =< 7$
                     $a = [0,1]$
                     (f<b>) -> f<b>
Operation: ~(f<b>) -> f<b>
Status Affected: None
Description: Bit 'b' in Register 'f' is toggled.
                     The 'a' bit selects which bank is accessed for
                     the operation. If 'a' is 1, the bank specified by
                     the BSR Register is used. If 'a' is 0, the access
                     bank is used.
Words: 1
Cycles: 1

Example 1:
```
        BTG      LATC, 7, 1    ; Toggle the value of
                               ; bit 7 in the LATC
                               ; Register
        Before Instruction:
            LATC = 0x0A
        After Instruction:
            LATC = 0x8A
```
Example 2:
```
        BTG         INDF0, 3, 1 ; Toggle the value of
                                ; bit 3 in the
                                ; register pointed to
                                ; by the value in the
                                ; FSR0 (FSR0H:FSR0L)
                                ; Register
        Before Instruction:
            FSR0 = 0xAC2
            Contents of Address
                (FSR0)= 0x20
        After Instruction:
            FSR0 = 0xAC2
            Contents of Address
                (FSR0)= 0x28
```

BZ Branch if Zero

Syntax:	[label] BZ n
Operands:	$-128 \le f \le 127$
Operation:	If zero bit is '1'
	(PC + 2) + 2n . PC
Status Affected:	None
Description:	If the Zero bit is '1', then the program will branch. The 2's complement number is added to the PC. Because the PC will have incremented to fetch the next instruction, the new address will be PC+2+2n. This instruction is then a two-cycle instruction.
Words:	1
Cycles:	1 (2)

Example 1:

```
        HERE    BZ   Z_CODE   ; If zero bit is
        ZERO      .            ; clear, execute
                               ; this code
                GOTO     MORE_CODE
        Z_CODE    .            ; Else, this code
                  .            ; executes
    Case 1: Before Instruction:
                PC = address HERE
                Z = 0
            After Instruction:
                PC = address ZERO
    Case 2: Before Instruction:
                PC = address HERE
                Z = 1
            After Instruction:
                PC = address Z_CODE
```

Example 2:

```
        HERE    BZ   $+OFFSET ; If zero bit is
                               ; set, branch to
                               ; HERE + OFFSET
        NZERO   GOTO     PROCESS_CODE
        PLUS0     .
```

```
          PLUS1    .
          PLUS2    .
          PLUS3    .
          PLUS4    .
          PLUS5    .
          PLUS6    .
Case 1:       Before Instruction:
                  PC = address HERE
                  Z = 0
              After Instruction:
                  PC = address NZERO
Case 2:       Before Instruction:
                  PC = address HERE
                  Z = 1
              After Instruction:
                  PC = address HERE + OFFSET
```

CALL   Call Subroutine

Syntax:	[label] CALL k
Operands:	k in range 0 to 2047
Operation:	(PC) + -> TOS,
	k-> PC<10:0>,
	(PCLATH<4:3>)-> PC<12:11>
Status Affected:	None
Description:	Call Subroutine. First, the 13-bit return address (PC+1) is pushed onto the stack. The eleven-bit immediate address is loaded into PC bits <10:0>. The upper bits of the PC are loaded from PCLATH<4:3>. CALL is a two cycle instruction.
Words:	1
Cycles:	2

Example:

```
HERE      CALL      THERE,1  ; Call subroutine
                             ; THERE. This is a
                             ; fast call so the
                             ; BSR, WREG, and
                             ; STATUS are forced
                             ; onto the Fast
                             ; Register Stack

Before Instruction:
    PC = AddressHere
After Instruction:
    TOS = Address Here + 1
    PC = Address There
```

CLRF   Clear f

Syntax: [label] CLRF   f
Operands: f in range 0 to 127
Operation: 00h ->f
                       1-> Z
Status Affected: Z
Description: The contents of Register 'f' are cleared
                       and the Z bit is set.
Words: 1
Cycles: 1

Example 1:

```
        CLRF      FLAG_REG,1
            Before Instruction:
                FLAG_REG = 0x5A
            After Instruction:
                FLAG_REG = 0x00
```

Example 2:

```
        CLRF      INDF0,1      ; Clear the register
                               ; pointed at by FSR0
                               ; (FSR0H:FSR0L)
        Before Instruction:
                FSR0 = 0xc2
                Contents of address
                FSR0= 0xAA
                z = x
        After Instruction:
                FSR0 = 0xc2
                Contents of address
                FSR0 = 0x00
                Z = 1
```

CLRWDT   Clear Watchdog Timer

Syntax:	[label] CLRWDT
Operands:	None
Operation:	00h -> WDT
	0 -.> WDT prescaler count,
	1 ->.~ TO
	1 -> ~. PD
Status Affected:	TO, PD
Description:	CLRWDT instruction clears the Watchdog Timer. It also clears the postscaler count of the WDT. Status bits TO and PD are set.
Words:	1
Cycles:	1

Example:

```
        CLRWDT          ; Clear the Watchdog
                            ; Timer count value
    Before Instruction:
        WDT counter = x
        WDT postscaler
            count = 0
        WDT postscaler = 1:128
        TO = x
        PD = x
    After Instruction:
        WDT counter = 0x00
        WDT postscaler
        count = 0
        WDT postscaler = 1:128
        TO = 1
        PD = 1
```

COMF   Complement f

Syntax: [*label*] COMF f, d, a
Operands: 0 ≤ f ≤ 255
                     d = [0,1]
                     a = [0,1]
Operation: ~(f) -> destination
Status Affected: Z, N
Description: The contents of Register 'f' are 1's
                     complemented. The 'd' bit selects the
                     destination for the operation.If 'd' is 1; the result
                     is stored back in the File Register 'f'. If 'd' is 0,
                     the result is stored in the WREG Register. The
                     'a' bit selects which bank is accessed for the
                     operation. If 'a' is 1, the bank specified by the
                     BSR Register is used. If 'a' is 0, the access
                     bank is used.
Words: 1
Cycles: 1

Example:
```
        COMF    REG1, 0, 1 ; Complement the
                           ; value in Register REG1 and
                           ; place the result in the WREG
                           ; Register

    Case 1:        Before Instruction:
                       REG1 = 0x13      ; 0001 0011
                       Z, N = x
                   After Instruction:
                       REG1 = 0x13
                       WREG = 0xEC      ; 1110 1100
                       Z = 0
                       N = 1
    Case 2:        Before Instruction:
                       REG1 = 0xFF      ; 1111 1111
                       Z, N = x
                   After Instruction:
                       REG1 = 0xFF
                       WREG = 0x00      ; 0000 0000
```

```
                    Z = 1
                    N = 0
    Case 3:     Before Instruction:
                    REG1 = 0x00        ; 0000 0000
                    Z, N = x
                After Instruction:
                    REG1 = 0x00
                    WREG = 0xFF        ; 1111 1111
                    Z = 0
                    N = 1
```

CLRF   Clear f

Syntax:	[label] CLRF f,a
Operands	$0 \leq f \leq 255$
	$a \, \varepsilon \, [0,1]$
Operation:	00h -> f
	1 -> Z
Status Affected:	Z
Description:	Register f is cleared and the Zero bit (Z) is set.
Words:	1
Cycles:	1

Example:

```
    CLRF FLAG_REG, 1 ; Clear Register FLAG_REG
Before Instruction:
    FLAG_REG = 0x5A
    Z = x
After Instruction:
    FLAG_REG = 0x00
    Z = 1
```

CLRWDT   Clear Watchdog Timer

Syntax:	[label] CLRWDT
Operands:	None
Operation:	00h -> WDT
	0 -> WDT prescaler count,
	1 -> TO
	1 -> PD
Status Affected:	TO, PD
Description:	CLRWDT instruction clears the Watchdog Timer. It also clears the prescaler count of the WDT. Status bits TO and PD are set. The instruction does not change the assignment of the WDT prescaler.
Words:	1
Cycles:	1

Example:

```
CLRWDT
Before Instruction:        WDT counter= x
                           WDT prescaler = 1:128

After Instruction:         WDT counter=0x00
                           WDT prescaler count=0
                           TO = 1
                           PD = 1
                           WDT prescaler = 1:128
```

COMF   Complement f

Syntax:	[label] COMF f,d
Operands:	f in range 0 to 127
	d is 0 or 1
Operation:	(f) -> destination
Status Affected:	Z,N
Description:	The contents of Register 'f' are 1's complemented. If 'd' is 0, the result is stored in WREG. If 'd' is 1, the result is stored back in Register 'f'.
Words:	1
Cycles:	1

Example 1:

```
        comf    reg1,0
        Before Instruction:         reg1 = 0x13
        After Instruction:          reg1 = 0x13
                                    WREG = 0xEC
```

Example 2:

```
        comf    INDF,1
        Before Instruction:         FSR = 0xc2
                                    [FSR]= 0xAA
        After Instruction:          FSR = 0xc2
                                    [FSR] = 0x55
```

Example 3:

```
        comf    reg1,1
        Before Instruction:         reg1= 0xff
        After Instruction:          reg1 = 0x00
```

CPFSEQ   Compare f with WREG, Skip if Equal

Syntax:	[*label*] CPFSEQ f, a
Operands:	$0 \leq f \leq 255$
	$a \, \varepsilon \, [0,1]$
Operation:	(f) - (WREG)
	skip if (f) = (WREG)
Status Affected:	None
Description:	Compares the contents of Register 'f' to the contents of WREG Register by performing an unsigned subtraction. If 'f' = WREG, then the fetched instruction is discarded and a NOP is executed instead, making this a two-cycle instruction. The 'a' bit selects which bank is accessed for the operation. If 'a' is 1, the bank specified by the BSR Register is used. If 'a' is 0, the access bank is used.
Words:	1
Cycles:	1 (2 or 3)

Example:

```
      HERE     CPFSEQ   FLAG,1   ; Compare the
                                 ; value in Register
                                 ; FLAG to WREG. Skip
                                 ; the porgram memory
                                 ; if they are equal

   Case 1:        Before Instruction:
                      PC = address HERE
                      FLAG = 0x5A
                      WREG = 0x5A  ; FLAG - WREG = 0
                  After Instruction:
                      PC = address EQUAL    ; The two
                                        ; values were equal
   Case 2:        Before Instruction:
                      PC = address HERE
                      FLAG = 0xA5
                      WREG = 0x5A  ; FLAG - WREG = 0x4B
                  After Instruction:
                      PC = address NEQUAL ; The two
                                        ; values were not
                                        ; equal
```

CPFSGT   Compare f with WREG, Skip if Greater

Syntax: [*label*] CPFSGT f, a
Operands: $0 \leq f \leq 255$
                   $a \, \varepsilon \, [0,1]$
Operation: (f) - (WREG)
                   skip if (f) > (WREG) (unsigned comparison)
Status Affected: None
Description: Compares the contents of Register 'f' to the
                   contents of WREG Register by performing an
                   unsigned subtraction. If 'f' > WREG, then the
                   fetched instruction is discarded and a NOP is
                   executed instead, making this a two-cycle
                   instruction. The 'a' bit selects which bank is
                   accessed for the operation. If 'a' is 1, the bank
                   specified by the BSR Register is used. If 'a' is
                   0, the access bank is used.
Words: 1
Cycles: 1 (2 or 3)
Example:

```
        HERE    CPFSGT  FLAG,1  ; Compare the
                                ; value in Register
                                ; FLAG to WREG. Skip
                                ; the porgram memory
                                ; if FLAG > WREG
 Case 1:     Before Instruction:
                 PC = address HERE
                 FLAG = 0x5A
                 WREG = 0x5A  ; FLAG - WREG = 0
             After Instruction:
                 PC = address NGT ; The two
                                  ; values were equal
 Case 2:     Before Instruction:
                 PC = address HERE
                 FLAG = 0xA5
                 WREG = 0x5A  ; FLAG - WREG = 0x4B
             After Instruction:
                 PC = address NEQUAL ; The two
                                     ; values were not
                                     ; equal
```

CPFSLT Compare f with WREG, Skip if Less

Syntax:	[*label*] CPFSLT f, a
Operands:	$0 \le f \le 255$
	$a \; \varepsilon \; [0,1]$
Operation:	skip if (f) < (WREG)
Status Affected:	None
Description:	Compares the contents of Register 'f' to the contents of WREG Register by performing an unsigned subtraction. If 'f'< WREG, then the fetched instruction is discarded and a NOP is executed instead, making this a two-cycle instruction. The 'a' bit selects which bank is accessed for the operation. If 'a' is 1, the bank specified by the BSR Register is used. If 'a' is 0, the access bank is used.
Words:	1
Cycles:	1 (2 or 3)

Example:

```
         HERE     CPFSLT   FLAG,1   ; Compare the
                                    ; value in Register
                                    ; FLAG to WREG. Skip
                                    ; the porgram memory
                                    ; if FLAG < WREG
  Case 1:     Before Instruction:
                  PC = address HERE
                  FLAG = 0x5A
                  WREG = 0x5A  ; FLAG - WREG = 0x00
              After Instruction:
                  PC = address NLT ; The two
                                   ; values were equal
  Case 2:     Before Instruction:
                  PC = address HERE
                  FLAG = 0xA5
                  WREG = 0x5A  ; FLAG - WREG = 0x4B
              After Instruction:
                  PC = address NEQUAL ; FLAG < WREG
                                   ; Skip the next
                                   ; instruction
```

DAW   Decimal Adjust WREG Register

Syntax: [*label*] DAW
Operands: None
Operation: If [WREG<3:0> >9] or [DC = 1] then
                    (WREG<3:0>) + 6 . WREG<3:0>;
                    else
                    (WREG<3:0>) . WREG<3:0>;
                    If [WREG<7:4> >9] or [C = 1] then
                    (WREG<7:4>) + 6 . WREG<7:4>;
                    else
                    (WREG<7:4>) . WREG<7:4>;
Status Affected: C

Description: DAW adjusts the eight-bit value in WREG
                    resulting from the earlier addition of two
                    variables (each in packed BCD format) and
                    produces a correct packed BCD result.
Words: 1
Cycles: 1
Example:

```
HERE DAW       ; Decimal Adjust WREG
Case 1:
    Before Instruction:
        WREG = 0x0F  ; 0x0f = 15 decimal
        C = x
    After Instruction:
        WREG = 0x15
        C = 0
Case 2:
    Before Instruction:
        WREG = 0x68  ; 0x68 = 104 decimal
        C = x
    After Instruction:
        PC = 0x04
        C = 1           ; Indicated decimal
                        ; rollover
Case 3:
    Before Instruction:
        WREG = C6    ; 0xc6 = 198 decimal
```

```
        C = x
After Instruction:
    PC = 98
    C = 1          ; Carry to indicate
                   ; decimal rollover
```

DECF   Decrement f

Syntax: [label] DECF f,d
Operands: f in range 0 to 127
                           d is either 0 or 1
Operation: (f) - 1 -> destination
Status Affected: Z
Description: Decrement Register 'f'. If 'd' is 0, the result is
                           stored in WREG. If 'd' is 1, the result is stored
                           back in Register 'f'.
Words: 1
Cycles: 1

Example 1:

```
        decf      count,1
        Before Instruction:        count = 0x01
                                   Z = 0
        After Instruction:         count = 0x00
                                   Z = 1
```

Example 2:

```
        decf      INDF,1
        Before Instruction:        FSR = 0xc2
                                   [FSR] = 0x01
                                   Z = 0
        After Instruction:         FSR = 0xc2
                                   [FSR] = 0x00
                                   Z = 1
```

Example 3:

```
        decf      count,0
        Before Instruction:        count = 0x10
                                   WREG = x
                                   Z = 0
        After Instruction:         count = 0x10
                                   WREG = 0x0f
```

DECFSZ   Decrement f, Skip if 0

Syntax: [label] DECFSZ f,d
Operands: f in the range 0 to 127
                   d is either 0 or 1
Operation: (f) - 1 -> destination; skip if result = 0
Status Affected: None
Description: The contents of Register 'f' are
                   decremented. If 'd' is 0, the result is placed
                   in WREG. If 'd' is 1, the result is placed
                   back in Register 'f'. If the result is 0,
                   then the next instruction (fetched during
                   the current instruction execution) is
                   discarded and NOP is executed instead,
                   making this a two-cycle instruction.
Words: 1
Cycles: 1(2)

Example:

```
here:
        decfsz  count,1
        goto    here
Case 1:
Before Instruction:         PC = $
                            count = 0x01
After Instruction:          count = 0x00
                            PC = $ + 2 (goto skipped)
Case 2:
Before Instruction:         PC = $
                            count = 0x04
After Instruction:          count = 0x03
                            PC = $ + 1 (goto executed)
```

DCFSNZ   Decrement f, Skip if Not 0

Syntax: [label]	DCFSNZ f, d, a
Operands:	$0 \leq f \leq 255$
	$d \, \varepsilon \, [0,1]]$
	$a \, \varepsilon \, [0,1]$
Operation:	(f) - 1 -> destination; skip if result $\neq 0$
Status Affected:	None
Description:	The contents of Register 'f' are decremented. If the result is not 0, then the next instruction (fetched during the current instruction execution) is discarded and an NOP is executed instead, making this a 2-cycle instruction.
	The 'd' bit selects the destination for the operation.
	If 'd' is 1, the result is stored back in the File Register 'f'.
	IIf 'd' is 0, the result is stored in WREG.
	The 'a' bit selects which bank is accessed for the operation.
	If 'a' is 1, the bank specified by the BSR Register is used. If 'a' is 0, the access bank is used.
Words:	1
Cycles:	1 (2 or 3)

Example:

```
        HERE      DCFSNZ   CNT, 1, 1 ; Decrement the
                                     ; register CNT. If CNT
                                     ; not equal to zero, then
                                     ; skip the next
                                     ; instruction
                  GOTO LOOP
            CONTINUE
                           .
                           .
    Case 1:       Before Instruction:
                  PC = address HERE
                  CNT = 0x01
```

```
                    After Instruction:
                        CNT = 0x00
                        PC = address HERE + 2
    Case 2:         Before Instruction:
                        PC = address HERE
                        CNT = 0x02
                    After Instruction:
                        CNT = 0x01
                    PC = address CONTINUE
```

GOTO

Unconditional Branch

Syntax: [label] GOTO k

Operands: $0 \le k \le 2047$

Operation: k -> PC<20:0>

                     0 -> PC<0>

Status Affected: None

Description: GOTO allows an unconditional branch
                     anywhere within the entire 2Mbyte
                     memory range. The 20-bit immediate value 'k'
                     is loaded into PC<20:1>. GOTO is always a
                     two-cycle instruction.

Words: 1

Cycles: 2

Example:

```
HERE    GOTO    THERE
After Instruction:
        PC = address of THERE
```

INCF   Increment f

Syntax:	[label] INCF f,d
Operands:	f in range 0 to 127
	d is either 0 or 1
Operation:	(f) + 1 -> destination
Status Affected:	Z
Description:	The contents of Register 'f' are incremented. If 'd' is 0, the result is placed in WREG. If 'd' is 1, the result is placed back in register 'f'.
Words:	1
Cycles:	1

Example 1:

```
INCF      COUNT,1
Before Instruction:          count = 0xff
                             Z = 0

After Instruction:           count = 0x00
                             Z = 1
```

Example 2:

```
INCF      INDF,1
Before Instruction:          FSR = 0xC2
                             [FSR] = 0xff
                             Z = 0

After Instruction:           FSR = 0xc2
                             [FSR] = 0x00
                             Z = 1
```

Example 3:

```
INCF      COUNT,0
Before Instruction:          count = 0x10
                             WREG = x
                             Z = 0

After Instruction:           count = 0x10
                             WREG = 0x11
                             Z = 0
```

INCFSZ   Increment f, Skip if 0

Syntax: [label] INCFSZ f,d
Operands: f in range 0 to 127
                     d is either 0 or 1
Operation: (f) + 1 -> destination, skip if result = 0
Status Affected: None
Description: The contents of Register 'f' are incremented. If
                     'd' is 0, the result is placed in WREG. If 'd'
                     is 1, the result is placed back in Register 'f'.
                     If the result is 0, then the next instruction
                     (fetched during the current instruction
                     execution) is discarded and NOP is executed
                     instead, making this a tqo-cycle instruction.
Words: 1
Cycles: 1(2)

Example:

```
    HERE:
            INCFSZ  COUNT,1
            GOTO    HERE
Case 1:
  Before Instruction:  PC = $
            count = 0x10
  After Instruction:   count = 0x11
            PC = $ + 1 (goto executed)
Case 2:
  Before Instruction:  PC = $
            count = 0x00
  After Instruction:   count = 0x01
            PC = $ + 2 (goto skipped)
```

INFSNZ   Increment f, Skip if Not 0

Syntax: [label] INFSNZ f, d, a

Operands: $0 \le f \le 255$

                     $d \, \varepsilon \, [0,1]$

                     $a \, \varepsilon \, [0,1]$

Operation: (f) + 1 -> destination, skip if result $\ne$ 0

Status Affected: None

Description: The contents of Register 'f' are incremented. If
                     the result is not 0, then the next instruction
                     (fetched during the current instruction
                     execution) is discarded and NOP is executed
                     instead, making this a two-cycle instruction.
                     The 'd' bit selects the destination for the
                     operation.
                     If 'd' is 1, the result is stored back in the File
                     Register 'f'.
                     if 'd' is 0, the result is stored in WREG.
                     The 'a' bit selects which bank is accessed for
                     the operation.
                     If 'a' is 1, the bank specified by the BSR
                     Register is used. If 'a' is 0, the access bank is
                     used.

Words: 1

Cycles: 1 (2 or 3)

Example:

```
         HERE      INFSNZ   CNT,1,0
                        ; Increment register CNT
                        ; if CNT not equal 0, skip
                        ; the next instruction
         ZERO      GOTO     LOOP
         NZERO       .
                     .
     Case 1:  Before Instruction:
                   PC = address HERE
                   CNT = 0xFF
               After Instruction
                   CNT = 0x00
                   PC = address ZERO
```

```
Case 2: Before Instruction:
            PC = address HERE
            CNT = 0x00
      After Instruction
            CNT = 0x01
            PC = address NZERO
```

IORLW   Inclusive OR Literal with WREG

Syntax: [label] IORLW k
Operands: k is in range 0 to 255
Operation: (WREG).OR. k -> WREG
Status Affected: Z
Description: The content of WREG is ORed with the eight
                     bit literal 'k'. The result is placed in WREG.
Words: 1
Cycles: 1

Example 1:

```
iorlw   0x35
Before Instruction:      WREG = 0x9a
After Instruction:       WREG = 0xbfF
                         Z = 0
```

Example 2::

```
iorlw   myreg
Before Instruction:      WREG = 0x9a
Myreg is a variable representing a
location
in PIC RAM.               [Myreg] = 0x37
After Instruction:       WREG = 0x9F
                         Z = 0
```

Example 3:

```
iorlw   0x00
Before Instruction:      WREG = 0x00
After Instruction:       WREG = 0x00
```

IORWF   Inclusive OR WREG with f

Syntax: [label] IORWF f,d
Operands: f is in range 0 to 127
                        d is either 0 or 1
Operation: (W).OR. (f) -> destination
Status Affected: Z
Description: Inclusive OR WREG with register 'f'. If 'd' is 0,
                        the result is placed in WREG. If 'd' is 1, the
                         result is placed back in Register 'f'.
Words: 1
Cycles: 1

Example 1:

                IORWF   RESULT,0
                Before Instruction: result = 0x13
                                         WREG = 0x91
                After Instruction: result = 0x13
                                         WREG = 0x93
                                         Z = 0
Example 2:

                IORWF   INDF,1
                Before Instruction: WREG = 0x17
                                         FSR = 0xc2
                                         [FSR] = 0x30
                After Instruction: WREG = 0x17
                                         FSR = 0xc2
                                         [FSR] = 0x37
                                         Z = 0
Example 3:

                IORWF   RESULT,1
          Case 1: Before Instruction:
                                result = 0x13
                                         WREG = 0x91
                    After Instruction: result = 0x93
                                         WREG = 0x91
                                         Z = 0
          Case 2: Before Instruction: result = 0x00
                                         WREG = 0x00
                    After Instruction: result = 0x00
                                         WREG = 0x00
                                         Z = 1

LFSR   Load 12-Bit Literal to FSR

Syntax:	[*label*] LFSR f,k
Operands:	$0 \leq f \leq 2$
	$0 \leq k \leq 4095$
Operation:	k –> FSRx
Status Affected:	None
Description:	The 12-bit literal 'k' is loaded into the File Select Register (FSR Register) pointed to by 'f':
	f = 00 –> FSR0
	f = 01 ->. FSR1
	f = 10 –> FSR2
	f = 11 ->. Reserved
Words:	2
Cycles:	2

Example:

```
LFSR 2, 0x123 ; Load the 12-bit FSR2
                ; with 123h
Before Instruction:
    FSR0H = 0x05
    FSR0L = 0xA5
    FSR1H = 0x05
    FSR1L = 0xA5
    FSR2H = 0x05
    FSR2L = 0xA5
After Instruction:
    FSR0H = 0x05
    FSR0L = 0xA5
    FSR1H = 0x05
    FSR1L = 0xA5
    FSR2H = 0x01
    FSR2L = 0x23
```

MOVLW   Move Literal to WREG

Syntax: [label] MOVLW k
Operands: k in range 0 to 255
Operation: k- > WREG
Status Affected: None
Description: The eight-bit literal 'k' is loaded into WREG.
                   The don't cares will assemble as 0's.
Words: 1
Cycles: 1

Example 1:

```
        movlw   0x5a
        After Instruction:          WREG = 0x5A
```

Example 2:

```
        movlw   myreg
        Before Instruction:         WREG = 0x10
                                    [myreg] = 0x37
        After Instruction:          WREG = 0x37
```

MOVF

Move f

Syntax:	[label] MOVF f,d
Operands:	f is in range 0 to 127
	d is either 0 or 1
Operation:	(f) -> destination
Status Affected:	Z
Description:	The content of Register 'f' is moved to a destination dependent upon the status of 'd'. If 'd' = 0, destination is WREG. If 'd' = 1, the destination is file register 'f' itself. 'd' = 1 is useful to test a file register because status flag Z is affected.
Words:	1
Cycles:	1

Example 1:

```
MOVF    FSR,0
Before Instruction:        WREG = 0x00
                           FSR = 0xc2
After Instruction:         WREG = 0xc2
                           Z = 0
```

Example 2:

```
MOVF    INDF,0
Before Instruction:        WREG = 0x17
                           FSR = 0xc2
                           [FSR] = 0x00
After Instruction:         WREG = 0x17
                           FSR = 0xc2
                           [FSR] = 0x00
                           Z = 1
```

Example 3:

```
             MOVF    FSR,1
Case 1:  Before Instruction:    FSR = 0x43
         After Instruction:     FSR = 0x43
                                Z = 0
Case 2:  Before Instruction:    FSR = 0x00
         After Instruction:     FSR = 0x00
                                Z = 1
```

MOVFF   Move f to f

Syntax: [*label*] MOVFF fs, fd
Operands: 0 ≤ fs ≤ 4095
                         0 ≤ fd ≤ 4095
Operation: (fs) ->. fd
Status Affected: None
Description: The contents of Source Register 'fs' are moved
                         to Destination Register 'fd'. Location of source
                         'fs' can be anywhere in the 4096 byte data
                         space (000h$_{to}$ FFFh), and location of
                         destination 'fd' can also be anywhere from
                         000h to FFFh. MOVFF is particularly useful for
                         transferring a data memory location to a
                         Peripheral Register (such as the transmit buffer
                         or an I/O port) without affecting the WREG
                         Register.
                         Note: The MOVFF instruction cannot use the
                         PCL, TOSU, TOSH, andTOSL as the
                         Destination Register

Words: 2
Cycles: 2

Example 1:

```
MOVFF    REG1, REG2 ; Copy the contents
                    ; of Register REG1 to
                    ; Register REG2
Before Instruction:
    REG1 = 0x33
    REG2 = 0x11
After Instruction:
    REG1 = 0x33
    REG2 = 0x33
```

Example 2:

```
MOVFF REG2, REG1 ; Copy the contents of
                 ; Register REG2 to
                 ; Register REG1
Before Instruction:
    REG1 = 0x33
    REG2 = 0x11
```

```
After Instruction:
    REG1 = 0x11
    REG2 = 0x11
```

:

MOVLB   Move Literal to low nibble in BSR

Syntax: [*label*] MOVLB k
Operands: $0 \leq k \leq 15$
Operation: k ->. BSR<3:0>
Status Affected: None
Description: The 4-bit literal 'k' is loaded into the Bank
                        Select Register (BSR).
Words: 1
Cycles: 1

Example:

```
        MOVLB 5        ; Modify Least Significant
                       ; nibble of BSR Register
                       ; to value 5
        Before Instruction:
            BSR = 0x02
        After Instruction:
            BSR = 0x05
```

MOVWF   Move WREG to f

Syntax:	[label] MOVWF f
Operands:	f in range 0 to 127
Operation:	(WREG) -> f
Status Affected:	None
Description:	Move data from WREG to Register 'f'.
Words:	1
Cycles:	1

Example 1:

```
movwf   OPTION_REG
Before Instruction:        OPTION_REG = 0xff
                           WREG = 0x4f
After Instruction:         OPTION_REG = 0x4f
                           WREG = 0x4f
```

Example 2:

```
movwf   INDF
Before Instruction:        WREG = 0x17
                           FSR = 0xC2
                           [FSR] = 0x00
After Instruction:         WREG = 0x17
                           FSR = 0xC2
                           [FSR] = 0x17
```

MULLW   Multiply Literal with WREG

Syntax:	[*label*] MULLW k
Operands:	0 ≤ k ≤ 255
Operation:	(WREG) x k -> PRODH:PRODL
Status Affected:	None
Description:	An unsigned multiplication is carried out between the contents of WREG and the 8-bit literal 'k'. The 16-bit result is placed in PRODH:PRODL Register Pair. PRODH contains the high byte. WREG is unchanged. None of the status flags are affected. Neither an overflow nor carry is possible in this operation. A zero result is possible but not detected.
Words:	1
Cycles:	1
Example :	

```
MULLW 0xC4 ; Multiply the WREG Register
                  ; with the constant value
                  ; C4h
        Before Instruction:
            WREG = 0xE2
            PRODH = x
            PRODL = x
        After Instruction:
            WREG = 0xE2
            PRODH = 0xAD
            PRODL = 0x08
```

MULWF   Multiply WREG with f

Syntax: [*label*] MULWF f,a
Operands: $0 \le f \le 255$
                     a ε [0,1]
Operation: (WREG) x (f) -> PRODH:PRODL
Status Affected: None
Description: An unsigned multiplication is carried out
                     between the contents of WREG and the value
                     in Register File Location 'f'. The 16-bit result is
                     placed in the PRODH:PRODL Register Pair.
                     PRODH contains the high byte.
                     Both WREG and 'f' are unchanged.
                     None of the status flags are affected.
                     Neither an overflow nor carry is possible in this
                     operation. A zero result is possible but not
                     detected.
                     The 'a' bit selects which bank is accessed for
                     the operation.
                     If 'a' is 1, the bank specified by the BSR
                     Register is used. If 'a' is 0, the access bank is
                     used.

Words: 1
Cycles: 1

Example:
```
        MULWF MYREG, 1 ; Multiple the WREG
                       ; Register with the value
                       ; in MYREG Register
        Before Instruction:
             WREG = 0xE2
             MYREG = 0xB5
             PRODH = x
             PRODL = x
        After Instruction:
             WREG = 0xE2
             MYREG = 0xB5
             PRODH = 0x9F
             PRODL = 0xCA
```

NEGF   Negate f

Syntax: [*label*] NEGF f,a
Operands: $0 \le f \le 255$
                     $a \, \varepsilon. \, [0,1]$
Operation: $\sim (f) + 1 \rightarrow (f)$
Status Affected: C, DC, Z, OV, N
Description: Location 'f' is negated using two's complement.
                     The result is placed in the data memory
                     location 'f'. The 'a' bit selects which bank is
                      accessed for the operation. If 'a' is 1; the bank
                     specified by the BSR Register is used.
                     If 'a' is 0; the access bank is used.
Words: 1
Cycles: 1

Example:

```
     NEGF MYREG, 1      ; 2's complement the value
                            ; in MYREG
     Case 1:  Before Instruction:
                MYREG = 0x3A
                C, DC, Z, OV, N = x
              After Instruction
                MYREG = 0xC6
                C =0
                DC = 0
                Z =0
                OV = 0
                N =1
     Case 2:  Before Instruction:
                MYREG = 0xB0
                C, DC, Z, OV, N = x
              After Instruction
                MYREG = 0x50
                C =0
                DC = 1
                Z =0
                OV = 0
                N =0
     Case 3:  Before Instruction:
                MYREG = 0x00
```

```
         C, DC, Z, OV, N = x
After Instruction
     MYREG = 0x00
     C =1
     DC = 1
     Z =1
     OV = 0
     N =0
```

NOP   No Operation

Syntax:	[label] NOP
Operands:	None
Operation:	No operation
Status Affected:	None
Description:	No operation.
Words:	1
Cycles:	1

Example:

```
nop
Before Instruction:      PC = $
fter Instruction:        PC = $ + 1
```

OPTION   Load Option Register

Syntax:	[label] OPTION
Operands:	None
Operation:	(WREG) -> OPTION_REG
Status Affected:	None
Description:	The contents of WREG is loaded in the OPTION_REG register. This instruction is supported for code compatibility with PIC16C5X products. Because OPTION_REG is a Readable/writable register, code can directly address it without using this instruction.

Words: 1
Cycles: 1

Example:

```
movlw   b'01011100'
option
```

POP   POP Top of Return Stack

Syntax:	[*label*] POP
Operands:	None
Operation:	(TOS) -> bit bucket
Status Affected:	None
Description:	The Top of Stack (TOS) value is pulled off the return stack and is discarded. The TOS value then becomes the previous value that was pushed onto the return stack. This instruction is provided to enable the user to manage the return stack to incorporate a software stack.
Words:	1
Cycles:	1

Example:

```
HERE POP ; Modify the Top of Stack
         ; (TOS). The TOS points to
         ; what was one level down
   Before Instruction
       TOS = 0x0031A2
       Stack (1 level down) = 0x014332
   After Instruction
       TOS = 0x014332
       PC = HERE + 2
```

PUSH

PUSH Top of Return Stack

Syntax:	[*label*] PUSH
Operands:	None
Operation:	PC -> (TOS)
Status Affected:	None
Description:	The previous Top of Stack (TOS) value is pushed down on the stack. The PC is pushed onto the top of the return stack. This instruction is provided to enable the user to manage the return stack to incorporate a software stack.
Words:	1
Cycles:	1

Example:

```
        HERE PUSH      ; PUSH current Program
                          ; Counter value onto the
                          ; hardware stack
        Before Instruction:
            PC = 0x000124
            TOS = 0x00345A
        After Instruction:
            PC = 0x000126
            TOS = 0x000124
            Stack (1 level down)
                = 0x00345A
```

RCALL   Relative Call

Syntax: [*label*] RCALL n
Operands: -1024 ≤ n ≤ 1023
Operation: (PC + 2) –> TOS,
                     (PC + 2) + 2n -> PC
Status Affected: None
Description: Subroutine call with a jump up to 1K from the
                     current location. First, the return address
                     (PC+2) is pushed onto the stack. Then the 2's
                     complement number '2n' is added to the PC.
                     Because the PC will have incremented to fetch
                     the next instruction, the new address will be
                     PC+2+2n. This instruction is a two-cycle
                     instruction.
Words: 1
Cycles: 2
Example:

```
HERE RCALL Sub1 ; Call a program memory
                    ; location (Sub1)
                    ; this location must be
                    ; < 1024 locations forward
                    ; or > 1025 locations
                    ; backward
      Before Instruction:
          PC = Address (HERE)
          TOS = 0x0031A2
      After Instruction:
          PC = Address (Sub1)
          TOS = Address (HERE + 2)
          Stack (1 level down)
              = 0x0031A2
```

RESET   Reset Device

Syntax:	[*label*] RESET
Operands:	None
Operation:	Force all registers and flag bits that are affected by a MCLR reset to their reset condition.
Status Affected:	All
Description:	This instruction provides a way to execute a software reset.
Words:	1
Cycles:	1

RETFIE   Return from Interrupt

Syntax:	[label] RETFIE
Operands:	None
Operation:	TOS -> PC,
	1 -> GIE
Status Affected:	None
Description:	Return from Interrupt. The 13-bit address at the Top of Stack (TOS) is loaded in the PC. The Global Interrupt Enable bit, GIE (INTCON<7>), Is automatically set, enabling Interrupts. This is a two-cycle instruction.
Words:	1
Cycles:	2

Example:

```
retfie
After Instruction:        PC = TOS
                          GIE = 1
```

RETLW   Return with Literal in WREG

Syntax:	[label] RETLW k
Operands:	k in range 0 to 255
Operation:	k -> WREG
	TOS -> PC
Status Affected:	None
Description:	WREG is loaded with the eight bit literal 'k'. The program counter is loaded 13-bit address at the Top of Stack (the return address). This is a two-cycle instruction.
Words:	1
Cycles:	2

Example:

```
        movlw   2       ; Load WREG with desired
                        ; Table offset
        call    table   ; When call returns WREG
                        ; contains value stored
                        ; in table
  Table:
        addwf   pc      ; WREG = offset
        retlw   .22     ; First table entry
        retlw   .23     ; Second table entry
        retlw   .24
        .
        .
        .
        retlw   .29     ; Last table entry
        Before Instruction:     WREG = 0x02
        After Instruction:      WREG = .24
```

RETURN   Return from Subroutine

Syntax:	[label] RETURN
Operands:	None
Operation:	TOS -> PC
Status Affected:	None
Description:	Return from subroutine. The stack is POPed and the top of the stack (TOS) is loaded into the program counter. This is a two-cycle instruction.
Words:	1
Cycles:	2

Example:

```
RETURN
After Instruction
        PC = TOS
```

RLCF   Rotate Left f through Carry

Syntax:	[label] RLCF f,d,a
Operands:	f in range 0 to 127
	d is either 0 or 1
	a is either 0 or 1
Operation:	See description below
Status Affected:	C,Z,N
Description:	The contents of Register 'f' are rotated one bit to the left through the Carry Flag. If 'd' is 0, the result is placed in WREG. If 'd' is 1, the result is stored back in Register 'f'.
Words:	1
Cycles:	1

Example 1:

```
RLF      REG1,0
Before Instruction:        reg1 = 1110 0110
                           C = 0
After Instruction:         reg1 = 1110 0110
                           WREG =1100 1100
                           C =1
```

Example 2:

```
          RLF      INDF,1
Case 1: Before Instruction:    WREG = xxxx xxxx
                               FSR = 0xc2
                               [FSR] = 0011 1010
                               C = 1
        After Instruction:     WREG = 0x17
                               FSR = 0xc2
                               [FSR] = 0111 0101
                               C = 0
Case 2: Before Instruction:    WREG = xxxx xxxx
                               FSR = 0xC2
                               [FSR] = 1011 1001
                               C = 0
        After Instruction:     WREG = 0x17
                               FSR = 0xC2
                               [FSR] = 0111 0010
                               C = 1
```

RLNCF   Rotate Left f (No Carry)

Syntax: [*label*] RLNCF f, d, a

Operands: $0 \leq f \leq 127$

                                         d ε [0,1]

                                         a ε [0,1]

Operation: See description below

Status Affected: Z, N

Description: The contents of Register 'f' are rotated one bit
                          to the left. The Carry Flag bit is not affected.
                          The 'd' bit selects the destination for the
                          operation.
                          If 'd' is 1, the result is stored back in the File
                          Register 'f'.
                          If 'd' is 0, the result is stored in WREG.
                          The 'a' bit selects which bank is accessed for
                          the operation. If 'a' is 1, the bank specified by
                          the BSR Register is used. If 'a' is 0, the access
                          bank is used.

Words: 1

Cycles: 1

Example:

```
        RLNCF REG1, 0, 1 ; Rotate the value in REG1
                         ; 1 bit position left and
                         ; bit 7 loads into bit 0.
                         ; Then place the result in
                         ; the WREG Register
        Before Instruction:
            REG1 = 1110 0110
            Z, N = x
        After Instruction:
            REG1 = 1110 0110
            WREG = 1100 1101
            Z = 0
            N = 1
```

RRCF   Rotate Right f through Carry

Syntax:	[label] RRF f,d,a
Operands:	f in range 0 to 127
	d is either 0 or 1
	a is either 0 or 1
Operation:	See description below
Status Affected:	C,Z,N
Description:	The contents of register 'f' are rotated one bit to the right through the Carry Flag. If 'd' is 0 the result is placed in WREG. If 'd' is 1 the result is placed back in register 'f'.
Words:	1
Cycles:	1

Example 1:

```
        RRCF REG1 , 0
        Before Instruction:   reg1= 1110 0110
                 WREG = xxxx xxxx
                 C = 0
         After Instruction:   reg1= 1110 0110
                 WREG = 0111 0011
                 C = 0
```

Example 2:

```
        RRF  INDF , 1
    Case 1:  Before Instruction:  WREG = xxxx xxxx
                 FSR = 0xc2
                 [FSR] = 0011 1010
                 C = 1
         After Instruction:   WREG = 0x17
                 FSR = 0xC2
                 [FSR] = 1001 1101
                 C = 0
    Case 2:  Before Instruction:  WREG = xxxx xxxx
                 FSR = 0xC2
                 [FSR] = 0011 1001
                 C = 0
         After Instruction:   WREG = 0x17
                 FSR = 0xc2
                 [FSR] = 0001 1100
```

RRNCF   C = 1
                     Rotate Right f (No Carry)

Syntax: [label] RRNCF f, d, a
Operands: k in range 0 to 255
Operation: k - (W) -> W
Status Affected: C, DC, Z
Description: WREG is subtracted (2's complement method)
                   from the eight-bit literal 'k'. The result is placed
                   in WREG.
Words: 1
Cycles: 1

Example:
```
        RRNCF REG1, 0, 1; Rotate the value in REG1
                        ; 1 bit position right and
                        ; bit 0 loads into bit 7.
                        ; Then place the result in
                        ; the WREG Register
    Before Instruction:
            REG1 = 1110 0110
            WREG = x
            Z, N = 1
    After Instruction:
            REG1 = 1110 0110
            WREG = 0111 0011
            Z = 0
            N = 0
```

SETF   Set f

Syntax: [*label*] SETF f, a
Operands: $0 \le f \le 255$
                     a ε. [0,1]
Operation: FFh ->. f
Status Affected: None
Description: The contents of the specified register are set.
                     The 'a' bit selects which bank is accessed for
                     the operation. If 'a' is 1, the bank specified by
                     the BSR Register is used. If 'a' is 0, the access
                     bank is used.
Words: 1
Cycles: 1

Example:

```
        SETF FLAG_REG, 1 ; Set all the bits in
                         ; Register FLAG_REG
     Before Instruction:
         FLAG_REG = 0x5A
     After Instruction:
         FLAG_REG = 0xFF
```

SLEEP   Enter SLEEP mode

Syntax:	[*label*] SLEEP
Operands:	None
Operation:	00h ->. WDT,
	0 -> WDT prescaler count,
	1 -> ~ TO,
	0 -> ~. PD
Status Affected:	TO, PD
Description:	The power-down status bit, PD is cleared.
	Time-out status bit, TO is set. Watchdog Timer
	and its prescaler count are cleared. The
	processor is put into SLEEP mode with the
	oscillator stopped.

Words: 1

Cycles: 1

Example:

```
        SLEEP    ; Turn off the device
                      ; oscillator. This is the
                      ; lowest power mode
    Before Instruction:
        TO = ?
        PD = ?
    After Instruction:
        TO = 1†
        PD = 0
        † If WDT causes wake-up, this bit is
          cleared
```

SUBFWB   Subtract f from WREG with borrow

Syntax: [*label*] SUBFWB f, d, a
Operands: $0 \le f \le 255$
                     $d \, \epsilon \, [0,1]$
                     $a \, \epsilon \, [0,1]$
Operation: (WREG) – (f) – (C) -> destination
Status Affected: C, DC, Z, OV, N
Description: Subtract Register 'f' and carry flag (borrow)
                     from WREG (2's complement method). The 'd'
                     bit selects the destination for the operation.
                     If 'd' is 1; the result is stored back in the File
                     Register 'f'. If 'd' is 0; the result is stored in the
                     WREG Register.
                     The 'a' bit selects which bank is accessed for
                     the operation.
                     If 'a' is 1; the bank specified by the BSR
                     Register is used.
                     If 'a' is 0; the access bank is used.
Words: 1
Cycles: 1
Example:

```
        SUBFWB MYREG, 1, 1 ; WREG - MYREG - borrow
                           ; bit
        Before Instruction
            MYREG = 0x37
            WREG = 0x10
            C, DC, Z, OV, N = x
            C = 0
        After Instruction
            MYREG = 0xA8
            WREG = 0x10
            C = 0
            DC = 0
            Z = 0
            OV = 0
                        N = 1
```

SUBLW   Subtract WREG from Literal

Syntax: [label] SUBLW k
Operands: k in range 0 to 255
Operation: k - (W) -> W
Status Affected: C, DC, Z
Description: WREG is subtracted (2's complement method)
                    from the eight-bit literal 'k'. The result is placed
                    in WREG.
Words: 1
Cycles: 1

Example 1:

```
              sublw   0x02
     Case 1:  Before Instruction:      WREG = 0x01
                                       C = x
                                       Z = x
              After Instruction:       WREG = 0x01
                                       C = 1 if result +
                                       Z = 0
     Case 2:  Before Instruction:      WREG = 0x02
                                       C = x
                                       Z = x
              After Instruction:       WREG = 0x00
                                       C = 1 ; result = 0
                                       Z = 1
     Case 3:  Before Instruction:      WREG = 0x03
                                       C = x
                                       Z = x
              After Instruction:       WREG = 0xff
                                       C = 0 ; result -
                                       Z = 0
```

Example 2:

```
              sublw   myreg
              Before Instruction:      WREG = 0x10
                                       [myreg] = 0x37
              After Instruction        WREG = 0x27
                                       C = 1 ; result +
```

SUBWF   Subtract W from f

Syntax: [*label*] SUBWF f,d,a

Operands: f is in the range 0 to 255
                    d is either 0 or 1
                    a is either 0 or 1

Operation: (f) - (WREG) -> destination

Status Affected: C, DC, Z, OV, N

Description: Subtract (2's complement method)
                    .WREG Register from Register 'f'.
                    The 'd' bit selects the destination for
                    the operation. If 'd' is 1, the result is
                    stored back in the File Register 'f'. If 'd' is
                    0, the result is stored in WREG.
                    The 'a' bit selects which bank is
                    accessed for the operation. If 'a' is 1, the
                    bank specified by the BSR Register is
                    used. If 'a' is 0, the access bank is
                    used.

Words: 1

Cycles: 1

Example:

```
        SUBWF REG1, 1, 1 ; Subtract the value in
                         ; WREG Register from REG1,
                         ; placing the result in REG1
    Case 1: Before Instruction:
        REG1 = 3
        WREG = 2
        C, DC, Z, OV, N = x
        After Instruction:
        REG1 = 1
        WREG = 2
        C =1
        DC = 1
        Z =0
        OV = 0
        N =0
        ; result is positive
```

SUBWFB   Subtract W from f with Borrow

Syntax: [label]   SUBWFB f, d, a

Operands: f is between 0 and 255
                         d is either 0 or 1
                         a is either 0 or 1

Operation: (f) - (WREG) - (C) -> destination

Status Affected: C, DC, Z, OV, N

Description: Subtract (2's complement method) WREG
                         Register from Register 'f' with borrow.
                         The d bit selects the destination for the
                         operation. If 'd' is 1; the result is stored back in
                         the File Register 'f'.If 'd' is 0; the result is stored
                         in WREG.
                         The a bit selects which bank is accessed for
                         the operation. If a is 1, the bank specified by
                         the BSR Register is used. If a is 0, the access
                         bank is used.

Words: 1

Cycles: 1

Example:

```
SUBWF REG1, 1, 1 ; Subtract the value
                 ; WREG from REG1, placing the
                 ; result in REG1
   Case 1:  Before Instruction:
                 REG1 = 3
                 WREG = 2
                 C =1
                 DC, Z, OV, N = x
           After Instruction:
                 REG1 = 1
                 WREG = 2
                 C =1
                 DC = 1
                 Z =0
                 OV = 0
                 N =0
                 ; result is positive
```

SWAPF   Swap Nibbles in f

Syntax: [label] SWAPF f,d
Operands: f in range 0 to 127
                   d is either 0 or 1
Operation: (f<3:0>) -> destination<7:4>,
                   (f<7:4>) -> destination<3:0>
Status Affected: None
Description: The upper and lower nibbles of Register 'f' are
                   exchanged. If 'd' is 0, the result is placed in
                   WREG. If 'd' is 1, the result is placed in
                   Register 'f'.
Words: 1
Cycles: 1

Example 1:

```
SWAPF   REG,0
Before Instruction:        reg1 = 0xa5
After Instruction:         reg1 = 0xa5
                           WREG = 0x5a
```

Example 2:

```
SWAPF   INDF,1
Before Instruction:
        WREG = 0x17
        FSR = 0xc2
        [FSR] = 0x20
After Instruction:
        WREG = 0x17
        FSR = 0xC2
        [FSR] = 0x02
```

Example 3:

```
SWAPF   REG,1
Before Instruction:
        reg1 = 0xa5
After Instruction:
        reg1 = 0x5a
```

TBLRD   Table Read

Syntax: [*label*] TBLRD[*, *+, *-, or +*]
Operands: $0 \leq m \leq 3$
Operation: if TBLRD *,
                         (Prog Mem (TBLPTR)) -> TABLAT;
                         TBLPTR - No Change;
                     if TBLRD *+,
                         (Prog Mem (TBLPTR)) -> TABLAT;
                         (TBLPTR) +1 –> TBLPTR;
                     if TBLRD *-,
                         (Prog Mem (TBLPTR)) –> TABLAT;
                         (TBLPTR) -1 -> TBLPTR;
                     if TBLRD +*,
                         (TBLPTR) +1-> TBLPTR;
                         (Prog Mem (TBLPTR)) -> TABLAT;

Status Affected: None
Description: This instruction is used to read the contents of
                    Program Memory. To address the program
                    memory, a pointer called Table Pointer
                    (TBLPTR) is used. The TBLPTR (a 21-bit
                    pointer) points to each byte in the program
                    memory.
                    TBLPTR has a 2Mbyte address range. The
                    LSb of the TBLPTR selects which byte of the
                    program memory location to access.
                    TBLPTR[0] = 0: Least Significant byte of
                    Program Memory Word TBLPTR[0] = 1: Most
                    Significant byte of Program Memory Word
                    The Program Memory word address is the
                    same as the TBLPTR address, except that the
                    LSb of TBLPTR (TBLPTR[0]) is always forced
                    to '0'.
                    The TBLRD instruction can modify the value of
                    TBLPTR as follows:
                         • no change
                         • post-increment
                         • post-decrement
                         • pre-increment

Words: 1
Cycles: 2

Example 1:

```
TBLRD*+        ; Read byte addressed by
                ; TBLPTR, then increment
                ; TBLPTR
        Before Instruction:
            TABLAT = x
            TBLPTR = 0x00A356
            Contents of Address (TBLPTR)= 0x34
        After Instruction:
            TABLAT = 0x34
            TBLPTR = 0x00A357
```

Example 2:

```
TBLRD+*   ; Increment TBLPTR, then
                ; Read byte addressed by
                ; TBLPTR
        Before Instruction:
            TABLAT = x
            TBLPTR = 0x00A357
            Contents of Address (TBLPTR)= 0x12
            Contents of Address (TBLPTR + 1)= 0x28
        After Instruction:
            TABLAT = 0x28
            TBLPTR = 0x00A358
```

TBLWT   Table Write

Syntax: [*label*] TBLWT[*, *+, *-, +*]

Operands: $0 \leq m \leq 3$

Operation: if TBLWT*,
                     (TABLAT) –> Prog Mem (TBLPTR) or Holding
                      Register1;
                     TBLPTR - No Change;
                 if TBLWT*+,
                     (TABLAT) . Prog Mem (TBLPTR) or Holding
                      Register1;
                     (TBLPTR) +1 ->. TBLPTR;
                 if TBLWT*-,
                     (TABLAT) . Prog Mem (TBLPTR) or Holding
                      Register1;
                     (TBLPTR) -1 ->. TBLPTR;
                 if TBLWT+*,
                     (TBLPTR) +1 ->. TBLPTR;
                     (TABLAT) . Prog Mem (TBLPTR) or Holding
                      Register1;

                 Note 1: The use of a Holding Register(s) is device
                 specific. Please refer to the Device Data Sheet for
                 information on the operation of the TBLWT
                 instruction with the Program Memory.

Status Affected: None

Description: This instruction is used to program the contents of
                 Program Memory. To address the program memory,
                 a pointer called Table Pointer (TBLPTR) is
                 used.
                 The TBLPTR (a 21-bit pointer) points to each byte
                 in the program memory. TBLPTR has a 2 MBtye
                 address range. The LSb of the TBLPTR selects
                 which byte of the program memory location to
                 access.
                 TBLPTR[0] = 0: Least Significant byte of Program
                 Memory Word
                 TBLPTR[0] = 1: Most Significant byte of Program
                 Memory Word
                 The Program Memory word address is the same as
                 the TBLPTR address, except that the LSb of
                 TBLPTR (TBLPTR[0]) is always forced to '0'.

The TBLWT instruction can modify the value of TBLPTR as follows:
- no change
- post-increment
- post-decrement
- pre-increment

Words: 1
Cycles: 2 (many if long write to internal program memory)

Example 1:
```
TBLWT*+  ; Write byte addressed by
         ;  TBLPTR, then increment TBLPTR
Before Instruction:
    TABLAT = 0x55
    TBLPTR = 0x00A356
    Contents of (TBLPTR) = 0x34
After Instruction:
    TBLPTR = 0x00A357
    Contents of (TBLPTR) = 0x55
```

Example 2:
```
TBLWT+*  ; Increment TBLPTR, then Write
         ;  byte addressed by TBLPTR
Before Instruction:
    TABLAT = 0xAA
    TBLPTR = 0x00A357
    Contents of (TBLPTR) = 0x12
    Contents of (TBLPTR + 1) = 0x28
After Instruction:
    TBLPTR = 0x00A358
    Contents of (TBLPTR) = 0x12
    Contents of (TBLPTR + 1) = 0xAA
```

TSTFSZ   Test f, Skip if 0

Syntax: [*label*] TSTFSZ f, a
Operands: 0 ≤ f ≤ 255
                     a is either 0 or 1]
Operation: Skip if (f) = 0
Status Affected: None
Description: If Register 'f' = 0, the next instruction fetched is
                     discarded and a NOP is executed (two NOPs if
                     the fetched instruction is a two-cycle
                     instruction).
                     The 'a' bit selects which bank is accessed for
                     the operation.
                     If 'a' is 1, the bank specified by the BSR
                     Register is used.
                     If 'a' is 0, the access bank is used.

Words: 1
Cycles: 1 (2 or 3)
Example:

```
  HERE  TSTFSZ REG1,1   ; If REG1 = 0 then
                              ; skip the next
                              ; program memory
                              ; address
          NZERO    .
          ZERO     .
          Before Instruction:
              REG1 = 0xAF
              PC = Address (HERE)
          After Instruction:
              PC = Address (NZERO)
```

TRIS   Load TRIS Register

Syntax:	[label] TRIS f
Operands:	f in range 5 to 7
Operation:	(W) -> TRIS register f;
Status Affected:	None
Description:	The instruction is supported for code compatibility with the PIC16C5X products. Because TRIS registers are readable and writable, code can address these registers directly.
Words:	1
Cycles:	1

Example:

```
MOVLW   B'00000000'
TRIS    PORTB
```

XORLW   Exclusive OR Literal with WREG

Syntax: [label] XORLW k
Operands: k in range 0 to 255
Operation: (WREG).XOR. k -> WREG
Status Affected: Z
Description: The contents of WREG are XORed with
                     the eight bit literal 'k'. The result is placed in
                     WREG.
Words: 1
Cycles: 1

Example 1:

```
XORLW   b'10101111'
Before Instruction:
        WREG = 1011 0101
After Instruction
        WREG = 0001 1010
        Z = 0
```

Example 2:

```
XORLW   MYREG
Before Instruction:
        WREG = 0xaf
        [Myreg] = 0x37
After Instruction:
        WREG = 0x18
        Z = 0
```

XORWF   Exclusive OR WREG with f

Syntax: [label] XORWF f,d
Operands: f in range 0 to 127
                   d is either 0 or 1
Operation: (W).XOR. (f) -> destination
Status Affected: Z
Description: Exclusive OR the contents of WREG with
                   Register 'f'. If 'd' is 0, the result is stored in
                   WREG. If 'd' is 1, the result is stored
                   back in register 'f'.
Words: 1
Cycles: 1

Example 1:

```
XORWF   REG,1
Before Instruction:
        WREG = 1011 0101
        reg = 1010 1111
After Instruction:
        reg = 0001 1010
        WREG = 1011 0101
```

Example 2:

```
XORWF   REG,0
Before Instruction:
        WREG =  1011 0101
        reg = 1010 1111
After Instruction:
        reg = 1010 1111
        WREG = 0001 1010
```

Example 3:

```
XORWF   INDF,1
Before Instruction:
        WREG = 1011 0101
        FSR = 0xc2
        [FSR] = 1010 1111
After Instruction:
        WREG = 1011 0101
        FSR = 0xc2
        [FSR] = 0001 1010
```

Appendix E

Number Systems and Data Encoding

This appendix presents the background material necessary for understanding and using the number systems and numeric data storage structures employed in digital devices as well as the data types used by electronic-digital machines.

E.1 Decimal and Binary Systems

The Hindu-Arabic numerals, or decimal system, has been adopted by practically all the nations and cultures of the world. This system was introduced into Europe by the Arabs who had been using it because 800 A.D. and probably had copied it from the Hindu. It is a ten-symbol positional system of numbers that includes the special symbol for 0. The Latin title of the first book on the subject of the so-called "Indian numbers" is *Liber Algorismi de Numero Indorum*. The author is the Arab mathematician al-Khowarizmi.

The most significant feature of the Hindu-Arabic numerals is the presence of a special symbol (0), which by itself represents no quantity. Nevertheless, the symbol 0 is combined with the other ones to represent larger quantities. The principal characteristic of decimal numbers is that the value of each digit depends on its absolute magnitude and on its position in the digit string. This positional characteristic, in conjunction with the use of the special symbol 0 as a placeholder, allows the following representations:

```
   1 = one
  10 = ten
 100 = hundred
1000 = thousand
```

E.1.1 Binary Number System

The computers built in the United States during the early 1940s used decimal numbers. It was in 1946 that von Neumann, Burks, and Goldstine published a trend-setting paper titled "Preliminary Discussion of the Logical Design of an Electronic Computing Instrument," in which they proposed the use of binary numbers in computing devices arguing that binary numbers are more compact and efficient.

Another advantage of the binary system is that, in digital devices, the binary symbol 1 can be equated with the electronic state ON, and the binary symbol 0 with the state OFF. Furthermore, the two symbols of the binary system can also represent conducting and nonconducting states, positive or negative, or any other bi-valued condition. It has been proven that in digital electronics, two steady states are easier to implement and more reliable than a ten-digit encoding.

E.1.2 Radix or Base of a Number System

We have seen that in a positional number system, the weight of each column is determined by the total number of symbols in the set, including zero. In the decimal system, the weight of each digit is a power of ten, and a power of two in the binary system. The number of symbols in the set is called the base or radix of the system. Thus the base of the decimal system is 10, and the base of the binary system is 2. Notice that the increase in column weight from right to left is purely conventional. You could construct a number system in which the column weights increase in the opposite direction. In fact, in the original Hindu notation, the most significant digit was placed at the right.

E.2 Decimal versus Binary Numbers

We have seen that the binary system of numbers uses two symbols, 1 and 0. This is the simplest possible set of symbols with which we can count and perform arithmetic. Most of the difficulties in learning and using the binary system result from this simplicity. Figure E.1 shows sixteen groups of four electronic cells each in all possible combinations of two states.

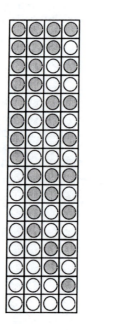 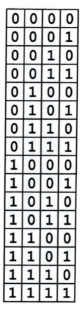

Figure E.1 *Electronic cells and binary numbers.*

Note that binary numbers match the physical state of each electronic cell. If we think of each cell as a miniature light bulb, then the binary number 1 can be used to represent the state of a charged cell (light ON) and the binary number 0 to represent the state of an uncharged cell (light OFF).

E.2.1 Hexadecimal and Octal

Binary numbers are convenient in digital electronics; however, one of their drawbacks is the number of symbols required to encode large values. For example, the number 9134 is represented in four decimal digits. However, the binary equivalent 10001110101110 requires fourteen digits. In addition, large binary numbers are difficult to remember.

One possible solution is to use other symbols to represent groups of binary digits. In this manner a group of three binary numbers allow eight possible combinations, thus we could use the decimal digits 0 to 7 to represent each possible combination of three binary digits. This grouping of three binary digits, called an octal representation, gives rise to the following table:

```
binary          octal
0 0 0             0
0 0 1             1
0 1 0             2
0 1 1             3
1 0 0             4
1 0 1             5
1 1 0             6
1 1 1             7
```

The octal encoding serves as a shorthand representation for groups of three-digit binary numbers.

Hexadecimal numbers (base 16) are used for representing values encoded in four binary digits. Because there are only ten decimal digits, the hexadecimal system borrows the first six letters of the alphabet (A, B, C, D, E, and F). The result is a set of sixteen symbols, as follows:

```
0 1 2 3 4 5 6 7 8 9 A B C D E F
```

Most modern computers are designed with memory cells, registers, and data paths in multiples of four binary digits. Table E.1 lists some common units of memory storage.

Table E.1

Units of Memory Storage

UNIT	BITS	HEX DIGITS	HEX RANGE
Nibble	4	1	0 to F
Byte	8	2	0 to FF
Word	16	4	0 to FFFF
Doubleword	32	8	0 to FFFFFFFF

In most digital electronic devices, memory addressing is organized in multiples of four binary digits. Here again, the hexadecimal number system provides a convenient way to represent addresses. Table E.2 lists some common memory addressing units and their hexadecimal and decimal range.

Table E.2

Units of Memory Addressing

UNIT	DATA PATH IN BITS	ADDRESS RANGE DECIMAL	HEX
1 paragraph	4	0 to 15	0-F
1 page	8	0 to 255	0-FF
1 kilobyte	16	0 to 65,535	0-FFFF
1 megabyte	20	0 to 1,048,575	0-FFFFF
4 gigabytes	32	0 to 4,294,967,295	0-FFFFFFFF

E.3 Character Representations

Over the years, data representation schemes have often been determined by the various conventions used by the different hardware manufacturers. Machines have had different word lengths and different character sets, and have used various schemes for storing characters and data. Fortunately, in microprocessor and microcontroller design the encoding of character data has not been subject to major disagreements.

Historically, the methods used to represent characters have varied widely, but the basic approach has always to choose a fixed number of bits and then map the various bit combinations to the various characters. The number of bits of the storage format limits the total number of distinct characters that can be represented. In this manner, the 6-bit codes used on a number of earlier computing machines allow representing 64 characters. This range allows including the uppercase letters, the decimal digits, some special characters, but not the lowercase letters. Computer manufacturers that used the 6-bit format often argued that their customers had no need for lowercase letters. Nowadays, 7- and 8-bit codes that allow representing the lowercase letters have been adopted almost universally.

Most of the world (except IBM) has standardized character representations by using the ISO (International Standards Organization) code. ISO exists in several national variants; the one used in the United States is called ASCII, which stands for *American Standard Code for Information Interchange*. All microcomputers and microcontrollers use ASCII as the code for character representation.

E.3.1 ASCII

ASCII is a character encoding based on the English alphabet. ASCII was first published as a standard in 1967 and was last updated in 1986. The first 33 codes, referred to as non-printing codes, are mostly obsolete control characters. The remaining 95 printable characters (starting with the space character) include the common characters found on a standard keyboard, the decimal digits, and the upper- and lowercase characters of the English alphabet. Table E.3 lists the ASCII characters in decimal, hexadecimal, and binary.

Table E.3

ASCII Character Representation

DECIMAL	HEX	BINARY	VALUE	
000	000	00000000	annual	(Null character)
001	001	00000001	SOH	(Start of Header)
002	002	00000010	STX	(Start of Text)
003	003	00000011	ETX	(End of Text)
004	004	00000100	EOT	(End of Transmission)
005	005	00000101	ENQ	(Enquiry)
006	006	00000110	ACK	(Acknowledgment)
007	007	00000111	BEL	(Bell)
008	008	00001000	BS	(Backspace)
009	009	00001001	HT	(Horizontal Tab)
010	00A	00001010	LF	(Line Feed)
011	00B	00001011	VT	(Vertical Tab)
012	00C	00001100	FF	(Form Feed)
013	00D	00001101	CR	(Carriage Return)
014	00E	00001110	SO	(Shift Out)
015	00F	00001111	SI	(Shift In)
016	010	00010000	DLE	(Data Link Escape)
017	011	00010001	DC1	(XON)(Device Control 1)
018	012	00010010	DC2	(Device Control 2)
019	013	00010011	DC3	(XOFF)(Device Control 3)
020	014	00010100	DC4	(Device Control 4)
021	015	00010101	NAK	(- Acknowledge)
022	016	00010110	SYN	(Synchronous Idle)
000	000	00000000	annual	(Null character)
023	017	00010111	ETB	(End of Trans. Block)
024	018	00011000	CAN	(Cancel)
025	019	00011001	EM	(End of Medium)
026	01A	00011010	SUB	(Substitute)
027	01B	00011011	ESC	(Escape)
028	01C	00011100	FS	(File Separator)
029	01D	00011101	GS	(Group Separator)
030	01E	00011110	RS	(Request to Send)
031	01F	00011111	US	(Unit Separator)
032	020	00100000	SP	(Space)
033	021	00100001	!	(exclamation mark)
034	022	00100010	"	(double quote)
035	023	00100011	#	(number sign)
036	024	00100100	$	(dollar sign)
037	025	00100101	%	(percent)
038	026	00100110	&	(ampersand)
039	027	00100111	'	(single quote)
040	028	00101000	(	(left/opening parenthesis)
041	029	00101001	)	(right/closing parenthesis)
042	02A	00101010	*	(asterisk)
043	02B	00101011	+	(plus)
044	02C	00101100	,	(comma)
045	02D	00101101	-	(minus or dash)
046	02E	00101110	.	(dot)
047	02F	00101111	/	(forward slash)
048	030	00110000	0	(decimal digits ...)
049	031	00110001	1	

(continues)

Table E.3

ASCII Character Representation (conitnued)

DECIMAL	HEX	BINARY	VALUE	
050	032	00110010	2	
051	033	00110011	3	
052	034	00110100	4	
053	035	00110101	5	
054	036	00110110	6	
055	037	00110111	7	
056	038	00111000	8	
057	039	00111001	9	
058	03A	00111010	:	(colon)
059	03B	00111011	;	(semi-colon)
060	03C	00111100	<	(less than)
061	03D	00111101	=	(equal sign)
062	03E	00111110	>	(greater than)
063	03F	00111111	?	(question mark)
064	040	01000000	@	(AT symbol)
065	041	01000001	A	
066	042	01000010	B	
067	043	01000011	C	
. . .				
090	05A	01011010	Z	
091	05B	01011011	[	(left/opening bracket)
092	05C	01011100	\	(back slash)
093	05D	01011101	]	(right/closing bracket)
094	05E	01011110	^	(circumflex)
095	05F	01011111	_	(underscore)
096	060	01100000	`	(accent)
097	061	01100001	a	
098	062	01100010	b	
099	063	01100011	c	
...				
122	07A	01111010	z	
123	07B	01111011	{	(left/opening brace)
124	07C	01111100	I	(vertical bar)
125	07D	01111101	}	(right/closing brace)
126	07E	01111110	~	(tilde)
127	07F	01111111	DEL	(delete)

E.3.2 EBCDIC and IBM

In spite of ASCII's general acceptance, IBM continues to use its one EBCDIC (Extended Binary Coded Decimal Interchange Code) for character encoding. Thus, IBM mainframes and mid-range systems such as the AS/400 still use a character set primarily designed for punched card technology.

EBCDIC uses the full 8 bits available to it, so parity checking cannot be used on an 8-bit system. Also, EBCDIC has a wider range of control characters than ASCII. EBCDIC character encoding is based on Binary Coded Decimal (BCD), which we discuss later.

E.3.3 Unicode

One of the limitations of the ASCII code is that an 8-bit representation is limited to 256 combinations, which is not enough for all the characters of languages such as Japanese and Chinese. This limitation of the ASCII led to the development of encodings that would allow representing large character sets. Unicode has been proposed as a universal character encoding standard that can be used for representation of text for computer processing.

Unicode attempts to provide a consistent way of encoding multilingual text and thus makes it possible to exchange text files internationally. The design of Unicode is based on the ASCII code but goes beyond the Latin alphabet to which ASCII is limited. The Unicode Standard provides the capacity to encode all the characters used for the written languages of the world. Like ASCII, Unicode assigns each character a unique numeric value and name. Unicode uses three encoding forms that use a common repertoire of characters. These forms allow encoding as many as a million characters.

The three encoding forms of the Unicode Standard allow the same data to be transmitted in byte, word, or double word format, that is in 8-, 16- or 32-bits per character. UTF-8 is a way of transforming all Unicode characters into a variable length encoding of bytes. In this format, the Unicode characters corresponding to the familiar ASCII set have the same byte values as ASCII. By the same token, Unicode characters transformed into UTF-8 can be used with existing software.

UTF-16 is designed to balance efficient access to characters with economical use of storage. It is reasonably compact, and all the heavily used characters fit into a single 16-bit code unit, while all other characters are accessible via pairs of 16-bit code units. UTF-32 is used where memory space is no concern, but fixed width, single code unit access to characters is desired. In UTF-32, each Unicode character is represented by a single 32-bit code.

E.4 Encoding of Integers

For unsigned integers there is little doubt that the binary representation is ideal. Successive bits indicate powers of 2, with the most significant bit at the left and the least significant one on the right, as is customary in the decimal system. Figure E.2 shows the digit weights and the conventional bit numbering in the binary encoding.

In order to perform arithmetic operations, the digital machine must be capable of storing and retrieving numerical data. The storage formats are designed to minimize space and optimize processing. Historically, numeric data was stored in data structures devised to fit the characteristics of a specific machine, or the preferences of its designers. It was in 1985 that the Institute of Electrical and Electronics Engineers (IEEE) and the American National Standards Institute (ANSI) formally approved mathematical standards for encoding and storing numerical data in digital devices.

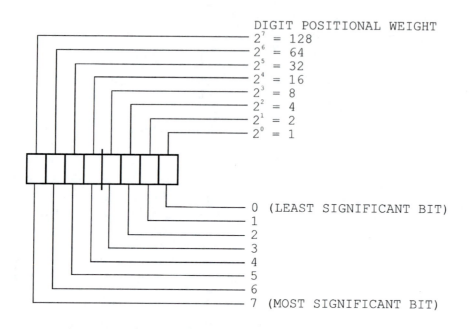

Figure E.2 *Binary digit weights and numbering.*

Data stored in processor registers, magnetic media, optical devices, or punched tape, is usually encoded in binary. Thus, the programmer and the operator can usually ignore the physical characteristics of the storage medium. This means that the bit pattern 10010011 can be encoded as holes in a strip of paper tape, as magnetic charges on a mylar-coated disk, as positive voltages in an integrated circuit memory cell, or as minute craters on the surface of the CD. In all cases 10010011 represents, the decimal number 147.

E.4.1 Word Size

In electronic digital devices, the bi-stable states are represented by a single binary digit, or bit. Circuit designers group several individual cells to form a unit of storage that holds several bits. In a particular machine, the basic unit of data storage is called the word size. Word size in computers often ranges from 8 to 128 bits, in powers of 2. Microcontrollers and other digital devices sometimes use word sizes that are determined by their specific architectures. For example, the mid-range PIC microcontrollers use a 14-bit word size.

In most digital machines, the smallest unit of storage individually addressable is 8 bits (one byte). Individual bits are not directly addressable and must be manipulated as part of larger units of data storage.

E.4.2 Byte Ordering

The storage of a single-byte integer can be done according to the scheme in Figure E.2. However, the maximum value that can be represented in 8 bits is the decimal number 255. To represent larger binary integers requires additional storage area. Because memory is usually organized in byte-size units, any decimal number larger than 255 will require more than one byte of storage. In this case, the encoding is padded with the necessary leading zeros. Figure E.3 is a representation of the decimal number 21,141 stored in two consecutive data bytes.

```
        machine storage
                              binary          decimal
                            = 01010010 10010101 = 21,141
```

Figure E.3 *Representation of an unsigned integer.*

One issue related to using multiple memory bytes to encode binary integers is the successive layout of the various byte-size units. In other words, does the representation store the most significant byte at the lowest numbered memory location, or vice versa? For example, when a 32-bit binary integer is stored in a 32-bit storage area, we can follow the conventional pattern of placing the low-order bit on the right-hand side and the high-order bit on the left, as we did in Figure E.3. However, if the 32-bit number is to be stored into four byte-size memory cells, then two possible storage schemes are obvious. as shown in Figure E.4.

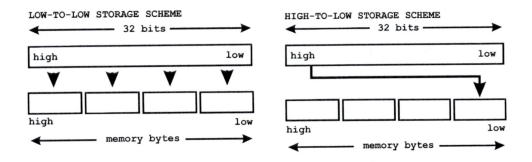

Figure E.4 *Byte ordering schemes.*

In the low-to-low storage scheme, the low-order 8 bits of the operand are stored in the low-order memory byte; the next group of 8-bits is moved to the following memory byte in low-to-high order, and so on. Conceivably, this scheme can be described by saying that the "little end" of the operand is stored first, that is, in lowest memory. According to this notion, the storage scheme is described as the little-endian format. If the "big-end" of the operand, that is, the highest valued bits, are stored in the low memory addresses, then the byte ordering is said to be in big-endian format. Some Intel processors (like those of the 80x86 family) follow the little-endian format. This is easy to remember because Intel rhymes with little. On the other hand, some Motorola processors (like those of the 68030 family) follow the big-endian format, while still others (such as the MIPS 2000) can be configured to store data in either format.

Often, the programmer needs to be aware of the byte-ordering scheme, for example, to retrieve memory data into processor registers so as to perform multi-byte arithmetic, or to convert data stored in one format to the other one. This last operation can be described as a simple byte swap. For example, if the hex value 01020304 is stored in four consecutive memory cells in low-to-high order (little-endian format), it appears in memory (low-to-high) as the values 04030201. Converting this data to the big-endian format consists of swapping the individual bytes so that they are stored in the order 01010304. Figure E.5 is a diagram of a byte swap operation.

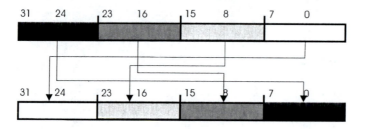

Figure E.5 *Data format conversion by byte swapping.*

E.4.3 Sign-Magnitude Representation

Representing signed numbers requires a way of differentiating between positive and negative magnitudes. One possible scheme is to devote one bit to represent the sign. Typically the high-order bit is set to denote negatives and reset to denote positives. Using this convention, the decimal numbers 93 and – 93 are represented as follows:

```
01011101 binary = 93 decimal
11011101 binary = -93 decimal
|
|------------ sign bit
```

This way of designating negative numbers, called a sign-magnitude representation, corresponds to the conventional way in which we write negative and positive numbers longhand; that is, we precede the number by its sign. Sign-magnitude representation has the following characteristics:

1. The absolute value of positive and negative numbers is the same.

2. Positive numbers can be distinguished from negative numbers by examining the high-order bit.

3. There are two possible representations for zero, one negative (10000000B) and one positive (00000000B).

But a major limitation of sign-magnitude representation is that the processing required to perform addition is different from that for subtraction. While this is not insurmountable, there are other numeric representations that avoid this problem. A consequence of sign-magnitude representation is that, in some cases, it is necessary to take into account the magnitude of the operands in order to determine the sign of the result. Also, the presence of a negative zero reduces the numerical range of the representation and is, for most practical uses, an unnecessary complication.

But the most important limitations of the sign-magnitude format are the complicated rules required for the addition of signed numbers. For example, considering two operands labeled x and y, the following rules must be observed for performing signed addition:

1. If x and y have the same sign, they are added directly and the result is given the common sign.

2. If x is larger than y, then y is subtracted from x and the result is given the sign of x.

3. If y is larger than x, then x is subtracted from y and the result is given the sign of y.

4. If either x or y is 0 or –0, the result is the non-zero element.

5. If both x and y are –0, then the sum is 0.

E.4.4 Radix Complement Representation

The radix complement of a number is defined as the difference between the number and the next integer power of the base that is larger than the number. In decimal numbers the radix complement is called the ten's complement. In the binary system, the radix complement is called the two's complement. For example, the radix complement of the decimal number 89 (ten's complement) is calculated as follows:

```
   100   = higher power of 10
 -  89
 -------
    11   = ten's complement of 89
```

The use of radix complements to simplify machine subtraction operations can best be seen in an example. Suppose the operation x = a – b with the following values:

```
    a = 602
    b = 353

          602
        - 353
        -----
  x =     249
```

Notice that in the process of performing longhand subtraction, we had to perform two borrow operations. Now consider that the radix complement (ten's complement) of 353 is

```
 1000 - 353 = 647
```

Using complements, we can reformulate subtraction as the addition of the ten's complement of the subtrahend, as follows:

```
        602
    +   647
        -----
       1249
       |_____ discarded digit
```

The result is adjusted by discarding the digit that overflows the number of digits in the operands.

In decimal arithmetic, there is little advantage in replacing subtraction with ten's complement addition. The work of calculating the ten's complement cancels out any other possible benefit. However, in binary arithmetic, the use of radix complements entails significant computational advantages — principally because binary machines can calculate complements easily and rapidly.

The two's complement of a binary number can be obtained in the same manner as the ten's complement of a decimal number, that is, by subtracting the number from an integer power of the base that is larger than the number. For example, the two's complement of the binary number 101 is:

```
     1000B  =   2^3 = 8 decimal (higher power of 2)
  -   101B  =         5 decimal
     _____             _____
     011B   =         3 decimal
```

And the two's complement of 10110B is calculated as follows:

```
     100000B  =  2^5 = 32 decimal (higher power of 2)
  -   10110B  =        22 decimal
     _____              _____
     01010B             10 decimal
```

You can perform the binary subtraction of 11111B (31 decimal) minus 10110B (22 decimal) by finding the two's complement of the subtrahend, adding the two operands, and discarding any overflow digit, as follows:

```
          11111B  =  31 decimal
       +  01010B  =  10 decimal (two's complement of 22)
          _____
          101001B
 discard_____|
          01001B  =   9 decimal (31 minus 22 = 9)
```

In addition to the radix complement representation, there is a diminished radix representation that is sometimes useful. This encoding, sometimes called the radix-minus-one form, is created by subtracting 1 from an integer power of the base that is larger than the number, then subtracting the operand from this value. In the decimal system, the diminished radix representation is sometimes called the nine's complement.

In the binary system, the diminished radix representation is called the one's complement. The one's complement of a binary number is obtained by subtracting the number from an integer power of the base that is larger than the number, minus one. For example, the one's complement of the binary number 101 (5 decimal) can be calculated as follows:

```
1000B   =   2^3 = 8 decimal

 111B   =   1000B minus 1 =   7 decimal
 101B                         5 decimal
_____                      _____

 010B   =                     2 decimal
```

An interesting feature of one's complement is that it can also be obtained by changing every 1 binary digit to a 0 and every 0 binary digit to a 1. In the above example 010B is one's complement of 101B. In this context the 0 binary digit is said to be the complement of the 1 binary digit, and vice versa. Most modern computers contain an instruction that inverts all the digits of a value by transforming all 1 digits into 0, and all 0 digits into 1. The operation is also known as a negation.

It is also interesting that the two's complement can be obtained by adding one to the one's complement of a number. Therefore, instead of calculating

```
    100000B
 -   10110B
    _____

     01010B
```

we can find the two's complement of 10110B as follows:

```
  10110B   = number
  01001B   = change 0 to 1 and 1 to 0 (one's complement)
+     1B     then add 1
  _____

  01010B   = two's complement
```

This algorithm provides a convenient way of calculating the two's complement in a machine equipped with a complement instruction. Finally, the two's complement can be obtained by subtracting the operand from zero and discarding the overflow.

E.4.5 Simplification of Subtraction

We have seen that the radix complement of a number is the difference between the number and an integer power of the base that is larger than the number. Following this rule, we calculate the radix complement of the binary number 10110 as follows

```
   100000B   =   2^5 = 32 decimal
 -  10110B   =         22 decimal
   _____             _____

    01010B             10 decimal
```

However, sometimes the machine calculation of the two's complement of the same value produces a different result; for example,

```
   100000000B   =   28 = 256 decimal
 -  00010110B   =         22 decimal
   _____             _____

   11101010B              234 decimal
```

The difference is due to the fact that in the longhand method, we have used the next higher integer power of the base compared to the value of the subtrahend (in this case, 100000B), while in the machine calculations we use the next higher integer power of the base compared to the operand's word size, which is normally either 8 or 16 bits. In the above example, the operand's word size is 8 bits and the next highest integer power of 2 is 100000000B.

In reality, the results of two's complement subtraction are valid as long as the minuend is an integer power of the base that is larger than the subtrahend. For example, to perform the binary subtraction of 00011111B (31 decimal) minus 00010110B (22 decimal), we can find the two's complement of the subtrahend and add, discarding any overflow digit, as follows:

```
          00011111B   =   31 decimal
      +   11101010B   =  234 decimal (two's complement of 22)
          _____
          100001001B
discard____|
          00001001B   =    9 decimal (31 minus 22 = 9)
```

In addition to the simplification of subtraction, two's complement arithmetic has the advantage that there is no representation for negative 0. While both the two's complement and the one's complement schemes can be used to implement binary arithmetic, system designers usually prefer the two's complement.

E.5 Binary Encoding of Fractional Numbers

In any positional number system, the weight of each integer digit can be determined by the formula:

$$P = d * BC$$

where d is the digit, B is the base or radix, and C is the zero-based column number, starting from right to left. Therefore, the value of a multi-digit positive integer to n digits can be expressed as a sum of the digit values:

$$dn*Bn + dn\text{-}1*Bn\text{-}1 + dn\text{-}2*Bn\text{-}2 + ... + d0*B0$$

where d is the value of the digit and B is the base or radix of the number system. This representation can be extended to represent fractional values. Recalling that

$$x^{-n} = \frac{1}{x^n}$$

we can extend the sequence to the right of the radix point, as in Figure E.6.

In the decimal system, the value of each digit to the right of the decimal point is calculated as 1/10, 1/100, 1/1000, and so on. The value of each successive digit of a binary fraction is the reciprocal of a power of 2; therefore the sequence is 1/2, 1/4, 1/8, 1/16, etc. Figure E.6 shows the positional weight of the integer and the fractional digits in a binary number.

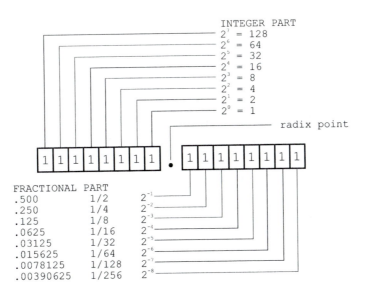

Figure E.6 *Positional weights in a binary fraction.*

E.5.1 Fixed-Point Representations

The binary encoding and storage of fractional numbers (also called real numbers) presents several difficulties. The first one is related to the location of the radix point. Because there are only two symbols in the binary set, and both are used to represent the numerical value of the number, there is no other symbol available for the decimal point.

One possible solution is to pre-format the field of digits that represent the integer part and the one that represents the fractional part. For example, if a real number is to be encoded in two data bytes, we can assign the high-order byte to the integer part and the low-order byte to the fractional part. In this case, the positive decimal number 58.125 could be encoded as shown in Figure E.7.

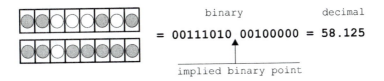

Figure E.7 *Binary fixed-point representation.*

In Figure E.7 we take for granted that the binary point is positioned between the eighth and the ninth digit of the encoding. Fixed-point representations assume that whatever distribution of digits is selected for the integer and the fractional part of the representation is maintained in all cases. This is the greatest limitation of the fixed-point formats. Suppose we want to store the value 312.250. This number can be represented in binary as follows:

```
312  = 100111000
.250 = .01
```

In this case the total number of binary digits required for the binary encoding is 11, which means that the number can be physically stored in a 16-digit structure (as the one in Figure E.7) leaving five cells to spare. However, because the fixed-point format we have adopted assigns eight cells to the integer part of the number, 312.250 cannot be encoded because the integer part requires nine binary digits. In spite of this limitation, the fixed-point format was the only one used in early computers.

E.5.2 Floating-Point Representations

An alternative to fixed-point is not to assume that the radix point has a fixed position in the encoding, but to allow it to float. The idea of separately encoding the position of the radix point probably originated in scientific notation, where a number is written as a base greater than or equal to 1 and smaller than 10, multiplied by a power of 10. For example, the value 310.25 in scientific notation is written

$$3.1025 \times 10^2$$

A number in scientific notation has a real part and an exponent part. Using the terminology of logarithms these two parts are sometimes called the mantissa and the characteristic. The following simplification of scientific notation is often used in computer work:

```
3.1025 E2
```

In the computer version of scientific notation, the multiplication symbol and the base are implied. The letter E, which is used to signal the start of the exponent part of the representation, accounts for the name "exponential form." Numbers smaller than 1 can be represented in scientific notation or in exponential form using negative powers. For example, the number .0004256 can be written as

$$4.256 \times 10^{-4}$$

or as

```
4.256 E-4
```

Floating-point representations provide more efficient use of the machine's storage space. For example, the numerical range of the fixed point encoding shown in Figure E.7 is from 255.99609375 to 0.00390625. To improve this range we can reassign the 16 bits of storage so that 4 bits are used for encoding the exponent and 12 bits for the fractional part, which is called the significand. In this case, the encoded number appears as follows:

```
0000 000000000000
+--+ +----------+
 ¦           |_____  12-bit fractional part
 ¦                        (significand)
 |_____  4-bit exponent part
```

If we were to use the first bit of the exponent to indicate the sign of the exponent, then the range of the remaining three digits would be 0 to 7. Notice that the sign of the exponent indicates the direction in which the decimal point is to be moved, which is unrelated to the sign of the number. In this example, the fractional part (or significand) could hold values in the range 1,048,575 to 1. The combined range of exponent and significand allows representing decimal numbers in the range 4095 to 0.00000001, which considerably exceeds the range in the same storage space in fixed-point format.

E.5.3 Standardized Floating-Point

Both the significand and the exponent of a floating-point number can be stored as an integer, in sign magnitude, or in radix complement form. The number of bits assigned to each field can vary according to the range and the precision required. For example, the computers of the CDC 6000, 7000, and CYBER series used a 96-digit significand with an 11-digit exponent, while the PDP 11 series used 55-digit significands and 8-digit exponents in their extended precision formats.

Variations, incompatibilities, and inconsistencies in floating-point formats created the need to develop a standard format. In March and July 1985, the Computer Society of the Institute of Electric and Electronic Engineers (IEEE) and the American National Standards Institute (ANSI) approved a standard for binary floating-point arithmetic (ANSI/IEEE Standard 754-1985). This standard establishes four formats for encoding binary floating-point numbers. Table E.4 summarizes the characteristics of these formats.

Table E.4

ANSI/IEEE Floating Point Formats

PARAMETER	SINGLE	SINGLE EXTENDED	DOUBLE	DOUBLE EXTENDED
total bits	32	43	64	79
significand bits	24	32	53	64
maximum exponent	+127	1023	1023	16383
minimum exponent	−126	1022	−1022	16382
exponent width	8	11	11	15
exponent bias	+127	...	+1023	...

Figure E.8 shows the IEEE floating-point single format.

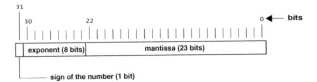

Figure E.8 *IEEE floating-point single format.*

If a floating-point encoding is to allow the representation of signed numbers, it must devote one binary digit to encode the number's sign. In the IEEE 754 single format in Figure E.8, the high-order bit represents the sign of the number. A value of 1 indicates a negative number. We prematurely end the discussion of binary floating-point encodings because binary floating-point calculations are out of the scope of this book.

E.5.4 Binary-Coded Decimals (BCD)

Binary floating-point encodings, as those defined in ANSI/IEEE 754, are the most efficient format for storing numerical data in a digital device and provide the fastest and most efficient way of performing numerical calculations. However, other representations are also useful and much easier to implement.

Binary-coded decimal (BCD) is a way of representing decimal digits in binary form. There are two common ways of implementing this encoding; one is known as the packed BCD format and the other one as unpacked. In the unpacked format, each BCD digit is stored in one byte. In the packed scheme, two BCD digits are encoded per byte. The unpacked BCD format does not use the four high-order bits of each byte, which is wasted storage space. On the other hand, the unpacked format facilitates conversions and arithmetic operations on some machines. Figure E.9 shows the memory storage of a packed and unpacked BCD number.

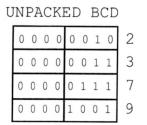

Figure E.9 *Packed and unpacked BCD.*

Floating-Point BCD

Unlike the floating-point binary numbers, binary-coded decimal representations and BCD arithmetic have not been explicitly detailed in a formal standard. Although ANSI/IEEE 758 covers numerical representation independent of radix, the standard does not provide specific formats like those in ANSI/IEEE 754. Therefore, each machine or software package stores and manipulates BCD numbers in a unique and often incompatible way. Some machines include packed decimal formats, which are sign-magnitude BCD representations. These integer formats are useful for conversions and input-output operations. For performing arithmetic calculations, a floating-point BCD encoding is required. This approach provides all the advantages of floating-point as well as the accuracy of decimal encodings.

Our own BCD floating-point format, which we call BCD12, is shown Figure E.10.

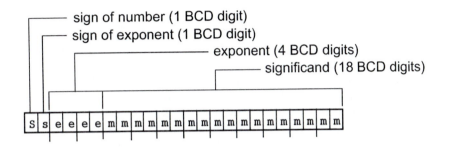

Figure E.10 Map of the BCD12 format.

The structure of the BCD12 format is described in Table E.5.

Table E.5

Field Structure of the BCD12 Format

CODE	FIELD NAME	BITS WIDE	BCD DIGITS	RANGE
S	sign of number	4	1	0 – 1 (BCD)
S	sign of exponent	4	1	0 – 1 (BCD)
E	exponent	16	4	0 – 9999
M	significand	72	18	0 – 99..99 (18 digits)
	Format size	96 (12 bytes)		

Notes:
1. The significand is scaled (normalized) to a number in the range 1.00..00 to 9.99..99.
2. The encoding for the value zero (0.00..00) is a special case.
3. The special value FFH in the sign byte indicates an invalid number.

BCD12 requires 12 bytes of storage and is described as follows:

1 .The sign of the number (S) is encoded in the left-most packed BCD digit. Therefore, the first 4 bits are either 0000B (positive number) or 0001B (negative number).

2. The sign of the exponent is represented in the 4 low-order bits of the first byte. This means that the sign of the exponent is also encoded in one packed BCD digit. As is the case with the sign of the number field, the sign of the exponent is either 0000B (positive exponent) or 0001B (negative exponent).

3. The following 2 bytes encode the exponent in 4 packed BCD digits. The decimal range of the exponent is 0000 to 9999.

4. The remaining 9 bytes are devoted to the significand field, consisting of 18 packed BCD digits. Positive and negative numbers are represented with a significand normalized to the range 1.00...00 to 9.00...99. The decimal point following the first significand digit is implied. The special value 0 has an all-zero significand.

5. The special value FF hexadecimal in the number's sign byte indicates an invalid number.

The BCD12 format, as is the case in all BCD encodings, does not make ideal use of the available storage space. In the first place, each packed BCD digit requires 4 bits, which in binary could serve to encode six additional combinations. At a byte level the wasted space is of 100 encodings (BCD 0 to 99) out of a possible 256 (0 to FFH). The sign field in the BCD12 format also is wasteful because only one binary digit is actually required for storing the sign. Regarding efficient use of storage, BCD formats cannot compete with floating-point binary encodings. The advantages of BCD representations are a greater ease of conversions into decimal forms, and the possibility of using the processor's BCD arithmetic instructions.

Appendix F

Basic Electronics

F.1 Atom

Until the end of the nineteenth century it was assumed that matter was composed of small, indivisible particles called atoms. At that time, the work of J.J. Thompson, Lord Rutheford, and Neils Bohr proved that atoms were complex structures that contained both positive and negative particles. The positive ones were called protons and the negative ones electrons.

Several models of the atom were proposed: the one by Thompson assumed that there were equal numbers of protons and electrons inside the atom, and that these elements were scattered at random, as in the leftmost drawing in Figure F.1. Later, Lord Rutheford's experiments led him to believe that atoms contained a heavy central positive nucleus with the electrons scattered randomly. So he modified Thompson's model as shown in the center drawing. Finally, Neils Bohr observed that electrons had different energy levels, as if they moved around the nucleus in different orbits, like planets around a sun. The rightmost drawing represents this orbital model.

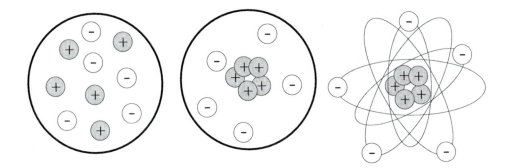

Figure F.1 *Models of the atom.*

Investigations also showed that the normal atom is electrically neutral and that the protons (positively charged particles) are 2,000 times more massive than the electrons (negatively charged particles). Furthermore, the orbital model of the atom is not actually valid because orbits have little meaning at the atomic level. A more accurate representation of reality is based on up to seven concentric spherical shells about the nucleus, The innermost shell has a capacity for 2 electrons and the outermost for 72.

The number of protons in an atom determines its atomic number; for example, the hydrogen atom has a single proton and an atomic number of 1, helium has 2 protons, carbon has 6, and uranium has 92. But when we compare the ratio of mass to electrical charge in different atoms, we find that the nucleus must be made up of more than protons. For example, the helium nucleus has twice the charge of the hydrogen nucleus, but four times the mass. The additional mass is explained by assuming that there is another particle in the nucleus, called a neutron, that has the same mass at the proton but no electrical charge. Figure F.2 shows a model of the helium atom with two protons, two electrons, and two neutrons.

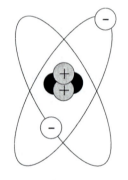

Figure F.2 *Model of the helium atom.*

F.2 Isotopes and Ions

But nature is not always consistent with such neat models. Whereas in a neutral atom, the number of protons in the atomic nucleus exactly matches the number of electrons, the number of protons need not match the number of neutrons. For example, most hydrogen atoms have a single proton, but no neutrons, while a small percentage have one neutron, and an even smaller one have two neutrons. In this sense, atoms of an element that contains a different number of neutrons are isotopes of that element. For instance, water (H_2O) containing hydrogen atoms with one or two neutrons is called "heavy water."

An atom that is electrically charged due to an excess or deficiency of electrons is called an ion. When the dislodged elements are one or more electrons, the atom takes on a positive charge. In this case, it is called a positive ion. When a stray electron combines with a normal atom, the result is called a negative ion.

F.3 Static Electricity

Free electrons can move about, or travel, through matter, or remain at rest on a surface. When electrons are at rest, the surface is said to have a static electrical charge which can be positive or negative. When electrons are moving in stream-like manner, we call this movement an electrical current. Electrons can be removed from a surface by means of friction, heat, light, or a chemical reaction. In this case, the surface becomes positively charged.

Six hundred years before our time, the Greeks discovered that when amber was rubbed with wool, the amber became electrically charged and would attract small pieces of material. In this case, the charge is a positive one. Friction can cause other materials, such as hard rubber or plastic, to become electrically charged negatively. Observing objects that have positive and negative charges we notice that like charges repel and unlike charges attract each other, as shown in Figure F.3.

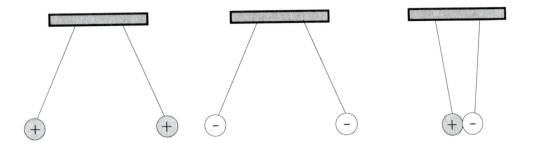

Figure F.3 *Like and unlike charges.*

Friction causes loosely held electrons to be transferred from one surface to the other. This results in a net negative charge on the surface that has gained electrons, and a net positive charge on the surface that has lost electrons. If there is no path for the electrons to take to restore the balance of electrical charges, these charges will remain until they gradually leak off. If the electrical charge continues building, it will eventually reach the point where it can no longer be contained. In this case, it will discharge itself over any available path, as is the case with lightning.

The point about static electricity is that it does not move from one place to another. This means that while some interesting experiments can be performed with it, it does not serve the practical purpose of providing energy to do sustained work.

Static electricity certainly exists, and under certain circumstances we must allow for it and account for its possible presence, but it will not be the main theme of these pages.

F.4 Electrical Charge

Physicists often resort to models and theories in order to describe and represent some force that can be measured in the real world. But very often these models and representations are no more than concepts that fail to represent the object physically. In this sense, no one knows exactly what is gravity, or an electrical charge. We call gravity the force between masses, which can be felt and measured. By the same token, bodies in "certain electrical condition" also exert forces on one another that can be measured. The term "electrical charge" was invented to explain these observations. Three simple postulates or assumptions serve to explain all electrical phenomena:

1. Electrical charge exists and can be measured. Charge is measured in Coulombs, a unit named for the French scientist Charles Agustin Coulomb.

2. Charge can be positive or negative.

3. Charge can neither be created nor destroyed. If two objects with equal amounts of positive and negative charge are combined on some object, the resulting object will be electrically neutral and will have zero net charge.

F.4.1 Voltage

We have seen that objects with opposite charges attract, that is, they exert a force upon each other that pulls them together. In this case, the magnitude of the force is proportional to the product of the charge on each mass. Like gravity, electrical force depends inversely on the distance squared between the two bodies; the closer the bodies, the greater the force. Consequently, it takes work and energy to pull apart objects that are positively and negatively charged — in the same manner that it takes work to raise a big mass against the pull of gravity.

The potential that separate objects with opposite charges have for doing work is called voltage. Voltage is measured in units of volts (V). The unit is named for the Italian scientist Alessandro Volta.

The greater the charge and the greater the separation, the greater the stored energy, or voltage. By the same token, the greater the voltage, the greater the force that drives the charges together.

Voltage is always measured between two points, which represent the positive and negative charges. In order to compare voltages of several charged bodies a common reference point is necessary. This point is usually called "ground."

F.4.2 Current

Electrical charge flows freely in certain materials, called conductors, but not in others, called insulators. Metals and a few other elements and compounds are good conductors, while air, glass, plastics, and rubber are insulators. In addition, there is a third category of materials called semiconductors, which sometimes seem to be good conductors but much less so other times. Silicon and germanium are two such semiconductors.

Figure F.4 shows two connected, oppositely charged bodies. The force between them has the potential for work; therefore there is voltage. If the two bodies are connected by a conductor, as in the illustration, the positive charges will move along the wire to the other sphere. On the other end, the negative charge flows out on the wire toward the positive side. In this case, positive and negative charges combine to neutralize each other until there are no charge differences between any points in the system.

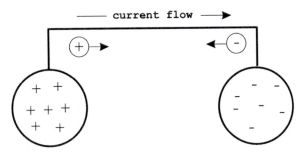

Figure F.4 *Connected opposite charges.*

The flow of an electrical charge is called a current. Current is measured in amperes (a), also called amps. Andre Ampere was a French mathematician and physicist. An ampere is defined as a flow of one Coulomb of charge in one second.

Electrical current is directional; therefore a positive current is the flow current from a positive point A to a negative point B. Actually, most current result from the flow of negative-to-positive charges.

F.4.3 Power

Current flowing through a conductor produces heat. In this case, the heat is the result of the energy that comes from the charge traveling across the voltage difference. The work involved in producing this heat is electrical power. Power is measured in units of watts (W), named after the Englishman James Watt who invented the steam engine.

F.4.4 Ohm's Law

The relationship between voltage, current, and power is described by Ohm's Law, named after the German physicist Georg Simon Ohm. Using equipment of his own creation, Ohm determined that the current that flows through a wire is proportional to its cross-sectional area and inversely proportional to its length. This allowed defining the relationship between voltage, current, power, as expressed by the equation:

$$P = V \times I$$

where P represents the power in watts, V is the voltage in volts, and I is the current in amperes. Ohm's law can also be formulated in terms of voltage, current, and resistance as shown in section F.6.2 of this appendix.

F.5 Electrical Circuits

An electrical network is an interconnection of electrical elements. An electrical circuit is a network in a closed loop, giving a return path for the current. A network is a connection of two or more simple elements, and may not necessarily be a circuit.

Although there are several types of electrical circuits, they all have at least some of the following elements:

1. A power source, which can be a battery, alternator, etc., produces an electrical potential.

2. Conductors, in the form of wires or circuit boards, provide a path for the current.

3. Loads, in the form of devices such as lamps, motors, etc., use the electrical energy to produce some form of work.

4. Control devices, such as potentiometers and switches, regulate the amount of current flow or turn it on and off.

5. Protection devices, such as fuses or circuit breakers, to prevent damage to the system in case of overload.

6. A common ground.

Figure F.5 shows a simple circuit that contains all of the above elements.

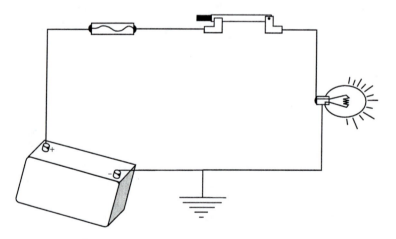

Figure F.5 *Simple circuit.*

F.5.1 Types of Circuits

There are three common type of circuits: series, parallel, and series-parallel. The circuit type is determined by how the components are connected. In other words, by how the circuit elements, power source, load, and control and protection devices are inter-connected. The simplest circuit is one in which the components offer a single current path. In this case, although the loads may be different, the amount of current flowing through each one is the same. Figure F.6 shows a series circuit with two light bulbs.

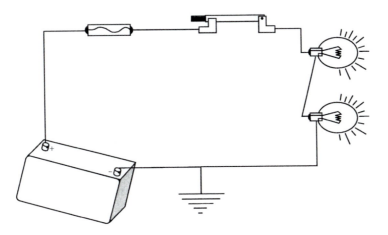

Figure F.6 *Series circuit.*

In the series circuit in Figure F.6, if one of the light bulbs burns out, the circuit flow is interrupted and the other one will not light also. Some Christmas light ornaments are wired in this manner; and if a single bulb fails, the whole string will not light.

In a parallel circuit, there is more than one path for current flow. Figure F.7 shows a circuit wired in parallel.

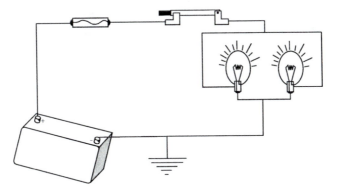

Figure F.7 *Parallel circuit.*

In the circuit of Figure F.7, if one of the light bulbs burns out, the other one will still light. Also, if the load is the same in each circuit branch, so will be the current flow in that branch. By the same token, if the load in each branch is different, so will be the current flow in each branch.

The series-parallel circuit has some components wired in series and others in parallel. Therefore, the circuit shares the characteristics of both series and parallel circuits. Figure F.8 shows the same parallel circuit to which a series rheostat (dimmer) has been added in series.

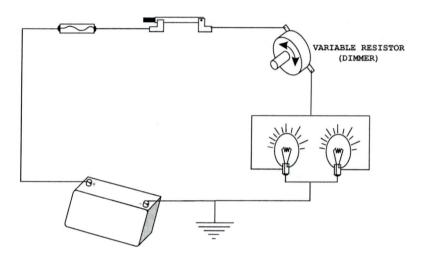

Figure F.8 *Series-parallel circuit.*

In the circuit of Figure F.8 the two light bulbs are wired in parallel, so if one fails, the other one will not. However, the rheostat (dimmer) is wired in series with the circuit, so its action affects both light bulbs.

F.6 Circuit Elements

So far we have represented circuits using a pictorial style. Circuit diagrams are more often used instead because they achieve the same purpose with much less artistic effort and are easier to read. Figure F.9 is a diagrammatic representation of the circuit in Figure F.8.

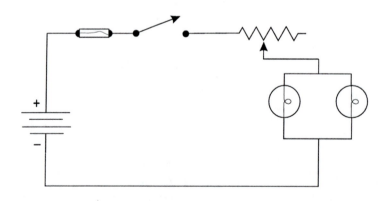

Figure F.9 *Diagram of a series-parallel circuit.*

Certain components are commonly used in electrical circuits; these include power sources (such as batteries), resistors, capacitors, inductors, and several forms of semiconductor devices.

F.6.1 Resistors

If the current flows from, say from a battery, is not controlled, a short-circuit takes place and the wires can be melted or even the battery may explode. Resistors provide a way of controlling the flow of current from a source. A resistor is to current flow in an electrical circuit like a valve is to water flow: both elements "resist" current flow. Resistors are typically made of materials that are poor conductors. The most common ones are made from powdered carbon and some sort of binder. Such carbon composition resistors usually have a dark-colored cylindrical body with a wire lead on each end. Color bands on the body of the resistor indicated its value, measured in ohms and represented by the Greek letter Ω. The color code for resistor bands can be found in Appendix E.

The potentiometer and the rheostat are variable resistors. When the knob of a potentiometer or rheostat is turned, a slider moves along the resistance element and reduces or increases the resistance. A potentiometer is used as a dimmer in the circuits of Figure F.8 and Figure F.9. The photoresistor or photocell is composed of a light-sensitive material whose resistance decreases when exposed to light. Photoresistors can be used as light sensors.

F.6.2 Revisiting Ohm's Law

We have seen how Ohm's Law describes the relationship between voltage, current, and power. The law is reformulated in terms of resistance so as to express the relationship between voltage , current, and resistance as follows:

$$V = I \times R$$

In this case, V represents voltage, I is the current, and R is the resistance in the circuit. Ohm's Law equation can be manipulated in order to find current or resistance in terms of the other variables, as follows:

$$I = \frac{V}{R}$$

$$R = \frac{V}{I}$$

Notice that the voltage value in Ohm's Law refers to the voltage across the resistor, in other words, the voltage between the two terminal wires. In this sense, the voltage is actually produced by the resistor, because the resistor is restricting the flow of charge much like a valve or nozzle restricts the flow of water. It is the restriction created by the resistor that forms an excess of charge with respect to the other side of the circuit. The charge difference results in a voltage between the two points. Ohm's Law can be used to calculate the voltage if we know the resistor value and the current flow.

A popular trick to help remember Ohm's Law consists of drawing a pyramid with the voltage symbol at the top and current and resistance in the lower level. Then, it is easy to solve for each of the values by observing the position of the other two symbols in the pyramid, as shown in Figure F.10.

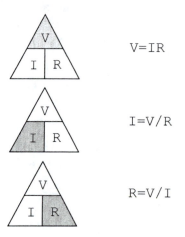

V=IR

I=V/R

R=V/I

Figure F.10 *Ohm's Law pyramid.*

Resistors are often connected together in a circuit, so it is necessary to know how to determine the resistance of a combination of two or more resistors. There are two basic ways in which resistors can be connected: in series and in parallel. A simple series resistance circuit is shown in Figure F.11.

F.6.3 Resistors in Series and Parallel

Resistors in a circuit are often connected in series. In this case, the total resistance for two or more resistors in series is equal to the sum of the individual resistances. The diagram in Figure F.11 shows two resistors ($R1$ and $R2$) wired in series in a simple circuit.

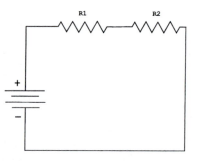

Figure F.11 *Resistors in series.*

In the case of Figure F.11, the total resistance (RT) is calculated by adding the resistance values of $R1$ and $R2$; thus, $RT = R1 + R2$. In terms of water flow, a series of partially closed valves in a pipe add up to slow the flow of water. On the other hand, resistors can also be connected in parallel, as shown in Figure F.12.

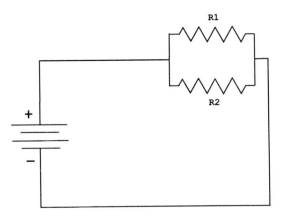

Figure F.12 *Resistors in parallel.*

Following the water pipe analogy, we can deduce that water will flow through multiple paths easier than it would through a single one. Thus, when resistors are placed in parallel, the combination will have less resistance than any one of the resistors. If the resistors have different values, then more current will flow through the path of least resistance. The total resistance in a parallel circuit is obtained by dividing the product of the individual resistors by their sum, as in the formula

$$RT = \frac{R1 \times R2}{R1 + R2}$$

If more than two resistors are connected in parallel, then the formula can be expressed as follows:

$$RT = \frac{1}{\dfrac{1}{R1} + \dfrac{1}{R2} + \dfrac{1}{R3} \ldots}$$

Also note that the diagram representations of resistors in parallel can have different appearances. For example, the circuit in Figure F.13 is electrically identical to the one in Figure F.12.

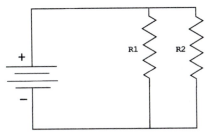

Figure F.13 *Alternative circuit of parallel resistors.*

Figure F.14 shows several commercial resistors. The integrated circuit at the center of the image combines eight resistors of the same value. These devices are convenient when the circuit design calls for several identical resistors. The color-coded cylindrical resistors in the image are of the carbon composition type.

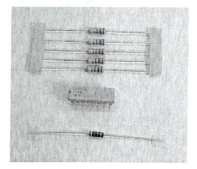

Figure F.14 *Resistors.*

F.6.4 Capacitors

An element often used in the control of the flow of an electrical charge is a capacitor. The name originated in the notion of a "capacity" to store charge. In that sense, a capacitor functions as a small battery. Capacitors are made of two conducting surfaces separated by an insulator. A wire lead is usually connected to each surface. Two large metal plates separated by air would perform as a capacitor. More frequently, capacitors are made of thin metal foils separated by a plastic film or another form of solid insulator. Figure F.15 shows a circuit that contains both a capacitor and a resistor.

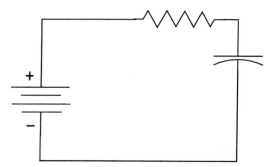

Figure F.15 *Capacitor circuit.*

In Figure F.15, charge flows from the battery terminals, along the conductor wire, onto the capacitor plates. Positive charges collect on one plate and negative charges in the other plate. Upon connecting a capacitor, the current is large, as there is no complete circuit of wire. The initial current is only limited by the resistance of the wires and by the resistor in the circuit, as in Figure F.15. As charge builds up on the plates, charge repulsion resists the flow and the current is reduced. At some point, the repulsive force from charge on the plate is strong enough to balance the force from charge on the battery, and current stops.

The existence of charges on the plates of the capacitor means there must be a voltage between the plates. When current stops, this voltage must be the same as the one on the battery. This must be the case, because the points in the circuit are connected by conductors, which means that they must have the same voltage, even if there is a resistor in the circuit. If the current is zero, then there is no voltage across the resistor, according to Ohm's Law.

The amount of charge on the plates of the capacitor is a measure of the value of the capacitor. This "capacitance" is measured in farads (f), named in honor of the English scientist Michael Faraday.

The relationship is expressed by the equation

$$C = \frac{Q}{V}$$

where C is the capacitance in farads, Q is the charge in Coulombs, and V is the voltage. Capacitors of 1 farad or more are rated. Generally, capacitors are rated in microfarads (μf), one millionth of a farad, or picofarads (pf), one trillionth of a farad.

Consider the circuit of Figure F.15 after the current has stabilized. If we now re-move the capacitor from the circuit, it will still hold a charge on its plates. That is, there will be a voltage between the capacitor terminals. In one sense, the charged capacitor appears somewhat like a battery. If we were to short-circuit its terminals, a current would flow as the positive and negative charges neutralized each other. But unlike a battery, the capacitor has no way of replacing its charge. So the voltage drops, the current drops, and finally there is no net charge left and no voltage differ-ences anywhere in the circuit.

F.6.5 Capacitors in Series and in Parallel

Like resistors, capacitors can be joined together in series and in parallel. Connecting two capacitors in parallel results in a bigger capacitance value, because there will be a larger plate area. Thus, the formula for total capacitance (CT) in a parallel circuit con-taining capacitors $C1$ and $C2$ is

$$CT = C1 + C2$$

Notice that the formula for calculating capacitance in parallel is similar to the one for calculating series resistance. By the same token, where several capacitors are connected in series the formula for calculating the total capacitance is

$$CT = \frac{1}{\dfrac{1}{C1} + \dfrac{1}{C2} + \dfrac{1}{C3} \cdots}$$

This formula is explained by the fact that the total capacitance of a series connection is lower than any capacitor in the series, considering that for a given voltage across the entire group, there will be less charge on each plate. Commercial capacitors are furnished in several types, including mylar, ceramic, disk, and electrolytic. Figure F.16 shows several commercial capacitors.

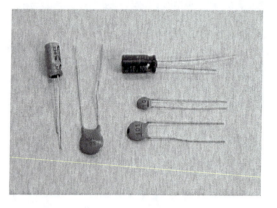

Figure F.16 *Assorted commercial capacitors.*

F.6.6 Inductors

The third and final type of basic circuit components are indictors. An inductor is a coil of wire with many windings. The wire windings are often made around a core of a magnetic material, such as iron. The properties of inductors derive from magnetic rather than electric forces.

When current flows through a coil it produces a magnetic field in the space outside the wire. This makes the coil behave just like a natural, permanent magnet. Moving a wire through a magnetic field generates a current the wire, and this current will flow through the associated circuit. Because it takes mechanical energy to move the wire through the field, then it is the mechanical energy that is transformed into electrical energy. A generator is a device that converts mechanical to electrical energy by means of induction.

The current in an inductor is similar to the voltage across a capacitor. In both cases, it takes time to change the voltage from an initially high current flow. Such induced voltages can be very high and can damage other circuit components, so it is common to connect a resistor or a capacitor across the inductor to provide a current path to absorb the induced voltage.

Induction is measured in henrys (h), but more commonly in millihenries, and microhenries. An electric motor is the opposite of a generator. In the motor, electrical energy is converted to mechanical energy by means of induction. In combination, inductors behave just like resistors: inductance adds in series. By the same token, parallel connection reduces induction.

F.6.7 Transformers

The transformer is an induction device that has the ability to change voltage or current to lower or higher levels. The typical transformer has two or more windings wrapped around a core made of laminated iron sheets. One of the windings, called the primary, receives a fluctuating current. The other winding, called the secondary, produces a current induced by the primary. Figure F.17 shows the schematics of a transformer.

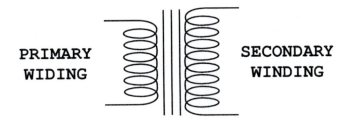

PRIMARY WIDING   SECONDARY WINDING

Figure F.17 *Transformer schematics.*

The device in Figure F.17 is a step-up transformer. This is determined by the number of windings in the primary and secondary coils. The ratio of the number of turns in each winding determines the voltage increase. A transformer with an equal number of turns in the primary and secondary transfers the current unaltered. This type of device is sometimes called an isolation transformer. A transformer with less turns in the secondary than in the primary winding is a step-down transformer and its effect is to reduce the primary voltage at the secondary winding.

Transformers require an alternating or fluctuating current as it is the fluctuations in the current flow in the primary that induce a current in the secondary winding. The ignition coil in an automobile is a transformer that converts the low-level battery voltage to the high-voltage level necessary to produce a spark.

F.7 Semiconductors

The term semiconductor stems from the property of some materials that can act either as a conductor or as an insulator, depending on certain conditions. Several elements are classified as semiconductors, including silicon, zinc, and germanium. Silicon is the most widely used semiconductor material because it is easily obtained.

In the ultra-pure form of silicon, the addition of minute amounts of certain impurities (called dopants) alters the atomic structure of the silicon. This determines that the silicon can then be made to act as a conductor or a nonconductor, depending upon the polarity of an electrical charge applied to it.

In the early days of radio, receivers required a device called a rectifier to detect signals. Ferdinand Braun used the rectifying properties of the galena crystal, a semiconductor material composed of lead sulfide, to create a cat's whisker diode that served this purpose. This was the first semiconductor device.

F.7.1 Integrated Circuits

Up to 1959, electronic components performed a single function, therefore, many of them had to be wired together to create a functional circuit. This meant that transistors were individually packaged in small cans. Packaging and hand wiring the components into circuits were extremely inefficient.

In 1959, at Fairchild Semiconductor, Jean Hoerni and Robert Noyce developed a process that made it possible to diffuse various layers onto the surface of a silicon wafer, while leaving a layer of protective oxide on the junctions. By allowing the metal interconnections to be evaporated onto the flat transistor surface, the process replaced hand wiring. By 1961, nearly 90% of all the components manufactured were integrated circuits.

F.7.2 Semiconductor Electronics

To understand the workings of semiconductor devices we need to reconsider the nature of the electrical charge. We have seen that electrons are one of the components of atoms, and atoms are the building blocks of all mater. Atoms bond with each other to form molecules. Molecules of just one type of atom are called elements. In this sense, gold, oxygen, and plutonium are elements because they all consist of only one type of atom. When a molecule contains more than one atom, it is known as a compound. Water, which has hydrogen and oxygen atoms, is a compound. Figure F.18 represents an orbital model of an atom with five protons and three electrons.

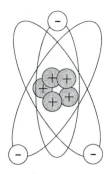

Figure F.18 *Orbital model of the boron atom.*

In Figure F.18, protons carry positive charge and electrons carry negative charge. Neutrons, not represented in the illustration, are not electrically charged. Atoms that have the same number of protons and electrons have no net electrical charge.

Electrons that are far from the nucleus are relatively free to move around because the attraction from the positive charge in the nucleus is weak at large distances. In fact, it takes little force to completely remove an outer electron from an atom, leaving an ion with a net positive charge. A free electron can move at speeds approaching the speed of light (approximately 186,282 miles per second).

Electric current takes place in metal conductors due to the flow of free electrons. Because electrons have negative charge, the flow is in a direction opposite to the positive current. Free electrons traveling through a conductor drift until they hit other electrons attached to atoms. These electrons are then dislodged from their orbits and replaced by the formerly free electrons. The newly freed electrons then start the process anew.

F.7.3 P-Type and N-Type Silicon

Semiconductor devices are made primarily of silicon. Pure silicon forms rigid crystals because of its four outermost electrons. Because it contains no free electrons, it is not a conductor. But silicon can be made conductive by combining it with other elements (doping) such as boron and phosphorus. The boron atom has three outer valence electrons (Figure F.18) and the phosphorus atom has five. When three silicon atoms and one phosphorus atom bind together, creating a structure of four atoms, there is an extra electron and a net negative charge.

The combination of silicon and phosphorous, with the extra phosphorus electron, is called n-type silicon. In this case the n stands for the extra negative electron. The extra electron donated by the phosphorus atom can easily move through the crystal; therefore, n-type silicon can carry and electrical current.

When a boron atom combines in a cluster of silicon atoms there is a deficiency of one electron in the resulting crystal. Silicon with a deficient electron is called p-type silicon (p stands for positive.) The vacant electron position is sometimes called a "hole." An electron from another nearby atom can "fall" into this hole, thereby moving the hole to a new location. In this case, the hole can carry a current in p-type silicon.

F.7.4 Diode

We have seen that both p-type and n-type silicon conduct electricity. In either case, the conductivity is determined by the proportion of holes or the surplus of electrons. By forming some p-type silicon in a chip of n-type silicon, it is possible to control electron flow so that it takes place in a single direction. This is the principle of the diode, and the p-n action is called a pn junction.

A diode is said to have a forward bias if it has a positive voltage across it from the p- to n-type material. In this condition, the diode acts rather like a good conductor, and current can flow, as in Figure F.19.

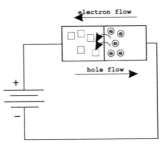

Figure F.19 *A forward biased diode.*

If the polarity of the voltage applied to the silicon is reversed, then the diode will be reverse biased and will appear nonconducting. This nonsymmetric behavior is due to the properties of the pn-junction. The fact that a diode acts like a one-way valve for current is a very useful characteristic. One application is to convert alternating current into direct current (DC). Diodes are so often used for this purpose that they are sometimes called rectifiers.

Index